SOLID STATE PHYSICS

VOLUME 9

£5
opp.

SOLID STATE PHYSICS

Advances in Research and Applications

Editors

FREDERICK SEITZ
Department of Physics
University of Illinois
Urbana, Illinois

DAVID TURNBULL
General Electric
Research Laboratory
Schenectady, New York

VOLUME 9

1959

ACADEMIC PRESS • NEW YORK AND LONDON

ACADEMIC PRESS INC.
111 FIFTH AVENUE
NEW YORK 3, N. Y.

United Kingdom Edition
Published by
ACADEMIC PRESS INC. (LONDON) LTD.
40 PALL MALL, LONDON S.W. 1

Library of Congress Catalog Card Number 55-12200

PRINTED IN THE UNITED STATES OF AMERICA

Contributors to Volume 9

H. P. R. FREDERIKSE, *Solid State Physics Section, National Bureau of Standards, Washington, D. C.*

ANDRE GUINIER, *University of Paris, Paris, France*

ERNST HEER, *University of Rochester, Rochester, New York, and Argonne National Laboratory, Lemont, Illinois*

A. H. KAHN, *Solid State Physics Section, National Bureau of Standards, Washington, D. C.*

DONALD S. MCCLURE, *RCA Laboratories, Princeton, New Jersey*

D. J. MONTGOMERY, *Michigan State University, East Lansing, Michigan*

THEODORE B. NOVEY, *University of Rochester, Rochester, New York, and Argonne National Laboratory, Lemont, Illinois*

W. W. SCANLON, *U. S. Naval Ordnance Laboratory, Silver Spring, Maryland*

H. C. WOLF, *II. Physikalisches Institut der Technischen Hochschule Stuttgart, Germany*

Preface

In the present volume seven articles are presented which cover quite a wide range of topics in solid state physics.

Two of the articles are devoted to electronic spectra; one of these, by Wolf, reviews aromatic crystals and the other, by McClure, reviews ionic crystals. An article by Scanlon continues the coverage, in this series, of important classes of semiconductors. Two fields that have come into greater prominence in solid state studies recently are reviewed in the articles of Heer and Hovey and of Kahn and Frederikse. New results and viewpoints concerning the long familiar field of static electrification are presented by Montgomery. The formation and structure of "zones," which are very small solute-rich regions that form with an extraordinarily high number density in certain supersaturated solid solutions, are reviewed by Guinier.

The editors and publishers deeply regret the tragic death of M. R. Schafroth. His article on "Theoretical Aspects of Superconductivity" was completed before his death and will appear in the next volume of this series.

FREDERICK SEITZ
DAVID TURNBULL

July, 1959

Contents

Oscillatory Behavior of Magnetic Susceptibility and Electronic Conductivity

A. H. Kahn and H. P. R. Frederikse

Heterogeneities in Solid Solutions

Andre Guinier

Electronic Spectra of Molecules and Ions in Crystals

Part II. Spectra of Ions in Crystals

Donald S. McClure

Contents of Previous Volumes

Volume 6

Volume 7

Volume 8

Articles Planned for Future Volumes

PHILLIP W. ANDERSON—CONYERS HERRING	Ferromagnetic and Antiferromagnetic Exchange Interactions
ALBERT C. BEER	Galvanomagnetic Effects
WERNER BRANDT—Y. H. PAO	Physics of High Polymers
B. N. BROCKHOUSE	Determination of the Normal Modes of Lattices by Neutron Spectroscopy
RICHARD H. BUBE	Electrical and Optical Properties of Cadmium Sulfide and Similar Materials
ELIAS BURSTEIN—G. PICUS	Infrared Spectra Arising from Foreign Atoms in Semiconductors
ELIAS BURSTEIN—MELVIN LAX	Infrared and Related Properties of Ionic Crystals
GEORGE A. BUSCH	Semiconducting Properties of Gray Tin
NICOLAS CABRERA	Theory of Crystal Growth
R. G. CHAMBERS	Determination of the Fermi Surface in Metals
ALAN H. COTTRELL	Subject to be announced
ROLAND DE WIT	The Continuum Theory of Stationary Dislocations
J. FRIEDEL	Theory of Solid Solutions
FAUSTO FUMI	Theory of Ionic Crystals
JOHN J. GILMAN	Dislocation Generation and Propagation in Lithium Fluoride
ROLFE GLOVER	The Properties of Thin Films
BARRY S. GOURARY—FRANK J. ADRIAN	Wave Functions for Electron-Excess Color Centers in Alkali Halide Crystals

A. V. Granato—Kurt Lücke	Internal Friction Phenomena Due to Dislocations
Paul Handler	Properties of Semiconductor Surfaces
Charles M. Herzfeld	Group Theory and Electronic Spectra of Solids
H. Inokuchi—H. Akamatu	Electrical Conductivity of Organic Semiconductors
Väinö Hovi	Thermodynamic and Physical Properties of Ionic Solid Solutions
Robert W. Keyes	Effects of Elastic Strain on the Electrical Properties of Semiconductors
C. Kittel	Cyclotron Resonance
Jan Korringa	Group Theoretical Discussions of the Spin Hamiltonian for Magnetic States in Crystals
Benjamin Lax—John Mavroides	Cyclotron Resonance
David Lazarus	Diffusion in Metals
G. Leibfried	Theory of Anharmonic Phenomena in Crystals
W. Low	Survey of Paramagnetic Resonance in Solids
R. J. Maurer	Transport Phenomena in Ionic Crystals
Elliott W. Montroll	Theory of Lattice Vibrations and Specific Heats
H. M. Rosenberg—K. Mendelssohn	Thermal Conductivity of Metals and Semiconductors at Low Temperatures
M. R. Schafroth	Theoretical Aspects of Superconductivity
Roland W. Schmitt	Resistivity of Metals and Alloys

The Electronic Spectra of Aromatic Molecular Crystals

H. C. Wolf

II. Physikalisches Institut der Technischen Hochschule, Stuttgart, Germany

I. Introduction

A schematic survey of the optical spectra of molecular crystals over the entire region of wavelength (Fig. 1) will encompass the following:

(1) The range of absorption associated with lattice vibrations in the long wavelength infrared (10–100 cm^{-1}).

(2) The range of absorption of the molecular vibrations in the short wavelength infrared (100–5000 cm^{-1}).

(3) The range of absorption associated with electron excitation in the visible and ultraviolet parts of the spectra. Lattice and molecular vibrations are usually stimulated simultaneously. The structure in this range of absorption is derived from a number of electron transitions having different energy which ultimately merge into the continuum associated with the ionization limit (50,000–100,000 cm^{-1}).

(4) A fluorescent range on the long wavelength limit of the lowest electronic absorption.

(5) A phosphorescent range somewhat farther toward the red than the range of fluorescence.

Of the five component parts of the over-all spectrum, only the first, namely the absorption by lattice vibrations, is associated purely with the crystal. All of the other ranges of absorption are found in the spectrum of the component molecules when observed in the free state and are modified only more or less when the molecules are bound into the crystal. An understanding of the spectra of the crystals requires a knowledge of the

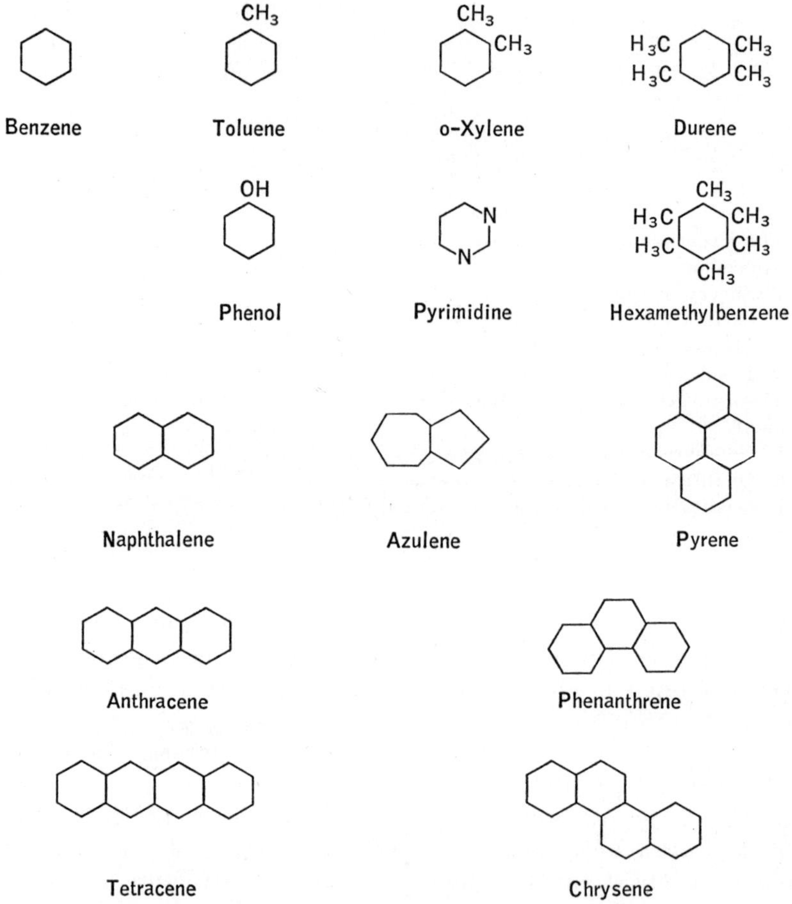

Structural formulas of organic molecules considered here.

spectroscopic behavior of the free molecules. We shall, as a result, give a brief survey of the spectra of organic molecules initially, particularly those of the aromatic type.

II. Survey of the Spectra of Aromatic Molecules

Although the interpretation of the spectra of polyatomic molecules faces very great difficulties, the understanding of the spectra of simple aromatic molecules has made great progress in recent years.[1–4] The π electrons of the rings are responsible for the essential form of the spectra. In the *absorption spectrum,* which begins in the visible or the near ultraviolet part of the spectrum, one observes a number of electronic excitation levels (0.0 transitions). These electronic levels are accompanied by the excitation levels of molecular vibrations and rotations; however, the latter can be observed only in dilute gases. As a result, they will not be discussed further in the following.

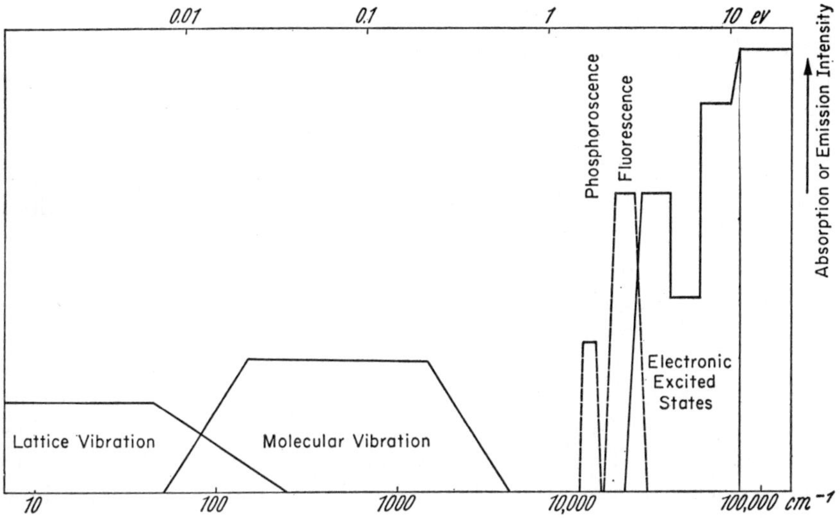

FIG. 1. Schematic representation of the optical spectral ranges for molecular crystals.

The molecular vibrational frequencies are also known from studies both in the infrared and by means of Raman spectra. In the region where the electron excitation spectra are observed, their intensities can be discussed chiefly in terms of two selection rules derived by Herzberg and Teller.[5]

(1) In the allowed transitions the totally symmetrical molecular vibrations are excited most intensely. The less the geometry of the molecules is altered during excitation, the more probable it is that there will be multiquantum excitations of such vibrations.

[1] H. B. Klevens and J. R. Platt, *J. Chem. Phys.* **17,** 470 (1949).
[2] J. R. Platt, *J. Chem. Phys.* **17,** 484 (1949).
[3] W. Moffitt, *J. Chem. Phys.* **22,** 320 (1954).
[4] R. Pariser, *J. Chem. Phys.* **24,** 250 (1956).
[5] G. Herzberg and E. Teller, *Z. physik. Chem.* (*Leipzig*) **B21,** 410 (1933).

In the case of simple aromatic molecules it is found experimentally that the intensity of certain totally symmetric vibrations exceeds that of all others by a considerable margin.

(2) Transitions which are forbidden on the basis of symmetry can be made "allowed" as a result of the presence of vibrations which are not totally symmetric. These vibrations impress their symmetry on the electronic transitions.

The electronic transitions can be distinguished from one another by their energy and symmetry, that is, by the position of the transition moment in the molecule. The higher transitions which lie in the vacuum ultraviolet converge toward the ionization limit in the manner of a Rydberg series. Very little is known about them because of the experimental difficulties encountered in carrying out investigations in these spectral regions.

The transitions of lower energy, in contrast, have been investigated extensively. In addition, they have been studied in considerable detail theoretically.[2–4]

At sufficiently low temperatures the absorption takes place only from the vibrationless ground state of the molecule, and terminates in the electronic and vibrational levels of the various excitation states of the molecule. With increasing temperature the various vibrational levels of the ground state are occupied in increasing proportion and, in accordance with a Boltzmann function, new absorption series originate in these vibrational levels. Moreover, the start of absorption is displaced toward longer wavelengths as a result. Thus, one obtains the term scheme shown in Fig. 2 in this manner. The rotational levels have been omitted for simplicity.

Many aromatic molecules exhibit fluorescence. Except in very dilute gases, the *fluorescence spectra* are independent of the excitation energy. In all condensed phases the energy in excess of that required to reach the first excited electronic state is dissipated very rapidly (in less than 10^{-12} sec) to the environment by radiationless processes. The fluorescence is produced by transitions from the first excited electronic state to the vibrational levels of the ground state regardless of the excitation. Thus, at low temperatures the absorption band of longest wavelength corresponds to the fluorescent band of shortest wavelength. The fluorescence-spectrum resembles closely the mirror image of the first absorption transition.

The decay time of fluorescence usually lies in the range of magnitude extending from 10^{-6}–10^{-9} sec.[6]

[6] T. Förster, "Fluoreszenz Organischer Verbindungen," Vandenhoek & Ruprecht, Göttingen, 1951.

Many aromatic molecules exhibit an additional *phosphorescence* at low temperatures when present in solid solution. That is, the lifetime for emission of light may be considerably longer, extending to 10 sec. Such emission usually lies at longer wavelengths than the fluorescent emission and can be ascribed to a triplet system. In such systems one can obtain a

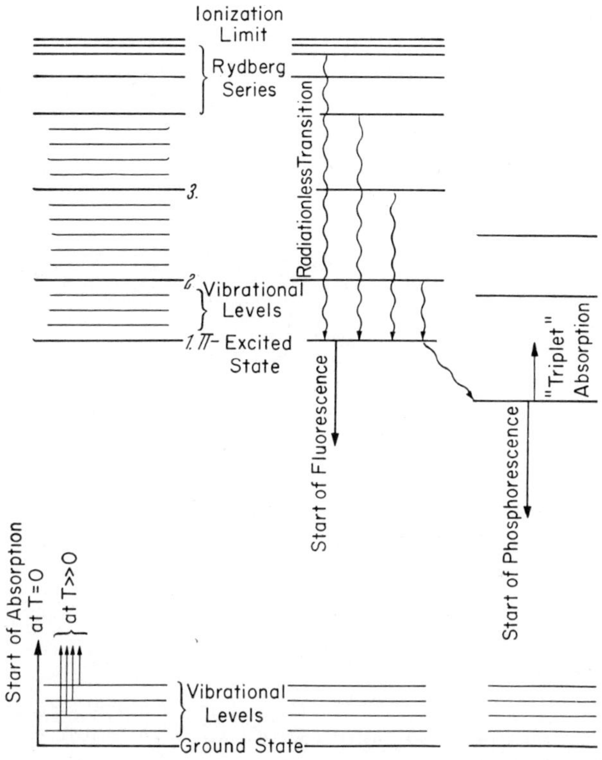

FIG. 2. Term scheme for absorption, fluorescence, and phosphorescence of organic molecules.

triplet absorption if the metastable states of excitation are populated sufficiently intensely as a result of the radiationless transitions from the singlet system.

The proof of the triplet character of the phosphorescing system was demonstrated in a few cases by magnetic measurements.[7,8] The experimental material relating to "triplet spectra" in pure molecular crystals

[7] G. N. Lewis and M. Kasha, *J. Am. Chem. Soc.* **66,** 2100 (1944).
[8] G. Kortüm and G. Littmann, *Z. Naturforsch.* **12a,** 395 (1957).

is still not plentiful and is not capable of a unique interpretation. In fact, it can be said that the triplet nature of the spectra is not completely proved in any case involving pure crystals.*

In the following we will generally discuss the "singlet spectra" exclusively.

The spectra obtained in different *states of aggregation* differ particularly in regard to the widths of the lines. The absorption and fluorescence spectra of gases consist of a large number of sharp lines which can be related to the various rotational, vibrational, and electronic levels and can be combined into band systems.

The lines undergo a broadening and a displacement when the molecules are in solution as a result of the interaction with the environment. Rotational lines can no longer be resolved. In order to obtain good resolution, the spectra of the solutions are usually investigated at low temperatures in solvents which are solidified into glass-like form and which preferably are composed of nonpolar molecules. The lines, however, usually are broadened strongly even at 77°K.

The *line width* depends upon the energy of interaction of the excited molecules with their environment. There is no general theory of the line or band width of molecular spectra at the present time. In the case of nonpolar molecules in nonpolar media, of interest to us here, the interaction consists principally in the coupling of the oscillating polarization of the electronic shells in the neighborhood of the excited molecule with the transition moment $\mathbf{m}$ of the excited molecule.[9] The interaction energy and, hence, the part of a line width determined by it is proportional to the square of the transition moment in the dipole approximation. The magnitude $|\mathbf{m}|^2$ is expressed in terms of the measurable quantity ϵ which is proportional to it and which appears in the exponent of the absorption relation $I = I_0 \cdot 10^{-\epsilon c d}$. Here I is the intensity, c is the concentration in moles per liter, and d is the layer thickness in centimeters. The quantity ϵ is defined as the molar decadal absorption coefficient.

The following rule of thumb is useful in discussing the band width in glass-like frozen solutions.

(1) In the case of weak transitions, for which ϵ is less than 10^3 cm^{-1} (mole/liter)$^{-1}$, the half-width of the bands lies between 30 and 100 cm^{-1}. In other words, the half-width definitely is smaller than the frequency

* *Note added in proof.* The paramagnetic character of the phosphorescent state has been demonstrated by electron spin resonance following a suggestion of K. H. Hausser by C. A. Hutchison and B. W. Mangum [*J. Chem. Phys.* **29,** 952 (1958)].

[9] N. S. Bayliss, *J. Chem. Phys.* **18,** 292 (1950); N. S. Bayliss and E. G. McRae, *J. Phys. Chem.* **58,** 1002 (1954); E. G. McRae, *ibid.* **61,** 562 (1957); J. Ferguson, *J. Chem. Phys.* **24,** 1263 (1956).

associated with typical molecular vibrations. Hence, one obtains spectra having a very intense structure.

(2) In the case of transitions of intermediate intensity, for which ϵ lies between 10^3 and 10^4, the half-width lies between 100 and 300 cm^{-1}. Thus, the half-width is of the same order of magnitude as the wave number of molecular vibrations. The vibrational bands overlap one another rather strongly and the spectra show relatively small amount of structure.

(3) In the case of intense transitions, for which ϵ is larger than 10^4, the half-width of the bands is so large that no vibrational bands can be resolved. The complete transitions spectrum consists of a broad maximum.

If one desires to compare the spectra of the crystal with those of the "free molecules," it is usually convenient and sufficient to work with the spectra of solutions instead of those of a gas in spite of the diffuse character of the former because the number of lines in the spectrum of the gas is tremendously large. The absorption and fluorescence spectra are complicated by the population of a number of vibrational states near the ground level as a result of thermal excitation.

One obtains sharper band or line spectra when the molecules are dissolved in an oriented way in crystals rather than in disordered solvents. The solute molecules then have somewhat specialized orientations relative to the molecules of the host with which they interact. Thus, the perturbation energy which determines the broadening is sharply defined. Line widths less than 5 cm^{-1} are observed at low temperatures in mixed crystals of this type; for example, in the case of naphthalene in durene. However, characteristic variations in the vibrational structure of the guest molecule is observed in the spectra of such mixed crystals as a consequence of the symmetry of the surrounding host medium. Thus, such mixed crystals cannot be looked upon simply as "oriented gases."

An additional broadening of the lines or bands which arises from the possibility of energy exchange between a molecule and its neighbors, and which may be designated *resonance broadening*,[10,11] can be observed in pure molecular crystals. Nevertheless, even here line widths which are less than 10 cm^{-1} can be observed at low temperatures in the most favorable cases.

Finally, it should be noted that a *displacement* of the *lines* or the *bands* toward the red can be observed when one makes the transition from the vapor or nonpolar solution to the crystal. This can also be interpreted as a consequence of the interaction of the nonpolar molecules in the excited state with the nonpolar molecules of the solvent.[9] The interaction

[10] H. C. Wolf, *Z. Physik* **143,** 266 (1955).
[11] D. P. Craig and P. C. Hobbins, *J. Chem. Soc.* pp. 2302–2309 (1955).

energy between the molecule and the environment differs in the various excited states. The interaction is proportional to $|\mathbf{m}|^2$ and, approximately, to the dielectric constant of the solvent. This interaction produces a diminution of the distance between the ground and the excited states, that is, leads to a displacement of the lines or bands toward the red. For a given solvent, the magnitude of the displacement is proportional

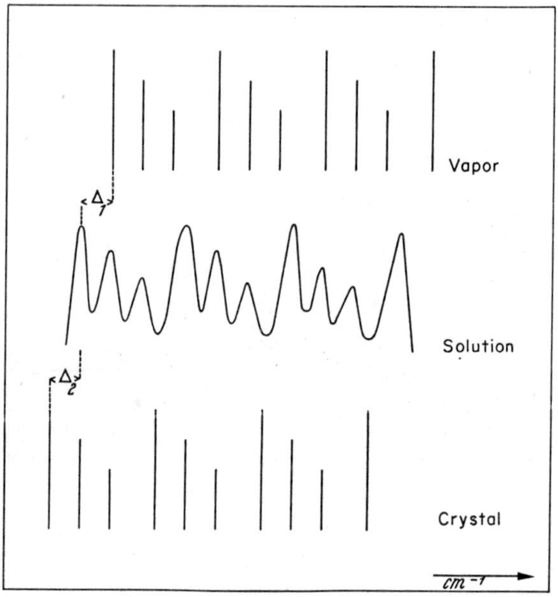

FIG. 3. Representation of the line width and line position of the spectrum in the transition from vapor to solution and crystal. The last two are indicated at low temperatures.

to $|\mathbf{m}|^2$, that is, to ϵ, and in general is largest for the crystal. The relationships governing the positions and widths of lines and the state of aggregation are shown schematically in Fig. 3.

This short review of the molecular spectra should provide a basis for better understanding of a discussion of the spectra of molecular crystals. For further information, it may be referred to the most recent comprehensive accounts of the absorption and fluorescence spectra.[6,12,13]

[12] M. Pestemer and D. Brück, *in* Houben-Weyl, "Methoden der Organischen Chemie," vol. III/2. Thieme, Stuttgart, 1955.

[13] H. Sponer, *Ann. Rev. Phys. Chem.* **6,** 193 (1955); M. Kasha and S. P. McGlynn, *ibid.* **7,** 403 (1956); T. Förster, *ibid.* **8,** 331 (1957); G. Herzberg, *ibid.* **9,** 315 (1958); J. Platt, *ibid.* **10,** to be published (1959).

III. The Experimental Investigation of the Spectra of Crystals of Aromatic Molecules

1. Survey of Older Work

The oldest detailed investigations of the absorption spectra of molecular crystals were made by Becquerel[14] in 1907. We are indebted to Obreimow and de Haas,[15] and the associated Russian school involving Prichotjko and Shabaldas for a systematic study. This work, carried out between 1925 and 1946, started with investigations of the crystals of alizarin, anthraquinone, acridine, azobenzene, as well as of some inorganic materials. Later work also involved the study of naphthalene, anthracene, and phenanthrene.[16,17] The application of the low temperatures (20°K and 4°K), of polarized light and high-resolution spectrographs yielded a wealth of observational material which subsequent work often has done no more than substantiate. The large numbers of lines have been organized into series without attempting to interpret the series. It was found that the spectra are displaced toward the red with decreasing temperatures.

The work of Pringsheim and Kronenberger[18,19] concerning the spectra of benzol and its derivatives at low temperatures is closely related to the work of the Russian school. As in the case of the Russian workers, they demonstrated the great sharpness of the spectral lines and the close relationship of the spectra to those of the free molecules.

Krishnan and Seshan[20] were the first to investigate the anisotropy of the molecular absorption in crystals with measurements on single crystals of chrysene. They confirmed the expectation that the absorption perpendicular to the plane of the benzol ring is smaller than that in the plane. The same result was obtained by Scheibe and co-workers[21] using single crystals of hexamethylbenzene. Krishnan and Seshan[20] determined the orientation of the guest molecules in mixed crystals of anthracene and tetracene.

[14] J. Becquerel, *Le radium* **4,** 328 (1907).

[15] I. W. Obreimov and W. J. de Haas, *Proc Roy. Acad. Amsterdam* **31,** 353 (1928); **32,** 1, 1324 (1929); *Comm. Leiden* No. 191a, 3 (1929).

[16] I. W. Obreimov and A. F. Prichotjko, *Physik. Z. Sowjetunion* **1,** 203 (1932).

[17] I. W. Obreimov and A. F. Prichotjko, *Physik. Z. Sowjetunion* **9,** 34, 48 (1936).

[18] P. Pringsheim and A. Kronenberger, *Z. Physik* **40,** 75 (1926).

[19] A. Kronenberger, *Z. Physik* **63,** 494 (1930).

[20] K. S. Krishnan and P. K. Seshan, *Proc. Indian Acad. Sci.* **8,** 487 (1939).

[21] G. Scheibe, S. Hartwig, and R. Müller, *Z. Elektrochem.* **49,** 372 (1943).

Above all, the new investigations of the Russian school[22–24] made it possible to distinguish between the crystal properties and molecular properties in the spectra of the solids for the first time. This, in turn, made it possible for Davydov[25,26] to construct a theory of the spectra of molecular crystals which can be looked upon as a foundation for all theoretical work even today.

In the following chapters, we shall survey the more recent work in which the observations have attained satisfactory agreement. In order to unify the presentation, the figures are selected from the work of the present author for the most part. Figures 20 and 21 provide examples of the beautiful spectra obtained in the work of McClure.

2. Goal of the Newer Work

The increasing interest which has been exhibited in the spectra of molecular crystals in recent years is a result of critical questions which have been raised concerning the physics of both solid bodies and molecules.

In the case of *molecular physics*, it is possible to use the modern theories of chemical binding and of the structure of molecules to calculate energies and symmetries of the excited states of molecules.[1–4] The calculated values of the energies are afflicted with large limits of error. In contrast, the symmetry properties, that is, the orientation of the transition moments in the molecule can be made the subject of rigorous comparison between theory and experiment. Such comparison is possible only if made with the use of measurements involving the spectra of crystals.

In the search for an interpretation of the spectra of crystals, one encounters typical solid state properties: the splitting of lines, lattice vibrations, and the displacement of the molecular spectra in the transition to the condensed phase. The understanding of other solid state properties, such as energy transfer, molecular interaction, and the influence of imperfections requires knowledge of the levels of optical excitation too. Thus, the investigation of the spectra of molecular crystals becomes a problem of *solid state physics*. Still further, the study of molecular crystals which have sharp spectral lines, whose separations can be identified with the energy levels of the constituent molecules, leads one to a better understanding of those crystals which do not possess sharp energy levels.

We should only mention in passing that in some cases the study of

[22] I. W. Obreimov and K. Shabaldas, *J. Phys. USSR* **7**, 167 (1943).
[23] I. W. Obreimov and K. Shabaldas, *J. Phys. USSR* **8**, 257 (1944).
[24] A. F. Prichotjko, *J. Exptl. Theoret. Phys.* (*U.S.S.R.*) **19**, 383 (1949).
[25] A. S. Davydov, *J. Exptl. Theoret. Phys.* (*U.S.S.R.*) **18**, 210 (1948).
[26] A. S. Davydov, *J. Exptl. Theoret. Phys.* (*U.S.S.R.*) **21**, 673 (1951).

the spectra of crystals may complete the results of analysis with the use of infrared and Raman spectroscopy.

3. Methods of Procedure

In order to achieve the goals mentioned above, it has proved necessary to carry out quantitative measurements of absorption, fluorescence, and eventually of the reflection spectra using single crystals, the highest resolution attainable, polarized light, and low temperatures. The results must be compared with the spectra derived from solutions. The methods developed to undertake these measurements will be described in the following paragraphs.

a. Choice of Material

To the extent possible, one searches primarily for materials which have sharp spectral lines, known crystal structures, possess spectra lying in an easily attainable region, and which are fluorescent. The fluorescence spectra makes it easier to interpret the absorption spectra. Up to the present time, the consistent pursuit of this line of reasoning has led to investigation of the simple aromatic hydrocarbons almost without exception.

b. Purification

One needs extremely pure substances if one wishes to avoid errors in interpretation of spectra which are falsified by impurities. Many examples of spectra which have been interpreted falsely because of the presence of impurities may be found in the literature. Although the absorption spectra usually are affected by impurities only in proportion to the amount of impurity present, the fluorescence spectra of crystals can be altered completely by impurities which are present even in minute amounts. In other words, the intrinsic fluorescence can be masked by that arising from the impurities. One of the most well-known cases occurs in the anthracene-tetracene system.[27,28] The fluorescence of anthracene is quenched by the addition of only about one part in a thousand of tetracene. A crystal of anthracene contaminated with tetracene in this manner displays the fluorescence of tetracene almost exclusively.

We shall not discuss the customary chemical methods of purifying substances. If one starts with "chemically pure" materials, the necessary degree of purity usually is obtained by two simple physical

[27] A. Winterstein, K. Schoen, and H. Vetter, *Naturwiss.* **22**, 237 (1934).
[28] H. C. Wolf, *Z. Physik.* **139**, 318 (1954).

methods, namely, chromatography and zone melting (or directed crystallization).[29–31]

c. Methods of Growing Crystals

Large single crystals from which sections with arbitrary boundaries can be cut may be seeded and grown by Bridgman's method. In this method, a crucible filled with a substance under investigation is lowered through a temperature gradient[32,33] (see Fig. 4). It is necessary to have a capillary at the end of the crucible where crystallization begins in order to furnish a unique seed for the single crystals.

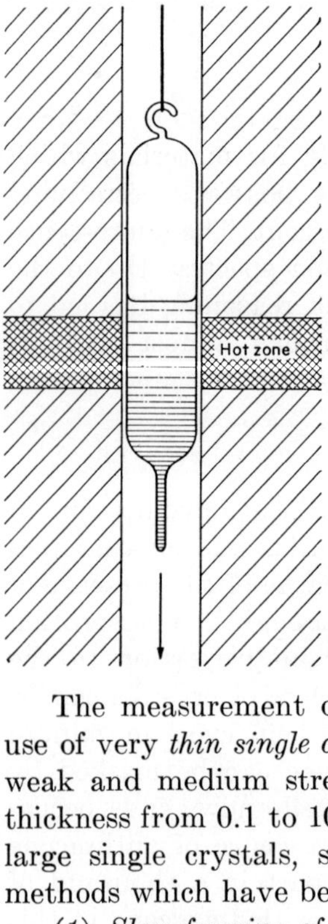

FIG. 4. Schematic representation of an apparatus for growing thick single crystals by slow directional crystallization.

The measurement of the absorption spectra requires, however, the use of very *thin single crystals*. If the absorption is in the range between weak and medium strength, single crystals should be in the range of thickness from 0.1 to 10 μ. Such crystals cannot be obtained by splitting large single crystals, so that one needs other methods. Some of the methods which have been developed for this purpose are as follows.

(*1*) *Slow freezing of the melt between two optically flat plates of quartz.* The thickness of the crystals obtainable in this way depends upon the

[29] R. C. Sangster and J. W. Irvine, *J. Chem. Phys.* **24,** 670 (1956).
[30] W. G. Pfann, *Solid State Phys.* **4,** 424 (1957).
[31] H. C. Wolf and H. P. Deutsch, *Naturwiss.* **41,** 425 (1954).
[32] H. Mette and H. Pick, *Z. Physik* **134,** 566 (1953).
[33] F. R. Lipsett, *Can. J. Phys.* **35,** 284 (1957).

surface or boundary tension of the melt. At pressures between zero and about two kp/cm^2, the thickness lies between 0.1 μ and 10 μ. The material between the quartz plates tends to grow in the form of a conglomerate of many, somewhat mutually disoriented crystals. In order to obtain larger single crystalline domains, one must take great care that the crystallization of the melt proceeds slowly and starts from a single point. The magnitude of the single crystalline domains seeded in this way is in the range between a few mm^2 and 0.5 cm^2. One can influence the perfection of the crystals and, in some cases, also the orientation of the individual domains by annealing.[34,35]

Generally, the quality of the crystals grown between the quartz plates does not attain that of crystals grown freely. The principal advantage of the method is that it provides a method of growing crystals of almost all materials quickly and simply. Moreover, the crystal is protected against sublimation and chemical decomposition to a large degree.

(*2*) *Sublimation.* Sublimation can be used to obtain very thin single crystals of outstanding quality and size for some substances which have a tendency to appear in leaflet form.

Anthracene provides the best example of a material which can be prepared in this way. The arrangement required is quite simple: anthracene powder which has been deposited at the bottom of a beaker covered with filter paper is brought rapidly to the melting point. If one takes care to reach a high vapor pressure inside the beaker, one obtains single crystals of the order of a few mm^2 in area and about 0.1 μ thick on the walls of the vessel and upon the filter paper in a few seconds. The largest crystals are thicker than 10 μ and of the order of a few cm^2 in area; and completely homogeneous. To avoid chemical transformations of the anthracene, the vessel should be freed of oxygen by introducing a pellet of solid CO_2. Similarly, light should be excluded.

Unfortunately, this method can be used with profit only for a few substances. Naphthalene, for example, gives crystals which are not as good and are smaller.

(*3*) A third method has been used successfully in the case of naphthalene, durene, and hexamethylbenzene. In it one deposits a solution of the substance in a light, liquid solvent *upon a water surface.* The solvent evaporates and the substance crystallizes in thin single-crystal layers or in the form of individual, thin single crystals. The thickness and the quality of the single crystals depends upon the solvent and upon the concentration when referred to the area of the surface of water employed.

[34] W. L. Broude and A. F. Prichotjko, *J. Exptl. Theoret. Phys.* (*U.S.S.R.*) **22,** 605 (1952).
[35] W. L. Broude and A. F. Prichotjko, *Kristallografia* **1,** 334 (1956).

Whereas petrol ether or hexane are the best solvents for naphthalene and durene, benzene is best for hexamethylbenzene. This method cannot be applied to chemically less stable materials, such as anthracene and pyrene, without taking special precautions to exclude oxygen and light.

(*4*) Crystals having a thickness between 10 and 100 μ and a surface area of the order of cm^2 can be obtained for many materials by slow crystallization from *supersaturated solutions.* A good example of this is provided by hexamethylbenzene in benzene solutions.

Crystals grown by the methods (2), (3), and (4) described in the foregoing, almost inevitably have a single orientation: the surface of greatest area of the thin crystals coincides with the cleavage plane. In some cases, one can obtain other orientations by using the first method, which is based on drawing the melt between quartz plates.[34–36] To achieve this, one can seed the melt, or one can allow a small region which is spontaneously grown and appropriately oriented to grow at the cost of the remainder of the crystal by suitable annealing.

d. Orientation

The orientation of the crystals can usually be obtained very easily. One determines the cleavage plane in thick crystals and uses the methods of crystal optics to determine and differentiate between the axes. In general, a konoscopic observation in the polarization microscope[37] is sufficient. The optical data have been collected by Winchell.[38] It frequently is necessary to refer to the older data of Groth.[39] Once the cleavage plane and the orientations in it are known, it is relatively easy to determine the orientations of other planes.

Using these simple optical methods, it is possible to determine the axes of the crystal to about 1 or 2 degrees of accuracy without difficulty. The crystal can then be mounted with the corresponding orientation in the crystal holder of the experimental arrangement.

e. Thickness

The thickness of the thin crystals can be determined in the following ways: for thickness greater than 10 μ, it is only necessary to use the *area* and the *weight;* for thickness between 1 and 10 μ, one can use the *doubly refracting interference colors* between crossed or parallel Nicols. If the

[36] W. L. Broude, *J. Exptl. Theoret. Phys.* (*U.S.S.R.*) **22,** 600 (1952).

[37] F. Rinne and M. Berek, "Anleitung zu Optischen Untersuchungen mit dem Polarisationsmikroskop." Schweizerbarth, Stuttgart, 1953.

[38] A. N. Winchell, "The Optical Properties of Organic Compounds." Academic Press, New York, 1954.

[39] W. Groth, "Chemische Kristallographie." Teubner, Leipzig, 1918.

double refraction is known, one can determine the thicknesses immediately from available color tables for polarization microscopy. The values of double refraction lie in the range between about 0.1 and 0.2 for most of the crystals investigated. The crystals appear gray when less than about 1 μ thick.

Crystal thicknesses less than 1 μ can be determined with the use of Newton's interference colors in transmitted or reflected light. Such crystals behave like interference filters as a result of the interference of light reflected from the front and back sides. The relation between crystal thickness d and the wavelength of light of maximum transmission is given by the equation $d = N\lambda/2n$. The relation for minimal transition is $d = (2N + 1)\lambda/4n$, in which N is the order of interference and n is the refractive index. This method can be used to determine crystal thicknesses down to the order of 0.1 μ.

Bree and Lyons[40] have provided a careful test of the double beam method and have described a similar accurate multiple beam method.

f. Mounting

Among the methods developed to mount the crystals in the crystal holders, the following has proved to be particularly good. In all cases one employs round plates of quartz glass of the same size which are optically flat and polished. Thick crystals are fastened to the plates with a template made of pasteboard or copper sheet which is mounted by pressing or gluing. If one desires to avoid the boundary between the quartz and the crystal, one can replace the quartz support with a metal disk of the same size having an appropriate opening over which the crystal is mounted.

Thin crystals sometimes adhere to the quartz disk partly on their own and arrange themselves smoothly. This is true, for example, of anthracene. One only need supply an appropriate diaphram which leaves the crystal free.

g. Low Temperatures

In order to carry out measurements, it is necessary to cool crystals to very low temperatures. The cooling agents normally employed are liquid oxygen (90°K), nitrogen (77°K), hydrogen (20°K), and helium (4°K). It is highly desirable to have an arrangement whereby arbitrary intermediate temperatures can be attained and stabilized. Broude[41] has described such an arrangement.

Usually, however, the experiments are limited to the extreme temperatures determined by the cooling media. In the case of liquid oxygen

[40] A. Bree and L. E. Lyons, *J. Chem. Soc.* p. 2658 (1956).

[41] W. L. Broude, W. S. Medwedjew, A. F. Prichotjko, *J. Opt. i Spektr.* **2,** 317 (1957).

and nitrogen, it is simplest to employ a cuvette holder in the form of a bulb which may be cooled and mounted in a Dewar of quartz glass for thermal isolation. The cooling fluid is placed in the inner part of the metallic bulb. The crystal itself does not come in contact with it. Such low-temperature bulbs were employed in the absorption spectroscopy of

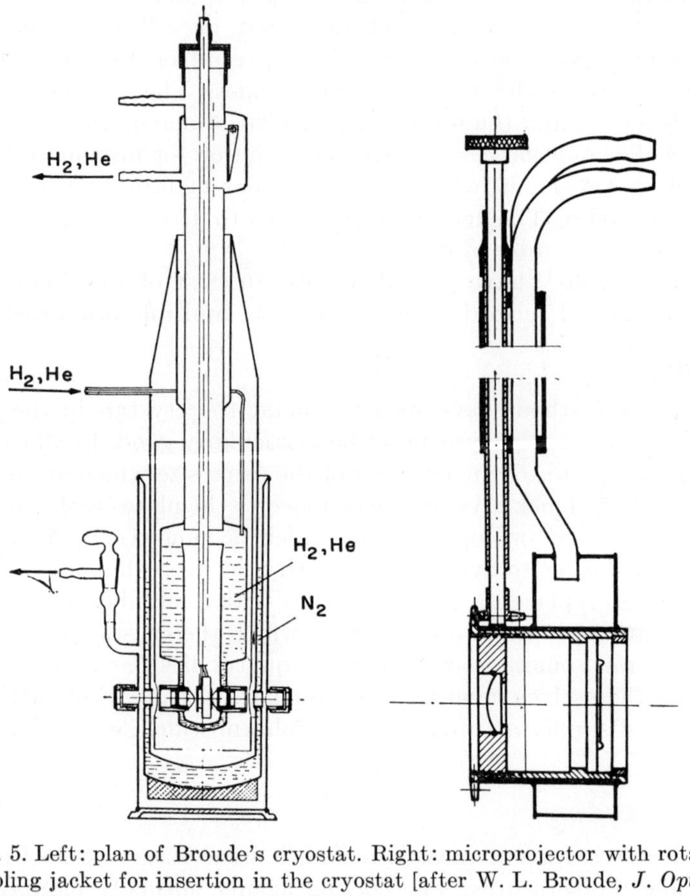

FIG. 5. Left: plan of Broude's cryostat. Right: microprojector with rotary device and cooling jacket for insertion in the cryostat [after W. L. Broude, *J. Opt. i Spektr.* **2**, 318 (1957)].

solutions in various arrangements which have been described in the literature.

It is necessary to employ more complex systems in going to lower temperatures. In general, double cooling, employing combinations of nitrogen and hydrogen or nitrogen and helium, is necessary. Apparatus that makes it possible to avoid placing the crystal in contact with the cooling fluid, but in which it is cooled indirectly, is regarded to have greatest merit. Appropriate arrangements have been described by Pesteil[42] and Broude.[41] The cryostat arrangement (Fig. 5) described by the latter has required a long and careful development. It involves the minimum consumption of helium for operation and, as a result of the presence of a lock, permits manipulation of the crystal without the need of warming the entire cryostat.

h. Absorption Spectra

The absorption spectra are usually measured photographically. The simplest arrangement is sketched in Fig. 6. The light of a continuous light source (tungsten ribbon lamp in the visible, high-pressure xenon lamp in the visible and ultraviolet to about 2400 A, hydrogen lamp below about 3000 A) passes through both the crystal and the polarization prism as a parallel beam and is focused upon the slit of the spectrograph in

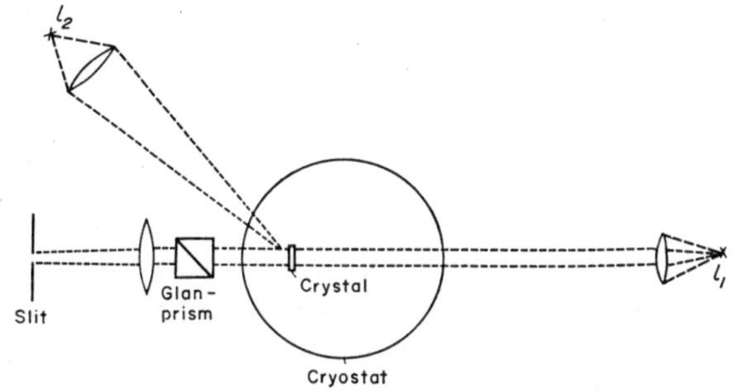

FIG. 6. Experimental arrangement for photographic measurements of polarized absorption and fluorescence spectra.

such a way that a somewhat diffuse and magnified image of the crystal is produced.

Frequently one is satisfied with a measurement of the position of the bands or lines and an approximate estimate of the intensity of absorption, that is, of the absorption coefficient. It is then only necessary to measure the darkening of the photographic plate with a recording microphotometer or with a comparator in the usual manner. A quantitative measurement of the absorption spectra of crystals involves very great difficulties. One must take account of the losses arising from reflection and scattering. The loss arising from reflection can be eliminated by comparing the absorption of a thin and a thick crystal. The loss due to scattering can be determined in part, namely, to the extent that it does not depend upon wavelength, by carrying out measurements in a region of the spectrum

[42] P. Pesteil and M. Barbaron, *J. phys. radium* **15**, 92 (1954).

free of absorption. Maier and Wimmel[43] have provided a good example of the quantitative measurement of absorption.

In each case it is necessary to test the validity of Beer's law by direct measurements on crystals of varying thickness in order to provide evidence that one is dealing with true absorption and not extraneous disturbances.

The spectrum lines of a low-pressure mercury lamp or of an iron arc can be used for calibration of the spectra. Equipment which places more demands upon technique, termed the "microprojector," has been employed by Broude[41] (Fig. 5). The crystal mount and the window of the Dewar contain a combination of quartz lenses which collect the light falling on the crystal and project an arbitrary portion of the crystal on the slit of the spectrograph.

i. Fluorescence

An arrangement essentially identical with that used for absorption can be employed in determining the fluorescence spectrum. The photographic method is still indispensable because of the very weak intensities of light. Quantitative measurements of the fluorescence spectrum, that is, measurements of the yield, are almost always avoided. Kortüm and Finckh[44] have provided an example of such a measurement of the yield. In general, the investigators limit themselves to an exact measurement of the position of the bands and to an estimate of the relative intensities.

In order to excite the luminescence, it is necessary to use sources of light having high intensity, particularly the high-pressure mercury and xenon lamps. The mercury line at 2537 A is particularly useful in the short wavelength ultraviolet, in spite of low luminous intensity which can be achieved. Metallic arcs can also be used, although the latter are not employed frequently. Kasha[45] has described a number of combinations of filters which can be used to separate out desired regions of the spectrum.

A method of fluorescence excitation "from the front," that is, on the crystal facing the slit of the spectrograph, is generally preferable since less exciting light enters the spectrograph. Moreover, one avoids much of the loss of fluorescent light by reabsorption in the crystal, as occurs in a simple transmission arrangement.

j. Polarization

In order to study polarization it is necessary to employ a uv transmitting prism of calcite or a Wollaston prism. The latter is particularly

[43] W. Maier and H. Wimmel, *Z. Elektrochem.* **59,** 876 (1955).

[44] G. Kortüm and H. Finckh, *Z. physik. Chem.* (*Leipzig*) **B52,** 263 (1942).

[45] M. Kasha, *J. Opt. Soc. Am.* **38,** 929 (1948).

useful for measuring simultaneously the two spectra which are polarized normally to one another. The spectrograph and the remaining optical apparatus should not introduce additional polarization or alter the plane of polarization of the light. Thus, quartz glass is to be preferred over crystalline quartz. Optical units made of crystal quartz can falsify the intensity distribution in the spectrum[46] as a result of rotational dispersion.

In each case one must make certain that the assembled optical system does not favor any of the polarization directions which are to be measured. This can be done best by making a polarization measurement on a crystal before and after rotations through 90°. The two photographs should give the same ratios of intensity for the two spectra which are polarized perpendicular to one another.

The optical properties of the crystal must be taken into account in each interpretation of a polarized crystal spectrum. One measures the three principal components of the polarized spectrum parallel to the three principal optical axes. If the direction of polarization of the light vector does not coincide with one of the principal optical axes, the light emerging from the crystal can be elliptically polarized in a manner depending upon the thickness of the crystal. Under these circumstances the intensity of light emerging from the crystal in two mutually perpendicular directions of polarization cannot be used to determine the absorption coefficients in these directions without further analysis. This difficulty does not occur when the absorption constant is measured in the customary way with the use of natural light.

In many cases the interesting crystal axes coincide with the principal optical axes. Thus, in monoclinic naphthalene and anthracene, the *b* axis is a principal axis. The two other principal axes must be perpendicular to it. Their projections upon the *ab* and *bc* planes fall upon the crystal axes *a* and *b* as a result. Light which is polarized parallel and perpendicular to the *b* axis and falls perpendicular to the planes *ab* or *bc* can, in consequence, not be elliptically polarized. There is no such simple relationship between the optical and the crystallographic axes in the *ac* plane. Little is known concerning the exact crystal-optical calculations of the polarized spectra when the direction of polarization is oriented arbitrarily in the crystal.

k. Optical Constants

Measurements of the *reflection spectra* and the *refractive indices* can represent a very valuable and meaningful addition to information concerning the absorption and fluorescence spectra.

According to dispersion theory, maxima of absorption are simultaneously maxima of reflection. As a matter of fact, the reflection spec-

[46] H. C. Wolf and D. Griessbach, *Naturwiss.* **42,** 206 (1955).

trum from clear cleavage faces duplicate the absorption spectrum. This method of determining the absorption spectra can be used to advantage when the absorption constant is exceedingly large. It is then possible to determine the relative intensity and the polarization, at least in an approximate manner, by making measurements of the reflection on thick crystals. The papers given in refs. 47–50 provide examples of the measurement of reflection spectra. The spectra obtained by reflection studies do not provide a faithful representation of absorption if the crystal surface is clouded and not smooth.

The refractive index and its dispersion can be measured with the elegant method of Fresnel[50] or by the method of channelized spectra[51] on good single crystals having plane parallel faces. The last method is primarily an application of the method described in Section 3e to determine the thickness. It requires no particular additions or modifications. Such channelized spectra can be obtained for each absorption and fluorescence measurement on good single crystals having plane parallel faces. Only a few maxima and minima can be seen in the case of very thin single crystals; however, a large sequence of striations often are visible in the case of thicker crystals. The striations tend to converge toward the wavelength at the absorption limit because of the high dispersion.

It is possible to use the foregoing method to determine the refractive index in the entire spectral range if one knows the thickness of the crystal and the value of the refractive index for a particular value of the wavelength in the range of the channelized spectra. An appropriate dispersion formula describing the dependence of the index upon the wavelength can then be used to determine the oscillator strength. This method of analysis provides an additional possible way of determining the relations between the oscillator strengths or the absorption coefficients in two different crystallographic directions.[47–50]

l. Photoconductive and Excitation Spectrum

The photoconductive spectrum, giving the photocurrent as a function of the wavelength of incident radiation, and the excitation spectrum, giving the intensity of fluorescence as a function of the exciting wavelength, can be used to obtain at least an estimate of the absorption spectra of the crystal in many cases. In general, these methods do not yield an absolute value for the absorption coefficient. However, one can

[47] A. Bree and L. E. Lyons, *J. Chem. Soc.* p. 2662 (1956).

[48] I. W. Obreimov, A. F. Prichotjko, and J. W. Rodnikowa, *J. Exptl. Theoret. Phys.* (*U.S.S.R.*) **18,** 409 (1948).

[49] A. Eitchiss, *J. Exptl. Theoret. Phys.* (*U.S.S.R.*) **20,** 471 (1950).

[50] Prichotjko *et al.*, *J. Opt. i Spektr.* **2,** 448 (1957).

[51] H. C. Wolf, *Z. Naturforsch.* **13a,** 414 (1958).

obtain the position and the polarization of the bands with reasonable precision, if high resolution is not required. The tetracene spectrum determined by Bree and Lyons[52] provides an example of a photoconductive spectrum.

m. Comparison with Other Spectra

It often is very valuable to compare the spectrum of a given crystal with other spectra in order to obtain a better understanding of the structure. Particularly valuable comparisons can be made with *spectra of solutions, of mixed crystals, and of deuterated substances.*

The measurements of the spectra of solutions are made so frequently that it is not necessary to discuss the methods in detail here. It usually is convenient to use a monochromator-multiplier arrangement because of the large band or line widths. One uses mixtures which solidify in a glass-like matter as solvents (for example, 96% ethanol) in making measurements at the temperatures of liquid oxygen or nitrogen.

The exact measurement of the spectra of mixed crystals was introduced into the spectroscopy of molecules and crystals by McClure.[53] In using the term "mixed crystal spectra" we shall imply spectra obtained from systems in which the molecule under investigation is dispersed as "guest" in a "host" crystal in an oriented manner. It is necessary that the guest G be soluble in the host crystal H. Moreover, H must be transparent for the spectrum of G under investigation. An anisotropy of the spectra indicates that G is incorporated in an oriented way in the crystal. The concentration of the guest can be determined by dissolving a crystal in a solvent and measuring the absorption. A comparison with the spectrum of the pure crystal of the guest molecule will indicate whether the molecules are incorporated in the host in a molecular manner or in the form of small crystalline domains[54] (Figs. 20 and 21).

A molecule which is incorporated in another crystal as a guest differs from a molecule of the pure crystal by the absence of a resonance interaction with its neighbors. Its environment is very similar in other respects if the molecules of the host crystal are closely related to it. Thus, by comparing the spectra of mixed and pure crystals, it is possible to obtain specific information concerning the influence of the resonance interactions upon the spectra.

The crystalline environment of the guest molecule is, however, also anisotropic. As a result, one cannot expect to regard the guest molecules in the mixed crystals simply as constituting an "oriented gas." In actual-

[52] A. Bree and L. E. Lyons, *J. Chem. Phys.* **22,** 1630 (1954).
[53] D. S. McClure, *J. Chem. Phys.* **22,** 1668 (1954).
[54] H. C. Wolf, *Z. Naturforsch.* **10a,** 244 (1955).

ity, the spectra of the guest molecules in mixed crystals also exhibit characteristic variations relative to the disordered solution which represent the symmetry character of the host crystal.

This, and the fact that one does not know the position of the guest molecule in the host, introduces problems in the evaluation of the spectra of mixed crystals. Usually it is possible to make plausible assumptions concerning the orientation of the incorporated molecules. If the position of the transition moment in the free molecule is known from other measurements, one can determine the orientation in which it is incorporated from the polarization of the spectra. Usually, however, one is in the reverse situation in which it is necessary to make assumptions concerning the orientation in order to determine the symmetry of the excitation state of the free molecule.

It is worth mentioning that the guest molecule in mixed crystals can exhibit strong phosphorescence, even when the pure crystal does not phosphoresce.

In general, one observes only the spectrum of the guest in mixed crystals. Observations of host fluorescence, which is weakened by the transfer of energy to the guest molecules, indicate that the widths of the lines are decreased[10] as a result of the incorporation of guest molecules in the crystal. To explain this, it is assumed that the guest molecules interrupt the propagation of energy in the crystal. Thus, the interaction with the lattice vibrations which determine the line broadening in the crystal is less.

The spectra of crystals of deuterated substances can be very important when it is not certain from observation of the band or line spacing whether one is dealing with a molecular or a crystal term. One can estimate the direction in which the molecular vibrational frequencies are displaced as a result of the change in mass and, thus, in many cases can distinguish whether the line separation can be explained in terms of the molecular vibrational frequencies.[55]

n. Sources of Error

Finally, we shall collect the sources of error which can falsify the interpretation of the spectra. Some of these have already been mentioned. They are responsible for the fact that the observations of different authors on spectra, which are investigated frequently, have not yet been interpreted uniquely in some cases.

(*1*) *Reabsorption.* Reabsorption decreases the intensity of fluorescent light in the region where the absorption and the fluorescence spectra

[55] D. S. McClure, *J. Chem. Phys.* **24,** 1 (1956).

overlap. This can happen to the 0–0 bands at low temperatures, for example.

If the 0–0 bands or lines overlap completely in both absorption and fluorescence, the principal effect is that the intensity of fluorescence of the 0–0 bands is weakened relative to that of the other spectra. In general, however, broad 0–0 bands overlap only at the edges in absorption and fluorescence. As a consequence, the apparent wavelength of the maximum of intensity of the 0–0 bands observed in fluorescence is shifted toward longer wavelengths (Fig. 7). Thus, a gap between absorption and fluorescence can be either generated or amplified. It is possible to avoid reabsorption to a considerable extent by exciting the fluorescence "from the front" on thin crystals. Furthermore, one can estimate the influence by calculation and, thus, correct the measured values.

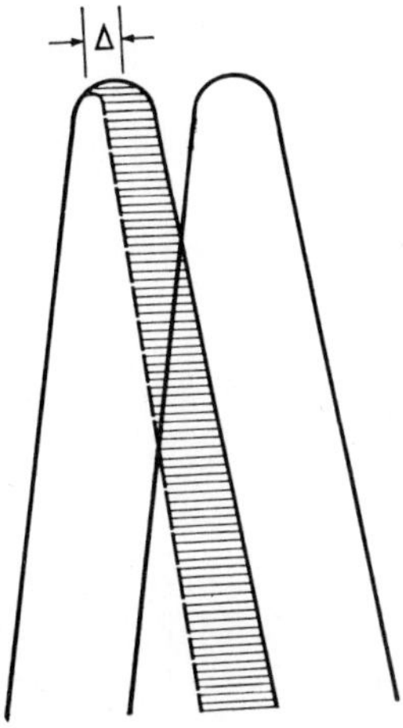

FIG. 7. Schematic representation of the falsification of a fluorescence band by reabsorption. The dashed part of the fluorescence band is removed as a result of reabsorption.

(*2*) *Phase change.* If the crystal undergoes a *phase change* during cooling, it may become polycrystalline and break into small fragments. This is observed in pyrene, for example. There are, however, examples in which the phase transition occurs and in which the specimen remains a single crystal. This is usually the case, for example, for hexamethylbenzene. In such instances one can detect the change in phase by observing the variation in double refraction between polarizing prisms or by observing the change in the spectrum. If a change in phase seems possible, it is advisable to measure the spectrum of the crystal at some intermediate temperatures as well as at room and low temperatures. One can then detect a discontinuous change in the temperature dependence of the spectrum resulting from the phase change.[56] Little is known regarding

[56] H. C. Wolf, *Z. Naturforsch.* **13a**, 336 (1958).

the nature of the lattice changes which accompany the phase changes that have been observed below room temperature to date. The variations in the optical properties display definite clues.[36]

(*3*) *Effect of attachment.* If the crystal becomes *rigidly attached* to quartz, which has a thermal expansion coefficient that is three powers of ten smaller than that of the organic material, the position, the sharpness, and even the polarization of the absorption bands can be altered as a result of different expansion.[57,58] The situation differs depending often on whether the crystal is bound to the quartz on one or two sides (Fig. 8). This effect has been studied very little to date. It is possible that it is responsible for the difference in the experimental results quoted by various authors.

(*4*) *Interference.* The interference by multiple reflection described in the foregoing can produce deceptive results, particularly for thin crystals where it is not easily recognized as such. It may seem to produce

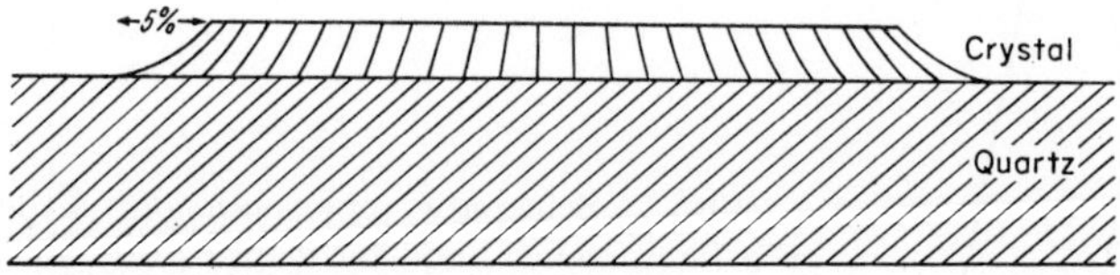

FIG. 8. The stresses induced during cooling in a crystal which is fastened on one side to a quartz mount [W. L. Broude, O. S. Pachamova, and A. F. Prichotjko, *J. Opt. i Spektr.* **2,** 323 (1957)].

sharp absorption and fluorescent lines. It is possible to distinguish this effect easily from that associated with true crystal or molecular lines by studying the spectra of crystals having different thicknesses and comparing the results, that is, by testing the validity of Beer's law.

It is necessary to take account of the loss in light resulting from reflection in the vicinity of the absorption edge where the refractive index rises very rapidly. The apparent wavelength of the peaks of the broader absorption and fluorescence bands can be altered somewhat as a result of this loss originating in reflection.

(*5*) *Errors in adjustment.* Small *errors in adjustment* of the crystals and the polarizing prisms can produce "false" light in a given direction of polarization and, hence, can falsify the absorption spectrum. This possibility is particularly strong when the crystal absorbs intensely in one direction and is completely transparent in the direction normal to it.

[57] W. L. Broude, O. S. Pachamova, and A. F. Prichotjko, *J. Opt. i Spektr.* **2,** 323 (1957).

[58] W. L. Broude and A. F. Prichotjko, *J. Opt. i Spektr.* **1,** 102 (1956).

For example, an error of adjustment of 2° will induce a "ghost structure" of the first absorption band near the absorption limit of an anthracene crystal. This "structure" can be removed by an appropriate rotation of the polarizing prisms.[51]

4. Results of Measurement

We shall survey the measured spectra in the following section, restricting attention to those crystals for which complete data concerning the absorption and fluorescence of crystals and solutions are available. The materials to be surveyed are as follows: a, benzene; b, hexamethylbenzene; c, naphthalene; d, anthracene; e, phenanthrene; f, durene and other derivatives of benzene and naphthalene; g, other aromatic hydrocarbons; h, other organic molecular crystals.

The experimental data concerning the crystal structure and the spectra of solutions and crystals of these substances will be compiled. Theoretical and spectulative considerations of the spectra will be omitted to the extent that this is possible, since they will be treated in Part IV of this article in a unified way. In some cases the principal details will be presented in a strongly schematic manner in the sense that the discussion will be restricted to the strongest bands or lines. The reader is referred to the original articles for detailed discussion of the weaker lines, the lattice frequencies, and similar matters. Likewise, details which are important for the physics of the molecules but are less interesting for the physics of the crystals will be omitted frequently.

a. Benzene

The spectra of benzene are relatively simple and easy to display. Thus, they serve as an appropriate introduction to the discussion.

The *crystal structure* of benzene has not yet been clarified in all details. The crystal is orthorhombic-bipyramidal with four molecules in the unit cell.[59,60] An approximate projection upon the cleavage plane *ac* is given in Fig. 9. The slope of the normal to the molecular planes relative to the *b* axis is 77°.

The *solution spectrum* is shown in Fig. 10. The fluorescence and absorption spectra resemble mirror images much as in the well-analyzed spectrum of the vapor.[61] The most intense bands of the first transition may be combined into a "molecular series," the *M* series, in which the

[59] E. G. Cox, *Proc. Roy. Soc.* **A135**, 491 (1932).

[60] E. G. Cox and J. A. S. Smith, *Nature* **173**, 75 (1954); E. G. Cox, D. W. J. Cruickshank, J. A. S. Smith, *Proc. Roy. Soc.* **A247**, 1 (1958).

[61] H. Sponer, G. Nordheim, E. Teller, and A. L. Sklar, *J. Chem. Phys.* **7**, 207 (1939).

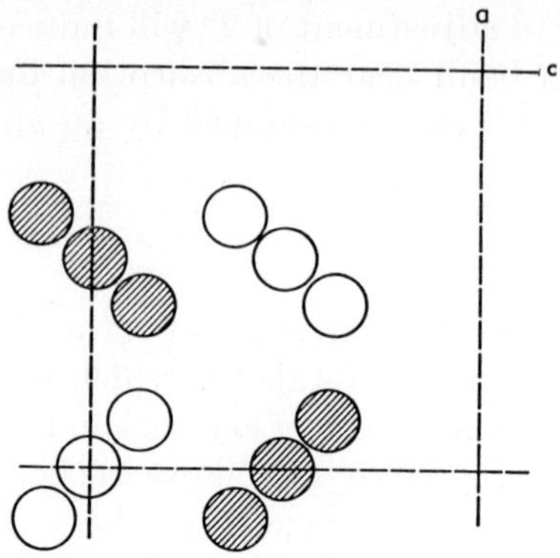

FIG. 9. Projection of the unit cell of the benzene crystal upon the plane *ac*. The cross-hatched molecules lie a distance $b/2$ higher than the two other [after E. G. Cox, *Proc. Roy. Soc.* **A135**, 491 (1932)].

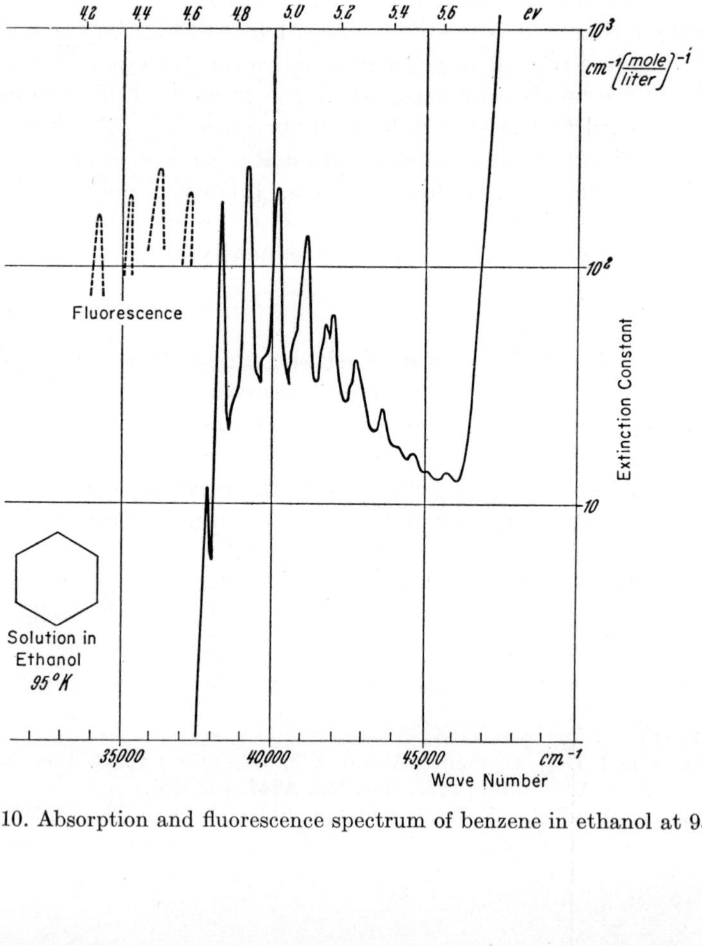

FIG. 10. Absorption and fluorescence spectrum of benzene in ethanol at 95°K.

0–0 band is absent:

$$M \text{ series} = 0\text{–}0 \text{ (forbidden)} \pm \nu_0 \pm n_1\nu_1 \pm n_2\nu_2$$

with

$n_1, n_2 =$ 0, 1, 2, 3

0–0 = 37,840 cm^{-1} in ethanol at 95°K.

$\nu_0 =$ Deformation vibration E_g with about 600 cm^{-1} in the ground state, 540 cm^{-1} in the excited state. This is the single excited vibration which provides the forbidden transition with its intensity.

$\nu_1 =$ Pulsation vibration A_{1g}(C—C) with 995 cm^{-1} in the ground state and 930 cm^{-1} in the excited state. This is a multiply excited vibration.

$\nu_2 =$ Pulsation vibration A_{1g}(C—H) with about 3050 cm^{-1} in the ground state and 2520 cm^{-1} in the excited state. This is usually multiply excited.

The upper sign is valid for absorption and the lower for fluorescence.

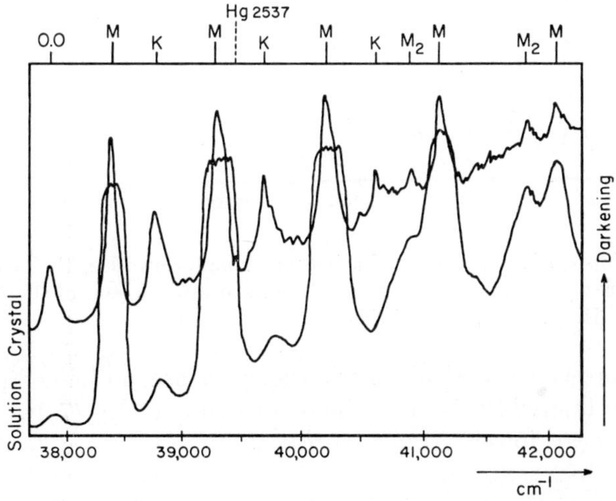

FIG. 11. Absorption spectrum of benzene in ethanol at 95°K. Also the unpolarized absorption spectrum of the crystal at 95°K. Curves obtained by a recording microphotometer. The K series is stronger in the crystal [H. C. Wolf, unpublished].

A second weak spectral series which is absent for the vapor appears even in the solution. It is intense in the crystal (Fig. 11) and for this reason is called the K series:

$$K \text{ series} = 0\text{–}0 \text{ (allowed)} \pm n_1\nu_1.$$

The *crystal spectrum* was investigated by a number of workers. The unpolarized form was studied by Kronenberger[19] at 20°K; the polarized form was studied by Broude and co-workers[34,41,62] at 77°K and 20°K. The latter was also studied by Zmerli and Pesteil[63] at 20°K and by Wolf[64] at 77°K and 90°K. Davydov[26] and Fox and Schnepp[65] have been concerned with the theoretical interpretations.

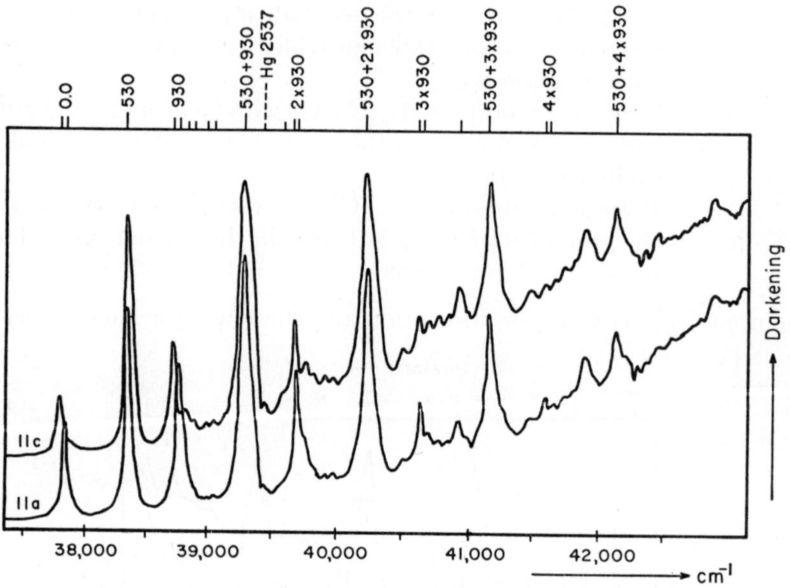

FIG. 12. Absorption spectrum of a benzene single crystal in the *ac* plane at 77°K taken with a recording microphotometer. The most intense lines of the *K* and *M* series are indicated [H. C. Wolf, unpublished].

The fluorescence and absorption spectra are no longer plain mirror images. The line width is only about 5 to 10 cm^{-1} at 77°K and becomes even smaller at lower temperatures, attaining values between 2 and 4 cm^{-1} at 20°K.

The *absorption spectrum* (Fig. 12) consists again of two series:

$$M \text{ series} = 0\text{–}0 \text{ (forbidden)} + \nu_0 + n_1\nu_1 + n_2\nu_2 + m(\nu_3 \cdots \nu_9)$$

in which $n_1, n_2 = 0, 1, 2, 3 \ldots; m = 0$ or 1. The intensity of this series is essentially the same in directions parallel to all three crystal directions.

[62] W. L. Broude, *J. Exptl. Theoret. Phys. (U.S.S.R.)* **21,** 665 (1951).
[63] A. Zmerli, H. Poulet, and P. Pesteil, *Colloque de Bellevue* (July 4, 1957).
[64] H. C. Wolf, unpublished (1958).
[65] D. Fox and O. Schnepp, *J. Chem. Phys.* **23,** 767 (1955).

The corresponding K series is

$$K \text{ series} = 0\text{--}0 \text{ (allowed)} + n_1\nu_1.$$

The terms of these series are split into two (or perhaps three) components. There is an intense component parallel to a and c and perhaps a weak component parallel to b.

The following numerical values are valid for the M series:

extrapolated position of 0–0

37,830 (77°K Wolf)
37,820 (20°K Pesteil and Zmerli)
37,800 (77°K Broude)
37,825 (20°K Broude)

two A_{1g} frequencies which form two periods

$$\nu_1 = 925 \text{ cm}^{-1}$$
$$\nu_2 = 2540 \text{ cm}^{-1}$$

an E_g frequency which makes the forbidden transition allowed

$$\nu_0 = 530 \text{ cm}^{-1}.$$

In addition to this, there are about six additional modes $\nu_3 \ldots \nu_9$, which are only simply excited and which have been not analyzed in detail. They have the approximate values 70, 550, 625, 700, 770, and 1200 cm^{-1}.

According to Broude,[41] the terms of the M series can be resolved into at least four components at 20°K. The frequency separations between these components are the same in all terms. It seems probable that this effect is a result of the superposition of lattice vibrations and molecular vibrations.

The K series is induced by the crystal field. One can draw conclusions about the crystal planes involved from the number of intense components (two) into which the lines are split. Normally, one obtains the ac plane. This assignment, suggested by Davydov, remains valid even when one takes into account the variations in crystal structure[60] which have been discovered since the work of Davydov. The ac plane is also expected to be the natural cleavage plane of benzene crystals. Finally, it is also possible to test this assignment by making measurements of the refractive indices.[63] The assignment of the two perpendicularly polarized spectra in the ac plane to the a and c directions is not as definite.

In the present article we shall follow the assignment proposed by Broude.[41] According to this analysis, the terms and components of the K

TABLE I. THE MOST INTENSE LINES OF THE K SERIES IN THE SPECTRUM OF CRYSTALLINE BENZENE

No.	Analysis	Measurement	Wave number of the components in cm^{-1}: c (strong)	a (strong)	b (weak)	Splitting a/c
1	0–0	90°K Wolf	37,800	37,845		45
		77°K Broude	37,786	37,811		25
		20°K Broude	37,803	37,839	37,846	36
		20°K Pesteil	37,798	37,842		44
2	923	90°K Wolf	38,730	38,775		45
		77°K Broude	38,702	38,730		28
		20°K Broude	38,724	38,764	38,772	40
3	2 × 923	99°K Wolf	39,680	39,710		30
		77°K Broude	39,635	39,656		19
		20°K Broude	39,665/72	39,687	39,696	20
4	3 × 923	90°K Wolf	40,600	40,620		20
		77°K Broude	40,581	40,586		5
		20°K Broude	40,600	40,615		15
5	4 × 923	90°K Wolf	41,535	41,545		10

series are those given in Table I. The measurements of Broude on the one hand, and of Pesteil and Wolf on the other, do not agree with one another exactly. The deviations exceed the limits of error considerably. It is clear, however, that the magnitude of the splitting is very different for the different lines and decreases in the direction of decreasing oscillator strength.

Only a doublet (a/c) is observed at 90° and 77°K. However, one can resolve an additional splitting of this doublet at 20°K. One of these weak components of the splitting was interpreted by Broude as the b component of 0–0. On the other hand, Pesteil and Zmerli regard all new resolved components as the result of lattice vibrations.

The *fluorescence spectrum* is more intense than in solution, perhaps because the quenching action of oxygen is less effective. A microphotometer tracing of the fluorescence spectrum of a single crystal at 77°K is shown in Fig. 13.

The most intense lines can be arranged into two series. One of these is described as follows:

$$M \text{ series} = 0\text{–}0 \text{ (forbidden)} - \nu_0 - n_1\nu_1$$

in which

$$n_1 = 0, 1, 2, 3 \ldots$$
$$0\text{–}0 = 37{,}800 \text{ cm}^{-1} \text{ (extrapolated)}$$
$$\nu_0 = E_g \text{ vibration } 600 \text{ cm}^{-1} \text{ simply excited}$$
$$\nu_1 = A_{1g} \text{ vibration } 995 \text{ cm}^{-1} \text{ multiply excited.}$$

The terms of the *M* series actually have two satellites at a distance of about 65 and about 130 cm^{-1}. These separations were not the same in all crystals investigated.

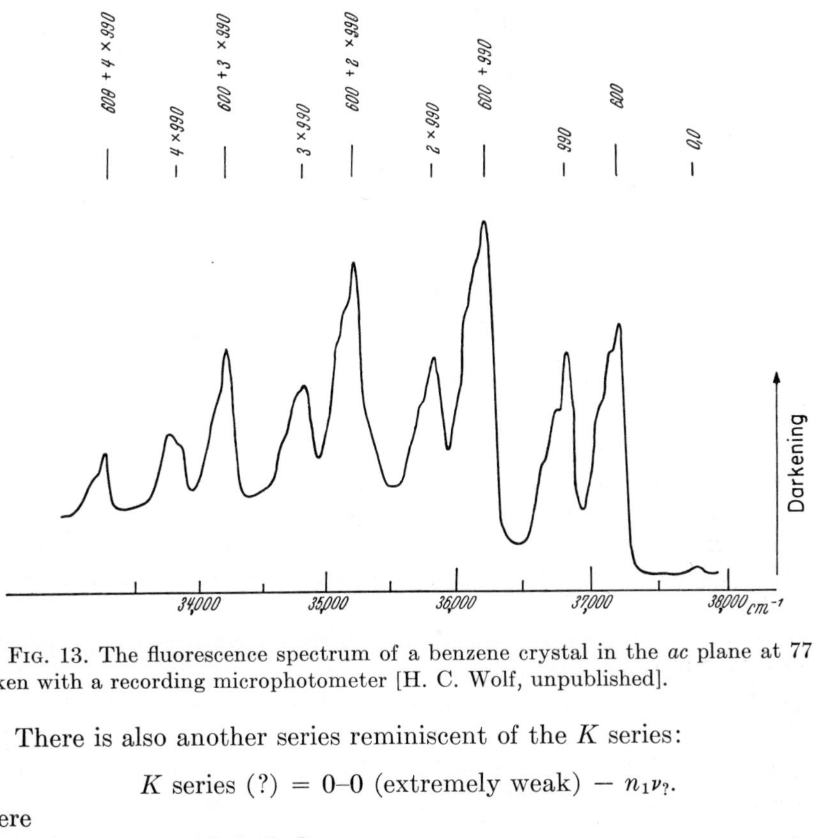

FIG. 13. The fluorescence spectrum of a benzene crystal in the *ac* plane at 77°K taken with a recording microphotometer [H. C. Wolf, unpublished].

There is also another series reminiscent of the *K* series:

$$K \text{ series (?)} = 0\text{–}0 \text{ (extremely weak)} - n_1\nu_?.$$

Here

$$n_1 = 0, 1, 2, 3 \ldots$$
$$0\text{–}0 = 37{,}800 \text{ cm}^{-1}.\ 0\text{–}0 \text{ is very weak, even when one takes into account reabsorption.}$$
$$\nu_? = 980 - 990 \text{ cm}^{-1}.$$

The members of these series usually have satellites similar to those of the *M* series.

There are two possible ways of interpreting this series: (a) $\nu_?$ is the A_{1g} frequency ν_1; the series is the K series. (b) The combination frequency $\nu_3 + \nu_4$ of the M series obtained from the Raman modes 607 and 406 cm^{-1} of the free molecule is very close to the frequency 995 cm^{-1} given in the foregoing. As a result of this accidental resonance excitation, those terms of the M series which coincide accidentally with the members of the missing K series are very strong. In this case, the second series belongs to the M series. It is very difficult to distinguish between these two possibilities. In both cases, the K series observed in fluorescence will appear different from that observed in absorption. In the first case the splittings are absent; in the second the relative intensities are different.

At first the measurements of Zmerli and Pesteil led to more hypothetical interpretations of the fluorescence of crystalline benzene.[63] Recently, these authors expressed the belief that they have found an explanation for the anomalous form of the fluorescence of benzene. According to Pesteil and Zmerli, the fluorescence spectrum of pure single crystals of benzene which have been degassed and subject to measurement in a vacuum is the complete mirror image of the absorption spectrum and is composed of a K series which is split, and whose 0–0 line coincides with that of absorption, and an M series which has not split. This situation is found only at 20°K. The "anomalous" fluorescence spectra to which the "satellites" visible in Fig. 13 clearly belong, and to which the "K" series of Fig. 13 perhaps belongs, is suggested to be produced by the adsorption of oxygen on benzene.

To complete the absorption measurements involving benzene crystals, Zmerli and Pesteil examined the spectrum of deuterated benzene.[63] The results correspond to those obtained from benzene if one takes into account a general displacement of the spectrum by about 200 cm^{-1} toward the violet.

The crystal properties associated with the benzene spectrum can be summarized in the following way.

(1) The displacement in going from solution to crystal is about 10 cm^{-1}.

(2) A K series appears in addition to the known M series observed for the free molecule. The K series has the structure of an allowed transition. Its terms split into two intense components parallel to the a and c axes. The magnitude of the splitting is different for the individual lines.

(3) The fluorescence spectrum exhibits deviations from the mirror image of the absorption spectrum. The principal cause of this is a displacement of the bands by amount between 500 and 800 cm^{-1} toward the red. The true fluorescence spectrum is obtained by excluding oxygen.

(4) The theory leads to the following additional consequences. One may conclude from the number and polarization of the components of splitting of the K series that the symmetry character of the states of molecular excitation is B_{2u}.

b. Hexamethylbenzene

Hexamethylbenzene (HMB) is one of the molecular crystals investigated most extensively because of its particularly simple *crystal structure.*

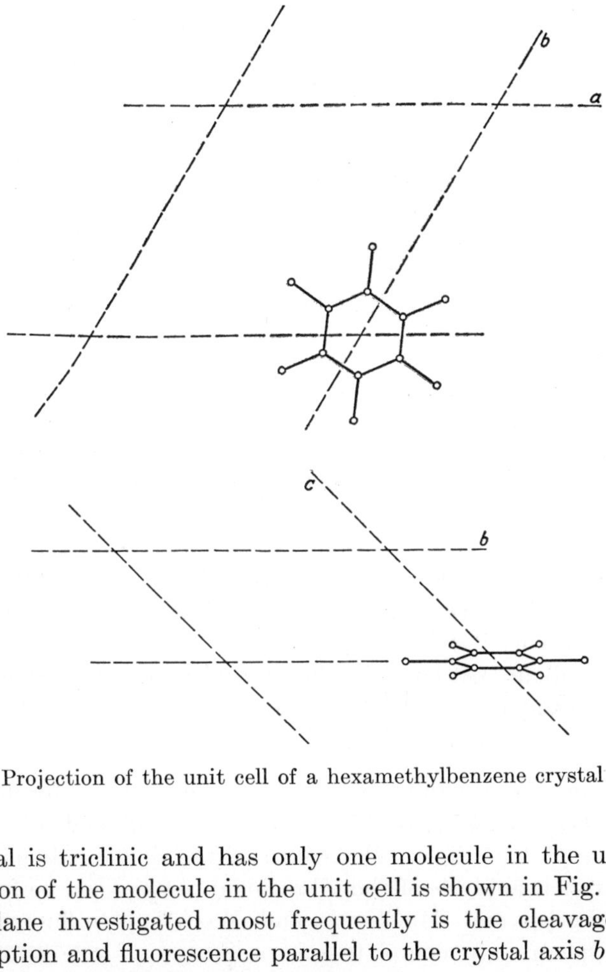

FIG. 14. Projection of the unit cell of a hexamethylbenzene crystal on the plane *ab* and *bc*.

The crystal is triclinic and has only one molecule in the unit cell.[66,67] The position of the molecule in the unit cell is shown in Fig. 14.

The plane investigated most frequently is the cleavage plane *bc*. The absorption and fluorescence parallel to the crystal axis *b* correspond

[66] K. Lonsdale, *Proc. Roy. Soc.* **A123,** 494 (1929).

[67] J. M. Robertson, "Organic Crystals and Molecules." Cornell Univ. Press, Ithaca, 1953.

essentially to processes involving the transition moment in the plane of the molecule. Conversely, measurements normal to b involve the transition moment normal to the plane of the molecule.

The absorption spectrum of crystals of HMB were measured first by Scheibe and his co-workers[21] at room temperature for the lowest transition. Additional measurements at low temperatures by Broude[36] demonstrated the existence of a second phase. The transition point lies at about 110°K. The low-temperature phase (LT phase) was explicitly investigated in absorption at 20°K by Schnepp and McClure.[68] Both phases were investigated at 77°K in absorption and fluorescence by Wolf.[56] Finally, higher transitions were obtained at lower resolution by Nelson and Simpson.[69] In the following, however, we shall discuss only the first transition.

The *spectrum of the solution* (Fig. 15) has the structure of a weak symmetry-forbidden transition which obtains its intensity as a result of the presence of a vibration (ν_5) which is not totally symmetric.

The following equation is valid for this series, which is completely analogous to the M series of benzene:

$$M \text{ series} = 0\text{–}0 \text{ (forbidden)} \pm \nu_5 \pm n_1\nu_1 \pm n_2\nu_2.$$

Here

$n_1, n_2 = 0, 1, 2, 3 \ldots$

0–0 = 35,400 cm^{-1} at 95°K in ethanol

ν_5 = about 520 ± 50 cm^{-1} in the ground state and 420 cm^{-1} in the excited state. The deformation vibration E_g(C—C) = 508 cm^{-1} of the infrared spectrum is the simple excited forelying vibration [or $E_g(C—CH_3)$ = 456 cm^{-1}].

ν_2 = about 1220 cm^{-1} in the ground state and 1200 cm^{-1} in the excited state. The pulsation vibration A_{1g}(C—C) is multiply excited.

ν_1 = about 570 cm^{-1} in the ground state and 580 cm^{-1} in the excited state. The pulsation vibration $A_{1g}(C—CH_3)$ is multiply excited.

Again the upper sign is valid for absorption and the lower for fluorescence.

The *spectrum of the crystal*, shown in Figs. 16 and 17, consists of numerous, very sharp lines. The width of the lines is about 5 cm^{-1} at 20°K. An explicit analysis of the vibrational pattern can be found in the work of Schnepp and McClure[68] and Wolf.[56] The fluorescence and absorption spectra are complete mirror images of one another in all crystallographic directions, the 0–0 frequency being the symmetry line.

[68] O. Schnepp and D. S. McClure, *J. Chem. Phys.* **2a,** 83 (1957).
[69] R. C. Nelson and W. T. Simpson, *J. Chem. Phys.* **23,** 1146 (1955).

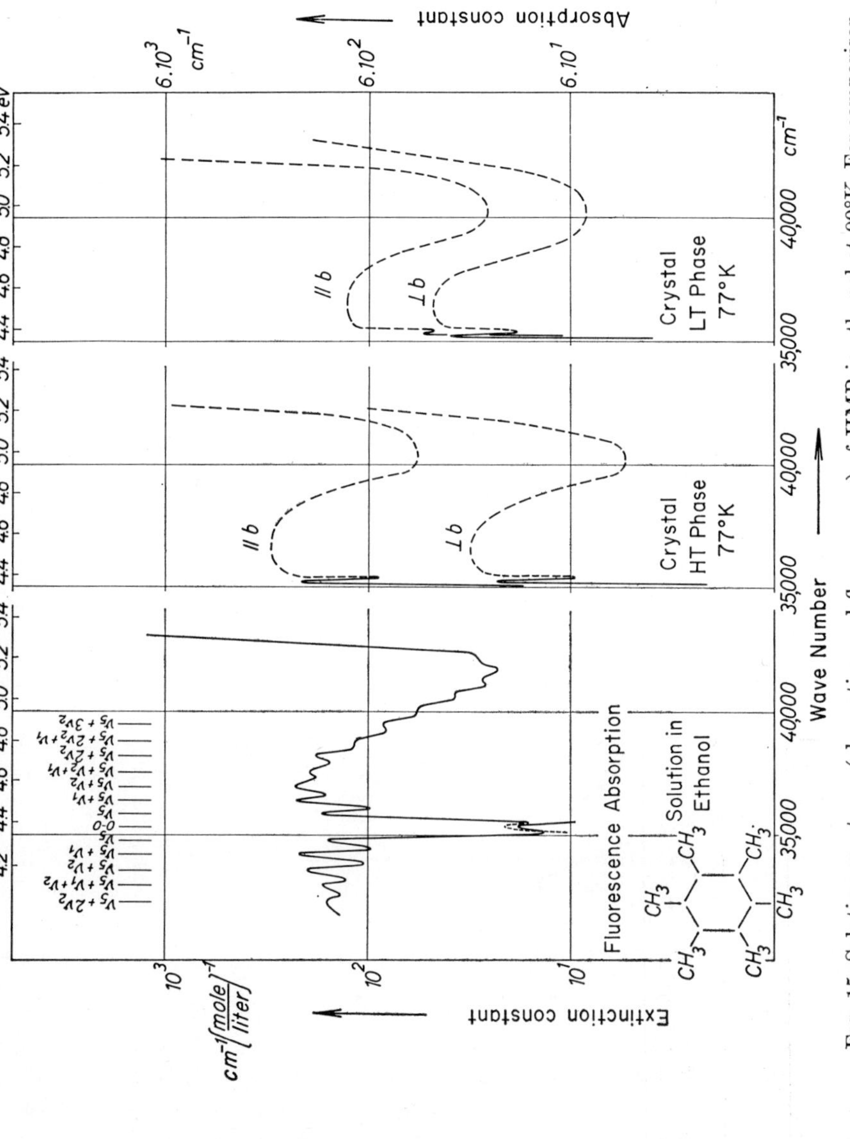

FIG. 15. Solution spectrum (absorption and fluorescence) of HMB in ethanol at 90°K. For comparison, the absorption spectra of the crystal in the high-temperature and low-temperature phases are given on the right. (0–0 line and the envelope of the remaining lines) [H. C. Wolf, Z. *Naturforsch.* **13a**, 336 (1958)].

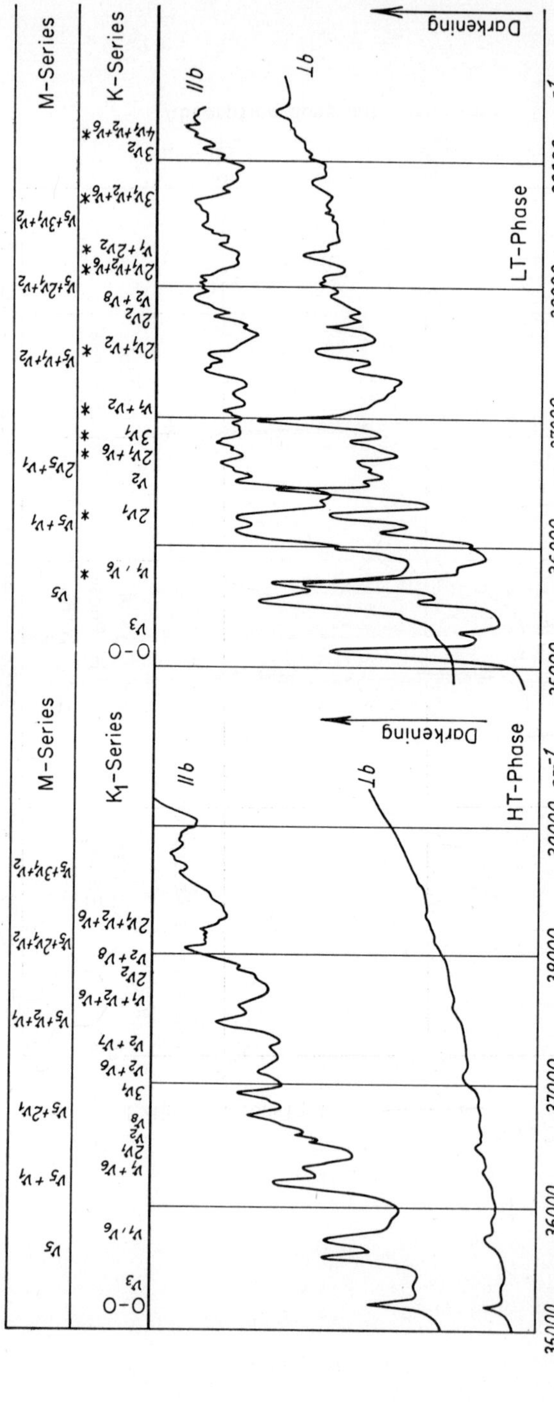

FIG. 16. The absorption spectrum of a HMB crystal in the *bc* plane. Both the high-temperature and low-temperature phases are shown at 77°K. Recording microphotometer curve. The numbering of the lines and divisions in the *K* and *M* series are based on work by H. C. Wolf [*Z. Naturforsch.* **13a**, 336 (1958)].

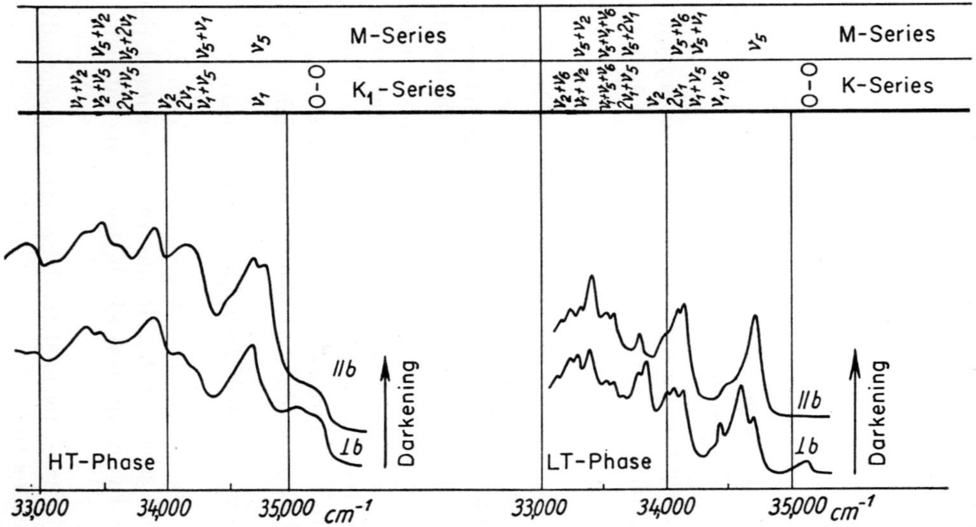

FIG. 17. The fluorescence spectra of HMB crystals in the *bc* plane. The curve for the high-temperature phase was obtained at about 110°K, that for the low-temperature phase at 77°K. The numbering of the lines and divisions in the *K* and *M* series are taken from work by H. C. Wolf [*Z. Naturforsch.* **13a**, 336 (1958)].

One can understand the spectra of the crystals best when attention is restricted to the most intense lines which are ordered into three series. As in solutions, one finds an M series of the form:

$$M \text{ series} = 0\text{–}0 \text{ (forbidden)} \pm \nu_5 \pm n_1\nu_1 \pm n_2\nu_2 \pm m(\nu_3 \cdots \nu_8).$$

Here

$$\begin{aligned} n_1, n_2 &= 0, 1, 2, 3 \ldots \text{ and } m = 0 \text{ or } 1 \\ 0\text{–}0 &= \left.\begin{array}{l} 35{,}200 \text{ cm}^{-1} \text{ in the HT phase} \\ 35{,}135 \text{ cm}^{-1} \text{ in the LT phase} \end{array}\right\} \text{at } 77^\circ\text{K}. \end{aligned}$$

There also is an equally intense series induced by the crystal which does not have a counterpart in solution:

$$K_1 \text{ series} = 0\text{–}0 \text{ (allowed)} \pm n_1\nu_1 \pm n_2\nu_2 \pm m(\nu_3 \cdots \nu_8).$$

The transition moment for the M series lies completely in the plane of the molecule in both phases. The transition moment for the K series lies primarily in the molecular plane in the high-temperature phase (HT phase), which is metastable at 77°K and perpendicular to the plane of the molecule in the LT phase. This is valid if one assumes that the position of the molecule in the crystal does not change substantially as a result of the phase transition. This matter is discussed in reference 56.

In addition, there is a third series parallel to the b axis, which can be seen clearly in the LT phase only. It obeys the relation:

$$K_2 \text{ series} = 0\text{–}0 \text{ (forbidden)} \pm \nu_? \pm n_1\nu_1.$$

Here $\nu_?$ is about 535 cm^{-1} in the ground state and 525 cm^{-1} in the excited state. K_2 is a partial series of K_1. It has the vibrational pattern of a transition forbidden by symmetry which derives its intensity from the vibration $\nu_?$.

The spectrum observed parallel to the a axis is essentially the same as that observed parallel to b.

The most important crystalline properties appearing in the spectrum of HMB crystals can be summarized in the following way:

(1) A displacement of about 200 cm^{-1} for the HT phase, and 265 cm^{-1} for the LT phase relative to solution in ethanol at 95°K.

(2) No splitting.

(3) The induction of an "allowed" crystal series whose transition moment lies primarily in the plane of the molecule in the HT phase and perpendicular to the molecule in the LT phase.

In addition, a "forbidden" series, probably related to the crystal series, is observed in the LT phase, the polarization being in the plane of the molecule.

(4) The molecular absorption and fluorescence (M series) also takes place only in the plane of the molecule in the crystal.

c. Naphthalene

An analysis of the spectrum of naphthalene is more difficult to attain than those of benzene and hexamethylbenzene for the following reasons.

(1) The number of totally symmetric vibrations which can form periods is larger.

(2) The lines, at least in part, are broader.

(3) The molecule is so oriented in the crystal that all three molecular axes possess large projections on all three axes of the crystal.

(4) The second transition in the absorption spectrum partly overlaps the first.

The *crystal structure* of naphthalene is well known.[70,71] The crystal is monoclinic, having two molecules in the unit cell. A model of the unit cell is given in Fig. 18 along with the relationships for projection of unit vectors parallel to the short or the long axes of the molecule on the three crystallographic directions.

The assignment of the band spacings to the vibrational modes of the molecule is obtained from the Raman spectra observed by Brandmüller[72] and the infrared measurements of Person, Pimentel, and Schnepp,[73] as well as those of Lippincott and O'Reilly.[74]

The spectrum of the solution in ethanol at 95°K is given in Figs. 19 and 23 (upper part). Three absorption transitions are observed in the ultraviolet above 2000 A.

(*1*) *Transition I*. For this transition $\epsilon \simeq 200$ cm^{-1} (mole/liter)$^{-1}$. The fluorescence spectrum is the mirror image of the absorption spectrum with 0–0 as the midpoint.[75,76]

The bands can be arranged in two series. The justification for this arrangement is based upon the different behavior of the two groups of bands in mixed crystals and pure crystals. The two series are as follows: an M series without a 0–0 band which is forbidden by symmetry and allowed by vibrations.

$$M \text{ series} = 0\text{–}0 \text{ (forbidden)} \pm \nu_0 \pm n_1\nu_1 \pm m(\nu_3 \cdots).$$

[70] S. C. Abrahams, J. M. Robertson, and J. G. Witte, *Acta Cryst.* **2,** 233 (1949).
[71] D. W. J. Cruickshank and A. P. Robertson, *Acta Cryst.* **6,** 698 (1953).
[72] J. Brandmüller and E. Schmid, *Z. Physik* **144,** 428 (1956).
[73] W. B. Person, G. C. Pimentel, and O. Schnepp, *J. Chem. Phys.* **23,** 230 (1955).
[74] E. R. Lippincott and E. J. O'Reilly, *J. Chem. Phys.* **23,** 238 (1955).
[75] R. Passerini and I. G. Ross, *J. Chem. Phys.* **22,** 1012 (1954).
[76] H. C. Wolf, *Z. Naturforsch.* **10a,** 3 (1955).

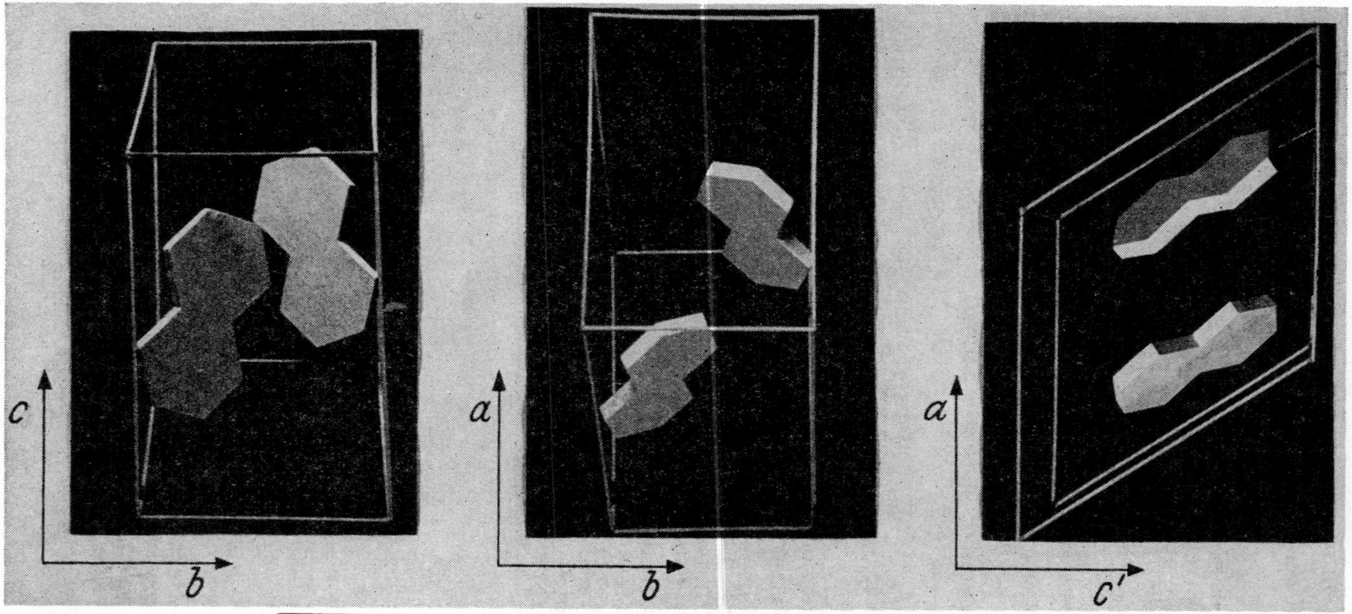

		Absorption	Emission		
Experimental		J_a / J_b	J_a / J_b	$J_b / J_{c'}$	$J_a / J_{c'}$
		1/3	1/3	3/1	1/1
Calculated	Short axis	1/7	1/7	6/1	1/1
	Long axis	4/1	4/1	1/17	1/4

FIG. 18. Model of the unit cell of the naphthalene crystal and the intensity ratios of the spectra parallel to the three crystal directions, both experimental and calculated for an "oriented gas" [from D. Griessbach, G. Will, and H. C. Wolf, *Z. Naturforsch.* **11a,** 791 (1956)].

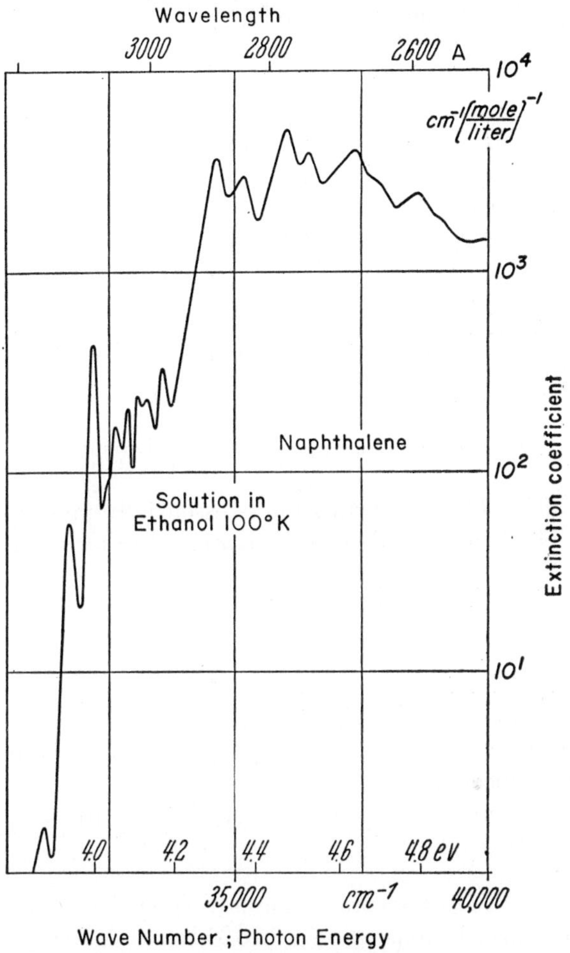

FIG. 19. Absorption spectrum of naphthalene in ethanol at 95°K [from H. C. Wolf, *Z. Naturforsch.* **10,** 3 (1955)].

Here

n_1 = 0, 1, 2, 3 . . . and m = 0 or 1

0–0 = 31,750 cm^{-1} in ethanol at 95°K

ν_0 = 500(±30) cm^{-1} in the ground state and 460(±30) cm^{-1} in the excited state as the forelying vibration which removes the selection rule forbidding the transition. The latter is only simply excited.

ν_1 = 1370 cm^{-1} in the ground state and 1390 cm^{-1} in the excited state. This is the A_{1g} vibration of the central carbon atoms.

Only about one and one-half periods of this series can be observed in absorption on account of the overlapping with the band arising from transitions II discussed in the following.

Moreover, there is an "allowed" series with a very small transition moment. It is completely missing in the spectrum of the vapor and is stronger in the crystal than in the solutions. The series is represented by the equation:

$$K \text{ series} = 0\text{–}0 \text{ (allowed)} \pm n_1\nu_1 \pm n_2\nu_2.$$

Here

$n_1,n_2 = 0, 1, 2, 3 \ldots$
$\nu_2 = 770(\pm 30)$ cm^{-1} in the ground state and $720(\pm 20)$ cm^{-1} in the excited state. (The A_{1g} Raman frequency is 760 cm^{-1}.)

Only a small part of the series appears in the absorption spectrum in this case also.

(*2*) *Transition II*. This transition is essentially more intense, ϵ being about 5000. The broad bands can be expressed in terms of a series:

$$M_2 \text{ series} = 0\text{–}0 + n_1\nu_1 + n_3\nu_3$$

in which

$n_1,n_3 = 0, 1, 2, 3 \ldots$
$0\text{–}0 = 34{,}680$ cm^{-1} in ethanol at 95°K
$\nu_1 = 1390$ cm^{-1}, multiply excited A_{1g} vibration (1370 cm^{-1} in the Raman spectrum)
$\nu_3 = 485$ cm^{-1}, multiply excited A_{1g} vibration (510 cm^{-1} in the Raman spectrum).

(*3*) *Transition III*. This transition is very strong, ϵ being larger than 100,000. In essence, it consists of a broad band showing little structure and possessing a maximum at 45,200 cm^{-1}. As a result of the very strong absorption, it has not yet been investigated in the crystal. It is not reproduced in Fig. 19.

Spectra of mixed crystals of naphthalene in durene have been investigated by McClure[53,55] and by Wolf.[54] The naphthalene molecules are dispersed in a molecular manner and are incorporated in an oriented way when not present in too large a concentration. The spectra are distinguished by particularly sharp lines and make it possible to obtain a clear separation of the *K* and *M* series (Figs. 20 and 21).

Unfortunately, the orientation of the guest molecules in the solvent crystal is not known, so that one is forced to make more or less plausible assumptions.

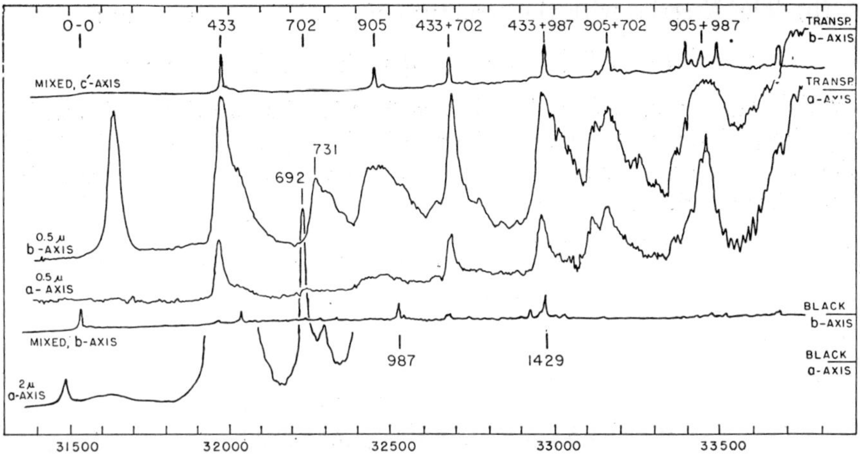

FIG. 20. Microphotometer tracings of the absorption spectra of naphthalene at 20°K. Reading from the top down, the various tracings are: (1) Mixed crystal spectrum of naphthalene in durene, c' axis (see Moffitt[3]), (2) b-axis of pure naphthalene crystal 0.5 micron thick, (3) a-axis of pure naphthalene crystal 0.5 micron thick, (4) mixed crystal spectrum along b-axis, (5) a-axis spectrum of pure naphthalene crystal 2 microns thick. The black and transparent refer to the 0.5-micron naphthalene crystal tracings. The first and fourth spectra demonstrate the similarity between the mixed crystal c' spectrum and the pure crystal spectrum, and the absence of much correlation with the mixed crystal b-spectrum. The mixed crystal spectra are shifted to the red by 10 cm^{-1}. The fifth spectrum is included to show the a-polarized component of the 0–0 band, which cannot be observed in 0.5-micron crystals, and to show the splitting of the 702 band [from D. S. McClure, *J. Chem. Phys.* **23,** 1575 (1955)].

If, to emphasize the essentials, one limits attention to the most intense lines, one can distinguish two series (Figs. 20, 21, and 23). These are as follows.

(a) The M series, which is forbidden by symmetry and allowed by vibration:

$$M \text{ series} = 0\text{–}0 \text{ (forbidden)} \pm \nu_0 \pm n_1\nu_1 \pm m(\nu_3 \cdots).$$

Here

$$n_1 = 0, 1, 2, 3, 4 \text{ and } m = 0 \text{ or } 1$$
$$0\text{–}0 = 31{,}555 \text{ (20°K, McClure)}$$
$$31{,}580 \text{ (95°K, Wolf)}.$$

Moreover, a second weaker M series also appears. This can be regarded as arising from the first by replacing ν_0 by 938 cm^{-1} in the ground state and by 905 cm^{-1} in the excited state.

The M series is polarized parallel to the c' axis of the durene crystal.

(b) The K series with the structure of an allowed transition:

$$K \text{ series} = 0\text{–}0 \text{ (allowed)} \pm n_1\nu_1 \pm n_2\nu_2 \pm m(\nu_3 \cdot \cdot \cdot)$$

in which $n_1,n_2 = 0, 1, 2, 3, 4$ and $m = 0$ or 1. The two A_{1g} modes which form the periods are 1380 and 760 cm^{-1} in the ground state and 1390 and 700 cm^{-1} in the excited state. There are also additional frequencies which do not form periods.

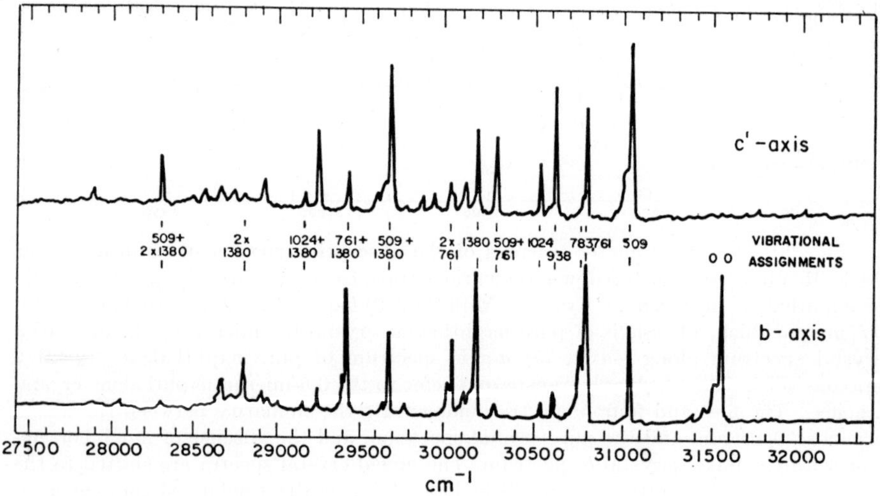

FIG. 21. Microdensitometer tracing of naphthalene fluorescence from α plane of durene crystal with about 0.1 per cent naphthalene in solid solution. The crystal temperature was 20°K, and since thermal equilibrium is established before emission, all transitions start from the vibrationless upper state. The short molecular axis practically coincides with the c' direction and thus the upper strip is essentially short axis fluorescence. The lower strip is long axis fluorescence [from D. S. McClure, *J. Chem. Phys.* **22,** 1668 (1954)].

The K series is polarized parallel to the b axis of the durene crystal.

The intensity ratios of the two series are very different in the mixed crystals of durene and in the solutions. The ratio ϵ_M/ϵ_K is between 6 and 10 in ethanol and less than unity in durene. The different polarizations for the two series are determined from the symmetry of the excited state of the molecule by superposition of the vibration ν_0 in the case of the M series and by the crystal field perturbation in the case of the K series.

In this connection, the absence of polarization in the fluorescence spectrum of deuterated naphthalene in the mixed crystal[55] with durene is difficult to understand.

The lattice modes are regarded to have the frequencies 39 and 98 cm^{-1} in the mixed crystal relative to the values 45 and 96 cm^{-1} in pure crystals.

The second transition in the mixed crystal definitely is polarized parallel to the c' axis of the durene crystal. The 0–0 band lies at 34,410 cm^{-1}.[53]

The spectrum obtained parallel to the a axis corresponds completely to that parallel to the b axis. One can conclude from this that the M series is polarized parallel to the short molecular axis of durene, whereas the K series is polarized parallel to the long axis. If the naphthalene molecule is incorporated in the direction "parallel" to the durene molecule, as McClure assumes, these assertions concerning the position of the transition moment in the molecule would also be valid for the molecules of naphthalene.

The *spectrum of the crystal* has been measured frequently. The results obtained by different authors are in considerable measure alike. The differences lie principally in the interpretation. The following work is given particular attention in this survey.

(1) The absorption spectra observed by Prichotjko[24] and by McClure[77] at 20°K. Small variations in the measured results obtained by these authors can be ascribed to differences in the methods of investigation, according to Broude.[50,58] Stresses introduced in the crystals during freezing presumably are the principal factors.

Related to these are the measurements of Wolf and co-workers[78] at 90°K with less resolution.

(2) The fluorescence spectra observed at 20°K by Obreimov and Shabaldas[22,23] and by McClure,[77] also the work at 95° and 77°K by Wolf.[78,79] The same remarks hold here as for the absorption spectra.

Transition I. The absorption (Fig. 22) and fluorescence spectra (Fig. 23) are mirror images.

Again one finds principally two series. The M series is the most intense. For it, we have:

(a)

$$M \text{ series} = 0\text{–}0 \text{ (forbidden)} \pm \nu_0 \pm n_1\nu_1 \pm n_2\nu_2 \pm m(\nu_3 \cdots).$$

Here

$$\begin{aligned} 0\text{–}0 &= 31{,}560 \text{ cm}^{-1} \text{ (extrapolated)} \\ \nu_1 &= 1380/1390 \text{ cm}^{-1} \text{ as in durene} \\ \nu_2 &= 520/495 \text{ cm}^{-1}. \end{aligned}$$

There is also a forelying vibration ν_0 which gives the transition its intensity. Its frequency value is 433 cm^{-1} in the excited state and in the range 470 to 500 cm^{-1} in the ground state.

[77] D. S. McClure and O. Schnepp, *J. Chem. Phys.* **23,** 1575 (1955).
[78] D. Griessbach, G. Will, and H. C. Wolf, *Z. Naturforsch.* **11a,** 791 (1956).
[79] H. C. Wolf, unpublished (1958).

Only the first one and one-half periods of the M series are visible in the absorption spectrum since the series runs into the second transition. The intensity ratios parallel to the three crystal axes are given in Fig. 18.

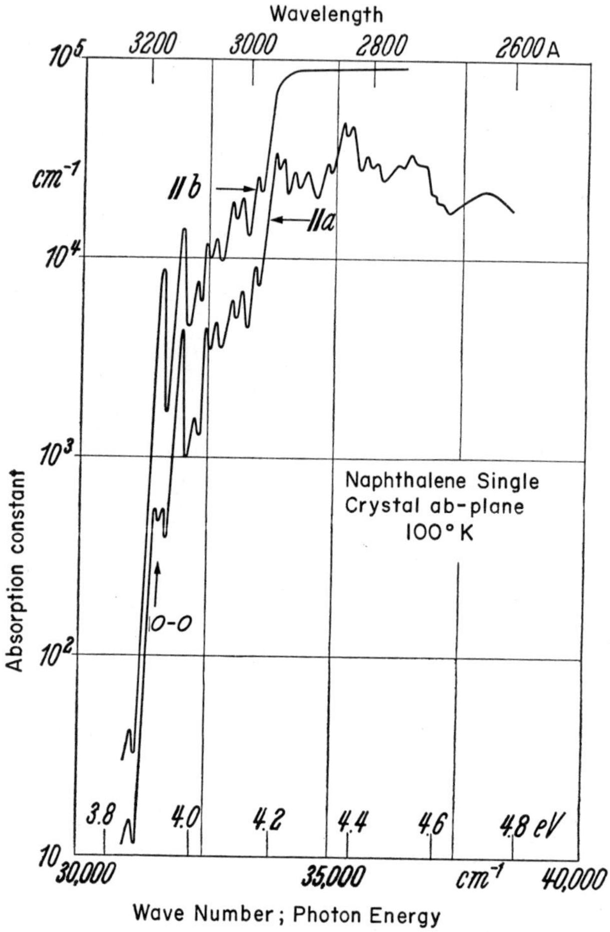

FIG. 22. Absorption spectrum of single crystals of naphthalene in the *ab* plane at 100°K [from D. Griessbach, G. Will, and H. C. Wolf, *Z. Naturforsch.* **11a**, 791 (1956)].

(b) In addition to the foregoing, only two pairs of lines have been observed in the crystal which can be ascribed to the K series with certainty and which split in the crystal. The series obeys the relation:

$$K \text{ series} = 0\text{–}0 \text{ (allowed)} \pm n_2\nu_2.$$

The first lines of this series correspond to 0–0 = 31,480 and 31,640 cm^{-1} with a Davydov splitting of 160 cm^{-1}. The second line corresponds

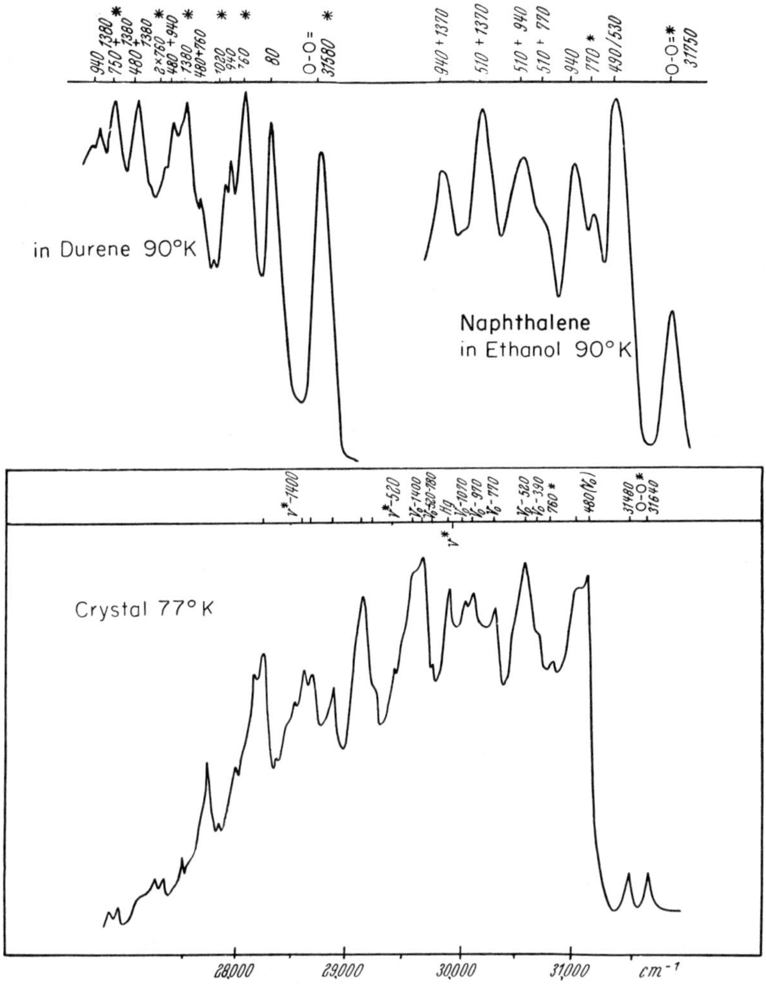

FIG. 23. Fluorescence spectrum of naphthalene taken with a recording microphotometer. Below: a crystal at 77°K; above left: durene mixed crystal, 90°K; above right: in ethanol, glass-like solidification, 90°K. An analysis is given. The lines indicated with a star belong to the *K* series. (The two upper spectra were obtained with a different spectrograph from that used in obtaining the lower curves; as a result the dispersions are not the same.) The weak line ν^* in the spectrum of the crystal coincides almost exactly with the Hg line at 3341 A.

to the superposition of the Raman frequency 760 cm^{-1} immediately up on 0–0. The splitting of these pairs of lines is in the range from 40 to 50 cm^{-1} in absorption. In fluorescence, the magnitude of the splitting has not yet been measured accurately. Additional members of the series cannot be observed because of the low intensities of the terms appearing in it. For example, the frequency $\pm n_1\nu_1$ is to be expected.

Among the members of the *K* series, the component of long wavelength is polarized in the *ac* plane, whereas that of short wavelength is polarized parallel to the *b* axis. The ratio of the intensity of the *M* series to that of the *K* series is similar to that in solution. The value is about 4 to 1.

The polarizations of the *K* and *M* series do not agree with one another in the pure crystal, much as in mixed crystals. In the case of the *K* series, the excited states of the molecules are perturbed by the crystalline field, whereas they are perturbed by a nontotally symmetric molecular vibration in the case of the *M* series.

The position of the *transition moment in the free molecule* can be determined from polarization measurements on the crystal with the use of the *M* series as well as the *K* series. The *M* series indicates polarization parallel to the short axis of the molecule. The polarization is determined by the symmetry of the molecular vibration ν_0 for this series which is not definitely known at present. As a result, it is not possible to infer the position of the purely electronic transition moment with certainty.

The *K* series, of which only the polarization relation for the 0–0 components is well known, also arises from molecular transition moments parallel to the short molecular axis, if one places reliance upon the intensity ratios along three axes of the crystals.[78] Wolf,[80] therefore, regards the short axis as the direction of the transition moment in the free molecule. In contrast, McClure[77] suggests that the measured polarization is produced by an interaction with the polarized transition II which actually is parallel to the short axis of the molecule, and that the moment of the free molecule for transition I lies along the long axis of the molecule.

It should be mentioned that there is an additional group of as yet unexplained observations which has been suppressed in the previous summary. Whereas, the characteristics of the spectrum described previously appear to be associated with the undistorted crystal, the following facts presumably are a consequence of imperfections in the crystals. The observations are as follows.

(1) The ratio of the intensities of the two 0–0 components which is about $b/a = 10$ at 90°K grows to a value over 100 at 20°K, according to McClure.[77]

[80] H. C. Wolf, *Z. Naturforsch.* **11a**, 797 (1956).

(2) Broude and his co-workers[57,58] have demonstrated that the manner in which the crystal is mounted upon the support through which it is cooled can have an essential influence upon the absorption spectrum. Similar phenomena have been observed in the fluorescence spectrum.[79] The sharp fluorescent lines of a crystal which is free of stress during cooling broaden to unsharp bands in crystals which are mounted in a rigid form. Polarization measurements which would indicate whether or not additional splitting occurs have not yet been made.

(3) Prichotjko and McClure[24,77] have observed the start of an additional series containing very sharp molecular lines at 29,945 cm^{-1}. The fluorescence spectrum of this series extends toward long wavelengths; the separations correspond principally to the molecular modes 520 and 1380 cm^{-1}. This group of lines is not always observed. According to McClure, its disappearance is coupled with the occurrence of green phosphorescence in many crystals. As a result of measurements of the author[79] (Fig. 23), it appears that this group of lines is strong only in crystals which are bound rigidly to the quartz mount and which suffer mechanical stressing during cooling. The origin of this group of lines is not yet clear.

(4) McClure has also observed a very small splitting of the vibration 1380 cm^{-1}, by about 8 cm^{-1}, in the M series. He associates this splitting with splitting of the fundamental vibration observed in the Raman spectrum of the crystal.[73]

(5) According to Wolf,[10] the width of the fluorescent lines of naphthalene crystals can be decreased by the addition of anthracene because the energy transfer is interrupted in such crystals.

Transition II. The second transition can be resolved into its components in the crystal only in a direction parallel to a. The absorption is too intense in a direction parallel to b. The polarization corresponds about to that of the M series in transition I.

The following series relations are valid for transition II.

$$M_2 \text{ series} = 0\text{–}0 + n_1\nu_1 + n_2\nu_2 + m(\nu_3 \cdots).$$

Here

$$n_1, n_2 = 0, 1, 2, 3, 4; m = 0 \text{ or } 1$$
$$0\text{–}0 = 33{,}840 \text{ according to Wolf}$$
$$33{,}780 \text{ according to McClure.}$$

The period forming A_{1g} frequencies are:

$$\nu_1 = ca\ 1390 \text{ cm}^{-1}, \quad \nu_2 = ca\ 500 \text{ cm}^{-1}.$$

There are also simple excited modes of 175, 725, and 1040 cm^{-1}.

This series has also been interpreted as a K series possessing a splitting of 175 cm^{-1} for all lines, the values of 0–0 being 33,840 and 34,015 cm^{-1}. The alternate interpretation is, however, less probable.

The most important characteristics of the spectra of naphthalene crystals are collected in Table II.

Table II. Data for the Spectra of Naphthalene (Wave Numbers in cm^{-1})

		ϵ	Line width 90°K	0–0 90°K	Splitting	Polarization
Transition I						
Solution in ethanol:	M series	*ca* 350	*ca* 80	31,750	. . .	. . .
	K series	*ca* 70				
Mixed crystals with durene:	M series	$\frac{M}{K} = 1$	20; 3–5 at 20°K			$\|c'$ durene
	K series			31,580	. . .	$\|b$ durene
Crystal:	M series	$\frac{M}{K} = 4$	50; 10–20 at 20°K	31,560	. . .	Short molecular axis
	K series			31,480 31,640	160 and 50	$\|b$ and in the *ac* plane
Transition II						
Solution in ethanol		*ca* 5000	150	34,700	. . .	. . .
Mixed crystal in durene	Similar solution		*ca* 10 at 20°K	34,400	. . .	$\|c'$ durene
Crystal	Similar solution		100	33,800	. . .(?)	Short molecular axis

d. Anthracene

The absorption and fluorescence spectra of anthracene crystals have been investigated by various authors in numerous studies. In spite of this, the results of the work are not yet completely unified, particularly in the area associated with fluorescence. Some of the differences in observations can be resolved by taking into account the perturbations of the spectrum which occur as a result of reabsorption and interference during reflection. However, other differences have endured investigation and will ultimately require an explanation.

An exact analysis of the spectrum of anthracene is made difficult by the very large band width. In this respect anthracene is essentially different from the crystals discussed previously, namely, hexamethylbenzene and naphthalene.

The *crystal structure* is well known.[81] Since it is very similar to that of naphthalene, we shall simply refer to the unit cell of naphthalene shown in Fig. 18.

The *spectrum of the solution* possesses two transitions in the region of wavelength of interest to us. Transition I (Fig. 24) consists of a small number of broad bands. The half-width is still about 250 cm^{-1} in ethanol at 90°K. The extinction coefficient is about 8000; the absorption and fluorescence spectra are mirror images. The 0–0 bands for absorption (26,400 cm^{-1}) and fluorescence (26,250 cm^{-1}) do not have identical maxima; however, they overlap strongly.

There are two A_g fundamentals which build periods attached to the 0–0 bands. They are about 400 and 1450 cm^{-1} in the excited state and about 400 and 1400 cm^{-1} in the ground state. Thus, the series formula for the allowed transitions is:

$$0\text{–}0 \pm n_1\nu_1 \pm n_2\nu_2 \text{ in which } n_1,n_2 = 0, 1, 2 \ldots .$$

Transition II is extremely intense, ϵ being 230,000, and is composed of a broad band at about 40,000 cm^{-1} showing little structure.

Transition I has been photographed by Sidman in the *mixed* crystal with naphthalene or phenanthrene at 20°K.[82] He has found the series observed in solution. However, it now has sharp lines and a larger number of vibrational frequencies. In the case of the most intense lines, the series equation for solutions is valid, namely:

$$\begin{aligned} 0\text{–}0 &= 25{,}856 \text{ for anthracene and naphthalene} \\ &= 26{,}080 \text{ for anthracene and phenanthrene.} \end{aligned}$$

The spectra of the mixed crystals exhibit polarization in absorption which is associated with the transition moment along the short axis of the anthracene molecule, provided the anthracene molecule is oriented in the mixed crystal in the same way as the molecules of the host.[43] The fluorescence spectrum apparently is depolarized.[82] The depolarization increases with decreasing temperature.[83]

Transition I was photographed first in *pure crystals* by Obreimov and Prichotjko.[17,84] Subsequently, the fluorescence spectrum was measured by

[81] V. C. Sinclair, J. M. Robertson, and A. M. Mathieson, *Acta Cryst.* **3**, 251 (1950).
[82] J. W. Sidman, *J. Chem. Phys.* **25**, 115 (1956).
[83] J. Ferguson and W. G. Schneider, *J. Chem. Phys.* **25**, 780 (1956).
[84] I. W. Obreimov, A. F. Prichotjko, and K. Shabaldas, *JETP* **6**, 1062 (1936).

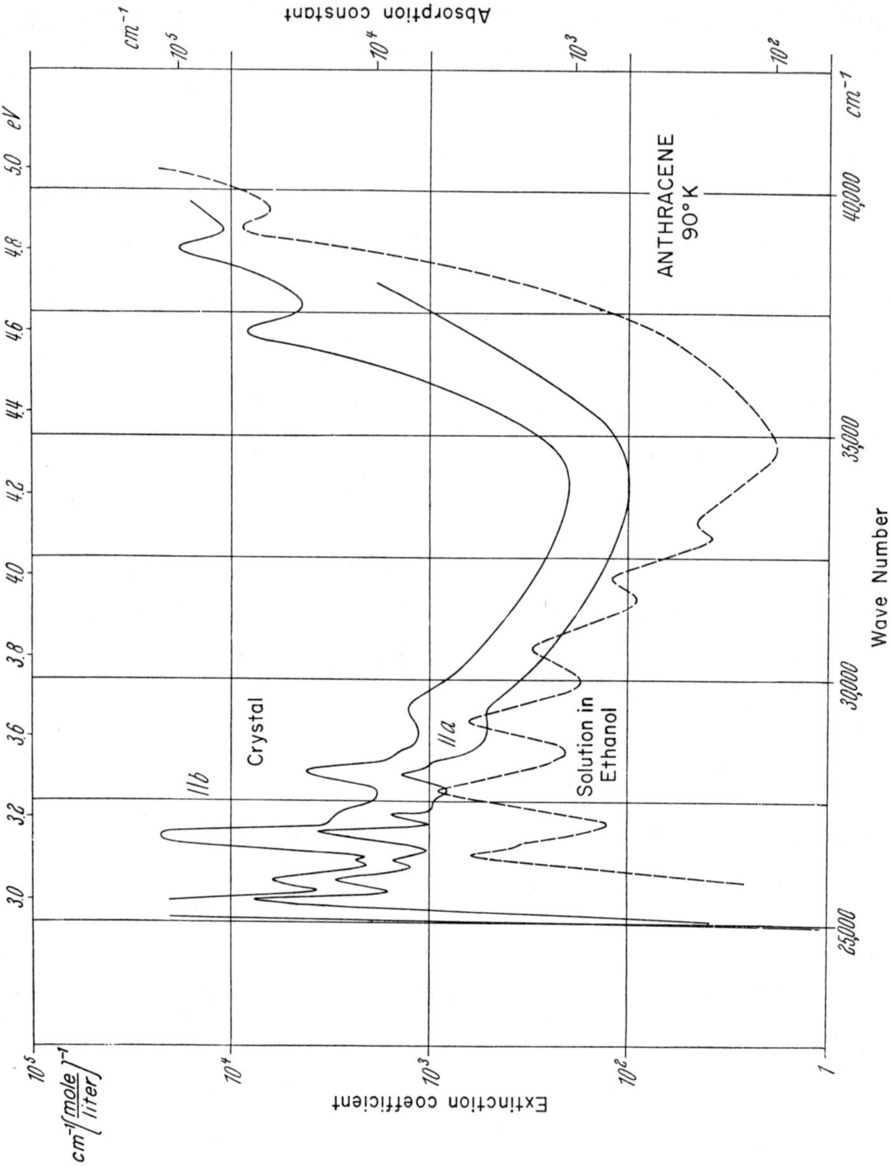

FIG. 24. The absorption spectrum of single crystals of anthracene in the *ab* plane (above) and of anthracene in ethanol at 90°K. (The dashed portion has an ordinate which is a factor 10 powers smaller) [from H. C. Wolf, *Z. Naturforsch.* **13a,** 414 (1958)].

Pesteil and Barbaron[42,85] at 14°K. The absorption and fluorescence spectra were measured at 4°K by Sidman and at 77° and 90°K by Wolf.[51]

The results observed in *absorption* are in essential agreement, apart from a displacement toward the red by about 90 cm^{-1}, in going from 90°K (Figs. 24 and 25) to 4°K.[86]

In this case the series formula is:

$$0\text{–}0 + n_1\nu_1 + n_2\nu_2$$

with

$$\begin{aligned} 0\text{–}0 &= \left.\begin{matrix} 25{,}490\ \text{cm}^{-1}\ \text{parallel}\ a \\ 25{,}270\ \text{cm}^{-1}\ \text{parallel}\ b \end{matrix}\right\}\text{mean } 25{,}380\ \text{cm}^{-1}\ \text{at } 90°\text{K} \\ \nu_1 &= 390\ \text{cm}^{-1} \\ \nu_2 &= 1410\ \text{cm}^{-1}. \end{aligned}$$

All bands indicate a Davydov splitting. This varies between about 200 cm^{-1} for the most intense and about 30 cm^{-1} for the weak bands. The *b* component lies at longer wavelengths than the *a* component. The value of the splitting can vary[51,57] by 25%, depending upon the manner in which the crystal is mounted in the cryostat.

According to the interpretation given previously, the 0–0 transition forms a *K* series in superposition with the totally symmetrical vibrations. This series splits in the crystal.

The *intensity ratio* of the absorption coefficients, ϵ_b/ϵ_a, is about 3.

The *fluorescence spectra* known to the present differ from one another and in their relationship to the absorption spectrum. The fluorescence spectra observed by Pesteil and Barbaron[42] at 14°K and by Sidman[86] at 4°K indicate, in the main, a number of broad bands with the same series structure observed in solution. A large number of sharp lines are superimposed. The separation of these lines can be identified with the frequencies of molecular vibrations. The weak intensity of these sharp lines relative to the background of broad bands leads to the supposition that they are the fluorescence lines derived from molecules which are more or less isolated in the crystal, perhaps at imperfections. It is also to be presumed that the principal emission of the crystal is, in contrast, to be found in the bands whose width, like that of the absorption band, is about 200 cm^{-1}. This broad structure is also observed at 77° and 90°K.

It should be noted, however, that even the positions of the broad bands vary with the experimental conditions in a way that is not yet sufficiently understood. The fluorescence spectrum at 90°K, as observed

[85] P. Pesteil, Thesis, Paris (1954).
[86] J. W. Sidman, *Phys. Rev.* **102,** 96 (1956).

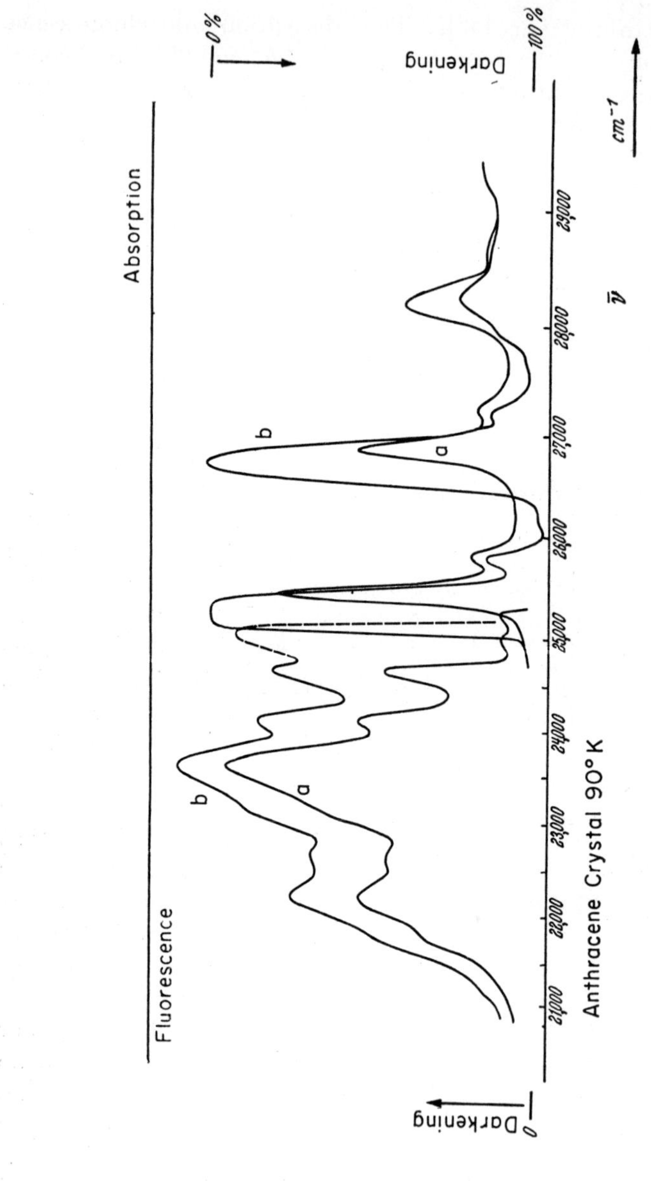

FIG. 25. Absorption and fluorescence spectrum of an anthracene single crystal in the *ab* plane at 90°K taken with a recording microphotometer. The dashed part of the fluorescence curve parallel to *b* is corrected [after H. C. Wolf, Z. *Naturforsch.* **13a**, 414 (1958)].

extensively in the largest number of investigated crystals, is shown in Fig. 25.

An "allowed" spectrum with the value of 0–0 at 25,100 cm^{-1} is found parallel to the *b* axis when corrections are made for reflection, interference, and reabsorption. The customary periods observed are about 400 and 1400 cm^{-1}. The 0–0 band is not observed parallel to the *a* axis. In other respects, the spectra are the same. A forelying fluorescence which is weak and which has not been given a unique interpretation is observed parallel to *a*. Deviations from these fluorescence spectra have been observed.[51]

A gap between the absorption and fluorescence bands is common to all observations, although, as is shown in Fig. 25, the 0–0 bands overlap at 90°K. An intensity ratio similar to that found in absorption is also observed in the directions *b* and *a*, namely, about 3/1.[51,83] On the other hand, the Davydov splitting possibly is absent in the fluorescence spectrum, that is, it behaves like an *M* series.

The variability of the fluorescence spectrum with experimental conditions, the absence of an immediate relationship with the 0–0 transition of the absorption spectrum, and the absence of a Davydov splitting imply that the observed spectrum is not that of the pure anthracene crystal. It appears most likely that the fluorescence arises from chemical agents such as oxygen or disturbed regions of the crystal, perhaps associated with stresses produced during cooling, whose lowest state of excitation lies below that of the pure anthracene molecule, and which, as a result, quench the true fluorescence of anthracene. According to this viewpoint, the largest part of the fluorescence spectra of crystalline anthracene observed to date can be ascribed to imperfections. Systematic investigations of this suggestion have not yet been made. The following summarizing comments can be made regarding transition I in crystalline anthracene.

An allowed *K* series is observed in absorption. It has a Davydov splitting between 10 and 220 cm^{-1}; there is a displacement relative to solutions in ethanol of 1000 cm^{-1}; the line width is about 200 − 250 cm^{-1}.

The fluorescence seems principally to be emission from imperfections whose origin varies with the experimental conditions in a way which has not yet been clarified.

Transition II has been measured by Craig and Hobbins,[11,87] by Bree and Lyons,[47] and Wolf.[51] Two bands, for which ϵ is about 9000, are observed parallel to *b* at about 37,300 and 38,500 cm^{-1}. In other words, they are much weaker than the second transition in solution. Craig, as well as Bree and Lyons, suggests that this corresponds to only one, namely, the weaker, of the Davydov components of the transition. They

[87] D. P. Craig, *J. Chem. Soc.* p. 539 (1955).

suggest that the other component, that is, the *a* component, lies at about 48,800 cm^{-1}. This would correspond to a Davydov splitting of about 11,000 cm^{-1}. The experimental situation is not yet sufficiently clear.

e. Phenanthrene

The spectroscopic behavior of the phenanthrene crystal is very similar to that of anthracene. Only the first transition has been investigated to date.

The *crystal structure*[88] has not yet been determined with certainty. Nevertheless, it is clear that it is very similar to that of anthracene.

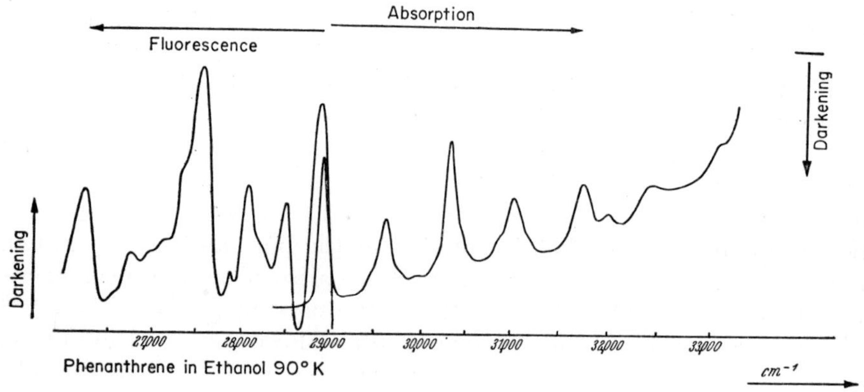

FIG. 26. The absorption and fluorescence spectrum of phenanthrene solution in ethanol at 90°K. Recording microphotometer curve [from H. C. Wolf, *Z. Naturforsch.* **13a,** 420 (1958)].

The *spectrum in solution* in ethanol at 90°K is represented in Fig. 26. It has the structure of an allowed transition. The vibrational structure in fluorescence is somewhat different from that in absorption. The following series formula is valid for the most intense lines:

$$\text{for absorption } 0\text{–}0 + n\nu_1 + m\nu_2$$

with

$$\begin{aligned} 0\text{–}0 &= 28{,}950 \text{ cm}^{-1} \\ n &= 0, 1, 2, 3 \\ m &= 0, 1 \\ \nu_1 &= 1410 \text{ cm}^{-1} \\ \nu_2 &= 685 \text{ cm}^{-1} \end{aligned}$$

$$\text{for fluorescence } 0\text{–}0 - n\nu_1 - m(\nu_{2,3,4})$$

[88] B S. Basak, *Indian J. Phys.* **24,** 309 (1950).

with

$$
\begin{aligned}
0\text{–}0 &= 28{,}920 \text{ cm}^{-1} \\
n &= 0, 1, 2, 3 \\
m &= 0 \text{ or } 1 \\
\nu_1 &= 1350 \text{ cm}^{-1} \\
\nu_{2,3,4} &= ca\ 400, 820, 1020 \text{ cm}^{-1}.
\end{aligned}
$$

The *crystal spectrum* at 90°K is shown[89] in Fig. 27. The *absorption spectrum* is easy to understand. The following series is valid for the most intense lines:

$$0\text{–}0 + n\nu_1 + m(\nu_{2,3,4,5,6})$$

with

$$
\begin{aligned}
0\text{–}0 &= 28{,}615 \text{ cm}^{-1} \text{ (mean between } a \text{ and } b) \\
n &= 0, 1, 2, 3 \\
m &= 0 \text{ or } 1 \\
\nu_1 &= 1405 \text{ cm}^{-1} \\
\nu_{2,3,4,5,6} &= 430, 550, 715, 1080, \text{ and } 1540 \text{ cm}^{-1}.
\end{aligned}
$$

The Davydov splitting for the most intense line (0–0) is about 50 cm^{-1} and decreases with decreasing absorption coefficient to a value near 0 for the weakest bands. The a component lies on the long wavelength side, in contrast to the situation for anthracene. The ratio of the absorption constants ϵ_b/ϵ_a is about 3.

The experimental results obtained by Obreimov and Prichotjko[17] and by McClure[90] at 20°K agree with this interpretation. In other words, the spectrum has the pattern of an allowed transition with a strong 0–0 line and multiply excited vibrational frequencies ("K series").

The *fluorescence spectrum* is less easy to understand. It exhibits a polarization effect only in the vicinity of the 0–0 band and is no longer simply a mirror image of the absorption spectrum.

Presumably, one is dealing with two superimposed spectra,[89] namely, a weak fluorescence spectrum of the unperturbed crystal whose 0–0 band coincides with the absorption and is split in the same way, and a strong unpolarized "imperfection spectrum" which starts at about 28,310 cm^{-1}. Only suppositions concerning the nature of the "imperfections" which overwhelm the fluorescence of the crystal are possible at present. Thus, there is no reliable interpretation of the gap of 300 cm^{-1} between absorption and emission. The imperfections may be of a crystallographic nature, originating in the stresses produced during cooling or of a chemical nature, arising from oxygen complexes.

89 H. C. Wolf, *Z. Naturforsch.* **13a**, 420 (1958).
90 D. S. McClure, *J. Chem. Phys.* **25**, 481 (1956).

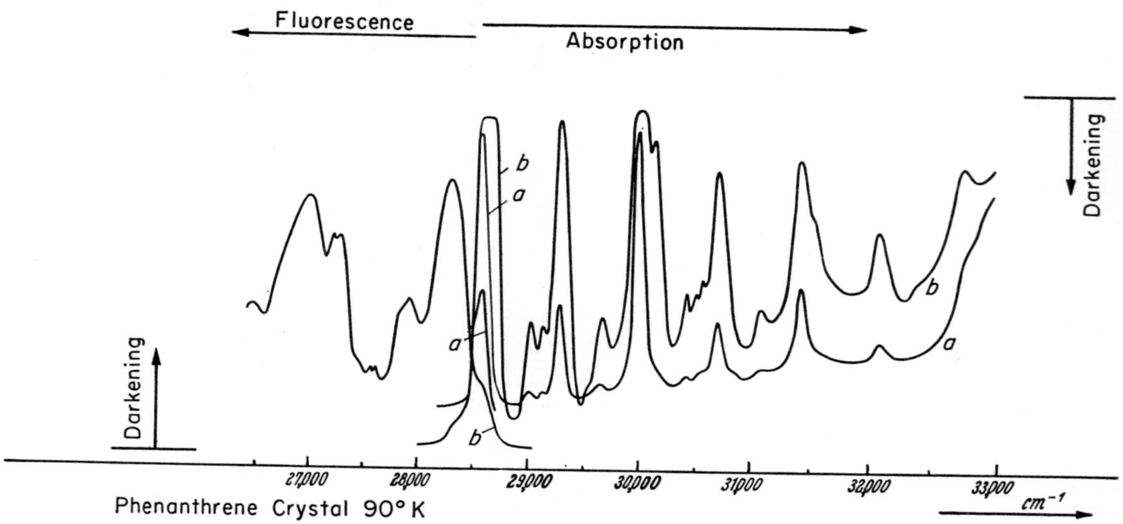

FIG. 27. Absorption and fluorescence spectrum of a single crystal of phenanthrene in the *ab* plane at 77°K. Recording microphotometer curve [H. C. Wolf, *Z. Naturforsch.* **13a,** 420 (1958)]. The fluorescence spectra are identical parallel to *a* and to *b* with the exception of 0–0.

In summary, the numerical values associated with the phenanthrene spectrum are as follows: (1) a "*K* series" with a Davydov splitting between 0 and 50 cm^{-1}; (2) a displacement between solution and crystal of 335 cm^{-1}; (3) a band width of about 100 cm^{-1}.

f. Simple Benzene and Naphthalene Derivatives

The substitution of weak polar groups on benzene or naphthalene rings alters the structure of the π-electron system only slightly. Only the first transitions appearing in the spectra of these derivatives have been examined to date. Naturally, they resemble those of the fundamental substances very closely.

The essential difference lies, in general, in the larger line or band width. Moreover, the first transition is no longer forbidden by symmetry whenever the substitution decreases the symmetry of the molecule. The crystal structures and, hence, the orientation of the single crystals under consideration are, for the most part, unknown.

Wolf[91] has made a thorough investigation of durene, which is symmetrically substituted tetramethylbenzene. This substance possesses a spectrum resembling benzene and a crystal structure resembling naphthalene. The differences between the characteristics of the basic substance and the derivatives are particularly clear.

Insofar as the large band width permits, the series formula for absorption and fluorescence can be expressed in the form:

$$0\text{–}0 \pm n\nu_1 \pm m(\nu_2, \nu_3, \nu_4)$$

with

$$\begin{aligned} n &= 0, 1, 2, 3 \ldots \text{ and } m = 0 \text{ or } 1 \\ 0\text{–}0 &= 35{,}875 \text{ in the crystal in absorption} \\ &= 35{,}680 \text{ in the crystal in fluorescence} \\ &= 36{,}250 \text{ in ethanol in absorption} \\ &= 35{,}920 \text{ in ethanol in fluorescence.} \end{aligned}$$

The period-forming frequency is $\nu_1 = 1250$ cm^{-1}. The simple excited frequencies are 700, 450, 160 cm^{-1}. (The values of the frequencies are approximately the same in the ground and the excited states.)

According to the designations introduced previously the series is to be regarded as a *K* series. Since the crystal structure is similar to that of naphthalene, one necessarily expects the same type of splitting. Experimentally, practically no polarization or splitting is observed in absorption in the *ab* plane or in fluorescence in *ab*, *ac*, and *bc* planes (Fig. 28). This surprising result presumably is to be explained by the fact that the CH_3

[91] H. C. Wolf, unpublished (1957).

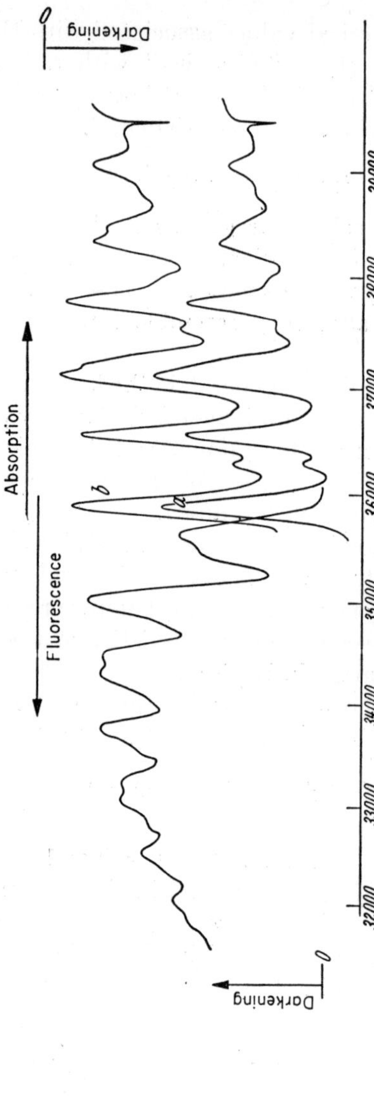

FIG. 28. The absorption and fluorescence spectrum of a durene single crystal in the *ab* plane at 77°K. Taken with a recording microphotometer [after H. C. Wolf, unpublished].

group is not completely frozen into position at 77°K. This will also explain the large band width.

Broude has investigated the absorption spectra of a large number of *mono-* and *dialkylbenzenes* at 77°K.[92,93] The results for all substances are very similar to those for benzene and hexamethylbenzene. They can be summarized in the following manner.

(1) Most of the substances possess a "high-temperature" and "low-temperature" modification (HTM, LTM). For the most part both phases were investigated.

(2) The spectra are very similar to those of benzene. However, the series which is designated as the K series in the latter case actually appears in the spectrum of the free molecule in the substituted components (allowed transitions).

(3) Again, one can distinguish between K and M series:

$$K \text{ series} = 0\text{–}0 \text{ (allowed)} + n\nu_1 + m(\nu_2, \nu_3, \nu_4 \cdots)$$

with $n = 0, 1, 2, 3 \ldots$ and $m = 0$ or 1;

$$M \text{ series} = 0\text{–}0 \text{ (weak or absent)} + \nu_0 + n\nu_1 + m(\nu_2, \nu_3, \nu_4 \cdots)$$

with $n = 0, 1, 2, 3$ and $m = 0$ or 1. Here ν_1 is the most intense totally symmetric vibrational mode of the molecule; ν_2, ν_3 . . . are other mainly total symmetric modes; ν_0 is a mode which is not totally symmetric and which converts the M series into an "allowed" type. The numerical values for 0–0 and the frequencies ν_0 and ν_1 are contained in Table III.

TABLE III. VALUES OF THE WAVE NUMBER OF THE 0–0 LINE AND OF THE FREQUENCIES ν_0 AND ν_1 FOR A NUMBER OF BENZENE DERIVATIVES

Substance	0–0 LTM	0–0 HTM	ν_1	ν_0
Ethylbenzene	37,180/215	37,235	945 ± 15	550
n-Hexylbenzene	37,115	37,225	945 ± 15	555
Toluene	37,065	37,145/175	945 ± 15	520
p-Dichlorbenzene	35,690	35,680		
o-Xylene		36,880	960	450
m-Xylene	36,830	36,780	950	460
p-Xylene		36,305	775	475
o-Ethyltoluene		37,060	950	485
p-Ethyltoluene	36,535	36,440	770	530
p-Cymol		36,670	780	560
p-Di-tertiary Butylbenzene		37,065	1060	460/540

[92] W. L. Broude, *J. Opt. i Spektr.* **1,** 387 (1956).
[93] W. L. Broude, *J. Opt. i Spektr.* **2,** 454 (1957).

It is necessary to assume that the K and M series are polarized normal to one another even in the free molecule. The K series corresponds to purely electronic excitation. Its transition moment lies in the plane of the molecule perpendicular to the principal direction of substitution. The M series is polarized in the plane of the molecule along the direction of principal substitution.

(4) Since the bands are broad in comparison with those for benzene, and have been investigated only at 77°K, the Davydov splitting is observed only in a few cases. This splitting applies only to the K series.

(5) One can draw conclusions about the position of the molecule in the crystal from the ratio of intensities of the components of the two series relative to the two crystal axes. In the main, the crystal structures are not known.

Broude[94] has examined the spectrum of *hexaethylbenzene* and has compared it with that of hexamethylbenzene. The spectra are similar.

Additional, less explicit investigations have been made for *phenol* and *bromphenol*,[95] for *1- and 2-methyl-naphthalene* in absorption,[96] and for all the *mono- and dimethylnaphthalenes*[97] in fluorescence. Thus far the measurements have indicated only that the 0–0 transition is allowed in all these substances and that the band width is very large. Davydov splitting or other essentially crystalline effects cannot be measured with certainty in these crystals.

g. *Additional Aromatic Hydrocarbons*

The absorption and fluorescence spectra of a number of other aromatic hydrocarbons have been investigated. In general, the results are not as complete as for the substances described previously and lead, in the main, to nothing new. They indicate only that the relation between the absorption and fluorescence spectrum usually is less simple in the large molecules.

Less evident, for example, are the relations in the case of pyrene.[98] The crystal structure is only approximately known in this case.[99] The absorption spectrum in solution (Fig. 29) begins with 0–0 at 26,950 cm^{-1} (log $\epsilon = 2.4$). The second transition begins at about 28,900 cm^{-1}. The fluorescence spectrum in solution begins with 0–0 at 26,790 cm^{-1} (Fig. 30).

[94] W. L. Broude, *Izvest. Akad. Nauk SSSR, Ser. Fiz.* **17,** No. 6, 699 (1953).
[95] S. B. Banerjee, *Indian J. Phys.* **30,** 353 (1956).
[96] S. B. Banerjee, *Indian J. Phys.* **30,** 106 (1956).
[97] H. C. Wolf, *Z. Naturforsch.* **10a,** 270 (1955).
[98] H. C. Wolf, unpublished (1957).
[99] J. M. Robertson and J. G. Witte, *J. Chem. Soc.* p. 358 (1947).

The spectrum displays a complicated vibrational structure which presumably is composed of at least three systems having origins at 26,790, 25,740, and 23,770 cm^{-1}. These correspond to the bands 1, 5, and 16 in Fig. 30. At higher concentrations, the well-structured fluorescence

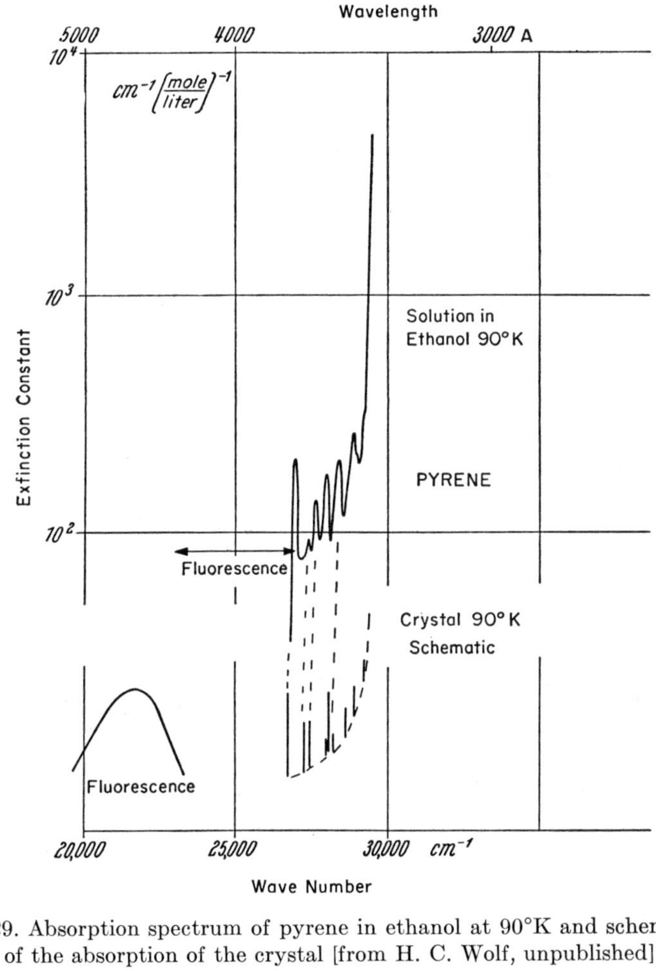

FIG. 29. Absorption spectrum of pyrene in ethanol at 90°K and schematic representation of the absorption of the crystal [from H. C. Wolf, unpublished].

spectrum undergoes a transition to a band with a broad maximum in the long wavelength region having no structure.[100] The *spectrum of the crystal* is represented schematically in Fig. 29 in directions parallel to the *a* and *b* axes. Transition I has the same structure in the directions parallel to *a* and *b*, the value of 0–0 being 26,720 cm^{-1}, that is, displaced by 230 cm^{-1}

[100] T. Förster and K. Kasper, *Z. Elektrochem.* **59**, 976 (1955).

relative to the value for the solution. To these are added vibrational bands which can be associated in part with bands which appear in the spectrum of the solution. No splitting is observed. The ratio of the intensities ϵ_a/ϵ_b is about 3. This ratio is about 10 in the second transition. No structure can be resolved in this case because of the intense absorption.

There is no simple relation between absorption and fluorescence in the crystal. The fluorescence spectrum[98,101] possesses a broad structureless maximum at about 21,700 cm^{-1} (Fig. 30). In solution, as in the crystal, it is necessary to assume a variation in the configuration or the association of the molecule when in the excited state, if one desires to understand the difference between the absorption and the fluorescence spectrum.

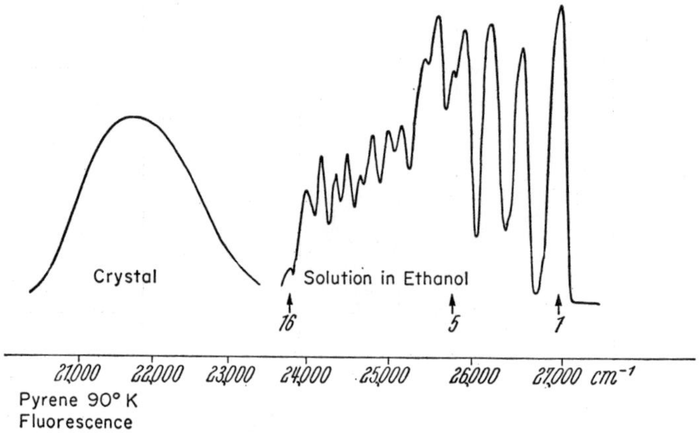

FIG. 30. The fluorescence spectrum of pyrene in ethanol and of a single crystal of pyrene at 90°K taken with a recording photometer [after H. C. Wolf, unpublished].

Sidman[107] has photographed and analyzed the absorption and fluorescence spectrum of the molecules *acepleiadiene and acepleiadylene* (with odd number rings). The observations were made on liquid solutions, on solid solutions in pyrene, and on crystalline specimens.

The absorption spectrum of single crystals of *tetracene* was measured by Bree and Lyons[52] with a photoconductive method. These observations, as well as preliminary absorption measurements made by Craig and co-workers[102] place the position of 0–0 for the first transition at about 19,200 cm^{-1} and lead to a Davydov splitting of about 600 cm^{-1}. The vibrational bands are separated by about 1400 cm^{-1}. The latter are multiply excited and possess smaller intensities and splitting. The transitions II and III lying towards shorter wavelengths cannot be analyzed

[101] S. C. Ganguly and N. K. Choudhury, *J. Chem. Phys.* **21,** 554 (1953).
[102] D. P. Craig, P. C. Hobbins, and J. R. Walsh, *J. Chem. Phys.* **22,** 1616 (1954).

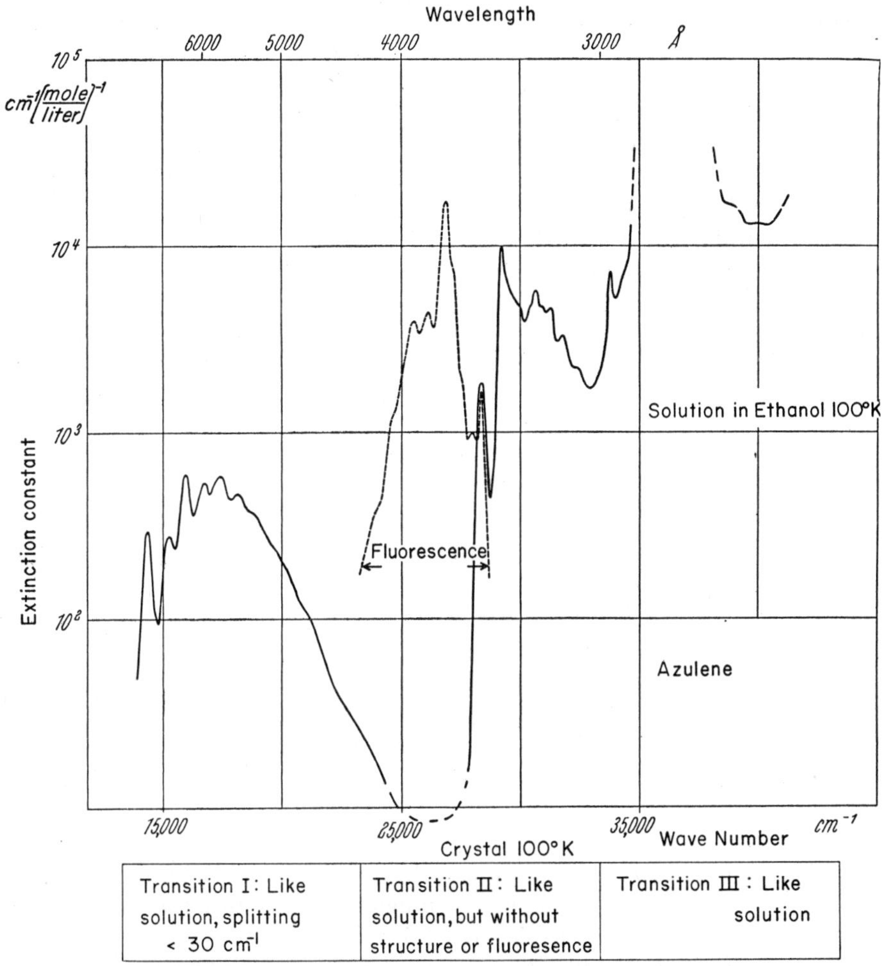

FIG. 31. Absorption and fluorescence spectrum of azulene in ethanol at 90°K and a specification of the crystal spectrum. The broken portion designates the fluorescence spectrum after H. C. Wolf, unpublished.

in a clear way at the present time. As in the case of transition I, the intensity ratios indicate that the polarization is parallel to the long axis of the molecule.

In addition, measurements on *azulene* crystals are available.[103] The structure is not yet known.[104] The spectrum of the solution is contained in Fig. 31.

[103] H. C. Wolf, unpublished (1958).

[104] J. M. Robertson and H. M. M. Shearer, *Nature* **177,** 885 (1956).

Absorption and fluorescence spectra obtained from the mixed crystals with naphthalene by Sidman and McClure[105] indicate that the first two transitions are polarized essentially perpendicularly to one another. (Sidman observed a depolarization of the fluorescence spectrum which is not well understood.) One obtains the same polarization for both transitions in the spectrum of the crystal, shown schematically in Fig. 31 with the ratio ϵ_a/ϵ_b about 3. Whereas the first transition consists of sharp lines or bands which coincide with the spectrum of the solution, the absorption in the second transition is diffuse and fluorescence is absent. Presumably the excited molecules have a different structure in the crystal and in solution.

Lyons[106] investigated the absorption of single crystals of two *pyrimidene* derivatives. It is possible to compare this with the spectrum of benzene. However, the lack of knowledge of the crystal structure makes a close analysis difficult.

h. *Other Molecular Crystals*

Without attempting to be complete, we shall mention the following investigations of the spectra of other materials.

Sidman[107] investigated the absorption spectra of *p-benzoquinone, anthraquinone,* and *naphthoquinone* at 4°K. McClure and Sidman[108] also investigated the absorption and fluorescence spectrum of *biacethyl* crystals $(CH_3CO)_2$ at 4°K.

A number of absorption spectra of different molecular crystals have been photographed at low temperatures by Deb, Sen, Swamy, and other Indian physicists.[109]

Peterson and Simpson[110] investigated the absorption bands of *myristamide* crystals in the short wavelength region of the ultraviolet.

All of these investigations are of much more interest for the physics of molecules than for the physics of solids, since, in the main, only a small amount of crystallographic data is known. They do indicate, however, that the principles gained from an analysis of the spectra of the simple aromatic hydrocarbons can be applied to other substances.

i. *Phosphorescent Spectra*

Several general items of interest have been gleaned from the studies at low temperature of numerous aromatic hydrocarbons employed in the

[105] J. W. Sidman and D. S. McClure, *J. Chem. Phys.* **24,** 757 (1956).

[106] L. E. Lyons, *J. Chem. Phys.* **20,** 1814 (1952).

[107] J. W. Sidman, *J. Am. Chem. Soc.* **78,** 2362, 4567 (1956); *J. Chem. Phys.* **27,** 820 (1957).

[108] J. W. Sidman and D. S. McClure, *J. Am. Chem. Soc.* **77,** 6461–6471 (1955).

[109] *Indian J. Phys.* (1950 to 1957).

[110] D. L. Peterson and W. T. Simpson, *J. Am. Chem. Soc.* **79,** 2375 (1957).

form of glass-like solidified solutions. One observes phosphorescent spectra different from the fluorescence spectra: (1) in respect to spectral position, lying toward longer wavelengths; (2) in respect to the lifetime, which is of the order of seconds; and (3) in respect to the intensity, which is normally weak. Moreover, the phosphorescence normally is interpreted as emission from a triplet state. Similar phosphorescence has been observed by Pesteil, Zmerli, and co-workers at 20°K in a large number of pure molecular crystals, namely, benzene and derivatives such as xylene and C_6Cl_6, naphthalene, and methyl derivatives of naphthalene.[111] The spectra can be analyzed in terms of series similar to those for the absorption and fluorescence spectra. The intensity and the structure of the phosphorescent spectra are strongly dependent upon such matters as the crystal structure and the presence of oxygen. Splitting has not yet been observed. One also encounters the point of view that the phosphorescence is only characteristic of impure crystals.[77] Additional investigations will be needed to clarify this matter.

j. Summary of the Experimental Results

One observes "*K*" and "*M*" series in crystals which differ with respect to polarization and the form of splitting.

The *K* series arise from the allowed transition of the free molecule, as well as from transitions which are forbidden in the free molecule, when the selection rule for the transition is lifted by the crystalline field.

The *M* series, which do not split, arise from forbidden transitions of the free molecules under conditions in which the selection rule forbidding the transition is lifted as a result of a molecular vibration which is not totally symmetric.

The *K* series display Davydov splitting. The polarization behavior is determined by the crystal symmetry.

The *M* series do not show splitting. The polarization behavior is determined by the molecular symmetry. The most important conclusions are summarized in Table IV.

IV. Theory of the Spectra of Molecular Crystals and Comparison with Experiment

We do not yet possess a complete theory of the spectra of molecular crystals. However, one can understand the most important properties in a qualitative and partly quantitative manner.

[111] P. Pesteil and A. Zmerli, *Ann. phys.* [12] **10,** 1079 (1955); *Cahiers phys.* **55,** 71–72 (1956); *Compt. rend.* **242,** 1876 (1956); L. Pesteil, P. Pesteil, and A. Zmerli, *ibid.* **242,** 2822 (1956); P. Pesteil and A. Zmerli, *ibid.* **243,** 1757 (1956).

TABLE IV. SUMMARY OF THE EXPERIMENTAL RESULTS

Crystal	Transition	ϵ cm^{-1} $\left(\frac{\text{mol}}{\text{liter}}\right)^{-1}$	Splitting cm^{-1}	Displacement cm^{-1}	Line width cm^{-1}	Polarization
HMB, HTM	I	200(0–0)	0	200	<10	M series in molecular plane, K series also perpendicular to plane
LTM	I	60(0–0)	0	265	<10	
Benzene	I	250	45 ÷ 0	10	<10	B_{2u}
Mono- and dialkyl-benzene	I	Similar to benzene			Larger than for benzene	2 mutually perpendicular polarized series in the molecular plane
Naphthalene	I	350	160	200	50	M and K series more intensive in the direction of the short axis of the molecule
	II	5000	?	900	100	Short axis
Methyl-naphthalene		Similar to naphthalene			Larger than for naphthalene	
Anthracene	I	8000	220 ÷ 10	1000	200	Short axis
	II	230,000	?			Short axis or long axis
			16,000	?	≫200	
Phenanthrene	I	300	50 ÷ 0	335	100	?
Durene	I	20,000	0	375	100	Almost unpolarized
Pyrene	I	250	0	230	<20	In the crystal both transitions are polarized similarly
	II	50,000	?	>500	>20	
Azulene	I	600	*ca* 30(?)		*ca* 50	In the crystal both transitions are polarized similarly
	II	2000 (0–0)	0		diffuse	
Tetracene	I	10,000	*ca* 600	*ca* 1000/1500	250	Long axis of the molecule
	II, III	>10,000	?	?	?	

5. Comparison with an "Oriented Gas"

To a first approximation one can attempt to regard a molecular crystal as an "*oriented gas*" in considering its spectroscopic behavior. In this limit one looks upon the system as if composed of oriented molecules which do not interact with one another. The transition moments $\mathbf{M}_a$, $\mathbf{M}_b$, and $\mathbf{M}_c$ of the three principal components of the spectra of the oriented gas can be obtained from those of the free molecule by projecting the transition moments of the molecules upon the optical axes of the crystal and summing over all molecules of the crystal

$$\mathbf{M}_a = \text{const } \Sigma_n \, \mathbf{m}_n \cdot \mathbf{a}$$

and correspondingly for the other directions. This approximation is too crude. As can be determined from the lattice energy of the typical molecular crystal (e.g., 2 kcal/mole for argon), one obtains a value of the binding energy per molecule of about 0.1 ev. This interaction energy of the unit of the crystal with its neighbors is substantially smaller than the value in ionic crystals (e.g., NaCl has a lattice energy of 180 kcal/mole). Nevertheless, it is sufficiently large to give the absorption and fluorescence spectra of aromatic molecular crystals, whose lattice energies lie between those of argon and NaCl, *typical crystalline properties*. These properties are: (1) a displacement relative to the gaseous state and a splitting of bands and lines; (2) a difference between the K and M series; (3) a variation of the selection rules for optical transitions; (4) a variation of the polarization relations; (5) a variation of the molecular vibration frequencies and a superposition of lattice vibrations; (6) absorption and fluorescence of distorted regions of the crystal. Of these six crystalline properties, the first, namely, the displacement and splitting of the lines, can be handled quantitatively on a theoretical basis. The other five properties will be discussed here only in a qualitative manner since, in the opinion of the author, they have not yet reached their terminal form at the present time.

In general, one must abandon the model of the oriented gas and deal with *weak coupling*. One can assume that the forces between molecules are smaller than the inner molecular forces.

6. Quantitative Treatment of Displacement and Splitting

a. *Quantum-Mechanical Calculations*

The basic development for understanding the spectra of molecular crystals was provided by Davydov.[25] It rests upon the investigations of

Frenkel and Peierls[112,113] concerning nonconducting excited states. The essential calculations can be found in the work of Davydov[25] and in the related work of Winston,[114] Craig,[87] and Fox and Schnepp.[65] We shall present the fundamental ideas and the results of the theory here.

In the theoretical determination of the displacement and splitting of the molecular terms in the crystal, one establishes the splitting pattern qualitatively with the help of the *theory of representations of symmetry groups,* particularly of the space groups.[115] In conjunction with this, one requires the results of *quantum-mechanical perturbation theory* in order to carry out approximate, quantitative calculation of the splitting.

A (hypothetical) state of excitation of the crystal in which the excitation energy is localized in one of the N molecules is N-fold degenerate, since there is equal probability that the energy will reside on any one of the N molecules in the crystal. Thus, one can construct the excited states of the entire crystal from those of the individual molecules. This is done in the following way. As a first step, one constructs the excited functions of the *unit cell* from the excited functions of the individual molecules of the *unit cell* by selecting linear combinations of the molecular functions. If these linear combinations are to be appropriate stationary states, they must belong to the same symmetry group as the unit cell. It is a problem of group theory to indicate those representations of the symmetry group of the unit cell which are compatible with a given representation of the symmetry group of the molecule. In general, *one* molecular excitation function corresponds to *a number* of functions of the unit cell and, indeed, at most to Z, where Z is the number of molecules in the unit cell. These are the components of the Davydov splitting.

A method of determining the number of components into which a molecular function splits, based on group theory, has been presented by Winston.[114]

In the second step, excitation functions for the *entire crystal* are constructed from those for the unit cell. The symmetry relations obeyed by these representations must correspond to representations of the three-dimensional space groups. If $M = M_a M_b M_c$ is the number of unit cells in the crystal, M_a, M_b, M_c being the number of cells along each of the three lattice directions, there will be M different excitation functions of this type. They can be distinguished by different values of the wave number vector of an "excitation wave," whose three components $\mathbf{k}_a$, $\mathbf{k}_b$, $\mathbf{k}_c$ have values ranging from $2\pi/2a$, $2\pi/2b$, and $2\pi/2c$ to $2\pi/M_a a$, $2\pi/M_b b$,

[112] J. Frenkel, *Phys. Rev.* **37,** 17, 1276 (1931); *Physik. Z. Sowjetunion* **9,** 158 (1936).
[113] R. Peierls, *Ann. Phys.* **13,** 905 (1932).
[114] H. Winston, *J. Chem. Phys.* **19,** 156 (1951).
[115] G. F. Koster, *Solid State Phys.* **5,** 174 (1957).

and $2\pi/M_c c$, respectively. Here a, b, c are the three axes of the unit cell. Since M_a, M_b, and M_c are large numbers, $\mathbf{k}_a$ extends for all practical purposes from 0 to π/a, and similarly for $\mathbf{k}_b$ and $\mathbf{k}_c$.

The M excitation functions of the crystal arise from each excitation function of the unit cell. This "energy band" merely has formal meaning for the spectra since only the optical transitions between states for which $\mathbf{k} = 0$ are allowed. As a consequence, the spectra of molecular crystals consist of sharp lines at low temperatures.

The selection rule $\mathbf{k} = 0$, that is, that the wavelength of the propagation wave be essentially infinite, means that the wave has the same phase of excitation in all unit cells of the crystal. From the particle standpoint, the selection rule implies that the momentum gained by the lattice from

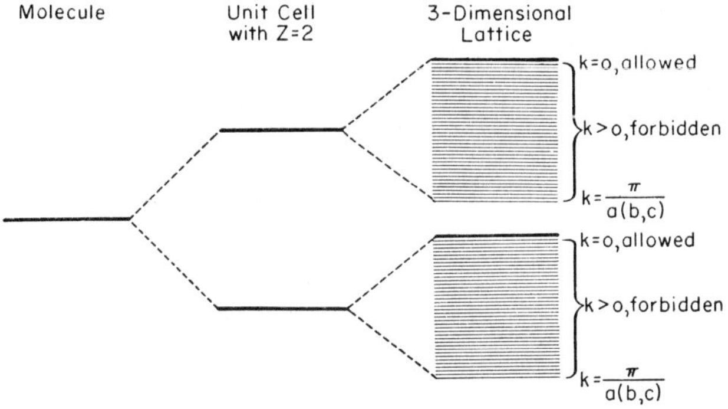

FIG. 32. Schematic means of constructing the excitation functions for the unit cell and the crystal from those of the individual molecules.

the radiation is negligibly small. This relation is valid only as long as the wavelength of the radiation is large compared to the lattice constant. Thus, the selection rule breaks down when one considers excitation with x-rays or by electrons with energies of the order of a volt or more. It also breaks down when lattice vibrations are considered. The transition from the excitation functions of the individual molecules to those of the total crystal is represented schematically in Fig. 32.

In deriving the *quantitative form of the energy states*, it is necessary to consider the interaction of the molecule m with all other molecules n of the crystal (n runs from 1 to N excluding n). Two different forms of the interaction are to be taken into account, namely, Coulomb and exchange interactions.

The *Coulomb interaction energy* in the ground state is identical with the van der Waals binding energy of the constituent molecules. An excited molecule possessing the transition moment $\mathbf{m}$ has a different, generally larger Coulomb interaction with its unexcited neighbors. This is roughly proportional to $|\mathbf{m}|^2$.[9] The difference

$$D_{mn} = \Sigma_n \{(\phi_m'\phi_n|V_{mn}|\phi_m'\phi_n) - (\phi_m\phi_n|V_{mn}|\phi_m\phi_n)\}$$

[Here is ϕ_m the function for which the mth molecule is in the ground state, ϕ_m' is the corresponding function for the excited state, and V_{mn} is the interaction energy operator.]

is designated as the "exciton Coulomb sum."[112] It produces a *displacement* of the lines or the bands of the molecules in the crystal relative to those of the free molecule, the displacement being toward longer wavelength. This displacement is different for each electronically excited state of the molecule and, for given crystalline geometry, is proportional to the square of the transition moment. It often is possible to limit the sum to the dipole interaction with the nearest neighbors in numerical calculations.

In addition to the foregoing, there is the *exchange interaction* energy which is associated with the ability of the excitation energy to wander from molecule to molecule. The important exchange sums are

$$I_{mn} = \Sigma_n (\phi_m'\phi_n|V_{mn}|\phi_m\phi_n').$$

It is often possible to restrict attention to nearest neighbors here as well. In the dipole approximation the sums I_{mn} can be expressed in the form

$$\frac{\mathbf{m}_m \cdot \mathbf{m}_n}{r^3}$$

in which $\mathbf{m}_m$ is the transition moment of the mth molecule and r_{mn} is the separation of the mth and nth molecules. The value of this expression depends not only on the magnitude but also on the relative orientation of the transition moment.

It is necessary to distinguish between the exchange interaction of molecules which are translationally equivalent and those which are not. The sum for the *translationally equivalent* molecules can be grouped into a term I_1. This contributes to the *displacement* of the spectrum relative to that of the free molecule. Its magnitude is proportional to $\mathbf{m}^2$.

The exchange interaction of molecules which are *not translationally equivalent*, that is, for the different molecules of the unit cell, depends upon the phase relations between the two interacting molecules (e.g., whether they are symmetrical or antisymmetrical relative to one another). In the case in which $Z = 2$, one obtains a second partial sum I_2 from I_{mn}. It is comprised of the interactions of the molecule m with all molecules which are not translationally equivalent to it. Its sign can be positive or negative. This term produces a *Davydov splitting*. In the case in which Z is larger than 2, other partial sums $\pm I_3 \cdots \pm I_{Z-1}$ can enter into the interaction energy between the molecule m and the other classes of molecules which are not translationally equivalent to it. Thus, the energy states of a molecular crystal can be expressed in the form

$$\Delta E_{\text{crystal}} = \Delta E_{\text{molecule}} + D + I_1 \pm I_2(\pm I_3 \cdots \pm I_{Z-1}).$$

The Davydov splitting differs from the known Bethe splitting in crystals in the sense that it occurs even for molecular states which are not degenerate. It arises from the wandering of the excitation energy in the crystal and, hence, is restricted to the excited states. The excitation states which show Davydov splitting can no longer be localized in a single molecule of the crystal.

b. *A Visualizable Vector Model*

One can determine the *number* and the *symmetry* of the possible crystalline states with a very simple vector model which contains the essential features of the Davydov theory. It also yields information concerning the relative intensities and the exact positions of the components of splitting in the crystal.[80]

One associates a unit vector lying in the direction of the transition moment of the individual molecule with each of the molecules which are coupled. The addition or subtraction of all these vectors in a unit cell yields a resultant which in magnitude and direction is the transition moment of the coupled vibration, that is, the Davydov component. This is illustrated in Fig. 33.

This construction has been applied to the spectrum of the naphthalene crystal,[80] a case for which $Z = 2$. It is easy to demonstrate that the results obtained for the benzene crystal ($Z = 4$) agree with the results of the group theoretical treatment.[26,65]

As a basis for this construction, one starts from the elementary cell and regards an excited molecule as an oscillating dipole. One such dipole interacts with the neighboring molecule which is unexcited initially. The coupled vibration which arises can be regarded as a superposition of a symmetric and an antisymmetric vibration of two coupled pendula. The difference in frequency is proportional to the energy of coupling.

The resulting periodic dipole moment of the symmetric vibration of two equivalent dipoles which are oriented parallel to one another is equal to the sum of the individual moments for weak coupling. The resulting moment of the antisymmetric vibration is just the difference of the individual dipole moments, which vanishes. In other words, the antisymmetric vibration of equivalent dipoles is optically inactive. In contrast, in coupling the dipole vibrations of two molecules which are not translationally equivalent, one obtains a resulting periodic dipole moment from both the symmetric and the antisymmetric motions. That is, one observes a splitting of the optical transitions into two components. The number of components of splitting is correspondingly greater in the case of the interaction of more than two molecules which are not translationally equivalent. One can go from the elementary cell to the crystal in the sense that the latter can be regarded as a superposition of as many lattices as there are molecules in the unit cell. Each of the sublattices can be regarded to consist of translationally equivalent molecules. In general, each will give rise to a band of levels.

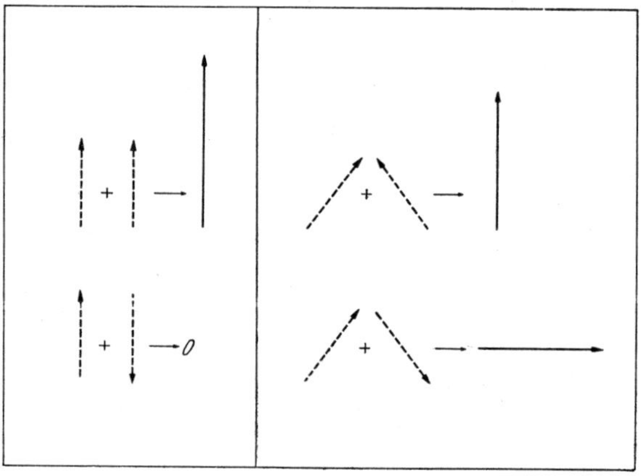

FIG. 33. Two-dimensional scheme for combining two transition moments for the symmetric and antisymmetric vibrations. In the case of translationally equivalent transition moments (left), one obtains only a single nonvanishing resultant. In the general case, one obtains two nonvanishing resultants.

The consequences of the theory described above can be summarized in the following way.

(1) The displacement of the crystal spectrum relative to that of the free molecule is proportional to the square of the transition moment, that is, proportional to ϵ, for similar geometry, in particular for the different electronic transitions in the same crystal.

(2) The magnitude of the splitting is exactly proportional to the second power of the transition moment. However, it depends on the scalar product of two vectors and, hence, upon the relative orientation of each.

(3) The number and position of the components which are split off is given by the vector construction described previously.

(4) One does not obtain any splitting in the case in which $Z = 1$. There is also no splitting in cases in which Z is larger than 1 if the different molecules of the cell are so oriented relative to one another that the projection of their transition moments upon one another vanishes.

An extension of these theoretical considerations is necessary in the case of *very intense transitions* for which ϵ is larger than 10^4 cm^{-1} (mole/liter)$^{-1}$. In this case the Davydov splitting will be between 1000 and 100,000 cm^{-1}. The deviations from the theory to be expected in this case have not been observed up to the present time because of the very great difficulty of measuring intense transitions in crystals. Perhaps the measurements on the second anthracene transition[47,49] represent a case of this type.

c. *Comparison with Experiment*

The *magnitude of the displacement and the splitting* would be determined in an absolute manner if one knew the interaction integrals described in Section 6a. Actually, these have been evaluated only approximately and not without arbitrary assumption to date.

As a result, one is restricted to a comparison of the *relative* numerical values of the experimental quantities. That is, one can compare the magnitude of the splitting and displacement in the different bands and transitions of the same crystal. This can also be done with different crystals; however, only in cases in which one takes account of the differences in crystal symmetry at the same time, that is, of the differences in separations and relative positions of the molecules.

The displacements of the crystal bands relative to those of solutions actually should be about proportional to the values of ϵ (Table IV). The magnitude of the splitting of different bands in the same crystal should, likewise, be greater for strong bands than for weak ones. Benzene (Fig. 12), anthracene (Fig. 25), and phenanthrene (Fig. 27) provide examples of this. The splitting is of the same order of magnitude as the displacement (Table IV).

The *number* and the *symmetry* of the components derived by splitting is determined by Z and by the symmetry of the molecular functions. Unfortunately, one does not know the functions initially and must infer them from the measured components of the crystal spectrum. In practice this is done by making a plausible assumption regarding their nature and testing it by seeing if it leads to results which are reasonable experimentally. This has been done for a number of crystals in Table V.

In general, the number and polarization of the components of splitting agree with the theory. The association with molecular functions is not always unique.

TABLE V. COMPARISON OF THE EXPERIMENTAL SPLITTING IN THE SPECTRA OF A NUMBER OF MOLECULES WITH THAT EXPECTED THEORETICALLY

Crystal	Z	Symmetry of the molecular function (assumed)	Splitting relations Theoretical	Splitting relations Experimental
Benzene	4	B_{2u}	A_u: inactive B_{1u}: $\parallel a$, strong B_{2u}: $\parallel b$, weak B_{3u}: $\parallel c$, strong	 $\parallel a$, strong $\parallel b$, ? $\parallel c$, strong
Hexamethyl-benzene	1		No splitting	No splitting
Naphthalene, anthracene	2	B_{2u} or	A_u: $\parallel b$, strong B_u: *ac* plane, weaker	$\parallel b$, strong $\parallel ac$ plane, weaker
		B_{3u}	A_u: $\parallel b$, weak B_u: *ac* plane, strong Splitting greater than for B_{2u}	The measured intensities thus support B_{2u}. The order of energy of the two components is the reverse for N and for A. The splitting in naphthalene is just as large as in anthracene in spite of the smaller transition moment. This supports B_{2u} (anthracene) and B_{3n} (naphthalene). The question remains open.[a]
Phenanthrene	2(?)	B_{2u} or B_{3u}	A_u: $\parallel b$ B_u: *ac* plane Likewise, however, the reversed order of energy, larger splitting	The reverse of the order of energies found in anthracene in spite of the smaller splitting. The question of whether the state in the molecule is B_{2u} of B_{3u} remains open.

[a] D. Griessbach, G. Will, and H. C. Wolf, *Z. Naturforsch.* **11a**, 791 (1956); H. C. Wolf, *ibid.* p. 797; **13a**, 414 (1958).

7. Qualitative Treatment of Additional Crystal Properties

a. Localized States: M Series

Along with the lines of the *K series* one finds the *M series*, whose terms do not split. The excitation energy must be localized in this case; that is, the excitation energy must not wander among the unit cells.

An explanation of this situation can be obtained from experiment. In all examples of the *M* series which have been observed to date, a vibration of the molecule which is not totally symmetric is superimposed upon the pure electronic excitation. The associated deformation of the molecule decreases the probability of a transfer of the energy to a neighboring molecule so severely that the probability of the transfer is smaller than that for a radiationless transfer of the vibration energy. Thus, a transfer of energy is suppressed in the states of the *M* series.

In this connection Winston[114] has noted that an analysis of the *lifetime* of the excitation energy can effect a decision concerning the localization. For a single molecule the lifetime is $\tau = h/\Delta E$ sec, in which ΔE is the energy of interaction. When $\Delta E = 100\ \text{cm}^{-1}$, τ is about 10^{-12} sec. The transfer of energy is not possible if one is dealing with molecular vibrations which will relax in a time very short compared to this.

b. Optical Selection Rules in the Crystal Field

A variation in the selection rules for optical transitions between the ground and the excited states of the molecules can occur as a result of the crystalline field when the *site symmetry* of the molecule in the crystal is less than the intrinsic symmetry of the molecule. This is the case in all of the crystals which have been examined up to the present time and leads to a relaxation of the selection rules forbidding vibrationless electron excitation (the first 0–0 band) in the case of benzene, naphthalene, and hexamethylbenzene. The pure electron transition I is forbidden in the free molecule of these materials, whereas it is intensified in the crystal. In fact, the transition probability for 0–0 also varies in mixed crystals such as naphthalene in durene.

Such variations in the transition probability for the free molecule when the molecule enters into the crystal have the effect of producing a large intensification of the *K* series relative to the *M* series, if the *M* series is the only allowed series in the free molecule, as is the case for benzene, naphthalene, and hexamethylbenzene.

Variations in the transition probability can also occur as a result of a change in the symmetry of the molecule which is incorporated in the crystal. This presumably is the case in the low-temperature phase of hexamethylbenzene.[56]

The boundary between "forbidden" and "allowed" transitions is much more flexible in molecular crystals than for solutions and gases.

c. Polarization Relations

The polarization relations observed in electronic transitions in molecular crystals, that is, the ratios of the oscillator strength parallel to the three principal axes of the crystal lattices are different from the values that would be expected for an oriented gas (see Table VI).

TABLE VI. MEASURED AND CALCULATED INTENSITY RATIOS IN DIFFERENT CRYSTALLOGRAPHIC DIRECTIONS FOR NAPHTHALENE AND ANTHRACENE
The values are for the *M* and *K* series in pure crystals and in mixed crystals.

	$a:b$	$b:c'$	$a:c'$	Ref.
Naphthalene				
Pure crystal, transition I	*M*: 1:3	3:1	1:1	78, 80
	K: 1:8	8:1	1:1	
Mixed crystal, transition I	*M*:	<1		
(durene)	*K*:	$\gg 1$		53
Pure crystal, transition II	$<1:2$			80
Mixed crystal, transition II				
(durene)		$\ll 1$		53
Anthracene				
Pure crystal, transition I	1:3			51
Mixed crystal, transition I (naphthalene)	1:3	1:1		43
Oriented gas, calculated				
Long axis (B_{3u})	4:1	1:17	1:4	78, 80
Short axis (B_{2u})	1:7	6:1	1:1	

The situation in a mixed crystal corresponds more closely to that in an oriented gas. Indeed, in some cases of this type[105,116] one actually measures a different polarization for different transitions of the gas molecule.

In contrast, differences in polarization in two different transitions have not been found up to this time in pure crystals, as a direct result of measurements of intensity. The experimental ratios of the oscillator strength parallel to two crystallographic directions which are perpendicular to one another are equal or similar for the two first transitions in materials, such as naphthalene, azulene, pyrene, and perhaps even in

[116] J. W. Sidman, *J. Am. Chem. Soc.* **78,** 4217 (1956).

anthracene, and are different from what one would expect for an oriented gas. One would expect the polarizations for the two directions to be orthogonal to one another.

The deviations from the relationships for the oriented gas are larger for the K series, in which the crystal symmetry can also determine the position of the transition moment in the molecule,[56] than for the M series.

The polarization relations in crystals can be determined by *intramolecular* and *intermolecular* interactions. In the first case, one is dealing with a partial mixing of two molecular excitation states which is induced by a molecular vibration of an appropriate symmetry.[5] This effect must occur in mixed as well as in pure crystals.

Since, however, the polarization ratios differ far more from those of an oriented gas in a pure crystal than in a mixed crystal, the second, intermolecular effect, must be the more important. One can take the viewpoint that there has been a "theft" of intensity as a result of the perturbation by the crystal field[11,77] induced by a resonance interaction between *different* excited states of *different* molecules in the crystal. These crystal forces mix the crystal wave functions for different molecular states. Thus, the intensities of transitions are distributed anew in the spectrum of the crystal. The amount of the new distribution is different in the different crystallographic directions and, thus, alters the intensity ratios relative to those for an oriented gas. The effect will have a particularly strong influence when a weak transition is separated from a stronger one by a small amount of energy.

McClure and Schnepp[77] have assumed that an extreme case occurs in naphthalene: transition I is weak and has a symmetry B_{3u}; transition II is strong and has a symmetry B_{2u}. Both transitions have a component with the crystal symmetry A_u and B_u in the crystal. The strong component A_u of transition II transfers so much of its character to the weak component A_u of transition I, as a result of mixing, that the A_u component of transition I in the resultant spectrum is stronger than the B_u component of the same transition. In this case the B_u component of the transition II is influenced less. Thus, the polarization relations of transition I are closer to those for a molecular state having a symmetry B_{2u} than for one having a symmetry B_{3u}.

The principle of theft of intensity can often be used in a detailed way to give a plausible explanation of the measured polarization ratios. Frequently, however, there are strong contradictions with experience.[11] It is evident that there is need to explore the question of the polarization ratios in crystals further.

d. Molecular and Lattice Vibrations

The vibrational modes appearing in the spectra of molecular crystals have not received a great deal of attention up to the present time. The *frequencies of the molecules* in crystals are very much the same as those of molecules in the gas or in solution.[117] *Lattice vibrational frequencies* have been observed in all measurements which were carried out below about 100°K. However, they have not been analyzed carefully. In many cases, in particular, in all transitions for which ϵ is larger than about 1000 cm^{-1} $(mole/liter)^{-1}$ no lattice vibrations are found even at the lowest temperatures because the line or band widths are still larger than about 50 cm^{-1}. They increase even more strongly with increasing transition moment in the spectra of molecular crystals than in solutions.

e. Absorption and Fluorescence of Disturbed Regions of the Crystal

The K series is associated with the wandering of the energy of excitation through the total crystal, that is, with moving excitons. It may be noted that the concept of the exciton is in a sense unnecessary in molecular crystals since the exciton states are not really new states of the system. Rather, the solid is always conscious of the individual molecules. One can regard the energy as if localized in a molecule at any given moment. This point of view is valid as long as complete periodicity of the crystal is maintained, that is, as long as all molecules of the crystal are on equivalent places.

The situation is different for molecules which reside near crystallographic or chemical imperfections near the surface or in stressed regions of the lattice, for their energy terms are displaced. This implies a perturbation in the periodicity of the lattice and, hence, a disturbance in the path of the moving exciton.

Several different cases can be distinguished.

(*1*) *The perturbation is the same for all molecules of the crystal.* Then, the total spectrum of the crystal must be displaced.

(*2*) *The perturbation affects all molecules but varies from one to another.* In this case the sharpness of the spectrum should be diminished.

These two effects have been observed by Broude and his co-workers in the absorption spectrum of benzene, naphthalene, and anthracene.[57,58] They have also been observed by Wolf in anthracene[51] and naphthalene.[79]

(*3*) *Only a few molecules are perturbed.* In this case weak "perturbation bands" corresponding to the small concentration of imperfections will be observed in absorption. A radical change in the fluorescence spectrum

[117] G. C. Pimentel, A. L. McClellan, W. B. Person, and O. Schnepp, *J. Chem. Phys.* **23**, 234 (1955).

can occur in this case, however, if the perturbed levels lie lower than those of the undisturbed molecules. In this case the disturbed regions act as traps for excitons, and the fluorescence of the crystals is quenched by the imperfections. This case, which can be regarded as one of sensitized fluorescence of the imperfections, occurs for instance when tetracene is present in anthracene.[28] A gap between the absorption and fluorescence spectrum is associated with the suppression of the excited state near the imperfection. There are a large number of experimental indications of such effects. We may call attention to the work of Sidman,[86] Wolf,[64] and of Zmerli and Pesteil.[63] Relatively little is known about the nature of such imperfections, however, for principal attention has been given to the spectrum of the pure crystals.

V. Survey

The difference between the measured spectra of crystals and the hypothetical spectra of an oriented gas can be understood to a considerable degree in terms of the material presented in Part IV, which focuses attention on the investigations dealing with transitions that are either weak or of medium intensity.

Additional experiments will be necessary if we hope to make further advances in understanding the spectra of crystals. The experiments must have the resolution of the following problems as their goal:

(1) Further analysis of the fine structure observed in the spectra at lowest temperatures, particularly the interaction with lattice vibrations.

(2) More exact analysis of the position of the Davydov components in the crystal.

(3) Analysis of the imperfections which are responsible for the imperfection-induced fluorescence.

(4) Extension of the study to higher transitions.

(5) Extension to other crystals, particularly those which are not aromatic.

(6) Measurements of the decay time of fluorescence and phosphorescence at low temperatures.

(7) Measurements of the intensities and band widths in mixed crystals as functions of temperature and concentration in order to investigate the radiationless interaction processes in crystals. The questions of line width and the polarization ratios should be investigated further on the basis of theory.

Polar Semiconductors

W. W. Scanlon

U.S. Naval Ordnance Laboratory, Silver Spring, Maryland

I. Introduction

Polar crystals, in which the main part of the binding energy is the result of Coulomb attraction between ions, such as is the case for the alkali halides, have been investigated intensively over the past few decades. Out of the enormous mass of experimental material have come certain correlations between optical, electrical, and chemical behavior. The knowledge gained from these studies provided the foundation for the theory of solids.

The behavior of conduction electrons and holes in polar crystals and their interaction with the crystal lattice has not been subject to the intensive experimental scrutiny which has been the case for the valence semiconductors, germanium and silicon. This is in part a result of the low electrical conductivity of the polar crystals generally used in the investigations, which makes the measurement of the electrical properties difficult. In addition, the electrical conductivity in the polar crystals contains a large component arising from the ionic conductivity. This complicates the problem of interpreting the measurements, particularly at high temperatures. Detailed studies of the processes and parameters involving electrons and holes, such as the scattering mechanisms, the mobility, the effective masses, and the other interactions with the lattice, have not been as well established experimentally in polar crystals as in valence crystals.

Polar crystals having smaller energy gaps, and hence a lower resistivity, than the alkali halides are known to exist. Using such crystals it is possible to make the various electrical and optical studies with the same degree of precision as for germanium crystals. Thus an experimental foundation for theories of various transport phenomena in polar crystals can be established on a firmer basis than might be possible with the classical polar crystals.

The possibility of making accurate electrical measurements, such as the Hall effect, on polar crystals opens up the study of the physical chemistry of polar compounds in the important range of composition near exact stoichiometry. Previously such studies have been inaccessible to analysis by ordinary chemical techniques. These studies provide information on the relationships between liquid, solid, and vapor phases which is important in the preparation of crystals of the semiconductors with desired electrical properties.

In this chapter we shall review the known characteristics of a family of polar semiconductors consisting of PbS, PbSe, and PbTe. The materials all have small energy gaps and low resistivities. The characteristics to be discussed will include electron-lattice interactions as well as information on structural and chemical properties which are related to the electrical behavior of the materials.

Of the three compounds, PbS has been studied in the greatest detail. Hence much of the experimental and theoretical information on polar semiconductors will be based upon the properties of PbS.

II. Crystal Properties

1. Cohesive Forces in Crystals

Cohesion in semiconducting crystals is due to two principal kinds of force. One is described by classical concepts and is the result of the electrostatic forces between charges, illustrated by the polar bond in ionic crystals. The other force has a quantum-mechanical explanation and is associated with the sharing of electrons. It is called the covalent bond and is present in valence crystals. The two types of force are present in most crystals; the crystal is called a polar or valence crystal depending upon which force is dominant.

Certain general rules govern the amount of polar or covalent character of a crystal. The most polar substance is composed of diatomic molecules of the form AB where A is an atom in column I of the periodic table and B is an atom from column VII. These atoms have the greatest difference in electronegativity. Molecules composed of atoms from columns II–VI, etc., are progressively less polar since their electronegativity difference is less.

Covalent bonding, on the other hand, is greatest in elemental crystals. Germanium and silicon are good examples of crystals having covalent bonding. In diatomic covalent crystals, differences in atom size as well as differences in nuclear charge tend to cause polarization of the atoms and provide a polar contribution to the bond.

It is difficult to make an estimate of the polar or covalent nature of bonds in crystals. One of the principal methods for determining the ionic character of a bond is to measure its dipole strength. The ratio of the measured dipole strength and the theoretical value is a measure of the ionic character which may be expressed in per cent. While these values are not generally available for crystals, they can be measured for the molecules constituting the crystal. Such values are likely to provide only a qualitative estimate of the ionic character in a crystal because of the effect of interactions with neighboring molecules. The general trend is that the ionic character of a material in the crystal form is substantially greater than in the free molecule. Pauling[1] has shown that the dipole moments of the free molecules can be calculated from the electronegativities of the ions. On the basis of this calculation, the NaCl molecule is about 50% ionic, PbS about 20%, PbSe about 18%, and PbTe about 12%. Actually x-ray studies show that NaCl crystals are almost 100% ionic; we may conclude that the lead salt crystals are considerably more ionic than is indicated by the ionic character of their free molecules.

2. Crystal Structure

In general, the crystal structure is determined by the dominant bonding force. In valence crystals each bond is effectively localized and directed in space just as, for example, in the methane molecule or in the

[1] L. Pauling, "The Nature of the Chemical Bond," p. 68. Cornell University Press, Ithaca, N.Y., 1945.

larger paraffin chain. Valence crystals are primarily huge molecules in which each atom is bound to its neighbors by covalent bonds. Many of the crystals have the tetrahedral crystal structure, as, for example, germanium or silicon.

The structure of polar crystals is determined by the electrostatic forces which are not spatially oriented. As a consequence the arrangement of ions is determined primarily by geometrical considerations. A given ion will prefer to have the largest possible number with opposite charges around it, with the restriction that electrical neutrality must be preserved.

At close atomic distances, large repulsive forces set in so that we may describe the atom or ion as a sphere with a given radius to a first approximation. This radius is maintained approximately constant in whatever compound the atom may be associated. A given atom has a different radius depending upon whether the bond is ionic or covalent. In cases of ionic bonding in which various ionic charges are possible, the ratio also depends upon the ion charge. In the series of semiconducting compounds PbS, PbSe, and PbTe, the nearest neighbor distances, which are obtained by x-ray diffraction measurements, are more like those predicted from ionic radii than from covalent radii. The lattice constants, calculated densities and measured densities of the three compounds are given in Table I.

TABLE I. LATTICE CONSTANT, THEORETICAL DENSITY, AND EXPERIMENTAL DENSITY OF PbS, PbSe, AND PbTe

Crystal	Lattice constant A	Density gm/cm³ (Calc.)	(Exptl.)
PbS	5.936	7.597	7.5
PbSe	6.124	8.273	8.10
PbTe	6.46	8.25	8.16

Packing in polar crystals is a function of the ionic radii. The coordination numbers may be 8, 6, or 4, depending on whether the ion radii are equal, somewhat different, or widely different. The crystalline form of binary polar compounds generally is cubic. Crystals of NaCl and the semiconductors PbS, PbSe, and PbTe have the coordination number 6 associated with the NaCl lattice.

3. Lattice Energy

The lattice energy in a polar crystal is particularly amenable to calculation and has been the subject of a great deal of work. This type of calculation was first made by Madelung[2] and Born.[3] The lattice energy may be defined as the change in energy incurred in the process of bringing the ions from infinity to the positions they occupy in the crystal. Such calculations have been made for a number of polar crystals including PbS and PbSe.

Unfortunately the lattice energy is not a quantity which can be measured directly. Born[3] and Haber[4] devised a thermochemical cycle by which the lattice energy can be obtained from experimentally measurable quantities. Sherman[5] gives these values for a number of crystals including PbS and PbSe. The values are listed in Table II. Corresponding information is not available for PbTe.

TABLE II. EXPERIMENTAL AND THEORETICAL CRYSTAL ENERGY IN PbS AND PbSe

Crystal	Structure	a_0	n	U_0	Q	I	S	$E-D$	U_{exptl}	U_{theoret}
PbS	NaCl	5.93	10.5	703.3	22	515.5	47.7	−146	732	705
PbSe	NaCl	6.12	11.0	682.5	12	515.5	47.4	−152	727	684

a_0 = Lattice constant
n = Repulsive exponent in the Born potential expression
U_0 = Crystal energy per mole
Q = Chemical heat of formation
I = Ionization potential
S = Heat of sublimation
E = Electron affinity
D = Heat of dissociation
U_{exptl} = Crystal energy from Born-Haber cycle
U_{theoret} = Crystal energy including energy due to pressure volume product

The agreement between the theoretical and experimental values for the lattice energy is within 4% for PbS and 6% for PbSe. This is somewhat less than the agreement in the case of the alkali halides, which are known to be of extreme ionic structure. It suggests that the polarity is weaker in the lead salts than in the alkali halide crystals.

4. Plasticity of Crystals

The lattice of an ionic crystal contains vacancies and dislocations which have an important bearing upon the behavior of electrons or holes in the crystal. These defects are incorporated in the crystal during growth,

[2] E. Madelung, *Gött. Nachr.* 100 (1909); 43 (1910); *Physik. Z.* **11,** 898 (1910).
[3] M. Born, *Handbuch d. Phys.* **24/2,** 623 (1933).
[4] F. Haber, *Verhandl. deut. physik. Ges.* **21,** 750 (1919).
[5] J. Sherman, *Chem. Revs.* **11,** 153 (1932).

and may play an important role in such growth.[6] They may also be produced afterward by plastic deformation of the crystal. A brief description of some of the mechanisms and results of deformation will be given in the following section.

The three crystals, PbS, PbSe, and PbTe cleave readily along {001} planes, just as is the case for NaCl crystals. The cleavage force, while difficult to define quantitatively, is influenced by the thickness of the crystal, among other factors; the thicker crystals require the greatest work of cleavage.[7]

Cleavage becomes more difficult with increasing temperature, disappearing at about 700°C in PbS. Above this temperature, crystals of PbS can be deformed without cracking.[8] PbTe may be deformed plastically above 300°C.[9]

Planes and directions of glide in these crystals are believed to be influenced by the distribution of ion charge, among other things. One principle governing glide is that ions of the same sign should never approach each other since this would lead to repulsive forces perpendicular to the glide plane. In the case of PbS, the glide takes place on {100} planes while the glide directions are $<110>$.[10]

The deformation of a PbS crystal as a result of glide is illustrated in Fig. 1. The figure shows the appearance of a cleaved plane under oblique illumination following the application of force on opposite corners of a crystal which is bound by cleavage planes. Here glide is seen to be along {001} planes. The etch-pit pattern on a cleaved face of such a crystal reveals an enormous increase in the density of dislocations.[11] Deformations of this type can be produced by relatively small forces.

Pressure applied by a point on a cleaved surface of a PbS crystal produces prismatic and nonprismatic punch patterns. Indentation patterns are produced easily through crystals of the order of centimeters in thickness. In PbS the prismatic and nonprismatic slip planes and direction have the forms [110](001) and [001](100).[12] Under the punch the crystal shows an enormous increase in dislocations as revealed by etch pits.

The thermal stress which results from the nonuniform cooling which occurs in a rapid quench from a high temperature also produces a high

[6] L. J. Griffin, *Phil. Mag.* [7] **41,** 196 (1950).

[7] W. H. Tertsch, *Z. Krist.* **87,** 326 (1934).

[8] G. Tammann and K. Dahl, *Z. anorg. u. allgem. Chem.* **126,** 106 (1923).

[9] W. A. Rachinger, *Acta Met.* **4,** 647 (1956).

[10] E. Schmid and W. Boas, "Plasticity of Crystals," p. 228. F. A. Hughes, London, 1950.

[11] B. B. Houston, this laboratory, private communication.

[12] J. W. Davisson, E. Burstein, and P. L. Smith, *Phys. Rev.* **98,** 1544A (1955).

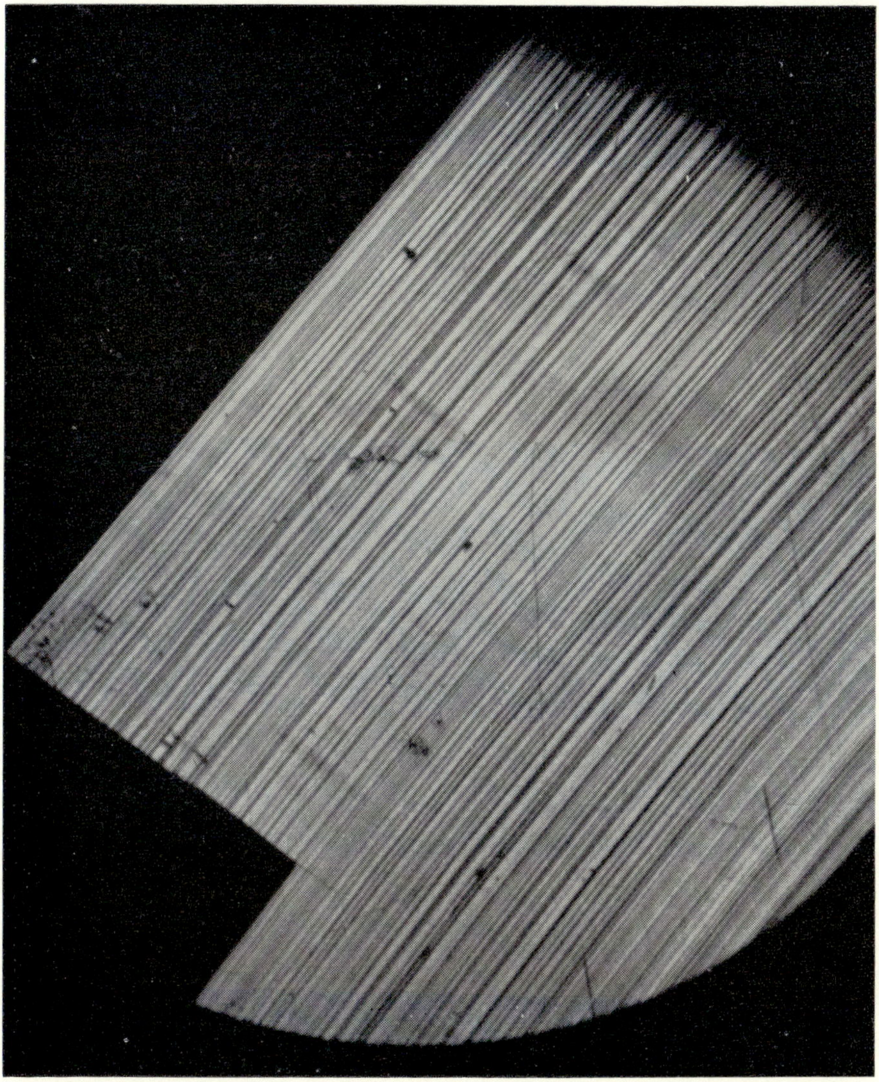

FIG. 1. Glide in a PbS crystal. (Courtesy B. B. Houston.)

density of dislocations in the crystals, as is illustrated in Fig. 2. The etch pit pattern typical of a natural PbS crystal is shown both before (a) and after (b) quenching from 500°C to room temperature.

We may conclude from these observations that the crystals of the PbS group may be deformed readily with the result that the concentration of lattice dislocations is increased substantially. These dislocations

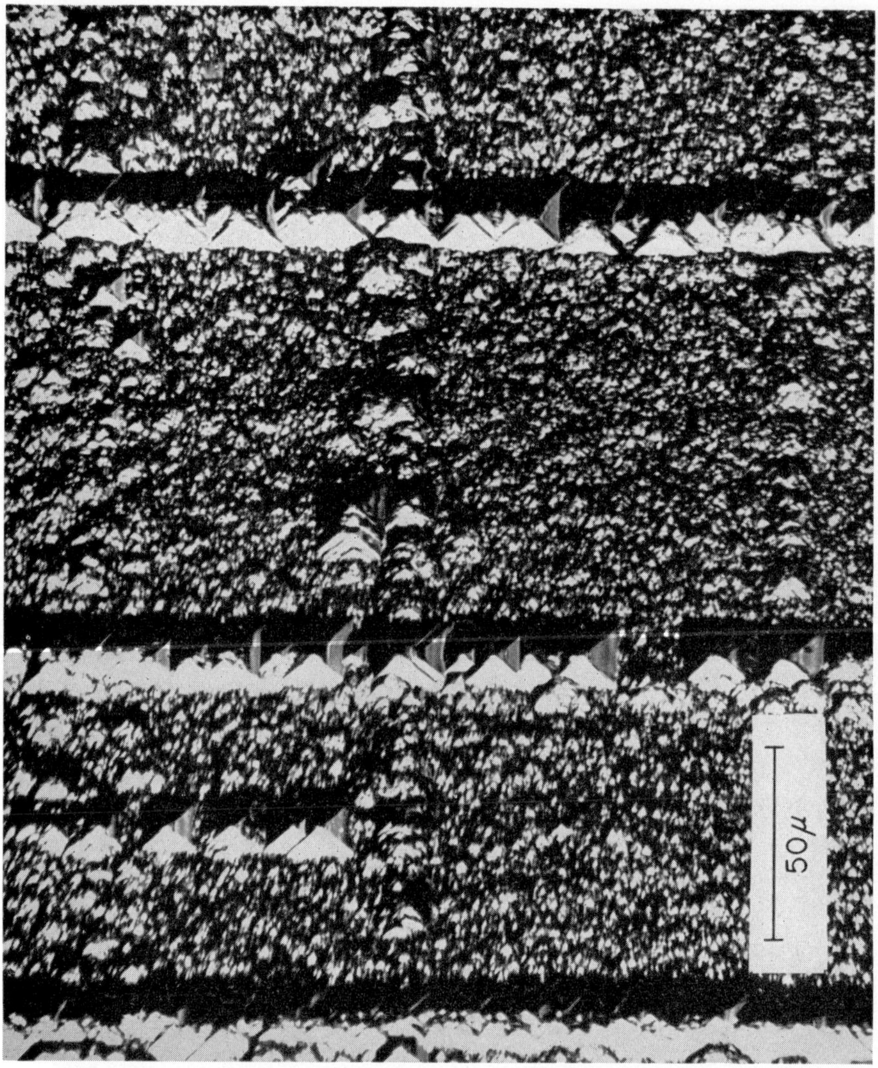

FIG. 2. Etch pit pattern on natural crystals of PbS (a) before heat treatment; (b) after quenching from 500°C to room temperature. [(a) Reproduced from R. F. Brebrick and W. W. Scanlon, *J. Chem. Phys.* **27**, 607 (1957).]

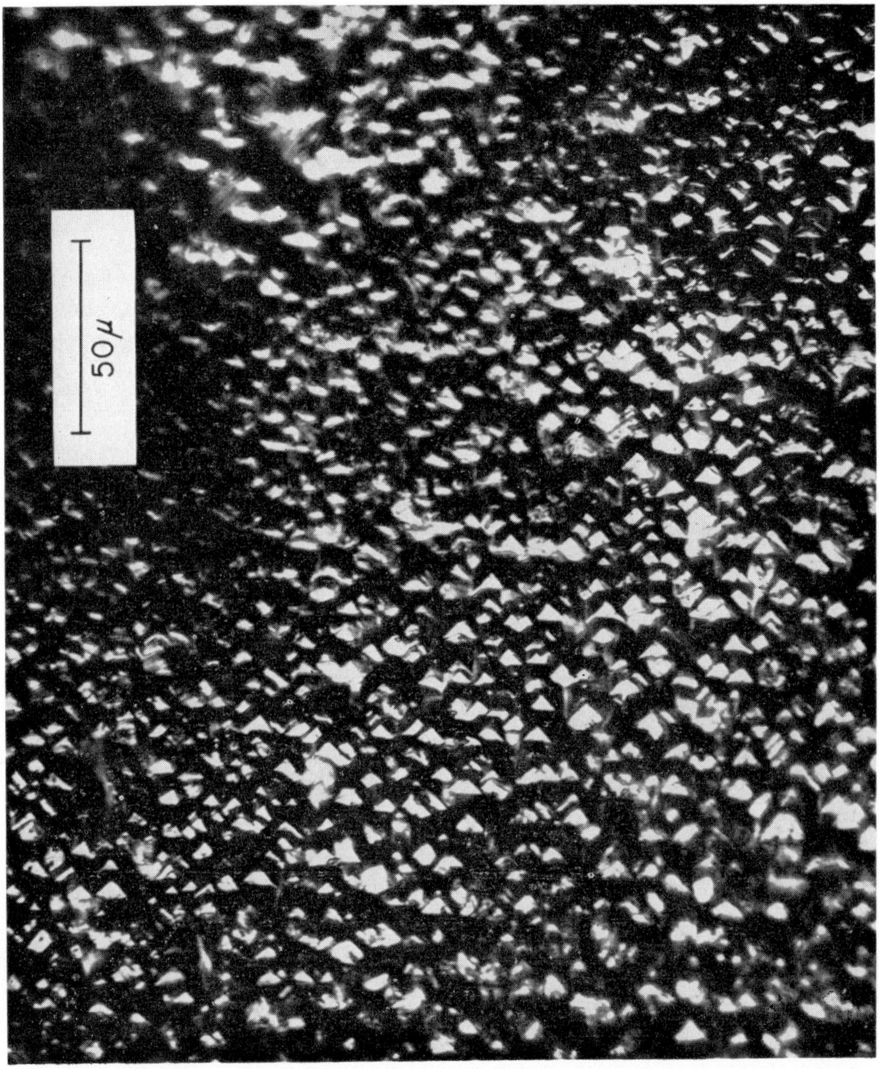

FIG. 2. (b)

play an important role in certain electrical properties of semiconductors, such as carrier lifetime, diffusion, and mobility. Further discussion of this problem will be made in the sections dealing with the electrical properties of the materials.

III. Physical Chemical Properties

5. Vacancies in a Polar Crystal Lattice

Compounds of polar or partially polar semiconductors present problems not generally met in dealing with the elementary valence semiconductors, germanium and silicon. The composition, and hence the electrical or optical properties, of polar crystals depend strongly upon the deviation from stoichiometric composition as well as upon the presence of foreign atoms in the lattice. Current statistical models predict that it is possible for all crystalline ionic compounds to show these deviations from stoichiometry.[13] The nonstoichiometric crystals can exist as a single phase over a range of composition, the additional or deficit atoms being represented by interstitial atoms and/or vacant lattice sites. In the pure compound PbS, an excess lead atom gives rise to a sulfur-ion vacancy. The lead atom, or cation, can dissociate in accordance with the relation

$$\mathrm{Pb} = \mathrm{Pb}^{++} + 2 \text{ electrons} \tag{5.1}$$

and the two electrons may be bound to the vacancy, which has a charge of $+2e$. The charge neutrality of the crystal is thus preserved. An excess sulfur atom or anion, on the other hand, produces a negative lead-ion vacancy that can attract the two holes created by the dissociation reaction

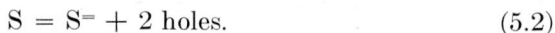

$$\mathrm{S} = \mathrm{S}^{=} + 2 \text{ holes}. \tag{5.2}$$

On the basis of considerations of size it seems unlikely that a sulfur atom can exist in interstitial positions in the PbS lattice, although, according to the same considerations, the lead atom may occupy such sites. An excess lead ion in an interstitial position would attract the two electrons created by the reaction corresponding to Eq. (5.1).

It is still possible for anion and cation vacancies to exist in a pure stoichiometric crystal although the numbers of each kind are equal. This concentration is a function of the temperature and may be quite high.

The anion and cation vacancies may exist in the crystal in a form in which they are uncharged, singly charged, or doubly charged. Bloem

[13] R. H. Fowler and E. A. Guggenheim, "Statistical Thermodynamics," Chapter 12. Cambridge University Press, London and New York, 1949.

defines the Schottky[14] equilibrium constants for the three cases[15]

$$\begin{aligned}[V_A][V_C] &= k_s' \\ [V_A^-][V_C^+] &= k_s \\ [V_A^{2-}][V_C^{2+}] &= k_s''\end{aligned} \tag{5.3}$$

in which the constants depend only on temperature. The subscripts A and C designate anion and cation, respectively.

From electrical data Bloem evaluates the temperature dependence of the dominant constant k_s as

$$k_s = 4 \times 10^{44} \exp \frac{-1.82}{kT}. \tag{5.4}$$

We may conclude that at any finite temperature a crystal contains anion and cation vacancies. In the pure, nonstoichiometric crystal, however, an excess of metal atoms produces an n-type semiconductor; an excess of the anion atom produces a p-type semiconductor.

The concentration of electrons or holes in the pure semiconductor is proportional to the concentration of positive or negative ion vacancies in the crystal, respectively. The ratio of charge carriers to ion vacancies generally is assumed to be unity.

6. Phase Relations

To prepare polar semiconductor crystals with desired electrical characteristics, it is necessary to understand the conditions which influence the number and kind of vacancies in the crystal lattice. These conditions are a complicated function of temperature and of the composition of liquid or vapor in contact with the solid. The equilibrium conditions for the liquid, solid, and vapor composition at various temperatures are described best by means of phase diagrams. In practice, crystal growing is a nonequilibrium process since large thermal gradients generally are required to promote crystallization from the liquid or vapor states. Hence it is difficult to define the thermodynamical conditions in a system undergoing crystallization. Once a crystal is formed, however, it is possible to apply the principles of thermodynamics in order to change the composition from one value to another in a reversible way. The theory governing the chemical equilibrium between a solid and ambient gas was first given by Schottky and Wagner.[14] The application of these principles to the preparation of polar semiconducting crystals was pointed out by Bloem[15] and Brebrick and Scanlon[16] for PbS and by

[14] W. Schottky and C. Wagner, *Z. physik. Chem. (Leipzig)* **11,** 467 (1930).
[15] J. Bloem, Thesis, *Philips Research Repts.* **11,** 273 (1956).
[16] R. F. Brebrick and W. W. Scanlon, *Phys. Rev.* **96,** 598 (1954).

Kröger[17] and co-workers for CdS. The ability of a polar crystal to exist as a single phase over a range of composition implies certain phase relationships between the crystal and its vapor. First, the two-phase, two-component system such as that of a crystal of lead sulfide in equilibrium with its vapor has two degrees of freedom, according to Gibbs' phase rule. Thus the composition of the crystal is fixed, for example, only when the temperature and vapor pressure of one of the components, are fixed. Second, the two components will, in general, have different proportions in the crystal and in the vapor. At equilibrium, therefore, a polar crystal can be maintained at a fixed composition for a range of temperature only by varying the vapor pressure of one of its components appropriately. This principle is important in cooling a crystal from its melting point, or in experiments in which certain properties are being studied over a range of temperature. The fact that large changes in crystal composition can occur when the second condition is not satisfied has been pointed out in studies of electrical properties in PbS.[18]

It is difficult to achieve the conditions favorable for crystal growth and for establishing a desired composition in the crystal simultaneously. In general, the requirements for the two processes are different. Hence it is convenient to separate the procedure of preparing crystals of polar semiconductors into two parts in practice. The first is concerned with the nonequilibrium process of forming the crystal from either the liquid or vapor state. The second centers about the thermal treatment in which the composition of the crystal is adjusted to the desired composition by controlling the pressure of one of the components of the vapor. We shall discuss the two processes separately.

7. Crystal Growing

Pliny, in the description of minerals, written about 70 A.D., gave the name galena to the natural crystalline form of lead sulfide. The crystals are found in a number of areas in the world; it is generally assumed that they were formed from solution. Crystals of PbS measuring almost a foot across have been found. The principal value of galena for many centuries has been the content of lead although the powdered mineral was also used by the women of ancient Egypt for painting their eye lids.

Natural crystals of lead selenide and lead telluride have been identified with the mineralogical names clausthalite and altaite respectively. These compounds are rare and only small crystals have been found.

[17] F. A. Kröger, H. J. Vink, and J. van den Boomgaard, *Z. physik. Chem. (Leipzig)* **203,** 1 (1954).

[18] W. W. Scanlon, *Phys. Rev.* **92,** 1573 (1953).

A large amount of experimental work has been carried out on natural PbS crystals; in some respects they are superior to synthetic crystals. Natural crystals having lower densities of dislocations than synthetic crystals have been found. The natural specimens may be more homogeneous and can have smaller stoichiometric deviations than typical untreated synthetic crystals. On the other hand, only PbS is available in reasonable quantities in nature so the problem of making synthetic crystals of the substances remains important for assuring a supply of the pure crystalline compounds. A review of various ways of making synthetic crystals of the compounds is given elsewhere.[19] The method most widely used today is to grow the crystals from the melt.

The compounds generally are prepared by a reaction between the purified elements. The anion elements, S, Se, or Te, can be purified by a vacuum distillation, or sublimation, process in which the first and last portions are discarded. The fact that this process removes metallic impurities effectively is demonstrated by spectroscopic analysis.[20] Lead which is 99.999% pure can be purchased. Since lead readily oxidizes in air at room temperature it usually is covered by a gray-colored oxide layer. Oxygen acts as an acceptor impurity in these semiconductors, so that steps must be taken to minimize the oxygen content of the lead. Most of the oxide can be eliminated by running molten lead through a hole 10 mils in diameter in a Pyrex funnel which is attached to the top of a six-foot long shot tower filled with hydrogen. The lead then is available in the form of silver-colored balls about 2-mm diameter. These can be stored in a dry atmosphere of hydrogen to prevent oxidation.

Stoichiometric proportions of lead and the appropriate anion are placed under vacuum in a sealed-off quartz container where they are made to react by heating the mixture gently with a flame. The reaction is exothermic and must be carried out with care to avoid excessive vaporization of the volatile anion element. Subsequently, the reacted material can be melted and formed into a crystal in the same container without exposing it to oxygen.

Another method[15] of preparing these compounds preliminary to growing crystals is based on the methods of wet chemistry. The compound PbS can be formed by the reaction of hydrogen sulfide with a solution of a lead salt such as lead nitrate or lead acetate. The precipitate is given a series of washings to yield a highly pure PbS powder.

A similar reaction can be used to form PbSe. H_2Se is unstable at room temperature, however, so that traces of unreacted selenium may be

[19] J. W. Mellor, "A Comprehensive Treatise on Inorganic and Theoretical Chemistry," Vol. 7, p. 779. Longmans, Green, New York, 1927.

[20] W. W. Scanlon, Thesis, Purdue University (1948).

present in the precipitate. The use of selenourea in place of H_2Se leads to a more stable reaction.

The corresponding tellurium compounds are very unstable so that the wet process for making PbTe is unsuitable.

Wet processes yield pure compounds which are nearly stoichiometric. One disadvantage of the wet method relative to the direct reaction of the elements is, however, that finely divided powders are formed. Oxygen reacts readily with the surface of these compounds.[21] Hence the large surface area of powders increases the chance of incorporating oxygen in crystals made by melting this form of the compound.

The melting temperature of the compounds is a function of the composition of the solid and the vapor surrounding the solid. In the case of lead sulfide, Bloem has found the melting temperature of the stoichiometric compound to be 1077°C whereas its maximum melting point is 1127°C. At the maximum temperature, the solid is n-type and has an excess lead ion concentration of about $6 \times 10^{18}/\text{cm}^3$.[22] The liquid and solid have about the same composition at this temperature. This point often is called the invariant melting temperature of the substance. Further details of the melting point and its dependence upon composition will be given later in connection with the phase diagrams for these compounds.

Less is known about the behavior of the melting temperature of the PbSe and PbTe systems. In the case of PbSe, Goldberg and Mitchell[23] give information on the melting temperature for a range of composition near the stoichiometric compound. They show that the liquid and solid have an invariant melting point when the composition corresponds to 0.005 atom per cent excess selenium. They assign the published maximum melting temperature of 1065°C[24] to this composition. Thus the first part of a near stoichiometric melt of PbSe to crystallize is p-type and has an excess of about $3 \times 10^{18}/\text{cm}^3$ selenium ions. The investigators give the melting temperature of the stoichiometric crystal as 1062°C.

In the case of PbTe, the crystal of the highest melting point is also p-type; therefore it has an excess of tellurium. The melting point is about 904°C.[25] Information obtained from experiments with crystal growing

[21] W. W. Scanlon, *in* "Semiconductor Surface Physics" (R. H. Kingston, ed.), p. 238. University of Pennsylvania Press, Philadelphia, 1957.

[22] J. Bloem and F. A. Kröger, *Z. physik. Chem.* (*Frankfurt*) [N.S.] **7**, No. 1/2, p. 1 (1956).

[23] A. E. Goldberg and G. R. Mitchell, *J. Chem. Phys.* **22**, 220 (1954).

[24] "International Critical Tables." McGraw-Hill, New York, 1928.

[25] M. Hansen, "Aufbau der Zweistofflegierungen," p. 998. Edwards Bros., Ann Arbor, Mich., 1943.

indicate that the excess Te[26] at the invariant melting point is at least 10^{18} atoms per cubic centimeter.

Single crystals of the materials generally are grown by a modification of the Bridgeman method.[27] An evacuated, sealed quartz tube with a conical bottom contains the material in a vertical double furnace. The top furnace is maintained about 50°C above the melting temperature and the bottom furnace about 50°C below the melting temperature. The quartz tube is moved downward at a rate of a few millimeters per hour. The crystal forms at the conical point first and grows upward.

The first solid to freeze from the stoichiometric melt has a composition close to that at the invariant melting point of the compound. This is n-type in the case of PbS and is p-type for PbSe and PbTe. As the crystal continues to grow, the melt becomes richer in one of the elements so that a change of sign of carrier eventually occurs, PbS becoming p-type, PbSe and PbTe becoming n-type. In the case of melts which are sufficiently rich in one or the other of the elements initially, it is possible to grow crystals having one sign of carrier over the entire length. Goldberg and Mitchell made a study of this behavior in PbSe.[23] Since the composition of the melt deviates from the stoichiometric proportion, the initial crystallization occurs at temperatures lower than the invariant melting temperature. This reduction in temperature could be a few hundred degrees. It depends upon the material and the composition of the melt.

After the solid has formed, one might expect composition changes to occur as a result of diffusion in the solid state. The diffusion proceeds in such a way as to reduce existing concentration gradients. For example, it is possible to wipe out a p-n junction in a small crystal by heating the specimen to a temperature near the melting point for a few minutes.

In the case of melts having large deviations in stoichiometric proportions, of the order of a few per cent, the solid which forms last generally shows segregation of a second phase. In PbTe, for example, segregation makes its appearance as a mixed phase of metallic lead and PbTe when there are large excesses of Pb. Similar results are observed for the other two compounds.

Inhomogeneities in the composition of crystals can be demonstrated by various electrical tests such as rectification, resistivity, thermoelectric power, or photoeffects. In the case of crystals of PbTe containing p-n junctions, we have been able to see the junctions directly because of differences in oxidation color of the n- and p-type material: n-type

[26] R. F. Brebrick, private communication, published in part: R. F. Brebrick, R. S. Allgaier, and G. L. Hammond, *Bull. Am. Phys. Soc.* [II] **4,** 134 (1959).

[27] W. D. Lawson, *J. Appl. Phys.* **22,** 1444 (1951); **23,** 495 (1952).

crystals exposed to air for a few months show a slight yellow tarnish whereas p-type crystals are unchanged in color.

We may conclude from the observations on growth of crystals from the melt that the preparation of large single crystals which have high resistivity and have homogeneous, controlled composition, is not easy. As will be discussed in the following section, this objective can be achieved in another way, namely, by adjusting the composition of the crystal from its initial arbitrary value to a desired one by the application of principles of thermodynamics to the solid-vapor system.

8. Diffusion

In principle, the composition of a polar crystal can be varied by use of heat treatment in a controlled atmosphere comprised of its components. In practice it is necessary to work at those temperatures at which the crystal-vapor equilibrium is established in a reasonable length of time. The equilibrium between the solid and vapor is established at the solid-vapor interface first. The concentration gradient in the solid leads to diffusion into or from the interior of the crystal until the entire specimen has the composition of the region near the surface. The time required to reach a certain percentage of the equilibrium composition in a crystal of given size can be calculated if the diffusion coefficient is known.

There are two types of diffusion coefficient of particular interest in solids. One, D^*, is obtained from radioactive tracer experiments in the absence of an over-all gradient in chemical composition and represents the rate at which a particular ionic species diffuses. The value of this intrinsic or self-diffusion coefficient depends on the jump frequency of the ion and the concentration of the most mobile point defects. The other diffusion coefficient, D_Δ, describes the rate at which a change in composition is propagated through the crystal and is designated the chemical diffusion coefficient, or the interdiffusion coefficient. If the electrical conduction in the crystal is primarily electronic, the second diffusion constant can be orders of magnitude larger than either of the self-diffusion coefficients.[28]

In diffusion experiments which are designed to alter the composition of the crystal one is interested in the chemical diffusion coefficient. One way of evaluating this quantity is to study the movement of p-n junctions in crystals having different initial compositions which are subjected simultaneously to a given heat treatment in a controlled atmosphere of the vapor phase. This method has been applied to PbS.[16]

A crystal which is n-type originally can be converted to p-type by exposing it to a sufficiently high temperature and sulfur pressure. If

[28] R. F. Brebrick, *J. Appl. Phys.* **30**, 811 (1959).

the crystal is quenched at an early stage in the change of composition, it will contain a *p-n* junction which divides its *n*-type interior from its *p*-type exterior.

If the composition at the crystal-vapor interface can be assumed to be a constant, independent of the initial composition of the crystal, and if the diffusion coefficient D is assumed to be independent of composition, the appropriate boundary value problem for a crystal of thickness h is:

$$\begin{aligned} \frac{\partial C}{\partial t} &= D\frac{\partial^2 C}{\partial x^2} \\ C(0,t) &= C(h,t) = C_S \\ C(x,0) &= C_i(0 < x < h). \end{aligned} \tag{8.1}$$

The analysis used here can be carried through in such a way that the concentration C refers either to the lead or the sulfide ion. In the following analysis, the concentration will be assumed to be that of the sulfide ion. Then C_S is the constant concentration of sulfide ions at the crystal surface, whereas C_i is the initial concentration in the crystal. The solution of the boundary value problem[29] is the rapidly converging series:

$$\begin{aligned} \frac{C(x,t) - C_i}{C_s - C_i} &= 1 - \frac{4}{\pi}\sum_{\nu=0}^{\infty} \frac{\sin\left[\frac{(2\nu+1)x}{h}\right]}{2\nu+1} e^{[(2\nu+1)^2\pi^2 Dt]/h^2} \\ &= S(x,Dt). \end{aligned} \tag{8.2}$$

In the case of practical interest in which the crystal is in the extrinsic range initially, a further assumption can be made, namely that

$$a(C_j - C_i) = n. \tag{8.3}$$

Here a is the number of conduction electrons per donor center, n is the initial concentration of electrons, and C_j is the sulfide ion concentration at the *p-n* junction.

By combining Eqs. (8.2) and (8.3), we obtain the following condition at the *p-n* junction,

$$n = [a(C_S - C_j) + n]S(x_j,Dt) \tag{8.4}$$

in which x_j is the distance of the *p-n* junction from the surface. This equation contains two unknowns, $a(C_S - C_j)$ and D. Thus measurements on two crystals having different initial compositions which are treated under identical conditions are required to determine the diffusion coefficient.

[29] W. Jost, "Diffusion in Solids, Liquids and Gasses," p. 37. Academic Press, New York, 1952.

An experimental apparatus for preparing crystals to be used in the foregoing analysis is shown in Fig. 3. The double furnace is used to establish the equilibrium crystal-vapor temperature and the vapor pressure of sulfur over the crystals.

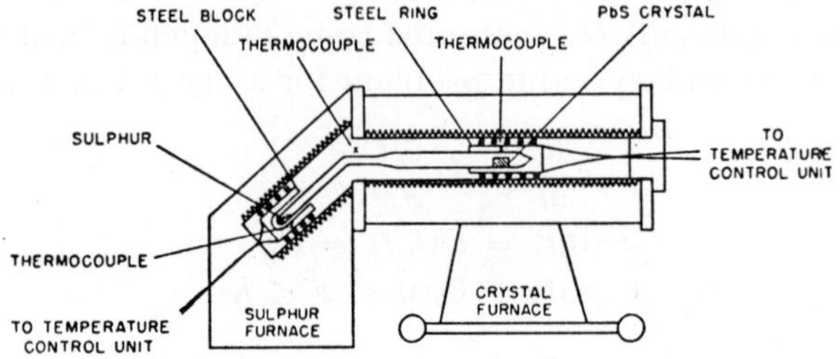

FIG. 3. Experimental apparatus used in vapor diffusion studies. [Reproduced from R. F. Brebrick and W. W. Scanlon, *Phys. Rev.* **96**, 598 (1954).]

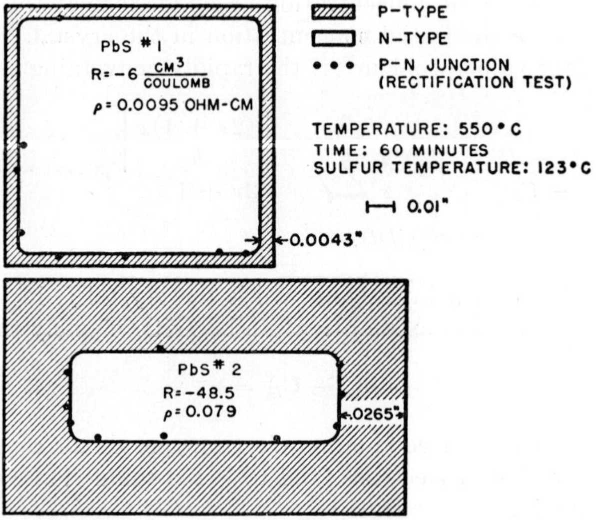

FIG. 4. Penetration of a *p-n* junction in two natural PbS crystals of different initial compositions. [Reproduced from R. F. Brebrick and W. W. Scanlon, *Phys. Rev.* **96**, 598 (1954).]

After the indicated time, the crystals are quenched to room temperature to freeze-in the composition established at the higher temperature. Figure 4 shows the penetration of the *p-n* junction in the two crystals. The experimental points locating the *p-n* junction were obtained by observing the sign of rectification as a sharpened tungsten point was moved across the surface of a cleaved midsection of the crystal. As a result of these measurements the diffusion coefficient for natural PbS crystals was found to be of the order of 10^{-6} to 10^{-7} cm²/sec at 500°C.

Using this value of D and Eq. (8.2), the time required for a crystal 2 mm thick to reach 99.8% of the equilibrium composition at its center is found to be about 20 hours at 500°C.

Bloem[15] gives equilibrium times for higher temperatures. These are shown in Table III.

TABLE III. TIME FOR PbS CRYSTALS 2-MM THICK TO COME TO EQUILIBRIUM WITH THE VAPOR AT VARIOUS TEMPERATURES
(Reproduced from J. Bloem, Thesis, University of Utrecht, 1956.)

T (°K)	Time for equilibrium
1200	20 min
1100	2 hr
1000	7 hr
900	20 hr
800	60 hr

It will be noted that the equilibrium time at a given temperature is somewhat larger for synthetic PbS crystals than for natural crystals. One possible explanation of this is that the natural crystals used generally have carrier concentrations of about 10^{17}/cm³, whereas synthetic crystals have concentrations in the range 10^{18} to 10^{19}/cm³. From theoretical considerations, Brebrick[28] predicted that diffusion in PbS crystals actually should depend upon composition.

Since the jump frequency for lead ions probably is higher than that for sulfur ions, diffusion should be more rapid in p-type materials than in n-type materials. Although this conclusion has not been verified numerically, Bloem[15] observes qualitatively that p-type crystals of PbS usually reach equilibrium sooner than n-type crystals.

The diffusion coefficient of radioactive lead atoms in pressed PbS powders is about 10^5 times smaller than the chemical diffusion coefficient obtained in the foregoing experiments. Anderson and Richards[30] found that $D = 2 \times 10^{-11}$ cm²/sec at 550°C and $D = 5 \times 10^{-11}$ cm²/sec at 600°C. Similar differences between the radioactive tracer diffusion coefficient and the chemical diffusion coefficient have been reported for Ag_2S.[31] According to Brebrick[28] the ratio of the two diffusion coefficients is proportional to the ratio of the density of cation vacancies to that of the total number of lattice sites. This ratio can be large.

Similar experiments concerning diffusion in PbSe have not been

[30] J. S. Anderson and J. R. Richards, *J. Chem. Soc.* p. 537 (1946).

[31] C. Wagner, *Z. physik. Chem.* (*Leipzig*) **B32,** 447 (1936); *J. Chem. Phys.* **21,** 1819 (1953).

carried out although we have observed, qualitatively, that the movement of a *p-n* junction in a PbSe crystal resembles the situation in PbS crystals.

The chemical diffusion rates in crystals of PbTe which are in contact with molten lead has been studied by Brady,[32] and by Boltaks and Mokhov.[33] They find that the value of the chemical diffusion coefficient in PbTe lies in the range from about 10^{-7} to 5×10^{-8} cm^2/sec at 500°C.

Thus the three crystals, PbS, PbSe, and PbTe appear to have somewhat similar chemical diffusion coefficients.

9. Crystal-Vapor Equilibrium Relationships

A study of the composition of a crystal which is in equilibrium with its vapor phase has been made only for lead sulfide. Crystals have been heated to various temperatures from about 770°K to 1270°K for periods of time sufficiently long to come into equilibrium with the vapor. The composition of the vapor has been varied over a wide range of values by controlling the partial pressure of one of the components of the vapor. Lead sulfide vapor is composed of molecules of Pb, S, and PbS. The partial pressure of S in a closed system can be regulated by controlling the temperature of solid or liquid sulfur in equilibrium with its vapors, provided that the temperature is the lowest one in the closed system. The experimental arrangement shown in Fig. 3 was used by Brebrick and Scanlon[16] for this purpose. A similar one was used by Bloem.[15]

The temperature of the oven containing the crystal was constant to ±2°C at about 500°C whereas the oven containing the sulfur was constant to within ±0.5°C during a 20-hour heat treatment. With this arrangement, sulfur pressures lying in the range from about 3×10^{-5} mm Hg to 3×10^{-1} mm Hg were obtained when the sulfur reservoir temperatures were in the range from 50°C to 160°C.

The temperatures in the sulfur reservoir are sufficiently low that the sulfur vapor is composed of various proportions of S_2, S_6, and S_8 molecules. However, the temperature is sufficiently high in the part of the closed system containing the PbS crystals that the sulfur molecules are almost completely dissociated into the S_2 molecular form. The total sulfur pressure above solid sulfur was, therefore, taken as the pressure of S_2 in the region surrounding the PbS crystal.

Bloem[15] used this technique for achieving sulfur pressures up to an atmosphere. The constants for sulfur at these pressures given by Brockmöller[34] were used to relate sulfur temperature and pressure.

[32] E. L. Brady, *J. Electrochem. Soc.* **101**, 466 (1954).

[33] B. I. Boltaks and Yu. N. Mokhov, *Zhur. Tekh. Fiz.* **26**, 2448 (1956).

[34] I. Brockmöller, *in* "A Comprehensive Treatise of Inorganic and Theoretical Chemistry" (J. W. Mellor, ed.), Vol. 10, pp. 58, 59. Longmans, Green, London, 1930.

An alternate method, used by Bloem[15] in order to obtain controlled partial pressures of sulfur in the range from 10^{-1} to 10^{-3} mm of Hg, makes use of the thermal dissociation of H_2S in H_2. The sulfur pressure associated with various mixtures of these gases was calculated from the data of Randall and Bichowski.[35] The equilibrium equation for the sulfur partial pressure is given by

$$P_{S_2}{}^{\frac{1}{2}} = \frac{1}{k_{H_2S}} \frac{P_{H_2S}}{P_{H_2}}. \tag{9.1}$$

The dependence of the equilibrium constant, k_{H_2S}, upon temperature, obtained by Bloem, is shown in Table IV. In Bloem's experiments, pressure ratios of P_{H_2S} to P_{H_2} in the range from about $\frac{1}{10}$ to 10 were used. A slow stream of the appropriate mixture of gases was passed through PbS powder first to establish the partial pressures of lead and sulfur in the system at the equilibrium temperature more quickly. The stream was then passed over the crystal.

TABLE IV. THE DEPENDENCE OF THE EQUILIBRIUM CONSTANT k_{H_2S} UPON TEMPERATURE

(Reproduced from J. Bloem, Thesis, University of Utrecht, 1956.)

Temp. (°K)	log k_{H_2S}
1200	1.35
1100	1.71
1000	2.12
800	3.26

The time to achieve equilibrium between the vapor and crystals about 2 mm thick, which was measured by Bloem using synthetic crystals at various temperatures, is given in Table III. Eisenmann[36] observed that four hours was sufficient to achieve equilibrium in synthetic crystals 1 mm thick at 873°K. This diffusion time is roughly equivalent to Bloem's observation made on thicker crystals.

Diffusion seems more rapid in natural PbS crystals. Brebrick and Scanlon[16] found that 20 hours was sufficient for natural crystals at about 773°K. This is to be compared with a time of 60 hours at approximately the same temperature in synthetic crystals observed by Bloem.

The equilibrium times vary somewhat for different crystals. Apparently they depend upon composition or other unknown factors. In general, p-type crystals have a shorter equilibrium time than n-type crystals.

[35] M. Randall and F. R. von Bichowski, *J. Am. Chem. Soc.* **40,** 368 (1918).

[36] L. Eisenmann, *Ann. Physik* **38,** 121 (1940).

For crystals having other thicknesses, the equilibrium time increases approximately as the square of the thickness according to the equation for the diffusion process given by (8.2).

After reaching equilibrium with the vapor, the crystals were quenched to room temperature to freeze-in the composition established at the higher temperature. Brebrick and Scanlon[16] quenched the crystal and container in air, requiring about 2 minutes to reduce the temperature; Bloem[15] immersed the container in cold water. A rapid quench is desirable in order to avoid changes in crystal composition which occur

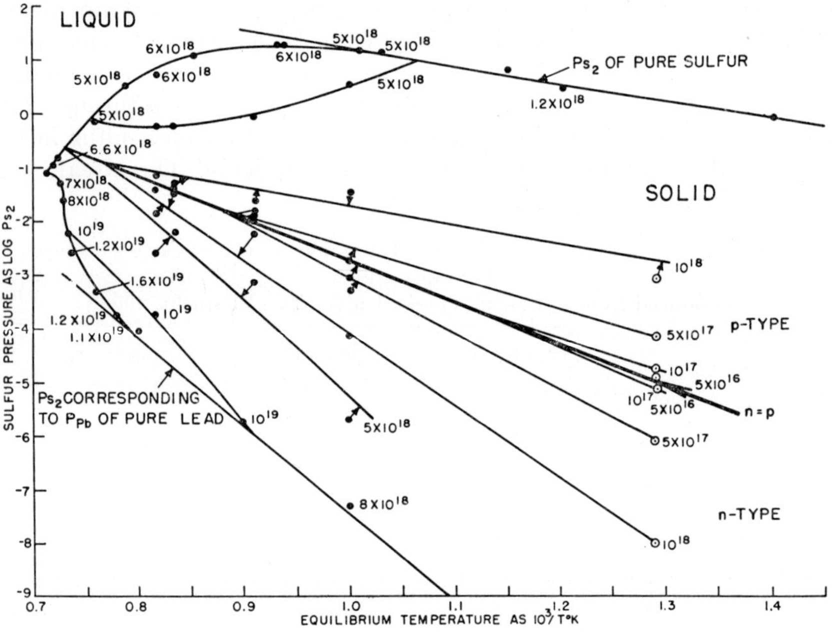

Fig. 5. Phase diagram for the PbS system.

primarily near the surface as the crystal-vapor system cools. The compositions of the crystals were then obtained from measurements of the Hall coefficient.

Data on composition and equilibrium vapor pressure of sulfur at temperatures 1223°K, 1200°K, and 1000°K were obtained by Bloem whereas Brebrick and Scanlon give data at 773°K. A useful phase diagram is obtained by combining these data with the observations of Bloem and Kröger[22] on the melting point composition relationships for PbS. This diagram, shown in Fig. 5, has the same form as the one published by Bloem and Kröger except that it contains additional data for equilibrium experiments at low temperature.

This diagram shows the pressure-temperature conditions necessary to establish a desired equilibrium composition in a PbS crystal. There generally is a great deal of interest in producing crystals having nearly stoichiometric compositions. This diagram suggests that such crystals are not made easily except at low equilibrium temperatures where the resolution in pressure seems more favorable. However, long times of treatment are required to reach equilibrium. The maximum electron concentration obtainable in PbS crystals appears to be about 10^{19}, whereas the maximum hole concentration is about $6 \times 10^{18}/\text{cm}^3$. The solid phase exists within a range extending about $\pm 0.1\%$ on either side of the stoichiometric composition. It will be noted that the intrinsic concentration for natural crystals (data at $10^3/T = 1.3$) falls on the intrinsic line for pure synthetic crystals. On the other hand, Bloem[15] shows that impurity concentrations of atoms like Bi and Ag as small as $10^{17}/\text{cm}^3$ shifts the position of the intrinsic line of S_2 pressure by several orders of magnitude at this temperature. Thus we may conclude that the natural crystals used in these experiments have a purity as high as that of the synthetic crystals.

The line representing stoichiometric composition is seen to intersect the liquidus line at a temperature near 1360°K, which is the melting point of the pure stoichiometric solid. The maximum melting point is seen to be 1400°K and corresponds to a composition having an excess of lead of about $6 \times 10^{18}/\text{cm}^3$.

Lines of equal composition appear to be distributed approximately symmetrically about the stoichiometric line. One can show from the statistical model of the crystal[26] that this symmetry is to be expected. Furthermore, the analysis suggests that the curve for the stoichiometric composition may be a straight line on a log P_{S_2} *vs* $1/T$ plot.

These curves also show that at a constant P_{S_2} the deviation of the stoichiometry of a crystal can vary by several orders of magnitude if it is heated slowly from about 500°C to 1000°C. This is an important consideration in any experiment in which the behavior of a property is being studied over a range of high temperatures. For example, the effects of such changes in composition upon the Hall effect and resistivity of PbS crystals[18] have been reported. Detailed phase diagrams of the crystal-vapor systems of PbSe and PbTe in the neighborhood of the stoichiometric compound are not available. The phase diagram of the liquid-solid system given by Hansen[25] shows only the stoichiometric compound of PbSe. Goldberg and Mitchell[23] give more details of the phase diagram of the liquid-solid system near the stoichiometric compound and indicate that the maximum melting point occurs for the composition with excess selenium.

The liquid-solid phase diagram of PbTe shown by Hansen[25] implies that the solid phase exists over a range of composition of about 10 or 15% relative to that of the ideal stoichiometric compound. Such a range is now believed to be excessive. A range of less than 0.1% probably is more accurate.[26] The composition having the maximum melting temperature contains at least 10^{18} atoms per cubic centimeter of excess tellurium.

Koval'chik and Maslakovets[37] report that the maximum solubility of excess lead and tellurium in PbTe are about 10^{19} and 4×10^{18} per cc respectively. Recent results of Brebrick[26] give the values 3×10^{18} and $10^{19}/\text{cm}^3$, respectively, for the maximum lead and tellurium excess concentrations.

10. Foreign Impurity Atoms in Polar Semiconductors

Thus far we have considered only the problems connected with the preparation of crystals of pure compounds. It is also possible to influence the properties of polar semiconductors by adding foreign elements to the crystals. These atoms may be incorporated in the lattice either in interstitial or at normal lattice sites.

We shall consider first the problem of impurities in normal lattice sites. The concentration of imperfections is related to the concentration of other defects by the laws of mass action and electroneutrality. Over certain limited ranges of composition, the situation may be described by the principle of controlled valency, formulated by Verwey and coworkers,[38] or by the principle of controlled vacancy, formulated by Wagner and Schottky.[39]

Each of these principles corresponds to different approximations of the electroneutrality law. In PbS, the composition range is so wide that neither of the principles covers the range exclusively. An extensive treatment of this subject is given by Kröger and Vink[40] and applied to PbS by Bloem[15] and others.[41]

The chemical equilibrium between the solid and vapor of impure PbS crystals was studied extensively by Bloem.[15] The experimental procedures used were the same as those used for the pure crystals except that various quantities of Ag and Bi ions were present in the crystals. The equilibrium results are summarized in Fig. 6 in which the pressure

[37] T. L. Koval'chik and Yu. P. Maslakovets, *Zhur. Tekh. Fiz.* **26**, 2417 (1956).

[38] E. Verwey, P. Haayman, F. Romeyn, and G. van Osterhout, *Philips Research Rept.* **5**, 173 (1950).

[39] C. Wagner and W. Schottky, *Z. physik. Chem. (Leipzig)* **B11**, 163 (1931).

[40] F. A. Kröger and H. J. Vink, *Solid State Phys.* **3**, 307 (1956).

[41] J. Bloem, F. A. Kröger, and H. J. Vink, *Rept. Bristol Conf. on Defects in Crystalline Solids, 1954* p. 273 (1955).

of the sulfur atmosphere corresponding to *p-n* transitions is plotted as a function of equilibrium temperature.

Bismuth moves the *p-n* transition point to higher pressures of sulfur whereas silver moves it to lower pressures at a given temperature. It will be noted that as few as 10^{17} ions/cc of Ag or Bi change the sulfur pressure required for the *p-n* transition by orders of magnitude at low equilibrium temperatures.

Foreign atoms can occupy interstitial sites in these crystals provided the ionic size does not exceed lattice interatomic distances greatly.

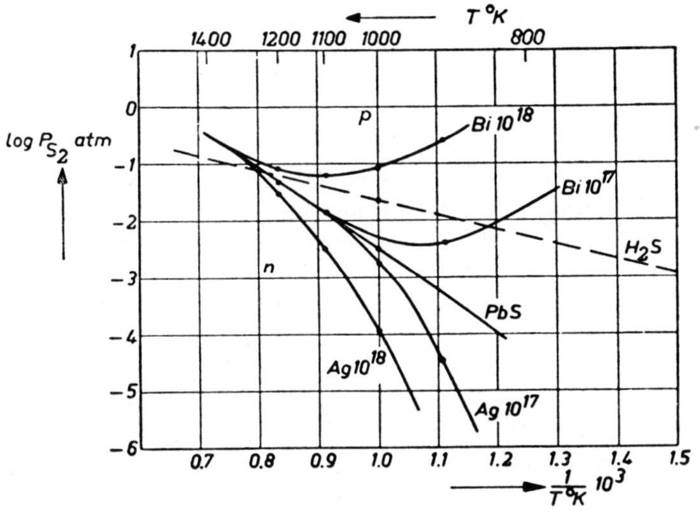

FIG. 6. The boundaries between *p*- and *n*-type regions for PbS and PbS with added foreign ions as a function of atmosphere and temperature. [Reproduced from J. Bloem, *Philips Research Repts.* **11,** 273 (1956).]

Bloem and Kröger[42] show that Cu^+ has sufficient space to migrate through the PbS lattice with a small activation energy. Using similar reasoning, they show that Ni^+ and Li^+ should also be mobile. Experiments with copper[42] and nickel[43] verify these observations; however, negative results were obtained for lithium.

In general, interstitial diffusion may be more rapid than diffusion by a vacancy mechanism. The diffusion coefficients of Cu^+ and Ni^+ are significantly higher than that of Pb in PbS. The diffusion coefficient of Cu^+ in PbS ranges from about 10^{-7} cm² sec⁻¹ at 100°C to 10^{-4} at 400°C; that for Ni^+ in PbS is somewhat smaller, ranging from about 10^{-9} $cm^2\ sec^{-1}$

[42] J. Bloem and F. A. Kröger, *Philips Research Repts.* **12,** 281 (1957).
[43] J. Bloem and F. A. Kröger, *Philips Research Repts.* **12,** 303 (1957).

at 200°C to 10^{-5} at 500°C. In the diffusion equation,

$$D = D_0 \exp(-Q/RT) \text{ cm}^2 \text{ sec}^{-1} \qquad (10.1)$$

the value of D_0 for Cu^+ is 5×10^{-3} cm² sec⁻¹; that for Ni^+ is 17.8 cm² sec⁻¹. Values of Q are 7130 cal/mole for Cu^+ and 22,000 cal/mole for Ni^+.

In the case of interstitial Cu^+, the lattice becomes saturated at about 3×10^{18} ions/cm³, whereas Ni^+ becomes saturated at a concentration near 5×10^{17} ions/cm³.

The ionization energy of interstitial Cu is 0.02 ev whereas that of Ni is 0.03 ev, as obtained from Hall data.

Hydrogen ions also diffuse rapidly in PbS crystals; however, it is believed to combine to form H_2 molecules which make no contribution to electrical conductivity, as seems to be the case for hydrogen in germanium.[44]

Sulfur atoms present at grain boundaries or other imperfections in PbS crystals have a marked influence on diffusion rates.[42] Crystals of PbS grown in a pressure of one atmosphere of sulfur have been found by chemical analysis to contain an excess of sulfur relative to the stoichiometric composition of 1.6%, whereas the electrical measurements indicate that the sulfur concentration is only 0.025%. The excess sulfur is believed to be concentrated on grain boundaries. The diffusion of Cu, for example, is almost completely retarded in such crystals. A *p-n* junction will not move across the boundaries until all the sulfur has reacted with the diffusing element.

A study of the effect of foreign metallic atoms and of the halogens on the electrical properties of PbTe was carried out by Koval'chik and Maslakovets.[37] The elements Sn, Ge, Fe, Co, Ni, Pt, Mg, Nb, and Bi acted as donors whereas groups II ‘I, and III elements, Au, Ag, Cu, In, Al, Zn, and Cd behaved as acceptors in the presence of Br atoms.

11. Alloys of the Pure Compounds

In consequence of the isomorphism of the three compounds PbS, PbSe, and PbTe, the common cation, and nearness of anion sizes, one may expect alloys of the compounds to exist in all proportions. This fact has been established experimentally at a number of laboratories. The problems of stoichiometry are very complicated for such alloys. However, single crystals of a number of alloys of PbS + PbSe and PbSe + PbTe have been made by the techniques used in growing crystals of the pure compounds.[45] These alloys range in composition

[44] H. Reiss, *Phys. Rev.* **100**, 1806A (1955).

[45] W. W. Scanlon, Intern. Conf. on Semiconductors, Rochester, N.Y., 1958, *J. Phys. Chem. Solids* **8**, 423 (1959).

from a few per cent to 50 atomic per cent. The cleavage properties and general appearances of the crystals are similar to those of the pure compounds. Their electrical, optical, and thermal characteristics, however, are dependent upon the composition and provide a continuous array of semiconductors possessing unique physical properties. These properties will be discussed in the sections to follow.

IV. Electrical and Optical Properties

We shall now consider the behavior of electrons and holes in polar semiconductors. One might expect the electron-lattice interactions in these crystals to differ in some respects from those in valence semiconductors as a result of the ionic character of the bonding. This difference is displayed most clearly by the mobility of electrons and holes since the mechanisms of scattering differ in polar and valence crystals. In general, however, the electrical and optical properties of the polar crystals obey essentially the same laws as those of valence crystals. In the discussion to follow, we shall review some of the semiconducting properties of the family of crystals represented by PbS, PbSe, and PbTe. It would be difficult to refer to all the published literature concerning the materials. Many excellent bibliographies covering the early research are given elsewhere.[46,47,48] We shall attempt to confine our attention to more recent studies.

12. Band Structure

The shape of the electronic energy bands in semiconductors is a subject of fundamental importance. The theoretical calculation of this structure is an extremely difficult task. It is interesting to point out that the first calculation of this type made for any semiconductor was that for PbS.[49] The analysis suggests that the bands are not simple paraboloids of revolution centered at the origin of the momentum axis but contain minima at values of momentum differing from zero. The calculations indicate that the minimum vertical separation between the valence and conduction bands occurs along the $\langle 110 \rangle$ directions and is about 1.3 ev, whereas the minimum nonvertical energy difference is about 0.3 ev.

Experimental evidence concerning the band structure of these materials has been obtained from studies of magnetoresistance and optical

[46] R. A. Smith, *Physica* **20,** 910 (1954).
[47] T. S. Moss, *Proc. I.R.E.* **43,** 1869 (1955).
[48] R. P. Chasmar, *in* "Photoconductivity Conference" (R. G. Breckenridge, F. J. Russell, and E. E. Hahn, eds.), p. 463. Wiley, New York, 1956.
[49] D. G. Bell, D. M. Hum, L. Pincherle, D. W. Sciama, and P. M. Woodward, *Proc. Roy, Soc.* **A217,** 71 (1953).

absorption. The magnitude of the magnetoresistance depends on the directions of the sample current and the magnetic field relative both to one another and to the direction of the symmetry axes of the sample.

The form of transport theory valid for small fields predicts that

$$\frac{\Delta\rho}{\rho} = M_{\text{abc}}^{\text{def}} \frac{\mu_H{}^2 H^2}{C^2} \tag{12.1}$$

where $\Delta\rho/\rho$ is the fractional change in the resistivity at zero field, $M_{\text{abc}}^{\text{def}}$ is the magnetoresistance coefficient, *abc* and *def* specify the directions of the current and magnetic field respectively, μ_H is the Hall mobility of the current carriers and H is the magnetic field. In emu, Gaussian, and practical units, the constant C has the values 1, 3×10^{10} cm/sec, and 1×10^8 cm^2-gauss/volt-sec respectively. Small fields imply that $\mu_H H/C \ll 1$.

In the case in which the energy surfaces are spherical and centered at $k = 0$, $M_{\text{abc}}^{\text{def}} = 0.27 \sin^2 \theta$ for nondegenerate samples in the lattice scattering range. Here θ is the angle between the current and magnetic field. If the electron gas is completely degenerate and if τ, the relaxation time, is a function of energy alone $M_{\text{abc}}^{\text{def}} = 0$.

In the case of a cubically symmetric model in which the surfaces of constant energy are ellipsoids of revolution along the $<100>$ direction in k space, the coefficient M_{100} is zero for all values of the ratio K of longitudinal to transverse effective mass which characterizes the ellipsoid. On the other hand, if the ellipsoids are oriented along the $<110>$ or $<111>$ directions, they may be distinguished by the value of the ratio M_{100}/M_{100}^{001}. This ratio is plotted as a function of K in Fig. 7 for the $<110>$ and $<111>$ oriented cubic models in the cases in which both classical and degenerate statistics apply. Note that the highest ratio is 2 for the $<110>$ model while that for the $<111>$ model is 4.

In a recent paper, Allgaier[50,51] reported values of magnetoresistance obtained on the lead salts at 77.4° and 4.2°K. The samples contained about 10^{18} carriers per cm^3. Consequently, classical statistics was a rather poor approximation at 77.4°K; all of the samples were highly degenerate at the lower temperature. The measurements were made only on cleaved crystals so that the direction of current flow was restricted to a $<100>$ axis.

Allgaier found that the longitudinal coefficient M_{100} generally was as large or larger than the transverse coefficient M_{100}^{001} for both n- and p-type material in all three lead salts. These results suggested that the conduction and valence bands in the lead salts may be similar, a situation which

[50] R. S. Allgaier, *Phys. Rev.* **112**, 828 (1958).

[51] R. S. Allgaier, Thesis, University of Maryland (1958).

is to be contrasted with that suggested by magnetoresistance data for germanium and silicon.

The relatively large longitudinal magnetoresistance which Allgaier found for all samples eliminates the simple band model immediately. It also eliminates the cubic model in which the ellipsoids are oriented along the <100> directions.

Allgaier's experimental data for PbS were not sufficiently precise or extensive to allow one to choose between the <110> and the <111> oriented models. The ratio M_{100}/M_{100}^{001} for PbSe was principally between

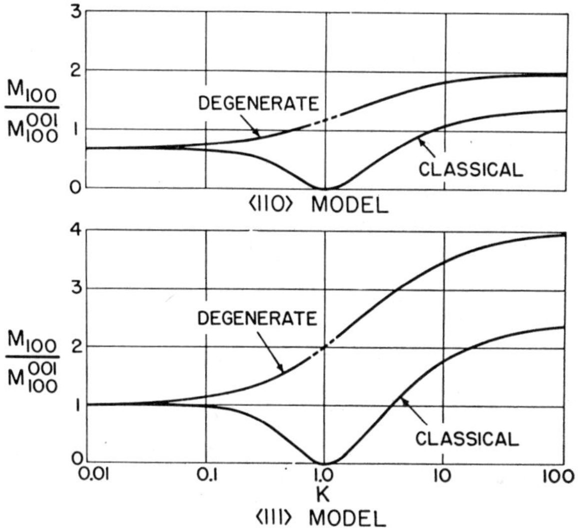

FIG. 7. The ratio M_{100}/M_{100}^{001} as a function of κ for the <110> and <111> models. [Reproduced from R. S. Allgaier, *Phys. Rev.* **112**, 828 (1958).]

1 and 2 so that either model is in agreement with the data if one assumes prolate ellipsoids ($K > 1$). Four of the five PbTe samples he studied had ratios between 2 and 5. Thus the <111> oriented model seems more appropriate for PbTe. This agrees with the conclusions of Shogenji and Uchiyama.[52]

13. The Intrinsic Energy Gap

Electronic transitions across the gap of forbidden energies between the filled band and the conduction band may be of two general types. The transition is called direct when the electron obeys the selection rule requiring no change of momentum. This selection rule may be relaxed if the electron interacts with the lattice so that the transition is accom-

[52] K. Shogenji and S. Uchiyama, *J. Phys. Soc. Japan* **12,** 1164 (1957).

panied by the emission or absorption of a phonon. The transition is then called indirect. Information on the transition energies can be obtained from certain thermal measurements such as those concerning the temperature dependence of resistivity or Hall effect, or from certain optical measurements such as those concerning transmission, reflectivity, or photoconductivity. Thermal measurements yield information on the indirect transition energy, often called the thermal energy gap, whereas optical measurements can be used to evaluate the energy for direct as well as indirect transitions. The energy for direct transitions often is designated the optical energy gap.

In ionic crystals there is an additional distinction between the thermal and optical energy gaps. This originates in the fact that a redistribution of the ionic charge, which changes the relative orientation of the bands, can occur during the relatively slow process of thermal excitation of an electron, but not for the faster optical excitation process. The difference appears as a difference between the excitation energies for the optical and thermal processes. It can be described in terms of the dielectric constants at low frequency, ϵ_s, and at high frequencies, ϵ_0, and is proportional to $(\epsilon_s - \epsilon_0)/\epsilon_s\epsilon_0$. If ϵ_s is appreciably larger than ϵ_0 as it could be in ionic crystals, the optical energy gap is significantly larger than the thermal gap. A simple band structure is assumed in this analysis. The foregoing conditions may be obscured in a multiple valley energy band system.

A theoretical analysis which relates the electron and hole concentrations to temperature for the simple energy band model has been carried out. In the absence of more detailed information concerning the energy band structure, this simplified treatment will be applied to the lead compounds in an attempt to evaluate the thermal energy gap approximately.

In the intrinsic semiconductor, the density of electrons n and holes p at thermal equilibrium is given by

$$np = N_c N_v \exp(-E_g/kT). \tag{13.1}$$

Here N_c and N_v are density of state terms for the conduction and valence bands:

$$\begin{aligned} N_c &= 2(2\pi m_n kT/h^2)^{\frac{3}{2}} \\ N_v &= 2(2\pi m_p kT/h^2)^{\frac{3}{2}} \end{aligned} \tag{13.2}$$

and m_n and m_p are the effective electron and hole masses. The introduction of these quantities into Eq. (13.1) reduces the expression for the product np to the form

$$\frac{np}{T^3} = A \exp(-E_g/kT) \tag{13.3}$$

where $A = 4(2\pi k/h^2)^3(m_n m_p)^{\frac{3}{2}}$.

Experimental values for n and p were obtained from Hall effect measurements on PbS[16]. The results are plotted in such a way that log (np/T^3) appears as a function of $1/T$ in Fig. 8. The slope of this curve is 0.34 ev and may be designated the thermal energy gap for PbS.

The measurements of the Hall effect used in this analysis were limited to temperatures below about 500°K. The crystals used in the studies had compositions which were adjusted by a high-temperature vapor diffusion process so that several cycles of data concerning the Hall effect could be

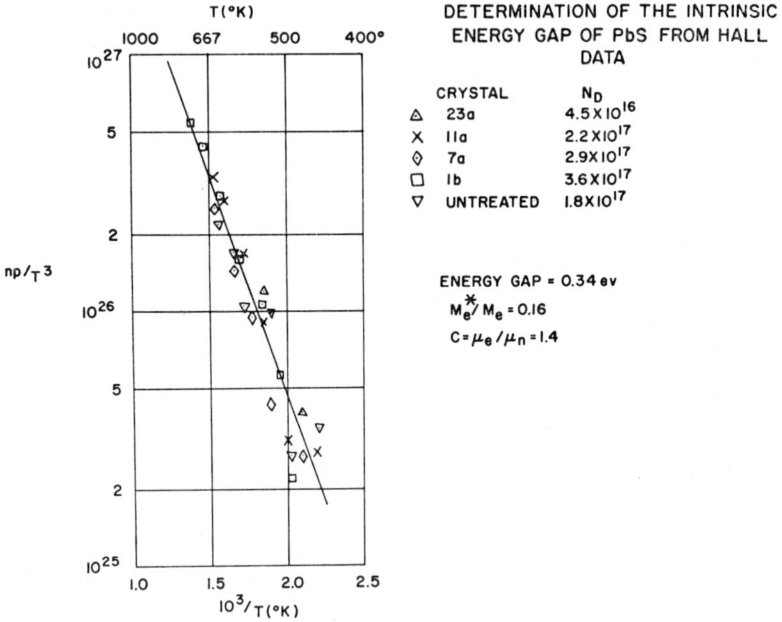

FIG. 8. Plot of log (np/T^3) *vs* $1/T$ for PbS.

obtained at temperatures below 500°K. Above this temperature, diffusion is sufficiently rapid to change the stoichiometric balance significantly in the time between Hall effect measurements. As a result the concentrations of carriers at high temperatures are determined largely by stoichiometric deviations, rather than by thermal excitation of carriers across the forbidden gap.[18] For these reasons the data concerning the Hall effect in the materials which extend to 1000°C cannot be interpreted with use of the equations for the intrinsic Hall effect to obtain a value for the energy gap.

The study of the optical absorption edge provides a more detailed picture of the transition of electrons from the filled to the conduction band. A theoretical analysis[53,54] predicts a somewhat different dependence of absorption on photon energy for the cases of direct and indirect transitions. For direct transitions, the absorption coefficients are large, being greater than about 5000 cm^{-1}, and depend upon the photon energy as $E^{\frac{1}{2}}$. The theoretical interpretation of the absorption associated with indirect transitions is not so clear. The absorption coefficient is low and may be a quadratic or cubic function of the photon energy.

The existence of direct and indirect transitions does not always imply that there is a minimum in the conduction band at points other than $k = 0$. Indirect transitions may also occur at the origin at temperatures above the Debye temperature θ, as a result of the emission or absorption of lattice phonons of energy $k\theta$. The effect of this type of transition is the production of a tail in the absorption curve at low energies. This tail should decrease at temperatures below θ. On the other hand, if indirect transitions are made to a valley in the conduction band away from the origin, the absorption tail should not be altered appreciably by cooling. A difference in the energy for direct and indirect transitions large compared to $k\theta$, however, provides good evidence for a minimum in the conduction band at a momentum different from that for the maximum of the valence band.

The first studies of the absorption coefficient in the family of compounds were made on PbS by Paul, Jones, and Jones.[55] Gibson[56] reported similar studies on PbS, PbSe, and PbTe. Absorption coefficients extending only to about 10^2 cm^{-1} were obtainable, on the crystals employed. Wide slits had to be used on the spectrometer in order to get sufficient energy through the crystals. This led to poor resolution of the absorption edge.

Avery[57] obtained data concerning the absorption coefficient in the range from 10^4 to 10^6 cm^{-1} with use of a reflection technique. Radiation polarized in planes at right angles was used in these measurements for two angles of incidence. This method yields results which are most accurate in regions of the spectrum where the absorption coefficients are large, of the order of 10^4 to 10^6 cm^{-1}. Thus the early studies of the absorption edge in these crystals were confined either to low or high absorption, leaving a gap in the data in the important region of absorption from

[53] "Photoconductivity Conference" (R. G. Breckenridge, F. J. Russell, and E. E. Hahn, eds.), J. Bardeen, F. J. Blatt, and L. H. Hall, p. 146; D. L. Dexter, p. 155; H. Y. Fan, M. L. Shepherd, and W. G. Spitzer, p. 184. Wiley, New York, 1956.

[54] H. Y. Fan, *Repts. Progr. in Phys.* **19,** 107 (1956).

[55] W. Paul, D. A. Jones, and R. V. Jones, *Proc. Phys. Soc.* **B64,** 528 (1951).

[56] A. F. Gibson, *Proc. Phys. Soc.* **B65,** 378 (1952).

[57] D. G. Avery, *Proc. Phys. Soc.* **B64,** 1087 (1951); **B65,** 425 (1952); **B66,** 133 (1953); **B67,** 2 (1954).

about 10^2 to 10^4 cm^{-1}. Since the regions of the spectrum covered by the two techniques overlapped in wavelength but not in absorption coefficients, there remained an uncertainty about the exact shape of the absorption edge. An analysis of these data in terms of direct and indirect transitions was not possible.

It was possible to reduce the thickness of the crystals to less than a micron by using a grinding and polishing technique. Transmission measurements made on crystals ranging in thickness from the order of millimeters to microns allowed determination of the entire absorption edge from about 10 to 10^5 cm^{-1} for PbS,[58] PbSe,[45] and PbTe.[45]

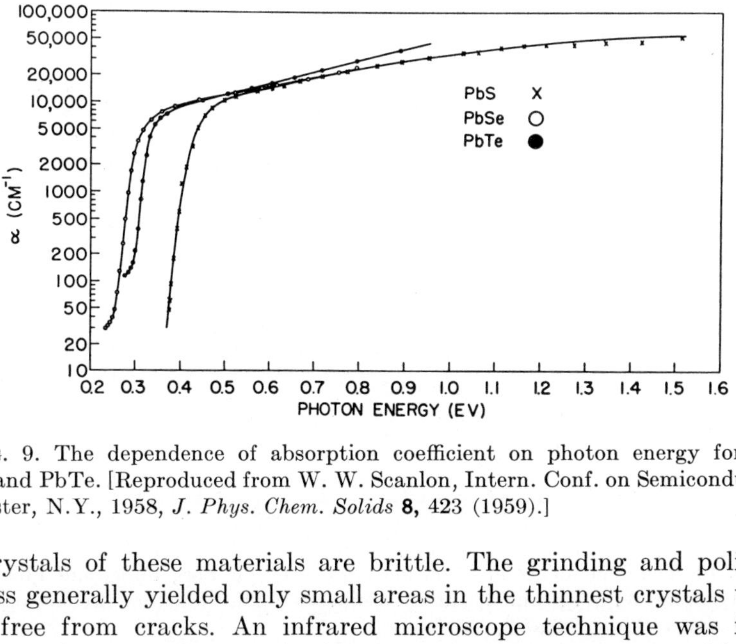

FIG. 9. The dependence of absorption coefficient on photon energy for PbS, PbSe, and PbTe. [Reproduced from W. W. Scanlon, Intern. Conf. on Semiconductors, Rochester, N.Y., 1958, *J. Phys. Chem. Solids* **8,** 423 (1959).]

Crystals of these materials are brittle. The grinding and polishing process generally yielded only small areas in the thinnest crystals which were free from cracks. An infrared microscope technique was found which made it possible to obtain absorption measurements with high resolution on areas as small as a few hundred microns on edge.

The absorption curves for the three materials are shown in Fig. 9. The resolution in the measurements ranged from about 0.04 μ at a wavelength of 4 μ to 0.06 μ at 2.5 μ. This was sufficiently small to reveal the details of the absorption edge.

The analysis of the absorption edge in terms of direct transition energies was obtained by plotting α^2 against photon energy for values of α greater than about 5000 cm^{-1} and taking the energy intercept at $\alpha = 0$. The curves are shown in Fig. 10. Indirect transition energies were

[58] W. W. Scanlon, *Phys. Rev.* **109,** 47 (1958).

obtained similarly by plotting $\alpha^{\frac{1}{2}}$ against photon energy, as shown in Fig. 11. The data are summarized in Table V. There is no evidence for the much larger values associated with direct energy transitions in these compounds, ranging up to 1.3 ev, which were suggested by several of the earlier workers in this field.

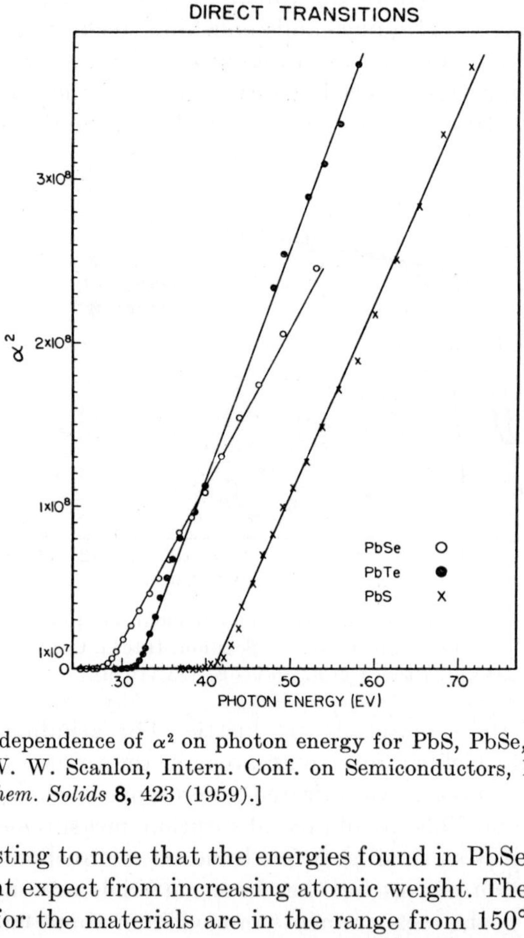

FIG. 10. The dependence of α^2 on photon energy for PbS, PbSe, and PbTe. [Reproduced from W. W. Scanlon, Intern. Conf. on Semiconductors, Rochester, N.Y., 1958, *J. Phys. Chem. Solids* **8,** 423 (1959).]

It is interesting to note that the energies found in PbSe are not in the order one might expect from increasing atomic weight. The characteristic temperatures for the materials are in the range from 150°K to 200°K.[59] Hence the maxima in the lattice phonon energies are of the order of 0.02 to 0.03 ev. As a result, the small differences observed in the direct and indirect transition energies could be accounted for without assuming a minimum in the conduction band away from the origin in k space.

While the energy gaps of most semiconductors have a negative temperature dependence which lies in the range from -3 to -5×10^{-4} ev/°K,

[59] D. H. Parkinson and J. E. Quarrington, *Proc. Phys. Soc.* **A67,** 569 (1954).

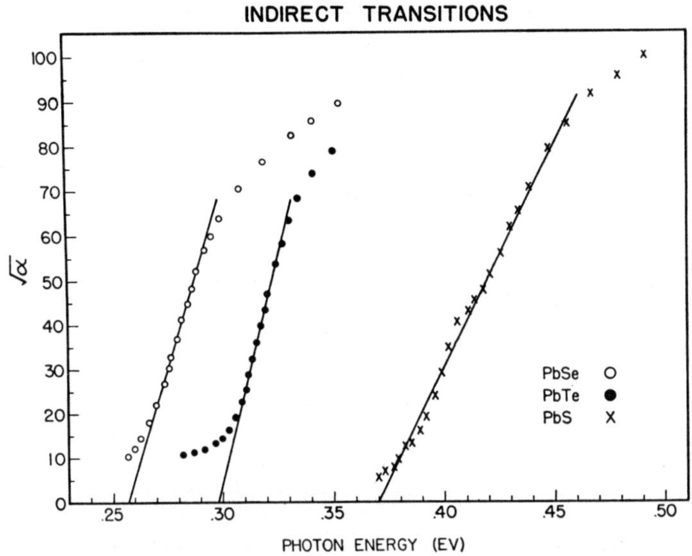

FIG. 11. The dependence of $\sqrt{\alpha}$ on photon energy for PbS, PbSe, and PbTe. [Reproduced from W. W. Scanlon, Intern. Conf. on Semiconductors, Rochester, N.Y., 1958, *J. Phys. Chem. Solids* **8,** 423 (1959).]

the PbS group have a positive temperature coefficient of about $+4 \times 10^{-4}$ ev/°K.[56]

Optical absorption studies have also been made on a group of single crystals of the alloys PbS + PbSe and PbSe + PbTe.[45] The purpose of the study was to see if an alloy with a minimum energy gap might exist between the compounds PbS and PbSe or PbSe and PbTe. The absorption curves were obtained for various alloys of the compounds. These display general features similar to those of the pure compounds. From the plots of α^2 and $\alpha^{\frac{1}{2}}$ as functions of photon energy, values for direct and indirect transition energies were obtained. These are summarized in Fig. 12,

TABLE V. SUMMARY OF TRANSITION ENERGY AND RADIATIVE RECOMBINATION DATA FOR PbS, PbSe, AND PbTe

Compound	Transition energy Direct	Indirect	R	n_i	τ_r
PbS	0.41	0.37	22.8×10^{18}	3×10^{15}	63
PbSe	0.29	0.26	14.6×10^{20}	2×10^{17a}	8.0
PbTe	0.32	0.29	8.9×10^{20}	6×10^{16a}	6.8

R = recombination rate, n_i = intrinsic carrier concentration, and τ_r = radiative lifetime.

[a] Calculated assuming $(m_e^*/m_e) = (m_h^*/m_h) = 1$.

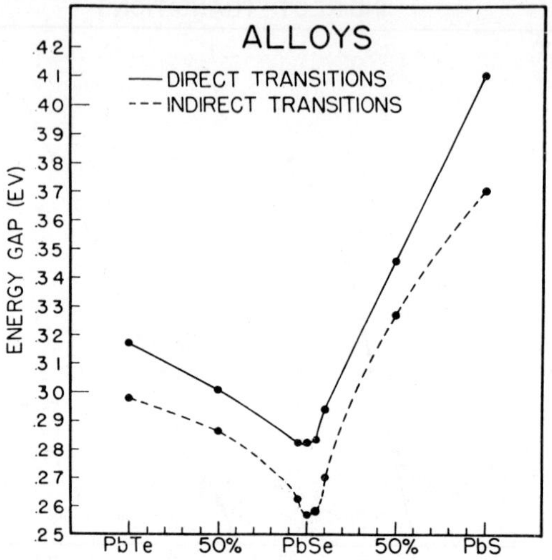

FIG. 12. Energy gaps in the alloys PbTe + PbSe and PbSe + PbS. [Reproduced from W. W. Scanlon, Intern. Conf. on Semiconductors, Rochester, N.Y., 1958, *J. Phys. Chem. Solids* **8**, 423 (1959).]

along with the corresponding data on the pure compounds. The energies are seen to vary continuously from one pure compound to another, the minimum energy gap occurring in pure PbSe.

14. Optical Constants

The semiconducting lead salts are characterized by large refractive indices. Avery[57] reports the following values at a wavelength of 3 μ from reflection studies: 4.10 for PbS, 4.59 for PbSe, and 5.35 for PbTe. We have found similar values for the materials from transmission studies. The refractive indices change little through the region of wavelength of the absorption edge.

There is evidence of direct interaction of the photons with the transverse polar vibrational modes of the lattice, the restrahlung frequency at very long wavelengths. Strong[60] gives curves showing that the process occurs at a wavelength of about 80 μ in PbS. Similar studies have not been made on PbSe and PbTe.

15. Transport Phenomena

We shall now consider the phenomena associated with the motion of electrons and holes when under the influence of applied electric or magnetic fields or temperature gradients. The solution of the Boltzmann

[60] J. Strong, *Phys. Rev.* **38**, 1818 (1931).

equation for the simple model and a general treatment of transport phenomena are given elsewhere.[61] The motion of electrons would be uninhibited by the lattice if the crystal were perfectly periodic. However, departures from a perfectly periodic structure produce scattering of the moving charges. These departures can be thermal vibrations of the lattice ions or atoms, lattice defects, impurities, both ionized and neutral, and other charge carriers. The influence of the various scattering mechanisms in valence crystals has been studied extensively and the results are summarized by Fan.[62] The same scattering mechanisms may be present in ionic crystals; however, their relative importance is believed to be different from that in valence crystals. These differences are revealed by the behavior of the Hall effect, resistivity, and mobility in the family of lead-salt semiconductors.

a. Hall Effect and Resistivity

Typical Hall effect and resistivity curves for crystals of PbS[16] and PbTe[63] having various concentrations of carriers are shown in Figs. 13 and 14 for the temperature range from about 80°K to 500°K. Measurements at temperatures in excess of about 500°K are difficult to make on these materials unless the composition of the atmosphere surrounding the crystal is adjusted suitably at each temperature to maintain a fixed crystal composition. Diffusion is sufficiently slow below about 500°K to maintain the crystal composition constant, regardless of the atmospheric composition.

A value for the intrinsic energy gap may be obtained from the Hall data for high temperatures. In the case of materials having a small gap, the transition from impurity conduction to intrinsic conduction may extend well into the linear portion of the log R *vs* $1/T$ curve at high temperatures. In other words, the Fermi level may still be shifting from near the conduction band to its intrinsic position. In this case the energy gap should be obtained from the relation $np = CT^3 \exp - E_g/kT$, which is independent of the position of the Fermi level. Such a plot of np/T^3 *vs* $1/T$, shown in Fig. 8, gives the energy gap 0.34 ev at absolute zero. This is in good agreement with the value obtained from optical measurements. Similar analysis of Hall data for PbTe gives a value for the intrinsic energy gap $E_g = 0.3$ ev.[64]

[61] A. H. Wilson, "Theory of Metals," Cambridge University Press, London and New York, 1953; F. Seitz, "Modern Theory of Solids," McGraw-Hill, New York, 1940.

[62] H. Y. Fan, *Solid State Phys.* **1,** 283 (1955).

[63] K. Shogenji and S. Uchiyama, *J. Phys. Soc. Japan* **12,** 252 (1957).

[64] K. Shogenji and S. Uchiyama, *J. Phys. Soc. Japan* **12,** 431 (1957).

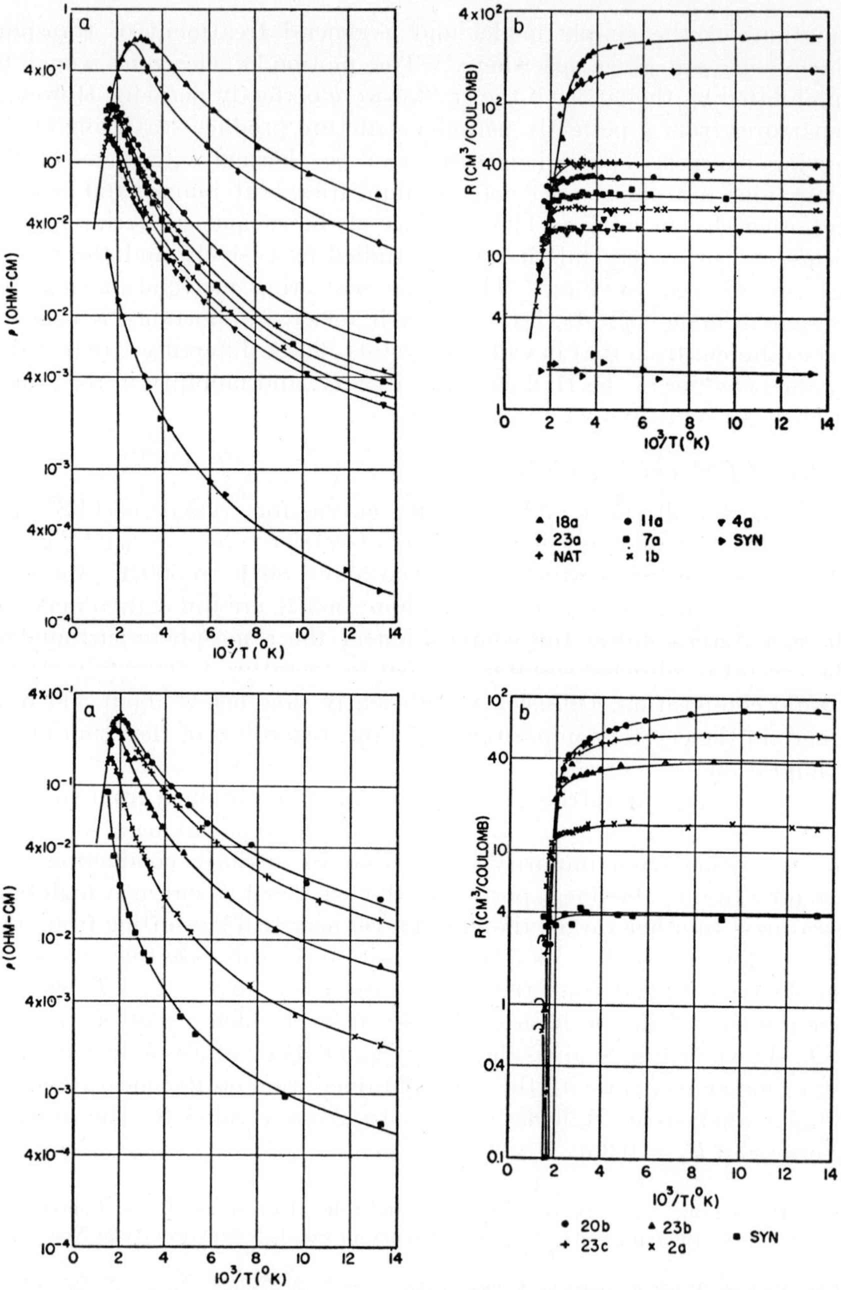

FIG. 13. Resistivity and Hall coefficient for *n*-type (top) and *p*-type (bottom) crystals of PbS. [Reproduced from R. F. Brebrick and W. W. Scanlon, *Phys. Rev.* **96,** 598 (1954).]

The intrinsic carrier concentration in PbS obtained from Hall data at room temperature is found to be $2.9 \times 10^{15}/\text{cm}^3$. Estimated values for PbSe and PbTe obtained assuming $(m_e^*/m_e) = (m_h^*/m_h) = 1$ are $6 \times 10^{16}/\text{cm}^3$ and $2 \times 10^{17}/\text{cm}^3$,[65] respectively. Extrapolation of the resistivity data indicates that PbS should have a resistivity of 2.5 Ω cm. Similar estimates, based upon published PbTe data,[63] indicate that intrinsic PbTe should have a resistivity of about 0.7 ohm-cm.

From the rectification measurements on a *p-n* junction in PbS, Moss[66] obtains an intrinsic carrier concentration of 2.7×10^{-15} cm^{-3} and an

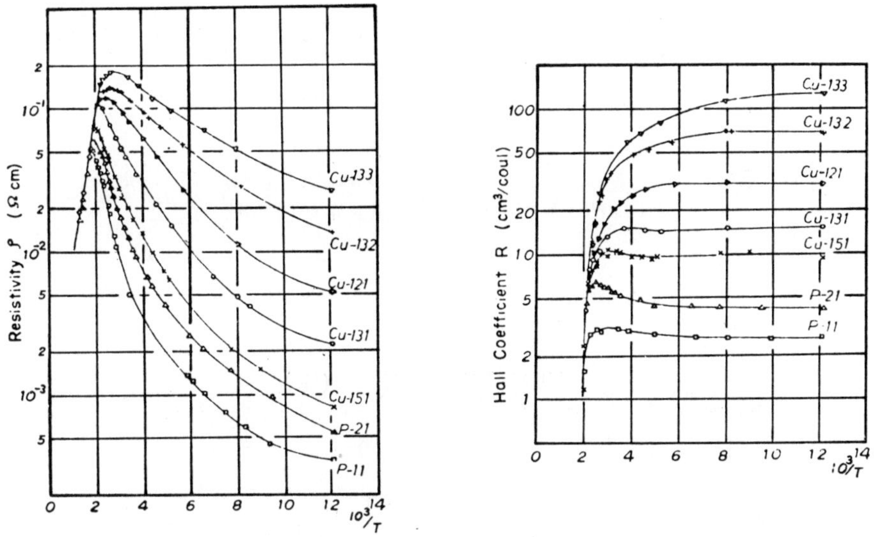

FIG. 14. Resistivity and Hall coefficient for pure PbTe (P) and PbTe doped with copper (Cu). [Reproduced from K. Shogenji and S. Uchiyama, *J. Phys. Soc. Japan* **12,** 252 (1957).]

intrinsic resistivity of 3.1 ohm-cm, which are in close agreement with the foregoing results.

Typical behavior of the Hall effect and resistivity at temperatures down to 4.2°K are shown in Figs. 15 and 16 for the three materials.[67] It is evident from these curves that the carrier concentrations vary only slightly in the temperature range from about 300°K to 4.2°K in crystals which have not been doped with foreign impurity atoms. Since the concentrations of carriers in many of these crystals are high, they become

[65] R. L. Petritz, *in* "Photoconductivity Conference" (R. G. Breckenridge, F. J. Russell, and E. E. Hahn, eds.), p. 62. Wiley, New York, 1956.

[66] T. S. Moss, *Proc. Phys. Soc.* **B68,** 697 (1955).

[67] R. S. Allgaier and W. W. Scanlon, *Phys. Rev.* **111,** 1029 (1958).

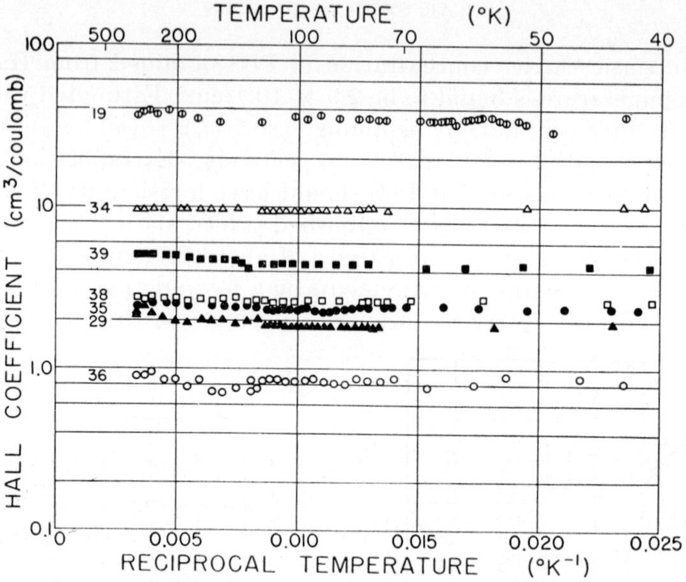

(a)

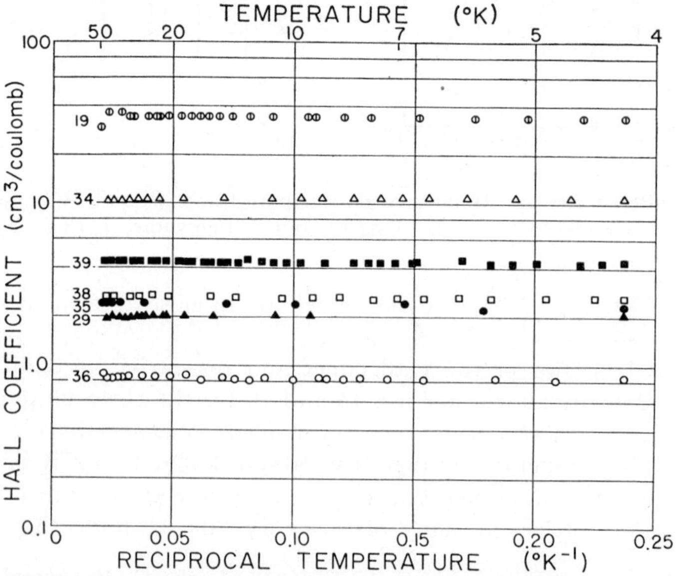

(b)

FIG. 15. Low temperature Hall coefficient in PbS, PbSe, and PbTe. (The symbols are identified in Fig. 16.) Curves (a) are in the temperature range 300° to 40°K, curves (b) 50° to 4.2°K. [Reproduced from R. S. Allgaier and W. W. Scanlon, *Phys. Rev.* **111,** 1029 (1958).]

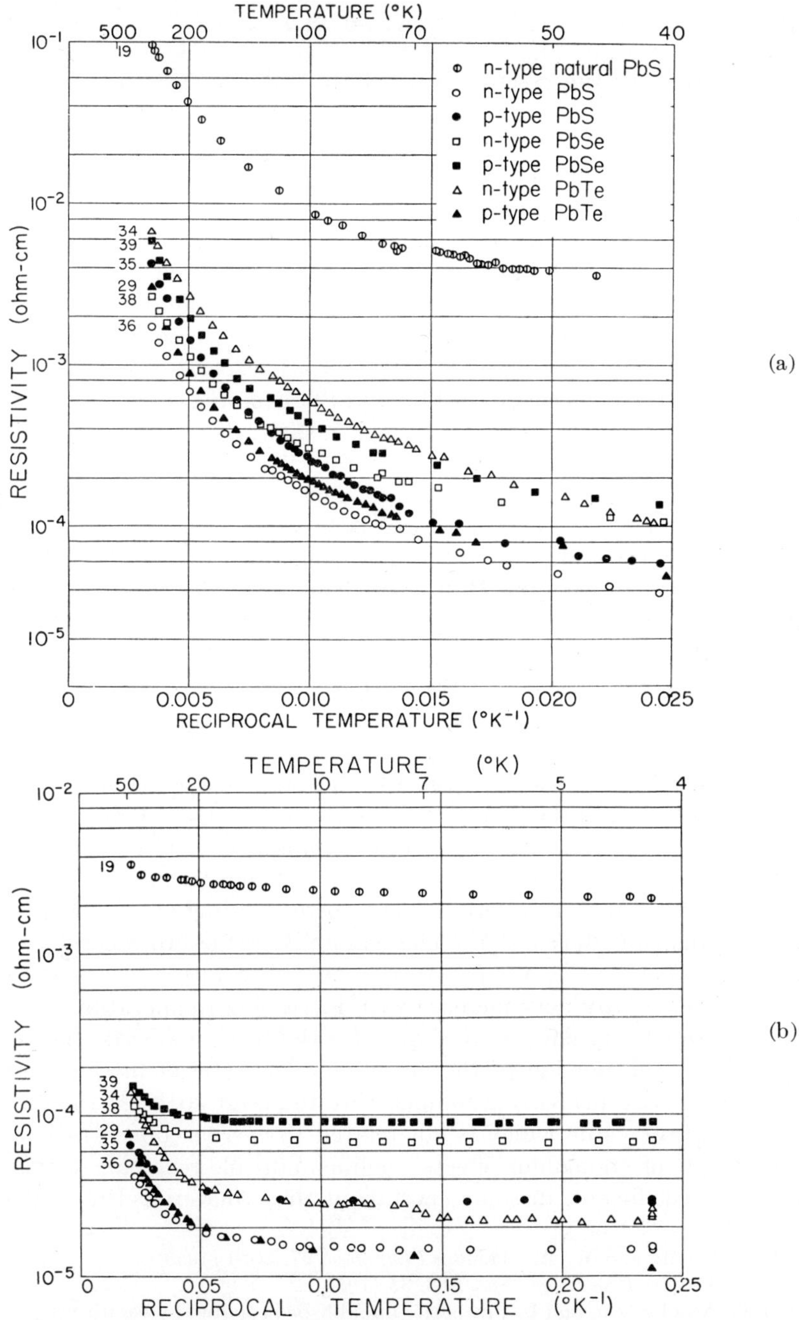

FIG. 16. Low temperature resistivity in PbS, PbSe, and PbTe, curves (a) 300° to 40°K and curves (b) 50° to 4.2°K. [Reproduced from R. S. Allgaier and W. W. Scanlon, *Phys. Rev.* **111,** 1029 (1958).]

degenerate and the essentially constant value of the Hall coefficient is not surprising. In the nondegenerate range, a constant carrier density may arise from shallow donor or acceptor levels, associated with deviations from stoichiometry, which have broadened and overlapped the conduction or valence bands.

Studies of the Hall effect have been reported for PbS crystals doped with Cu[43] and Ni[44] in which carrier concentrations were as high as 10^{18} per cm^3. These measurements indicate that the donor energy levels arising from Cu and Ni in PbS lie somewhat deeper in the forbidden energy gap, namely at 0.02 ev and 0.03 ev, respectively.

In the undoped crystals of PbS and PbSe having carrier concentrations of about 10^{18} per cm^3, Fig. 15, there is a slight decrease in the Hall coefficient as the temperature decreases to 4.2°K. This decrease may be a consequence of changing statistics. The degeneracy temperature of a conductor with 10^{18} carriers per cm^3 is $42/m^*$ °K where m^* is the effective mass. Most estimates for m^* range from 0.1 to 0.5.[41,46,68–70] Thus the observed decrease in the Hall coefficient occurs in an appropriate temperature range.

A slight increase in the Hall coefficient is observed at low temperatures in many crystals of PbTe. However, it is difficult to estimate activation energies because of the interference associated with changing statistics.

At high temperatures, Fig. 13, the sign of the Hall coefficient of *p*-type samples of these materials shows a reversal. This suggests that the mobility of holes is less than the mobility of electrons. Using the temperature at which the sign of the Hall coefficient changes, the mobility ratio in PbS is found to be 1.4.[16] The corresponding value given for PbSe is 1.4;[71] that for PbTe is 2.5.[63] These mobility ratios are generally assumed to be independent of temperature, although Smith[46] has evidence that the ratios actually may increase with increasing temperature.

The resistivity curves are similar for the three materials. The resistivity follows the portion of the curve characteristic of intrinsic material, decreasing with decreasing temperature to about 10°K. It then levels off to a constant value which is not related to carrier concentration. There are reports of anomalous effects, such as multiple reversals of the sign of the Hall coefficient, in some crystals of these materials.[64,72,73] These are

[68] R. L. Petritz and W. W. Scanlon, *Phys. Rev.* **97,** 1620 (1955).
[69] E. H. Putley, *Proc. Phys. Soc.* **B68,** 35 (1955).
[70] G. G. MacFarlane and L. Pincherle, unpublished, quoted by Smith.[46]
[71] E. Hirahara and M. Munakami, *J. Phys. Soc. Japan* **9,** 671 (1954).
[72] E. H. Putley, *Proc. Phys. Soc.* **B68,** 22 (1955).
[73] Y. Kanai and R. Nii, *J. Phys. Soc. Japan* **12,** 441 (1957).

offered as evidence of impurity band conduction. Another explanation of the effects exists. The author has been able to produce anomalies of this type in PbS crystals by heating them in an appropriate partial pressure of sulfur vapor so that they have a thin skin of *p*-type material over the entire surfaces of an originally *n*-type crystal as in Fig. 4. The opposite situation can be produced in *p*-type specimens. Such crystals may appear to be homogeneous when subject to various probing tests of the surface, such as those associated with rectification, resistivity, or thermoelectric power. The Hall curves, however, display such anomalies as multiple reversals of sign whereas the resistivity curves have unusual inflections not found in homogeneous crystals. The data give values for the Hall mobility which deviate considerably from the values found in homogeneous crystals. Upon grinding away the thin surface layer, the crystals exhibit the normal Hall effect and resistivity curves once more, as well as mobilities of the usual size.

These experiments suggest that the anomalous curves for Hall effect and resistivity reported in the literature may be the result of the presence of *p*-type and *n*-type regions in the crystals under investigation rather than a consequence of impurity band conduction.

b. *Mobility*

The mobility of electrons and holes is a measure of their velocity of drift in a unit electric field. It is a useful quantity for studying interactions of the moving carriers with the lattice since the temperature dependence of mobility reveals the nature of the scattering mechanisms present in the crystal.

The scattering of electrons or holes by the acoustical vibrational modes of the lattice is associated with a $T^{-\frac{3}{2}}$ mobility law, whereas scattering by ionized impurities in the lattice leads to a $T^{\frac{3}{2}}$ law.[74] A somewhat different expression has been obtained by Brooks[75] and Herring[76] by considering screening of the ionized centers. Scattering by the optical vibrational modes of the lattice leads to an expression for the mobility which varies exponentially with temperature.[68,77–80]

Gross defects, such as dislocations in the crystal lattice, may also

[74] E. M. Conwell and V. F. Weisskopf, *Phys. Rev.* **77,** 388 (1950).

[75] H. Brooks, *Phys. Rev.* **83,** 879 (1951).

[76] C. Herring, unpublished work, discussed by P. P. Debye and E. M. Conwell, *Phys. Rev.* **93,** 693 (1954).

[77] D. J. Howarth and E. H. Sondheimer, *Proc. Roy. Soc.* **A219,** 53 (1953).

[78] H. Fröhlich and N. F. Mott, *Proc. Roy. Soc.* **A171,** 496 (1939).

[79] H. Fröhlich, *Proc. Roy. Soc.* **A160,** 280 (1937).

[80] B. Davydov and I. Shushkevitch, *J. Phys.* (*U.S.S.R.*) **3,** 359 (1940).

scatter electrons and holes.[67,81–83] Dislocations may trap electrons, holes, or ions and behave like charged lines. These lines will be surrounded by cylindrical regions of opposite charge. Under these conditions the dislocations can have a relatively large effect in reducing the mobility, particularly at low temperatures.

The region of space charge around a charged dislocation also may lower the apparent mobility by distorting lines of current flow. Such an effect may be active even at high temperatures.

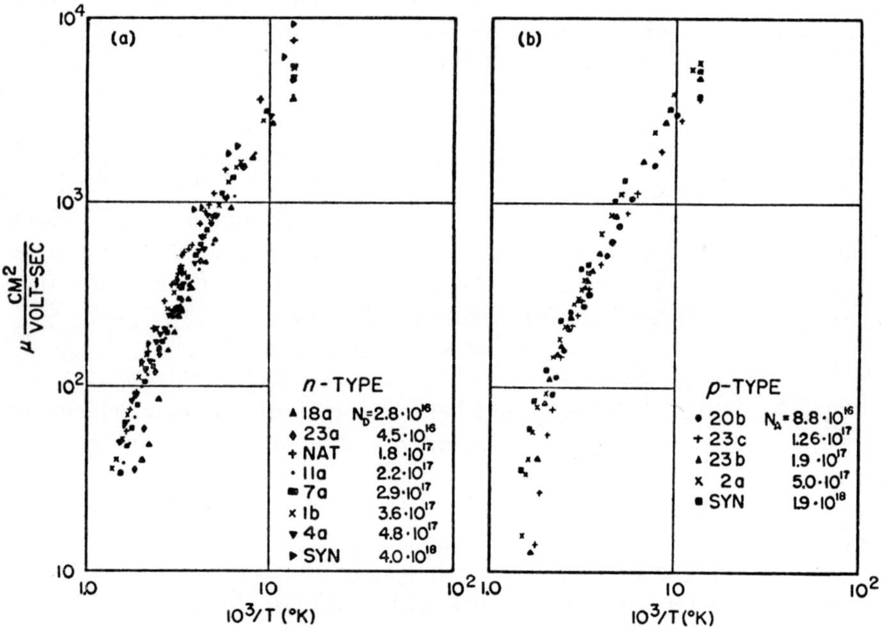

FIG. 17. Mobility of electrons and holes in PbS from 500° to 80°K. [Reproduced from R. L. Petritz and W. W. Scanlon, *Phys. Rev.* **97**, 1620 (1955).]

The behavior of mobility typically observed in the PbS family of semiconductors is shown in Figs. 17 and 18 which are obtained from measurements of Hall effect and resistivity on single crystals. Similar behavior has been observed in the compounds by a number of investigators.[84–86] A remarkable feature is that the mobility appears to be an

[81] G. L. Pearson, W. T. Read, Jr., and F. J. Morin, *Phys. Rev.* **93**, 666 (1954).
[82] W. T. Read, Jr., *Phil. Mag.* [7] **45**, 775, 1119 (1954); **46**, 111 (1955).
[83] D. L. Dexter and F. Seitz, *Phys. Rev.* **86**, 964 (1952).
[84] E. H. Putley, *Proc. Phys. Soc.* **B65**, 388 (1952).
[85] S. J. Silverman and H. Levenstein, *Phys. Rev.* **94**, 871 (1954).
[86] D. M. Finlayson and D. Greig, *Proc. Phys. Soc.* **B69**, 796 (1956).

intrinsic property of the crystals over a wide range of temperature extending from about 500°K to 80°K and is essentially independent of carrier concentration in the range extending from 10^{16} to 10^{19} carriers/cm^3. The behavior of mobility can be approximated by the formula $\mu = \mu_0 T^{-n}$ in which n is a constant which lies between 2 and 3. The most recent data give the value $n = 2.2$.[67] There appears to be no significant difference in the value of this number for crystals of PbS, PbSe, and PbTe, or for specimens which are n- and p-type.

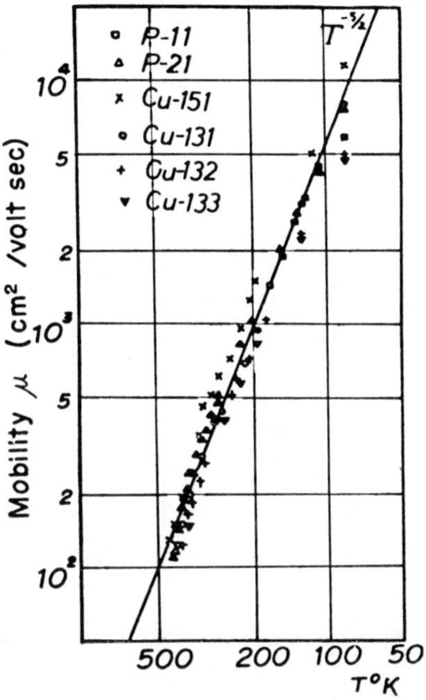

FIG. 18. Mobility of electrons in PbTe from 500° to 80°K. [Reproduced from K. Shogenji and S. Uchiyama, *J. Phys. Soc. Japan* **12,** 252 (1957).]

Attempts have been made to explain the observed behavior of mobility in terms of the intrinsic scattering mechanisms associated with the optical and acoustical modes of vibration of the lattice.[68] The agreement was found to be fair below the Debye temperature θ; however, the experimental values are much lower than those predicted by the theory at higher temperatures.

Various theories of polar scattering[77,78,87] predict a mobility which is proportional to the quantity $\exp(-\theta/T) - 1$. This function is concave upward when plotted on a log-log scale, whereas the experimental data

[87] F. E. Low and D. Pines, *Phys. Rev.* **98,** 414 (1955).

lead to a curve which is concave downward. Ehrenreich[88] demonstrated, however, that the mobility arising from polar scattering did become concave downward in InSb if a nonparabolic band model was used. Another recent analysis leads to a theoretical mobility which is proportional to $T^{-2.5}$.[89]

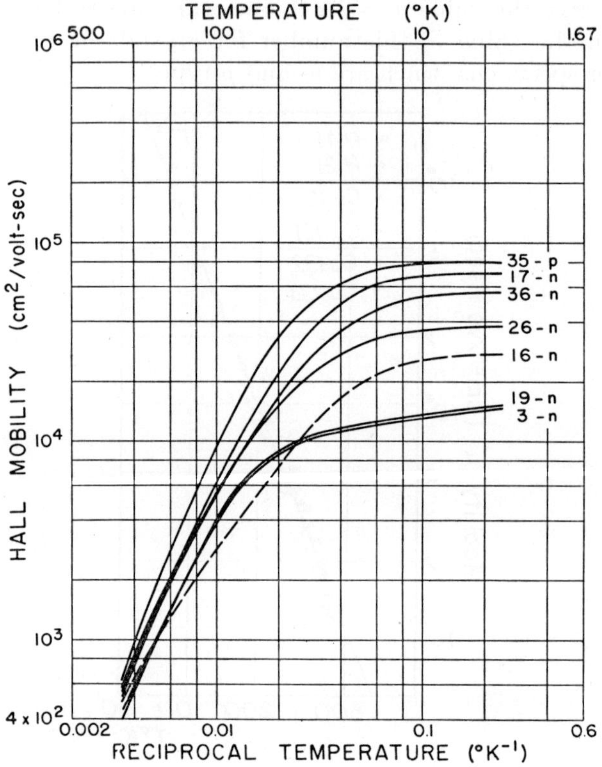

FIG. 19. Electron and hole mobilities in PbS from 300° to 4.2°K. [Reproduced from R. S. Allgaier and W. W. Scanlon, *Phys. Rev.* **111,** 1029 (1958).]

Studies of the mobility at lower temperatures reveal additional facts. Below about 50°K, the mobility curves shown in Figs. 19 to 21 for all three compounds gradually turn horizontally and become constant at 4.2°K in most crystals.[67] The mobilities in a few crystals, such as PbTe, continue to rise even at 4.2°K. A decrease in mobility is not observed in any case as the crystals are cooled to 4.2°K. This indicates that the

[88] H. Ehrenreich, *J. Phys. Chem. Solids* **2,** 131 (1957).

[89] T. A. Kontorova, *J. Tech. Phys.* (*U.S.S.R.*) **24,** 2217 (1955); K. B. Tolpygo and H. M. Fedorchenko, *J. Exptl. Theoret. Phys.* (*U.S.S.R.*) **31,** 845 (1956), transl., *Soviet Phys. JETP* **4,** 56 (1957).

scattering effects arising from ionized impurity centers are small in these crystals. The mobility varies widely from sample to sample at liquid helium temperature: 26,800–80,000 (PbS), 38,000–140,000 (PbSe), and 66,000 to 800,000 cm²/volt-sec (PbTe). These values are remarkably large if one considers the relatively large effective masses and the fact

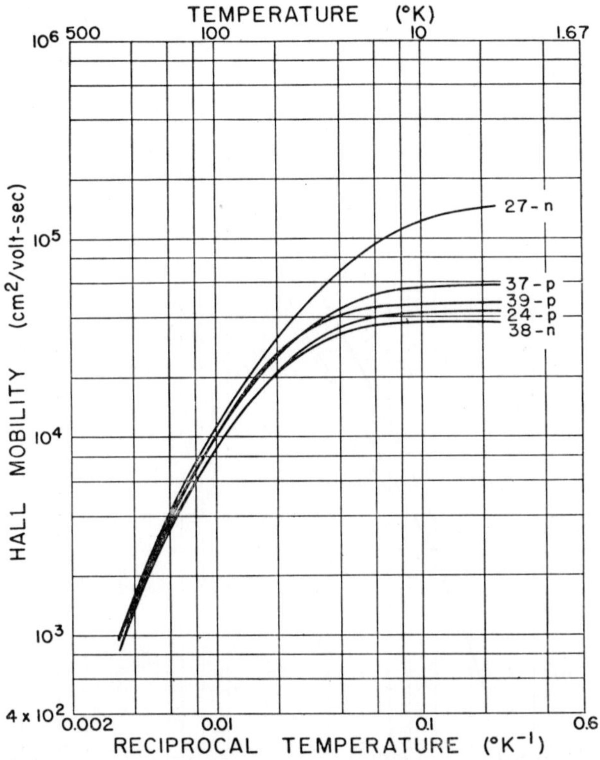

FIG. 20. Electron and hole mobilities in PbSe from 300° to 4.2°K. [Reproduced from R. S. Allgaier and W. W. Scanlon, *Phys. Rev.* **111,** 1029 (1958).]

that carrier concentrations as high as 10^{19} per cm³ were present in some cases.

The behavior of the mobility in the liquid helium range is characteristic of the phenomena associated with the residual resistance observed in metals. The mobilities become temperature-independent and their values vary from one sample to the next in an unsystematic manner.

If one used the ionized impurity scattering mobility formula of Dingle[90] and Mansfield,[91] one obtains mobilities of 50,000 and 500,000

[90] R. B. Dingle, *Phil. Mag.* [7] **46,** 831 (1955).
[91] R. Mansfield, *Proc. Phys. Soc.* **B69,** 76 (1956).

cm²/volt-sec providing the dielectric constants of the material are assumed to be of the order of 100 and 400, respectively. These are unusually large values for nonferroelectric materials. The published values of the static and high-frequency dielectric constants for PbS are 17.9 and 15.3, respectively.[24] Larger values of the static dielectric constant of the lead salts[92] have been suggested on the basis of a formula derived by Szigeti.[93]

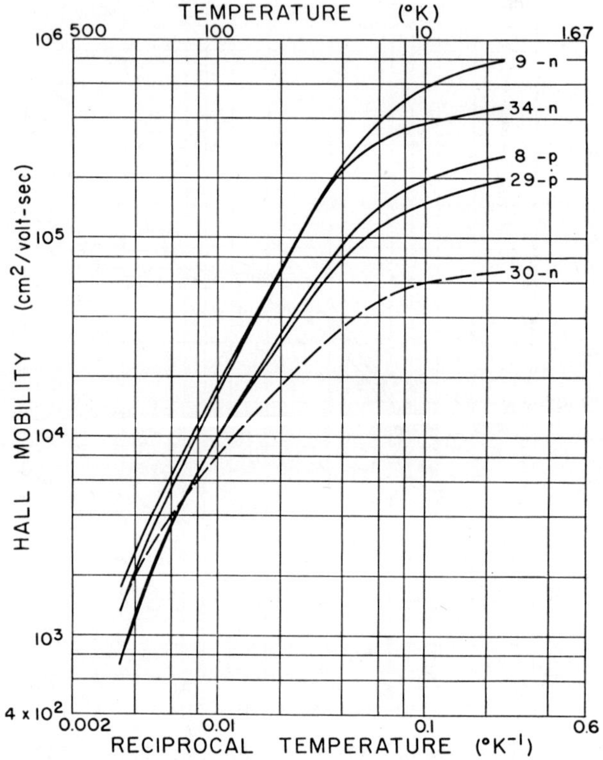

FIG. 21. Electron and hole mobilities in PbTe from 300° to 4.2°K. [Reproduced from R. S. Allgaier and W. W. Scanlon, *Phys. Rev.* **111,** 1029 (1958).]

On the other hand it should be pointed out that metals, which have much higher carrier densities than the lead salts, often have mobilities at low temperature which are several thousand cm²/volt-sec in spite of small values of the dielectric constants.

The wide range of mobilities found in a given material at low temperatures may be the result of different densities of dislocations.[67] An experiment was performed in which the density of dislocations was

[92] E. Burstein and P. H. Egli, *Advances in Electronics and Electron Phys.* **7,** 56 (1955).
[93] B. Szigeti, *Proc. Roy. Soc.* **A204,** 51 (1950).

increased in a given crystal. The Hall coefficient remained unchanged but the resistivity and mobility decreased significantly at low temperatures as a result. Similar results have been observed in deformed germanium.[81]

To summarize, the mobility of electrons and holes observed in the polar crystals PbS, PbSe, and PbTe in the range from 1000°K to 50°K appears to be an intrinsic property of the crystal and possibly may be explained in terms of a combination of optical and acoustical scattering mechanisms. The very large mobilities observed in these crystals at low temperatures even when the carrier concentrations extend to 10^{19} per cm^3 suggest unusually large static dielectric constants or a metal-like behavior in the highly degenerate semiconductors. The wide range in mobilities in a given material at low temperatures may be the result of variations in dislocation densities from one crystal to another.

c. *Mobility in Alloys*

The substitution of one anion for another in a given semiconducting lead salt introduces an additional scattering mechanism which generally reduces the carrier mobility, particularly at low temperatures. The behavior of mobility in these alloys has been studied in greatest detail in the PbTe + PbSe system by Kolomoets *et al.*[94] They find that the carrier mobility decreases with increasing departure from the pure compound. There is a maximum in the mobility *vs* composition curves near the 50% composition at low temperatures. This value is still well below the mobility in the pure compounds. The maximum is interpreted in terms of the onset of ordering in alloys possessing a 1:1 composition.

d. *Thermoelectric Power*

If two different conductors are joined at their two ends and the two junctions are kept at different temperatures, an electromotive force, which is proportional to the temperature difference (for small ΔT), is established. The thermoelectromotive force per degree is called the thermoelectric power α. In general, the thermoelectric power depends upon the scattering mechanism, carrier density, carrier sign, effective mass, mobility ratio, and other aspects of the band structure. Reviews of the theoretical expressions for thermoelectric power in semiconductors have been given by V. A. Johnson,[95] Herring,[96] and ter Haar.[97]

[94] N. V. Kolomoets, T. S. Stavetskaya, and L. S. Stil'bans, *Zhur. Tekh. Fiz.* **27**, 73 (1957).
[95] V. A. Johnson, *Progr. in Semiconductors* **1,** 65 (1956).
[96] C. Herring, *Phys. Rev.* **96,** 1163 (1954).
[97] D. ter Haar and A. Neaves, *Phil. Mag.* Suppl. 5, 241 (1956).

If carriers of one sign are present, the thermoelectric power has the form

$$\alpha = -\frac{1}{eT}\frac{\int_0^\infty E^2 l \frac{\partial f_0}{\partial E}\,dE}{\int_0^\infty El\frac{\partial f_0}{\partial E}\,dE} + \frac{\zeta}{eT}. \tag{14.1}$$

Here E is the kinetic energy of the carrier, l is the mean free path, f_0 is the unperturbed distribution function for the carriers and ζ is the Fermi level. If classical statistics may be applied and if the scattering is of simple lattice type, this expression becomes

$$\alpha = -\frac{k}{e}\left\{2 - \ln n + \frac{3}{2}\ln T + \ln\left[\frac{2(2\pi mk)^{\frac{3}{2}}}{h^3}\right]\right\}. \tag{14.2}$$

Here m is the effective mass for the density of states.

Expressions for the thermoelectric power in polar semiconductors have been derived by Howarth and Sondheimer[77] in the nondegenerate and degenerate cases.

Measurements of the thermoelectric power as a function of temperature are given by Putley[69] for PbSe and PbTe in the range of temperature from about 100°K to 1000°K, and by Kolomoets *et al.*[94] in the range from about 100°K to 700°K. Similar behavior was observed in PbS by Devyatkova *et al.*[98] The curves for the three compounds show the general feature of the thermoelectric power observed in valence semiconductors. There is agreement between experiment and the calculations based upon classical semiconductor theory except at very high carrier concentrations, where degeneracy effects are present. From the thermoelectric power Putley[69] finds that the effective mass of electrons in PbSe is about $0.3m$. Kolomoets *et al.*[94] find that it is about $0.4m$ in PbTe. Similarly Bloem[15] finds that the effective mass is about $0.2m$ in PbS, a value which is consistent with that obtained from mobility measurements.[68]

Bloem[15] finds that the room temperature value of the thermoelectric power in PbS crystals obeys the following emperical formula over a range of composition extending from 10^{17} to $10^{19}/\text{cm}^3$:

$$\alpha = \pm 200(19.4 - \log n)\mu V/°\text{C}. \tag{14.3}$$

Here the positive sign is valid for p-type crystals and the negative sign for n-type crystals; n refers to the carrier density.

[98] E. D. Devyatkova, Yu. P. Maslakovets, and M. S. Sominskii, *Bull. Acad. Sci. U.S.S.R., Ser. Fiz.* **5**, 409 (1941).

e. Thermal Conductivity

The transport of thermal energy is closely associated with electronic transport phenomena. In some materials, such as metals, thermal conduction is determined predominantly by electrons, whereas in others, such as insulators, it is determined essentially by lattice phonons. Semiconductors may exhibit both types of thermal conduction so that the total conductivity κ can be expressed as the sum of lattice conduction κ_l and electronic conduction κ_e:

$$\kappa = \kappa_l + \kappa_e. \tag{14.4}$$

Information concerning thermal conduction in semiconductors is meager partly because the experimental measurements are difficult. In PbS, however, Dunaev[99] was able to measure the thermal conductivity in a series of crystals having the comparatively large carrier concentration lying in the range from 10^{18} to 10^{19} per cm^3. Since the absolute value of the thermal conductivity was small, lying in the range from 6 to 8×10^{-3} cal/cm-sec-deg, he was able to separate the lattice and electronic components. The results showed an increase in thermal conduction with increasing carrier concentration. By extrapolating the thermal conductivity to zero carrier concentration one obtains a lattice conductivity $\kappa_l = 5.7 \times 10^{-3}$ cal/cm-sec-deg. In the case in which $n = 2.18 \times 10^{19}$ electrons/cm^3, the electronic component of the thermal conductivity is $\kappa_e = 1.47 \times 10^{-3}$ cal/cm-sec-deg. Using this value of κ_e Dunaev finds that the Wiedemann-Franz ratio is comparable to that in metallic conductors.

Similar studies have been made for PbTe by Devyatkova.[100] Crystals with carrier concentrations lying in the range from 10^{17} to 10^{18} and conductivities lying between 1740 and 60 ohm^{-1}-cm^{-1} were investigated at temperatures between 80 and 450°K. In PbTe, the thermal conductivities ranged from 5 to 7×10^{-3} cal/cm-sec-deg. By extrapolating the curve representing κ as a function of σ to zero σ, he obtains the value $\kappa_l = 5.25 \times 10^{-3}$ cal/cm-sec-deg. The ratio κ_e/σ is 1.10×10^{-6} and agrees with the Wiedemann-Franz law.

According to Busch and Schneider,[101] the electronic part of the thermal conductivity in a semiconductor can be represented by an equation of the form,

$$\kappa_e = \text{const } e^{-B/2kT}. \tag{14.5}$$

[99] Yu. A. Dunaev, *J. Tech. Phys.* (*U.S.S.R.*) **16,** 1101 (1946).
[100] E. D. Devyatkova, *Zhur. Tekh. Fiz.* **27,** 461 (1957).
[101] G. Busch and M. Schneider, *Physica* **20,** 1084 (1954).

In PbTe Devyatkova[100] finds that the quantity B appearing in Eq. (14.5) is 0.24 ev. This energy agrees closely with that for indirect electron transitions in PbTe, namely 0.29 ev, obtained from optical absorptions studies.[45]

The lattice thermal conductivity observed in PbSe is about 4×10^{-3} cal/cm-sec-deg. Thus all three compounds exhibit unusually low thermal conductivities relative to those found among semiconducting materials.

Further reduction in thermal conductivity can be accomplished by alloying the compounds. Ioffe[102] pointed out that the distortion of the crystalline lattice produced by alloying is relatively ineffective in scattering electron waves but is quite effective in scattering phonons which have wavelengths comparable to the lattice constant. Hence appropriate alloys of the compounds should have lower thermal conductivities than the pure compounds do. Experimental studies bear out these conclusions.[94]

16. Recombination and Trapping

The decay toward equilibrium of an excess density of electrons or holes in a semiconductor may occur by a number of recombination mechanisms. First, there is the possibility that a one-step process in which the electron in the conduction band combines directly with a hole will occur. The energy liberated in the process is given out as radiation. This recombination process is characterized by a radiative time constant τ_r.

A second volume recombination process occurs as the result of a three-body collision mechanism termed the Auger process. An electron in the conduction band combines with a hole in the valence band, giving up the energy to a third carrier. In this process the lifetime is inversely proportional to the square of the carrier density.

A third important recombination process is a two-step mechanism in which a hole recombines with an electron which has fallen into an energy level associated with a crystal imperfection. According to the Shockley-Reed theory,[103] the lifetime is independent of resistivity in strongly *n*- or *p*-type material when recombination takes place at lattice dislocations. The lifetime depends upon the density of dislocations in the crystal in this case.

Trapping provides another mechanism which influences the recombination of electrons and holes. A carrier may fall into a center in the crystal from which it must be re-excited before it can recombine with a

[102] A. F. Ioffe, "Semiconductor Thermoelements and Thermoelectric Cooling." Infosearch Limited, London, 1957.

[103] W. Shockley and W. T. Read, Jr., *Phys. Rev.* **87**, 835 (1952).

hole by one of the mechanisms discussed in the foregoing. There is much evidence to show that this process is an important one in photoconductive films of the lead salts. The carrier lifetime associated with trapping is different for electrons and holes. Usually one is concerned only with the trapping of minority carriers. An analysis of trapping in PbSe films is given elsewhere.[104] The role of trapping in crystals of PbS, PbSe, and PbTe has not been investigated.

The rate of radiative recombination may be calculated from the data on the optical absorption edge using the analysis of van Roosbroeck and Shockley.[105]

$$R = 32\pi^2 c \left(\frac{kT}{ch}\right)^4 \int_0^\infty \frac{n^3 \kappa u^3 du}{e^u - 1} \tag{16.1}$$

in which $u = h\nu/kT$, n is the refractive index, κ is the absorption index $= \alpha c/4\pi n\nu$, and α is the absorption coefficient. The integration can be performed numerically.

The radiative recombination lifetime τ_r then is obtained from the relation

$$\tau_r = n_i/2R \tag{16.2}$$

where n_i is the intrinsic carrier concentration. This analysis has been made for PbS,[58] PbSe,[45] and PbTe.[45] The results at 300°K are summarized in Table V, along with the values for direct and indirect transition energies.

The lifetimes measured experimentally in these crystals are much smaller than the radiative limit.[106] It is, therefore, reasonable to believe that other recombination mechanisms are acting. Two which have been considered are the Auger and the Shockley-Read mechanisms. They may be distinguished from one another by their dependence upon resistivity and upon the density of lattice dislocations.

Moss[107] found that the measured values of lifetime in a series of PbS crystals having different resistivity varied approximately as the square of the resistivity, suggestive of the Auger recombination mechanism. He did not specify the densities of dislocations in his crystals so that it is not possible to distinguish effects arising from the Shockley-Read recombination.

With the development of a suitable etch for revealing pits on PbS,[108] it became possible to compare the experimental dependence of lifetime

[104] J. N. Humphrey and R. L. Petritz, *Phys. Rev.* **105,** 1736 (1957).
[105] W. van Roosbroeck and W. Shockley, *Phys. Rev.* **94,** 1558 (1954).
[106] W. W. Scanlon, *Phys. Rev.* **106,** 718 (1957).
[107] T. S. Moss, *Physica* **20,** 989 (1954).
[108] R. F. Brebrick and W. W. Scanlon, *J. Chem. Phys.* **27,** 607 (1957).

on the density of dislocations with the results of the theory of Shockley and Read. The etch consists of a 3:1 mixture of a 100 g/l thiourea solution and concentrated HCl. It is used at a temperature of 60° to 80°C for a few minutes. The effectiveness of this etch on PbS:PbSe:PbTe is in the ratio of $1:10^{-5}:10^{-8}$.

The lifetime was found to vary from about 20 to 0.08 μsec in a group of PbS crystals which had the same resistivity but dislocation densities varying from about 10^5 to $10^7/cm^2$. On the other hand, a group of crystals which had the same high density of pits but varying resistivity had essentially the same short lifetime.[106] These results suggest that recombination in PbS crystals occurs principally at dislocation centers by the Shockley-Read mechanism.

V. Conclusions

The family of semiconductors PbS, PbSe, and PbTe provide an interesting group of materials to study for a number of reasons. They have sufficient polar character and low resistivity that it is possible to study the details of chemical composition in the immediate vicinity of the stoichiometric compound with a precision not possible with the usual methods of analytical chemistry. The materials also make studies of solid-vapor phase relationships possible to a degree of detail not previously obtainable in polar compounds.

The physics of these materials is also interesting since the crystals possess some of the properties of the classical ionic solids, the alkali halides, as well as being amenable to study by the techniques developed for the valence semiconductors.

Although a large number of electrical, optical, and thermal characteristics can be explained by conventional semiconductor theory, there are some exceptions. Among these is the high mobility, extending to about 10^6 cm²/volt-sec at low temperatures. The high mobility is found even in the presence of a large carrier concentration, in the range from 10^{18} to $10^{19}/cm^3$, and in spite of the relatively large effective masses, which are a few tenths the free mass of the electron.

The explanation of the approximately $T^{-2.2}$ behavior of the mobility in terms of optical and acoustical scattering of electrons and holes is not completely satisfactory particularly at temperatures above the characteristic temperature, even if one uses the most recent theoretical treatment of these scattering mechanisms in polar crystals.

The nature of the energy bands in the materials is now understood better. The problem of the energy gap has been clarified considerably as a result of the recent studies of the optical absorption edge. It is apparent that direct and indirect electronic transitions between the valence and conduction bands do not differ greatly in energy, being within an amount $k\theta$ of one another. This suggests that indirect transitions may not involve excitation to a minimum in the conduction band which occurs at a value of K different from that for the maximum in the valence band, as is the case in germanium and silicon. Studies of magnetoresistance favor a model with considerable anisotropy of mass. The cubic model having ellipsoids along the $<111>$ directions appears to be the most appropriate, at least in the case of PbTe, for both the conduction and the valence bands. Thus the conclusions derived from the optical and magnetoresistance studies appear to be consistent.

VI. Acknowledgments

The author wishes to express gratitude for the many discussions with Dr. R. F. Brebrick and Dr. R. S. Allgaier which occurred during the course of this writing, and to the Naval Ordnance Laboratory for making it possible to prepare this manuscript.

Static Electrification of Solids

D. J. MONTGOMERY
Michigan State University, East Lansing, Michigan

I. Foreword

Regrettably, the subject of static electrification has not reached as mature a state as most of its sister topics in the physics of surfaces and solids. The complexity of the phenomena is sufficiently formidable that the field has not appealed to many scientists. Furthermore, the technological problems arising from inadequate understanding of the phenomena have not yet become sufficiently severe to motivate much support for intensive or extensive programs. Yet there seems to exist no evidence that the subject will not yield to persistent attack, as have say magnetism, friction, and phosphorescence. It is the purpose of this chapter to bring features of this problem to the attention of workers in other fields and perhaps thereby enlist their aid.

Our discussion will be limited to the transfer of charge between solids, in view of the scope of the present series. It is a moot question whether the study of solid-solid contacts is the wisest way to make a beginning in understanding the complexities of static electrification. The structure of solids is known far better than the structure of liquids; but the nature of the contact between two solids is much more imperfectly understood than that between a solid and a liquid. In any event, in those cases where solid-liquid electrolytic processes have been taken as the point of departure in explaining static electricity, the extension to solid-solid phenomena has not been generally successful. Hence no contraindication exists that the study of solid-solid contacts will not constitute an avenue to fundamental understanding of static electrification. No matter what type of study ultimately leads us in the right direction, the road will be long, and there will be many turnings. In referring to the development of another branch of physical science, Sir Cyril Hinshelwood, President of the Royal Society of London in the 1957 Anniversary Address said[1]

" . . . Knowledge is pursued for two ends, for power and for understanding. Fortunately for those who believe with the preacher that wisdom is better than strength, the two ends are inseparable. Nobody who has witnessed the developments of the plastics industry, of internal combustion, the production of organic fine chemicals and drugs, or who has seen the many ways in which knowledge of the working of living cells is of value in practical investigations, can fail to realize how fruitful has been the analysis of chemical mechanisms. Nor can those who have studied such mechanisms fail to be impressed and perhaps disconcerted at their luxuriant diversity.

"Yet what the true seeker after knowledge in his heart desires is some simple design which he feels must underlie the facts. And nearly a century has gone to the seeking of it. The search for principles which are aesthetically satisfying seems often frustrated by the complexity of nature, and the conflict between imagination and austere regard for truth seems often to result in the passage of scientific theories through three stages. The first is that of gross over-simplification, reflecting partly the need for practical working rules, and even more a too enthusiastic aspiration after elegance of form. In the second stage the symmetry of the hypothetical systems is distorted and the neatness marred as recalcitrant facts increasingly revel against conformity. In the third stage, if and when this is attained, a new order emerges, more intricately contrived, less obvious, and with its parts more subtly interwoven, since it is of nature's and not of man's conception."

In static electrification we have not yet left the first stage; moreover, one man's notion of elegance is likely to blind him with regard to another's. Activity at the second stage will open the eyes of all. Achievement of the third stage we cannot promise; but it will not come without the first.

[1] Sir Cyril Hinshelwood, *Proc. Roy. Soc.* **A243,** v (1958).

II. Historical and Technical Background

When two objects are placed in contact and subsequently separated, what is the sign and what is the amount of charge transferred? This is the question that a theory of static electrification is supposed to answer. Despite the ancient history of electrostatic phenomena,[2] the question could not even be stated until two centuries ago. For it was not often realized in ancient times that contact is necessary in producing electrification. Nor was the commonness of the phenomenon realized; only a few semiprecious or rare substances—amber, jet, diamond perhaps—were believed capable of being electrified. It was not until after medieval times that the discovery of the electrification of more or less ordinary materials—glass, sulfur, wax—made the commonness of the phenomenon apparent (Gilbert, 1600). More than a century elapsed before the existence of two kinds of electricity was demonstrated (du Fay, 1733). The way was then open to investigations of ordering relations among electrifiable materials, with the idea of establishing a *triboelectric series*, that is, a list of materials so ordered that, upon rubbing any two of them together, the one listed higher acquires say a positive (vitreous) charge, and the other a negative (resinous) charge. The first such list was published just over two hundred years ago (Wilcke, 1757).

Investigations on the magnitude of charge could not come before development of theoretical concepts (conservation of charge, Watson, 1746; Franklin, 1747; law of interaction of charges, Coulomb, 1785) and experimental tools (electrometers and multipliers, e.g., Henley, 1772; Bennet, 1787). But surprisingly little work on the amount of charge transferred in static electrification was reported in the following years. Apparently only one substantial work (Péclet, 1834) appeared before the beginning of the twentieth century, when sporadic activity commenced. Even then very few investigators made static electrification their major

[2] A full survey of the early history of electrostatics cannot be attempted here; a quick review may be obtained from Whittaker or Roller.[3] For source material of all periods, Mottelay[4] is invaluable. For ancient times, Martin[5] is exhaustive; for the eighteenth century, Cohen[6] is valuable. Benjamin[7] has written a good general history.

[3] E. Whittaker, "A History of the Theories of the Aether and Electricity," Chapters I and II. Thomas Nelson & Sons, London, 1910, 1951; D. Roller and D. H. D. Roller, *Am. J. Phys.* **21**, 343 (1953).

[4] Paul Fleury Mottelay, "Bibliographical History of Electricity and Magnetism." Charles Griffin & Co., London; J. B. Lippincott Co., Philadelphia, 1922.

[5] Th. Henri Martin, "La Foudre, l'Electricité, et le Magnétisme chez les Anciens." Didier et Cie, Paris, 1866.

[6] I. Bernard Cohen, "Franklin and Newton." The American Philosophical Society, Philadelphia, 1956.

[7] Park Benjamin, "A History of Electricity." John Wiley and Sons, New York, 1898.

interest (among the outstanding are Coehn,[8] 1900–1925; Shaw, 1915–1930), the other workers in the early part of the century contenting themselves with only brief excursions into the field. But within the past decade or two, several groups have begun to attack the problem with tenacity, if not with unqualified success.

The renewal of interest stems from both technological and scientific considerations. The more prominent technological considerations are those of motivation, coming largely from the desire to avoid static electricity. For example, in the textile industry the increasing speeds of processing bring difficulties even with the natural, hydrophilic fibers, not to mention the synthetic, hydrophobic ones. In the photographic-film industry, the brush or spark discharges which dissipate high concentrations of static electricity result in fogging and blackening of the emulsion. From the standpoint of the consumer, static phenomena often become annoying, as with plastic seat covers, and sometimes dangerous, as in atmospheres containing ether, gasoline, or combustible dust. On the positive side, fibers, paint, and insecticides can be applied with the aid of static electricity, and certain copying processes are based on the adhesion of charged particles to unexposed portions of a charged photoconductive surface. Even in nuclear physics there are electrostatic generators depending in part on frictional generation of charge.

The scientific considerations leading to the renewed interest in static electrification of solids derive primarily from the improved understanding of the solid state within the past few decades. It is true of course that instrumentation has improved also, and that new materials—principally synthetic polymers—of fairly well-known composition and of high resistivity have become available. But it is the application of quantum mechanics to the solid state, so fruitful in related fields, that gives some hope of understanding the complicated events in charge transfer. True enough, only rudimentary theories of static electricity have been put forth. But even these suggest new experiments and techniques, and at this stage much more can hardly be expected.

As has been implied at the beginning, charge is normally transferred from one body to another upon touching and separation. As a rule this transfer produces no noteworthy effects. In the case of metal-metal contacts, the net charge retained is usually very small, and not detected by sense of touch, sight, or hearing. In the case of metal-insulator or insulator-insulator contacts, the charge transferred while the objects are touching may be large, but during separation it usually decreases to a

[8] Alfred Coehn, Berührungs-und Reibungselektrizität, *Handbuch d. Physik* **13,** 332 (1928).

very small value. At the speeds encountered in day-to-day events, namely 0.1 to 1000 cm/sec, the charge leaks back too fast to give noticeable effects whenever the resistivity of the more poorly conducting material falls below some value of the order of 10^9 ohm-cm^2/cm.[9] At higher speeds, static effects become noticeable even with more conductive materials, and at lower speeds, higher resistivities must occur before static effects can be observed. Thus we arrive at our first general statement: *Charge is ordinarily transferred between bodies when they are placed in contact; the amount that remains, and the duration of its stay on each body, depend on the electrical conductivity of the body, among other things.*

Now let us consider the magnitude of the charge transferred. Standard air near regions of low curvature of a body cannot sustain an electric field intensity greater than about 30 kv/cm, corresponding to a surface charge density of nearly 10 esu. If the excess or deficiency of charge is assumed to occur on singly charged atoms, 2×10^{10} atoms in each square centimeter are charged. But the total number of surface atoms in this area is perhaps 2×10^{15}. Hence only one out of 10^5 atoms is charged even at maximum electrification. So we arrive at our second general statement: *The fraction of the number of atoms which have to undergo gain or loss of an electronic charge is very small, of the order of 10^{-5}, even for the maximum electrification observed; hence almost no mechanism proposed can be either invoked or excluded by direct evidence.*

Our next point also is concerned with the amount of charge transferred. When the breakdown strength of the surrounding medium is exceeded, charge is lost by processes not relevant to static electrification proper. These extraneous effects may then mask the electrification phenomena, and what is intended to be a solid state study may actually become a gaseous-discharge study. Hence our third statement: *It is important to know whether the observed charge is that determined by the electrification processes proper, or whether it is the limiting charge determined by the occurrence of electrical breakdown.*

The next general statement has to do with the reproducibility of experimental results. It is a commonplace that the phenomena of static electrification are erratic. There is no denying that the great variability usually observed with these phenomena is far from being eliminated. But, as in any other field, the reasonable choice of material for experimentation and the careful control of variables go far in reducing the variability to manageable proportions. Apart from the obvious points to be made in choosing materials and in controlling conditions, two fac-

[9] In the following, ohm-cm^2/cm will be written simply as ohm-cm, despite the loss in explicitness.

tors need special attention. The first is that the chemical composition of a material must be specified carefully. In particular, when high polymers are investigated, composition may differ from source to source of materials, and even from batch to batch from the same manufacturer. The second point is that the mechanical state of a material must be specified carefully. In particular, this specification may have to include the strains, and in certain cases the history. Our fourth general statement is then: *Meaningful results can be hoped for if and only if the chemical composition and the mechanical state of materials are accurately specified, and if the ambient conditions and mechanical variables are carefully controlled.*

Finally, the complexity of the processes must be appreciated. Even if the materials were pure and homogeneous up to the outermost layer of atoms, and if the experiments were carried out in a perfect vacuum, the mechanical and electrical phenomena occurring would be complicated and imperfectly understood. The mechanical processes involve elastic and plastic deformation, flow and rupture, relaxation and creep. The electrical processes involve electronic conduction, ionic conduction, polarization, electron trapping, in fact, most of the processes of photoconductivity and luminescence, with departure from charge neutrality superimposed. Besides these points, materials depart far from the ideal, and most phenomena of interest occur in air containing some moisture. Thus our final general statement: *Great complexity is inherent in the problem, and painstaking experimentation and patient analysis must be expected to precede understanding.*

III. Experimental Studies

1. Relevant Variables

For practical purposes, the phenomena of interest in static electrification are the forces of attraction or repulsion resulting from electrification; the occurrence of sparks; the possible changes in frictional force; the time required for dissipation of charge; and so on. From a fundamental point of view, these phenomena are simply consequences of the transfer of excess charge and its subsequent behavior. Hence the dependent variable of primary interest in the present treatment will be the net charge retained upon separation—finite or infinite—of the objects that have been in contact. This quantity is to be considered as a function of numerous independent variables.

With some arbitrariness, we shall group the independent variables into sets pertaining to the materials, to the ambient conditions, and to the mechanical variables characterizing the contact process. A listing of the more important variables follows.

A. Material properties

Chemical composition of bulk of object
Chemical composition of surface of object
Molecular structure: orientation, crystallinity
State of strain
Size
Shape
State of electrical charge

B. Ambient conditions

Temperature
Atmosphere: composition, pressure
Electromagnetic fields

C. Mechanical variables

Type of contact (i.e., touching, impacting, rubbing, rolling, twisting)
Orientation of bodies during contact
Area of contact (including path length)
Duration of contact
Relative velocity of bodies
Force between bodies: normal, tangential

Certain other variables of obvious importance, such as resistivity, dielectric constant, pressure between contacting regions, are considered to be determined by specification of the variables listed above.

2. Experimental Techniques

a. Selection and Preparation of Samples

The choice of material for study is determined by the purposes of the inquiry. At least one of the materials must be an insulator, that is to say, with ordinary apparatus it must have a resistivity not less than 10^9 ohm-cm, to within a power of ten. The other material may be a conductor or an insulator. For technological studies, the choice of material usually is dictated by the problem demanding attention; for scientific studies, the choice of materials is wider. For simplicity and reproducibility, chemical purity and homogeneity are desirable. Stability with respect to absorption of, or reaction with, components of the atmosphere is helpful, but it is seldom obtainable. A geometry amenable to calculation is advantageous. Beyond these obvious points, the selection is controlled only by the scientific promise of the expected data.

The preparation of the samples involves removal of contamination, particularly at the surface, and conditioning of the material. Various washing and extracting procedures for removing contamination have been adopted by different workers, the details depending to some extent on the material but also on the kind of experiment; for example, if the contact is fairly severe, the phenomena often are not particularly sensitive to surface contamination. Materials are usually brought to standard conditions of temperature and moisture content by storage in a test chamber or other controlled atmosphere for a long enough time for equilibration to be achieved. For very fine filaments, a few minutes may suffice; for thick sheets of film, blocks of plastic, or large assemblies of fibers, several weeks may not be enough. Again, the type of experiment affects the time required; since the outside layers of the sample are affected most rapidly, they may get close to the desired conditions while the interior regions are still far from equilibrium. If the experiments require only a short period of time, the effect of diffusion from the interior regions may affect the results unimportantly.

Sometimes mechanical conditioning is desired. Ordinarily this involves nothing more profound then bringing the samples to some standard configuration—for example, straightening a rolled film or a coiled yarn—or applying a tension or compression. At other times one wishes to prepare the surface mechanically by polishing it or by roughening it, or to put the material through a specified stress-strain cycle. Methods of attaining these ends are described by the individual experimenters.

b. Provision of Suitable Ambient Conditions

(*1*) *Temperature.* Familiar methods of thermostating are used to maintain fixed temperature.

(*2*) *Atmosphere.* The composition of the atmosphere is controlled in standard ways. There has been little work with gases other than air, its content of water vapor and (infrequently) carbon dioxide being varied. Standard methods of controlling the composition are used, saturated-solution trays being common for small-scale experiments, and refrigeration-and-spray for large-scale ones. It is important to monitor the relative humidity with a sensing element located near the samples; even then it must be remembered that conventional elements have appreciable time lag and hysteresis. Claims of accuracy in relative humidity better than 1% are to be viewed critically.

The pressure of the atmosphere can be varied in standard ways. The complication inflicted on the rest of the apparatus has not made this variable a popular one. Moreover, the primary effect seems to be change of the breakdown voltage; hence the study becomes one of gaseous-discharge phenomena rather than solid state.

There is some possibility that the impact of photons might affect the movement of charge in certain materials. Hence provision must be made in certain cases for the exclusion or admission of radiation. So far, simply constructed light-tight boxes have been adequate to keep light out. Straightforward methods have been used for admitting radiation.

c. *Provision of Mechanical Controls*

Producing the desired manner of contact requires mechanical design of some kind, the details depending on the kind of contact, the ambient conditions, and usually the speed range required. The simplest solid-solid contact is the touching of spheres (as carried out by Harper,[10] for example, through flotation on mercury or by a lever arrangement). The most complicated contacts are perhaps those occurring in tumbling powder particles over a solid surface (as carried out, for example, by Debeau and by Gill and Alfrey[11]). Cases of intermediate complexity are those of a sphere rolling in a rotating cylinder (see, e.g., Peterson[12]) and yarns rubbing against a cylinder (see, e.g., Gonsalves and van Dongeren[13]). In selecting a type of contact for study, the experimenter is limited at one extreme by the intractability of analyzing a complicated contact, and at the other by simplifying the process so much that many of the phenomena of interest no longer appear.

Once the kind of contact has been selected by the experimenter or imposed by the phenomena, the means of controlling the variables follows standard practice. The bodies may be held in various clamps or jigs, which in turn are supported in adjustable mountings. As examples, there are shown in Fig. 1 a diagram of the apparatus of Harper[10] for the touching of spheres, and in Fig. 2 a diagram of that of Hersh[14] for the rubbing of filaments. Figure 3, a photograph of the apparatus of Cunningham,[15] shows an apparatus similar to that shown in Fig. 2. The area of contact may be controlled by orientating filaments so as to produce a certain rub type (see, e.g., Grüner[16]), by force between bodies (see, e.g., Medley[17]),

[10] W. R. Harper, *Proc. Roy. Soc.* **A205,** 83 (1951).
[11] D. E. Debeau, *Phys. Rev.* **66,** 9 (1944); E. W. B. Gill and G. F. Alfrey, *Nature* **163,** 172 (1949).
[12] J. W. Peterson, *J. Appl. Phy.* **25,** 907 (1954).
[13] V. E. Gonsalves and B. J. van Dongeren, *Textile Research J.* **24,** 1 (1954).
[14] S. P. Hersh, Ph.D. thesis, Princeton University (1954).
[15] R. G. Cunningham and D. J. Montgomery, *Textile Research J.* **28,** 971 (1958).
[16] H. Grüner, *Faserforsch. u. Textiltech.* **4,** 249 (1953).
[17] J. A. Medley, *J. Textile Inst.* **45,** T123 (1954); **48,** P112 (1957).

by path length of rub (see, e.g., Hersh and Montgomery[18]) or of roll (see, e.g., Wagner[19]). The effect of duration of the contact in the case of mere touching has not been studied much; in the studies of rubbing or rolling it appears implicitly in the analysis of the moving point of con-

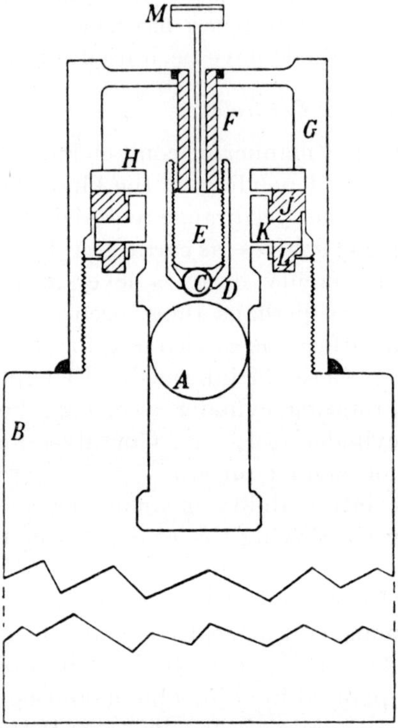

FIG. 1. Floating-ball apparatus of Harper for measuring the charge transferred upon controlled contact between spheres. The $\frac{1}{2}$-inch metal ball *A*, which slides freely in the cylinder turned out of *B*, the body of the apparatus, is floated on mercury (not shown) whose level may be altered by external means (not shown). The $\frac{5}{32}$-inch ball *C* is held by the cap *D* screwed on to the block *E* supported by the glass insulator *F*. The hood *G* carrying this assembly screws on to the body *B*. The Perspex rings *J* and *L* serve to insulate *K*, the central electrode of a standard cylindrical condenser, from the grounded guard ring *H*. Cap *D*, which forms the opposing electrode of the standard condenser, is connected via the button *M*, the fixed contact of the grounding key, to the electrometer, the rest of the key, and shielding, not shown. [After W. R. Harper, *Proc. Roy. Soc.* **A205**, 83 (1951).]

tact, where it is estimated from the extent of contact and the relative speed of the bodies. The speed is controlled by various mechanisms, for

[18] S. P. Hersh and D. J. Montgomery, *Textile Research J.* **25**, 279 (1955).
[19] P. E. Wagner, *J. Appl. Phys.* **27**, 1301 (1956).

example, those in standard industrial equipment (e.g., Ballou[20]) or by specially built research apparatus (e.g., Zaukelies[21]). The normal force between bodies may be controlled more or less directly by spring- or gravity-loading (see, e.g., Hersh and Montgomery[18]), or indirectly by

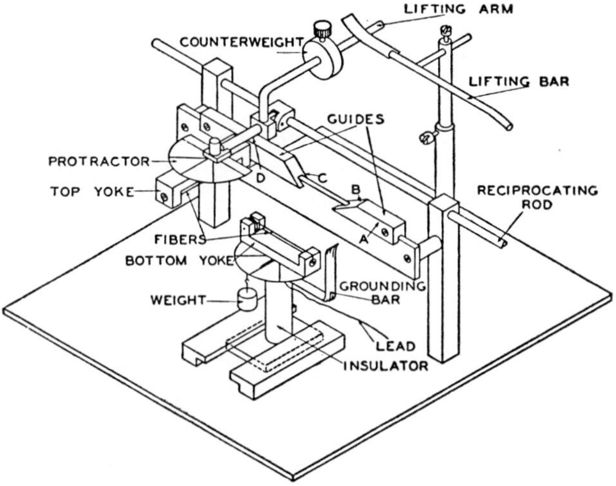

FIG. 2. Sketch of apparatus of Hersh for rubbing single fibers against each other under controlled conditions. One fiber is fixed in the grounded top yoke, which can be rotated in a horizontal plane to control the angle of rub. The other fiber is fixed similarly in the bottom yoke, which is mounted on a polystyrene stand-off insulator, and connected to an electrometer. The arm supporting the top yoke is fastened to a reciprocating rod driven by an electric motor. As the rod is driven, an adjustable guide keeps the top fiber above the bottom fiber at position A, lowers the top yoke and its fiber onto the bottom fiber as the arm reaches the depression B in the guide, and separates the fibers at the end of the depression in the guide at C. On the return journey the lifting arm engages the flat brass spring at the end of the lifting bar, which keeps the top-yoke above the lower fiber and returns it to position A in readiness for the next stroke. The source of ionizing radiation for discharging the fibers is not shown. [S. P. Hersh and D. J. Montgomery, *Textile Research J.* **25,** 279 (1955).]

tension and geometry (see, e.g., Gonsalves and van Dongeren[13]). The tangential force is not varied directly, but is determined by the normal force and the frictional properties of the material. In a few cases it has been measured (see, e.g., Levy *et al.*[22] and Quintelier *et al.*[22]).

[20] J. W. Ballou, *Textile Research J.* **24,** 146 (1954).
[21] D. A. Zaukelies, *Textile Research J.* **29,** in press (1959).
[22] J. B. Levy, J. H. Dillon, J. H. Wakelin, and J. H. Dusenbury, *Textile Research J.* **28,** 897 (1958); G. Quintelier, M. Warzée, and R. Sioncke, *J. Textile Inst.* **48,** P26 (1957).

FIG. 3. Photograph of part of the apparatus of Cunningham for rubbing single filaments under controlled conditions. The principle of operation is the same as that of the apparatus described in Fig. 2, with a slave selsyn replacing the reciprocating rod, and rotary solenoids actuating the lifting arm. A slot in the table underneath the lower filaments allows detection of sparks by a photomultiplier. The apparatus is contained in a chamber permitting control of temperature and pressure or humidity. [R. G. Cunningham and D. J. Montgomery, *Textile Research J.* **28,** 971 (1958).]

d. Detection and Measurement of Charge

The occurrence of excess charge in quantity can be detected directly by sight when it leads to a mechanical displacement or an electrical discharge, by hearing when it generates large sparks, by touch when a spark passes through the body or when mechanical force is sensed by the skin. In modern scientific studies, visible mechanical displacements are the only direct effects used to indicate presence of charge. For qualitative studies, a technique of value has been the use of "Lichtenberg figures," so-named after the worker who discovered them in 1777. In this technique a well-triturated mixture of powdered materials, differing in electrifiability and in color, is dusted over an electrified surface. The materials charge oppositely, each then being attracted to the charged region of appropriate sign, and thereby rendering it visible. Traditionally a mixture of sulfur and red lead (Pb_3O_4) has been used. Better results can be obtained with a mixture of three powders, such as cinnabar (HgS), red lead, and lycopodium. The use of other mixtures is described by Bŭrker[23] and by Hull.[24] Figures 4 and 5 show some results with this method. As an example of another technique based on mechanical displacement,[25] Fig. 6 shows a group of textile filaments that have been electrified by rubbing with a metal rod. The extent of separation and the rate of discharge give a semiquantitative measure of the electrifiability of the filaments, or as is usually the question of interest, of the efficacy of antistatic coating applied to it.

For quantitative work, charge may be detected by mechanical motion directly in string or quadrant electrometers, or indirectly following electronic amplification. In most cases these electrostatic methods must be used. In principle the charged region is presented to an electrometer probe in definite spatial relation to the probe and the surrounding objects, which should be conductors maintained at fixed potential. A certain fraction of the lines of electric flux emanating from the charged region terminate on the probe, and thereby affect the potential read on the electrometer. To translate this reading into charge carried by the region, the electrometer capacitance must be determined by direct calibration, and the fraction of the lines terminating on the probe at the time of measurement must be determined theoretically or experimentally.

If this fraction is significantly less than unity, and the body is an insulator, the precise determination of charge is extremely difficult, since a given total charge may be distributed in different ways on an insulator and thereby induce different potentials on the probe. In practice the error in estimating the fraction usually can be shown to be minor. Another complication arises from the fact that the electrometer probe cannot be insulated perfectly. Hence the measurement must be completed in a sufficiently short time that the charge on the probe does not change appreciably. It is primarily this factor that controls the complexity of actual instruments.

In case the charged region comprises an entire body, which for measurement may conveniently be isolated within a probe that largely or completely encloses it, a simple dc electrometer-tube instrument may suffice. For example, consider an object carrying a charge of 100 $\mu\mu$coul, a typical figure for some of the experiments to be discussed later. The fraction of the lines terminating on the probe will not be much less than

[23] K. Bürker, *Ann. Physik* **1,** 474 (1900).
[24] H. H. Hull, *J. Appl. Phys.* **20,** 1157 (1949).
[25] R. G. Cunningham and D. J. Montgomery, *Am. J. Phys.* **24,** 54 (1956).

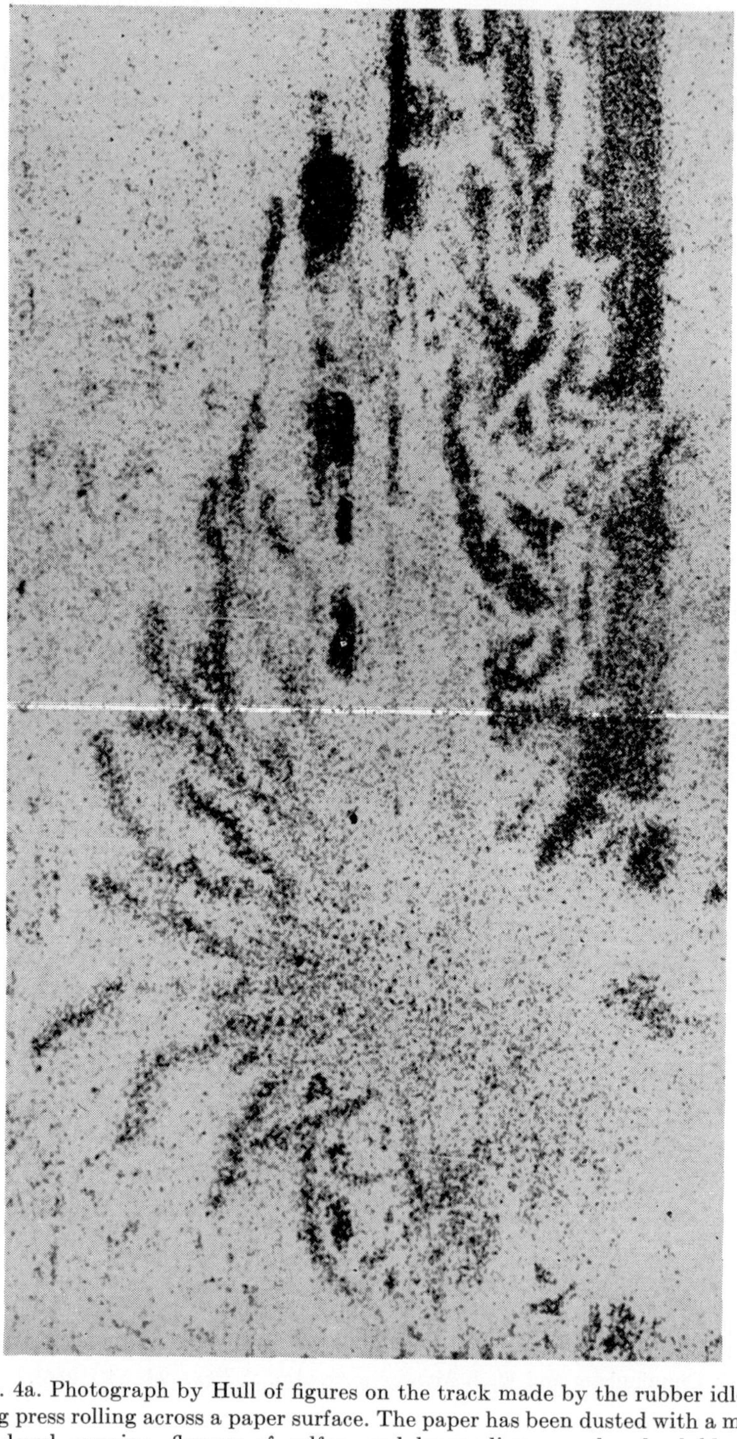

FIG. 4a. Photograph by Hull of figures on the track made by the rubber idler of a printing press rolling across a paper surface. The paper has been dusted with a mixture of powdered carmine, flowers of sulfur, and lycopodium powder dyed blue with methyl violet.

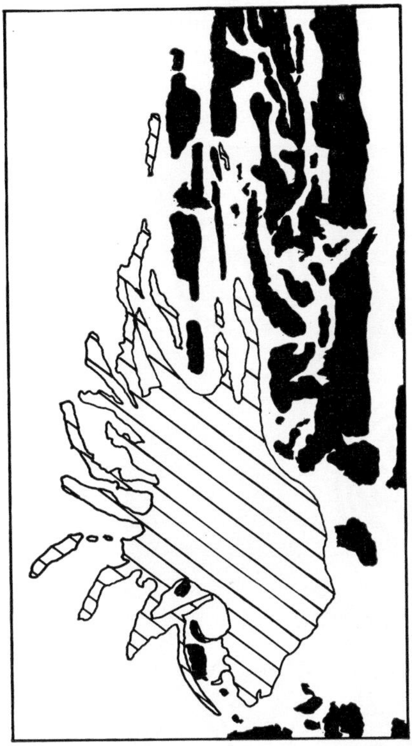

FIG. 4b. Diagram of Fig. 4a showing charge distribution. The cross-hatched areas are red (positive), the solid areas are dark blue (negative), and the remaining areas are light blue (more or less neutral). The peculiar star-shaped designs appear to be caused by electrical discharges onto the paper. [After H. H. Hull, *J. Appl. Phys.* **20**, 1157 (1949).]

unity. The electrometer itself may have an input capacitance of 10 $\mu\mu$f, with the probe and perhaps a loading capacitor increasing it to 100 $\mu\mu$f. The charge $q = 100$ $\mu\mu$coul gives a voltage $v = 1$ v, an amount easily measured to 1%. The electrometer may have a leakage resistance $R \sim 10^{13}$ ohms, and a grid current $i \sim 10^{-13}$ amp. The rate of change of voltage due to leakage current is equal to the voltage itself divided by the time constant $RC = 1000$ sec, that is, 10^{-3} v/sec, initially; and the rate arising from grid current is about the same, namely,

$$dv/dt = d(q/C)/dt = i/C = 10^{-3} \text{ v/sec}.$$

Hence the inertia of an indicating or recording instrument with a natural period of even several seconds will not introduce an error as large as 1%.

Much smaller charges can be measured with a quadrant electrometer. With a capacitance of 3 $\mu\mu$f, a leakage resistance of 10^{15} Ω, and no grid current, a charge of 3 $\mu\mu$coul will give 1 volt on the needle, with a time constant of 3000 sec.

In many cases, however, it is inconvenient or impossible to isolate the charged region for measurement. For instance, it may be desired to measure the charge on only a portion of a body; or to observe the time variation of the charge on an object connected to ground through a high resistance; or to monitor the charge continuously on material during processing. Then more complicated instrumentation is required. In such

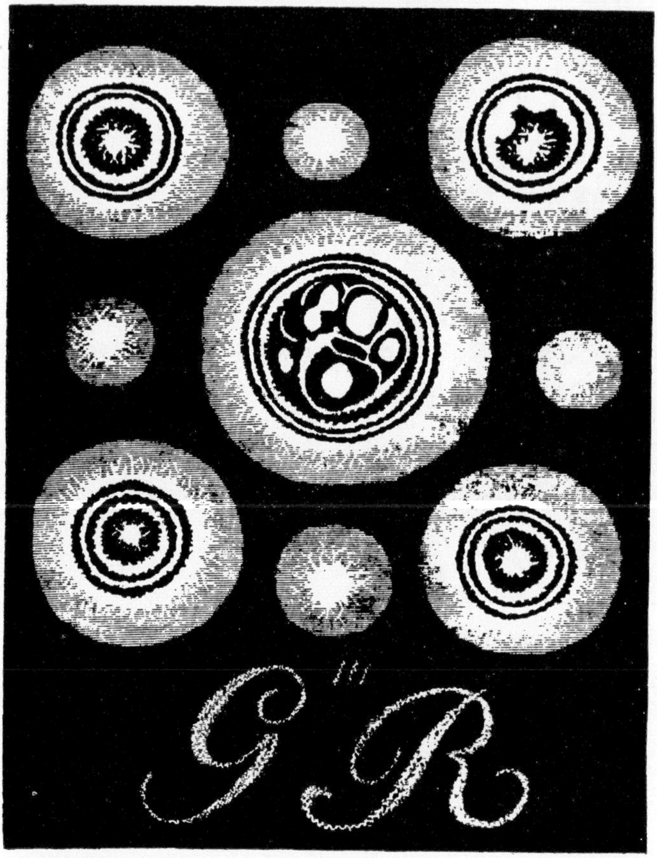

FIG. 5. Traditional Lichtenberg figures as popular in the eighteenth century. A glass plate is coated with tin foil on one side, and with gum-lac varnish on the other. The knob of a charged Leyden jar is drawn over the varnished side in the pattern desired, and the surface is then dusted with powdered rosin, chalk, or dye. [Encyclopaedia Britannica, 3rd ed., Edinburgh, 1798.]

cases the measurement is made by altering the number of lines terminating on the electrometer probe at the time of measurement. This alteration is effected by moving the charged region, the probe, or a neighboring object. In the simplest type of instrument, a grounded shutter is opened manually between the charged region and the probe of an electrometer, quadrant or electronic. This device is suitable only for rough estimation.

A more sophisticated form of this type of instrument utilizes a mechanically driven chopper to interrupt the electric lines from the charge to the probe (see, e.g., references 20, 26–36). Then ac amplification with its attendant high gain and stability can be used. This device is suitable for precise quantitative work. In another type of instrument the probe is moved; ordinarily the probe is a metal reed kept in vibration by an electromagnet. Again ac amplification is possible, and extreme sensitivity and stability may be attained. In certain rather specialized applications, the charged region itself is moved; for example when a body has nonuniformly charged regions, the varying charge induced on the probe as the body moves generates a signal (see, e.g., Sprokel[37]).

FIG. 6. Nylon filaments showing mutual repulsion upon electrification by rubbing. Several filaments have been suspended loosely between insulators and rubbed once with the back of a scissors blade. The distance between supports is approximately 12 in. The potential of the filaments may be shown to be several thousand volts. [R. G. Cunningham and D. J. Montgomery, *Am. J. Phys.* **24,** 54 (1956).]

In a few cases involving continuous production of static electrification, the rate of charge generation is large enough that galvanometric methods can be used. Even here, it is usually more convenient to send the cur-

[26] R. Gunn, *Phys. Rev.* **40,** 307 (1932).
[27] G. P. Harnwell and S. N. van Voorhis, *Rev. Sci. Instr.* **4,** 540 (1933).
[28] R. Penoir, *Compt. rend.* **205,** 1369 (1937).
[29] H. Prinz, *Arch. tech. Messen* J763–3, June 1939; J763–4, July 1939; J763–5, July 1942.
[30] R. S. Havenhill, H. C. O'Brien, and J. J. Rankin, *J. Appl. Phys.* **15,** 731 (1944); **17,** 338 (1946).
[31] R. C. Waddel, *Rev. Sci. Instr.* **19,** 31 (1948).
[32] H. W. Cleveland, *J. Soc. Motion Picture Television Engrs.* **55,** 37 (1950).
[33] M. Hayek and F. C. Chromey, *Am. Dyestuff Reptr.* **40,** 164 (1951).
[34] H. Dahl, "On the Sensitivity of Generating Electrostatic D. C. Voltmeters," Beret. XIV, p. 3. Chr. Michelsens Institutt, Bergen, 1951.
[35] A. S. Cross, *Brit. J. Appl. Phys.* Suppl. No. 2, p. S47 (1953).
[36] C. J. Young and H. G. Greig, *RCA Rev.* **15,** 469 (1954).
[37] G. J. Sprokel, *Textile Research J.* **27,** 501 (1957).

rent through a high resistance and measure the voltage drop with an electrometer.

e. Detection of Electrical Breakdown

Among the vast and complicated phenomena occurring in electrical discharge in gases, the existence of a breakdown voltage for air is the most important for interpretation of experiments on static electrification. It is certain that when the transferred charge is large enough, return of charge occurs by conduction through air upon separation. It is not clear what happens at small and intermediate charges. The dependence of the breakdown voltage on electrode size and air pressure is known well enough for fixed electrodes at macroscopic spacings, but the behavior is poorly understood for regions of breaking contact.

To detect the existence of breakdown experimentally, several investigators (e.g., Macky[38] and Cunningham[39]) have approached the problem indirectly by varying the ambient pressure. It would appear that the limiting charge attained in a given kind of experiment would decrease with decreasing pressure down to the minimum predicted by Paschen's law, and then increase again. Such behavior is indeed found, but the results do not lend themselves to quantitative interpretation. Consequently a direct approach was made by Cunningham and Montgomery,[15] who placed a photomultiplier directly below the region of contact of two filaments. Their results show the existence of sparks at high normal forces and rubbing speeds, but it is difficult to be sure that under less severe conditions the tube was not screened from very small sparks by the filaments themselves. This technique is adaptable to only a limited class of geometries. A probe that could pick up the electromagnetic radiation in the radio-frequency region, however, could be led to a receiver, and might well constitute a more sensitive and adaptable detector.

From the theoretical point of view, it is of course essential to determine how important the effect of breakdown is. Otherwise the testing of a quantitative prediction becomes complicated and perhaps impossible.

f. Dissipation of Unwanted Charge

The charge borne by an insulator ultimately will flow away by ionic or electronic conduction, or will be neutralized by normal atmospheric ionization. For experiments in which it is necessary to have the sample in a neutral state, it may be advantageous, however, to accelerate neutralization by supplying ions artificially. In the laboratory alpha, beta, and gamma or x-rays have been used, most investigators letting

[38] W. A. Macky, *Proc. Roy. Soc.* **A119,** 107 (1928).
[39] R. G. Cunningham, Ph.D. thesis, Michigan State University (1957).

convenience be their guide. Alpha rays provide a high concentration of ions; radium-beryllium can be used if the stray gamma radiation does not disturb the measuring circuit, and otherwise polonium-beryllium is satisfactory. Beta rays, usually from artificial sources, are effective over regions of greater depth. Gamma rays normally are used only for special purposes, as when ionization deep within the body of the material is desired.[22] X-rays, despite their low ionizing power in air, afford certain conveniences in that they can be switched on and off electrically.[38]

Of course, the study of dissipation is a large field in itself; we shall have occasion to refer to it later. Not only is it of extremely great technological importance, but it promises to be helpful in understanding the basic processes of static electrification (see, e.g., Fukada and Fowler[40]).

g. Measurement of Auxiliary Variables

By auxiliary variables are meant those parameters related to the properties of materials that are presumably of importance in determining the electrostatic behavior of materials, but are fixed by specification of the material and the ambient conditions. Dielectric constant and electrical resistivity are the two properties of this type which have received primary attention. For most insulators of interest, neither quantity is easy to measure or define, because of "polarization." It is simple to measure the dielectric constant as a function of frequency, but the extrapolation to zero frequency usually is of little meaning (cf. the data of Baker and Yager[41] on nylon). Moreover, it is not at all clear that the value at zero frequency is the quantity of interest. Perhaps a value associated with a frequency equal to the reciprocal of some characteristic time for the process is a more appropriate choice. Nonetheless, comparative values between materials at some arbitrarily chosen frequency are often of use. In the case of substances not obtainable as dense solids, various schemes must be developed for getting the dielectric constant of the solid material. Application to certain insulators, particularly textile fibers, has been made by Hearle.[42,43]

The factors peculiar to the measurement of resistance in dielectrics arise from the magnitude of the resistivity, the possible difference between surface resistivity and volume resistivity, nonohmic behavior, and time dependence.[44–46] Materials showing large electrostatic effects

[40] E. Fukada and J. F. Fowler, *Nature* **181,** 693 (1958).
[41] W. O. Baker and W. A. Yager, *J. Am. Chem. Soc.* **64,** 2171 (1942).
[42] J. W. S. Hearle, *Textile Research J.* **24,** 307 (1954).
[43] J. W. S. Hearle, *Textile Research J.* **26,** 108 (1956).
[44] J. W. S. Hearle, *J. Textile Inst.* **44,** T155 (1953); **48,** P40 (1957).
[45] G. E. Cusick and J. W. S. Hearle, *Textile Research J.* **25,** 563 (1955).
[46] S. P. Hersh and D. J. Montgomery, *Textile Research J.* **25,** 566 (1955).

usually have volume resistivities above 10^9 ohm-cm, rising to values as high as 10^{16} ohm-cm. We frequently are concerned with films or filaments of these materials as well as with blocks. Hence the actual resistances to be measured may run from $10^8\ \Omega$ to $10^{20}\ \Omega$. In any method the problem resolves itself into measurement of a small current. For applied voltages not in excess of 1000 v, the current may be as low as 10^{-17} amp. Normally the measurements must not require too much time, because of the time dependence of the resistance. Furthermore, leakage resistances of measuring instruments are hard to keep above $10^{15}\ \Omega$. Hence in practice electronic-electrometer methods are employed, with the consequence that currents smaller than 10^{-13} amp–10^{-14} amp cannot be measured easily with any degree of accuracy. The limit of resistance readily measured then turns out to be about 10^{16}–$10^{17}\ \Omega$; this limit can be attained only by minimizing leakage currents.

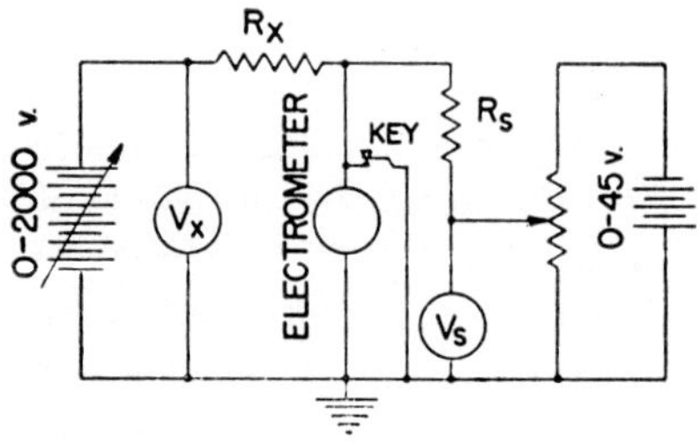

FIG. 7. Schematic diagram of circuit for measuring resistances up to 10^{16} ohms. [S. P. Hersh and D. J. Montgomery, *Textile Research J.* **22,** 805 (1952), following R. H. Norman, *J. Sci. Instr.* **27,** 200 (1950), and J. S. Townsend, *Phil. Mag.* [6] **6,** 598 (1903).]

A suitable circuit[47] is shown in Fig. 7. A Geiger-Mueller power supply impresses a specified voltage V_x as high as 1000 v across the unknown resistance R_x. The voltage V_s is adjusted until an electronic electrometer indicates zero potential difference across the terminals. In the steady state the current through R_x is equal to the current through R_s, whence $R_x = (V_x/V_s)R_s$. The circuit has the following advantages:

(1) The only path for the current which leaves R_x is through R_s and the electrometer insulators. Since the potential difference across the electrometer is zero at balance, the leakage current is small even if the leakage resistance should happen to be low.

(2) The electrometer is used only as a null indicator; therefore, only the roughest of calibrations is necessary, and deviations from linearity of response are of no importance.

[47] S. P. Hersh and D. J. Montgomery, *Textile Research J.* **22,** 805 (1952).

(3) Only low-impedance instruments, with their attendant ruggedness and stability, are necessary for the quantitative readings.

In practice, the ratio V_x/V_s should not exceed 1000, and it is necessary to have a set of standard resistors up to 10^{12} or $10^{13}\ \Omega$. These resistors are available commercially. In the higher ranges they need to be checked regularly because of possible instability and surface contamination. In the event that suitable standard resistors are not available, R_s may be replaced with a known capacitance. The measurement is then based on the time rate of change of voltage necessary to maintain zero potential difference across the electrometer (cf. Norman[48] and Townsend[49] for details).

For control purposes and for certain research activities, the resistance may be brought into a reasonable range by testing specimens in parallel. This method is particularly useful with filaments, which can be wound with many turns around suitable electrodes.

When the material is inhomogeneous, in particular, when the surface region has a resistivity much lower than the interior, it is often useful to introduce the concept of surface resistivity. With large specimens, for example, glass blocks or celluloid film, the concept may be defined and measured in a more or less standard way. (See, e.g., ASTM procedures,[50] which contain further references, and also AATCC procedures.[51]) In the case of small specimens, for example textile fibers, a direct measurement is not feasible. In principle the conductance may be separated into a surface and a volume conductance by studying the variation of total conductance with some dimension, say diameter. In practice this procedure is seldom possible because the composition of the material differs from one sample to another. In case of an antistatic coating, the conductance can be measured before and after the application. Then if the thickness of the coating can be considered negligible, the surface conductance can be taken as the increase in total conductance upon coating.

For routine measurements and certain research purposes, several megamegohmmeters are available commercially. Their operation and limitations are described in the manufacturers' literature. The convenience—as well as the cost—is high, but the highest resistance measurable reliably is usually less than 10^{15} ohms.

[48] R. H. Norman, *J. Sci. Instr.* **27,** 200 (1950).

[49] J. S. Townsend, *Phil. Mag.* [6] **6,** 598 (1903).

[50] ASTM Standards 1955, Part 6, p. 471, D257–54T, issued 1949; revised, 1952, 1954. Am. Soc. Testing Materials, Philadelphia.

[51] Tentative Test Method 76–54, Am. Assoc. Textile Chemists and Colorists, *Tech. Manual and Yearbook* **31,** 140 (1955).

3. Results—Qualitative and Semiquantitative

The most significant result of the various qualitative and semiquantitative studies is that charge transfer is to be expected whenever two objects are placed in contact and then separated. The two objects need not even be of different materials, if any difference in inhomogeneity, anisotropy, strain, temperature, etc., is present. Whether the transfer produces noticeable static electrification depends on the speed of neutralization. In the absence of external sources of ionization, appreciable effects are observed only when the specific resistivity of the better insulator rises about 10^8–10^9 ohm-cm or so.[17,18] This "threshold" of course depends on the phenomenon of interest and on the sensitivity of the instruments. Moreover, the threshold decreases as the speed increases for given phenomena and detectors. Neither the sign nor the order of magnitude of the charge seems to depend on whether the material of lower resistivity is a metal or an insulator.

a. *Charge Generated between Similar Materials*

When materials that are very similar are placed in contact under symmetrical conditions, the charge transferred is small and of random sign. However, materials that are apparently similar or even identical in chemical composition may have the chemical similarity or the contact

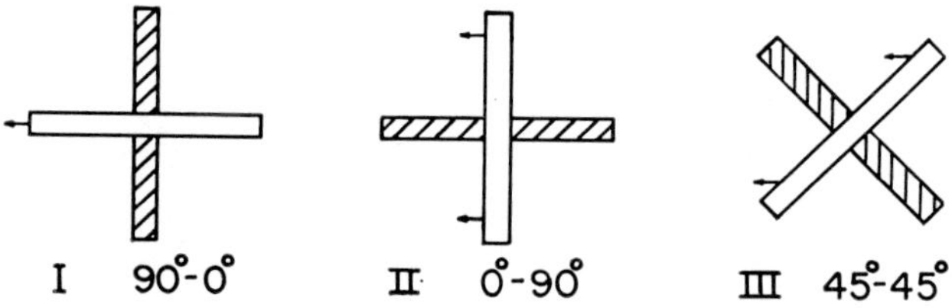

FIG. 8. Diagram to illustrate three major types of rub between crossed cylinders.

symmetry destroyed by differences in surface structure. Thus it was known in early days[52] that different surface treatments, for example, polishing or etching, might render bodies of the same material triboelectric with respect to one another. When the contact involves asymmetric rubbing, static electrification can occur even between samples having identical surface treatment. As an example, consider the rubbing of one cylinder against another, in the manner indicated in Fig. 8. Two asymmetric types of rubs, 90°–0° and 0°–90° are illustrated. Here the same region of one body is exposed to different regions on another. Also

[52] J. C. Wilcke, "Disputatio Physica Experimentalis de Electricitatibus Contrariis," Academy of Rostock, 1757.

shown is a symmetric type of rub, 45°–45°, in which new regions on each body are continually presented to new regions on the other. Shaw and Hanstock,[53,54] and later Grüner[16] and Henry,[55] studied the effect of asymmetrical rubbing of identical materials. Henry states that separation of charge occurs between two identical surfaces which are rubbed together in an asymmetric manner, say 90°–0°, so that the friction occurs mainly at one spot on one of the surfaces, but is spread over a large area of the other. The sign of charges is reversed if the two samples are interchanged (0°–90° rub). For his materials and conditions of rubbing Shaw found that the object on which the larger area has been rubbed always becomes positive. He explained the results on the basis of different surface strains. Grüner, in a related experiment, confirmed this result and interpreted it as a consequence of a higher mobility of the H^+ ions than of OH^- ions. Hersh[14] found for numerous materials, however, that the reverse rule was likely to hold. He was unable to reconcile the discrepancy, even after additional experimentation designed to test various explanations. Henry makes no general statement, and it is likely that no rule of wide applicability exists.

Another way in which chemically identical materials can differ is through molecular orientation. Hersh and Montgomery[18] studied charge generation upon symmetric rubbing (45°–45°) between nylon filaments drawn to different extents from the same melt. Filaments drawn from two to six times their original length were always positive to an undrawn one in a symmetric rub, but their was no significant electrification between filaments drawn to different ratios.

The effect of anisotropy in identical materials was discovered in wool by Martin[56] and examined quantitatively in both wool and hair by Hersh and Montgomery.[57] The observation shows that when one animal fiber of the kind mentioned is rubbed against another, the two fibers being placed with root ends pointing in the same direction, the sign of charge taken on by a given fiber depends on the direction of rub. On the other hand, when the root ends point in opposite direction, the charge transfered is negligible. This different behavior is reasonable in view of the scale structure of the animal fibers; however, the detailed explanation is difficult.

In summary, even though two insulators have the same chemical

[53] P. E. Shaw, *Proc. Phys. Soc.* **39,** 449 (1927).

[54] P. E. Shaw and R. F. Hanstock, *Proc. Roy. Soc.* **A128,** 480 (1930).

[55] P. S. H. Henry, *Brit. J. Appl. Phys.* Suppl. 2, S31 (1953); *J. Textile Inst.* **48,** P5 (1957).

[56] A. J. P. Martin, *Proc. Phys. Soc.* **53,** 186 (1941).

[57] S. P. Hersh and D. J. Montgomery, *Textile Research J.* **24,** 755 (1954).

composition, charge transfer may be expected if the rubbing is asymmetric, or the materials are anisotropic or oriented.

b. Charge Generated between Dissimilar Materials upon Symmetric Rubbing

In the case of symmetric rubbing of dissimilar materials, the chemical natures of the materials would be expected to determine the sign of the transferred charge. To a large extent, if not completely, this observation appears to be valid.[58]

The polarity of electrification might be expected to be a transitive property among materials. A *triboelectric series* could then be established in which any material becomes positive when rubbed against any material lower in the list. Wilcke[52] published the first such list (1757).

Positive
- Polished glass
- Wool flock
- Writer's quill
- Wood
- Paper
- Sealing wax
- White wax
- Unpolished glass
- Lead
- Sulfur
- Other metals

Negative

Two representative modern lists with emphasis on textiles[59] are given in Table I.

Ballou[20] states further that human skin lies in the region of the cellulosic materials, and that chrome-plated metallic surfaces are near ramie and acetate. Hersh[18] gives data to show that aluminum and magnesium stand above nylon in his series, and perhaps above wool; iron lies between nylon and acetate, and platinum between cellulose, acetate, and polyethylene.

The incompleteness of such triboelectric series is striking, as is the disagreement. Until a standard procedure is agreed upon, however, there is little motivation for an experimenter to include materials not relevant to his particular topic of investigation.

[58] The apparent exceptions, such as the behavior of polished *vs* matte glass, are probably a result of the difference in chemical composition of the surface layers, or of rubbing anisotropy of the type discussed above for wool fibers and animal hair.

[59] The chemical nature of the various materials is given in Appendix A.

TABLE I. REPRESENTATIVE TRIBOELECTRIC SERIES (POSITIVE AT TOP)

Ballou[a] (15% R.H., ~27°C)	Hersh[b] (33% R.H., 30°C)
Wool	Wool
Nylon	Nylon
Silk	Viscose rayon
Viscose rayon	Cotton
Cordura rayon	Silk
Cotton	Cellulose acetate
Fiberglas	Lucite
Spun ramie	Polyvinyl alcohol
Cellulose acetate	Dynel
Dacron yarn	Velon
Orlon yarn	Polyethylene
Polyethylene	Teflon
Saran	

[a] J. W. Ballou, *Textile Research J.* **24**, 146 (1954).
[b] S. P. Hersh and D. J. Montgomery, *Textile Research J.* **25**, 279 (1955).

Several workers, in particular Shaw and his colleagues,[60] have questioned the existence of a triboelectric series. In fact, they showed that it is possible to find substances *A*, *B*, and *C*, which have the characteristic that even though *A* is positive to *B* and *B* is positive to *C*, *A* might be negative to *C*. This case is rather exceptional, and remains unresolved.

c. Charge Generated between Dissimilar Materials upon Asymmetric Rubbing

The chemical dissimilarity swamps any effects of asymmetric rubbing in the majority of cases. With certain materials close together in a triboelectric series, however, the direction of charge transfer may be reversed by changing the conditions of rubbing, as discussed by Henry.[55] Hersh[18] has noted the same effect, citing its occurrence with polyethylene and Velon on Dynel.

4. RESULTS—QUANTITATIVE

Short of a compilation of all the numerical results and graphs occurring in the literature, it is impossible to describe the quantitative findings on static electricity without distortion. Most studies are not comparable with one another, in view of the differences in procedure and the variety of materials. Moreover, it is not justifiable to expect values obtained by one experimenter to apply to the findings of another. A representative

[60] P. E. Shaw and C. S. Jex, *Proc. Roy. Soc.* **A118**, 108 (1928).

picture can be obtained, nevertheless, from the description of the work of certain types.

In discussing the amount of charge generated, it is necessary to keep the conditions to which the stated charge refers clearly in mind. To illustrate this point, consider the behavior of two specimens which are to be rubbed against each other in a static-electrification experiment. By suitable chemical and electrical conditioning, for example by washing and discharging, specimens are brought to some reference state. They are then rubbed together once under specified conditions in such a way that a new surface on at least one specimen touches the surface of the other. We shall refer to the course of the charge variation in this case as the *initiatory* behavior. If this process is repeated with the same specimens, without cleaning of any kind but perhaps with discharging, it usually is found that the transferred charge changes, but soon levels off and attains a nearly constant value. In these circumstances we shall refer to the *steady-state* behavior. If now the rubbing is repeated a great number of times, the steady-state value may change slowly; we shall then talk about the *secular* behavior.

Each region of behavior has characteristic claims for study. Although the initiatory behavior would seem most susceptible to rational attack at first glance, few investigators have attempted it. The primary reason is that such experiments are inordinately time-consuming if the reference state is to be defined and controlled suitably, for the specimens must be in the virgin state before conditioning. Secondary reasons are that the initiatory behavior is not of interest in many applications, and that it may resist analysis more obdurately than the steady-state and the secular behavior. Ultimately the complete behavior must be understood; however this goal is indefinitely far removed. The majority of investigations have been of more or less steady-state behavior. We shall consider these in the main.

a. Reproducibility

In experiments involving pure touching of spheres without rubbing, Harper[10,61] found good reproducibility when sliding was eliminated. Harper worked with both metals and insulators. Bredov and Kshemianskaia[62] reported no difficulty in attaining reproducibility in similar experiments with metals and semiconductors of various shapes when brought against flat plates.

[61] W. R. Harper, *Proc. Roy. Soc.* **A218,** 111 (1953); **A231,** 388 (1955).

[62] M. M. Bredov and I. Z. Kshemianskaia, *Zhur. Tekh. Fiz.* **27,** 921 (1957); translated in *Soviet Phys. Tech. Phys.* **2,** 844 (1957).

Péclet[63] found deviations from run to run not greater than a few per cent in rubbing pads against a revolving cylinder; Hersh[14,18] and later Cunningham,[39,15] in rubbing single filaments against single filaments, reported that deviations from run to run did not exceed ±5%, while deviations from sample to sample were less than ±20%. Even with plates pressed together with negligible rubbing, in which the area of contact is difficult to control, Henry[55] found that the run-to-run variations were only about ±20%.

In summary, we may state that the reproducibility from run to run is ordinarily ±5% provided good control of mechanical and ambient conditions is maintained; the reproducibility is not so good from sample to sample, but for most experiments does not lie outside ±20%.

b. *Effects of Materials*

Until an adequate theory of static electrification is worked out, it is not clear what parameter or combination of parameters is to be used to characterize the static-electrification behavior of a given material quantitatively. Hence only a semiquantitative relation can be established at best. Almost all the observers who have reported triboelectric series have stated that the amount of charge transferred between materials in a series remains the same or increases only slightly with separation in the list.

Coehn[64] attempted to relate position in a triboelectric series with dielectric constant for solid dielectrics against liquids, those substances with higher dielectric constant being higher in the series. The law was extended[65] later to the statement that the charge produced is proportional to the difference in dielectric constants. The work of Shaw and Jex[66,60,53] and of Macky[38] showed that such a relation did not apply to solid-solid contacts. In recent times the law has been almost abandoned, for ac measurements of dielectric constants give values that show no correlation of the type required by Coehn's law, and dc measurements are almost meaningless because of polarization.

c. *Effect of Molecular Structure*

Few studies which attempt the formulation of quantitative relations have been made, and no quantitative relations have been established.

[63] E. Péclet, *Ann. chim.* (*Paris*) [2] **57,** 337 (1834).
[64] A. Coehn, *Ann. Physik* **64,** 217 (1898); **66,** 1191 (1898).
[65] A. Coehn, *Ann. Physik* **30,** 777 (1909).
[66] P. E. Shaw and C. S. Jex, *Proc. Roy. Soc.* **A118,** 97 (1928).

The available experiments on the effect of draw ratio show insufficient dependence on degree of orientation to permit any statement other than that orientation changes the static behavior; however, a draw ratio greater than two has little additional effect.

d. Effect of State of Strain

Hersh and Montgomery[18] showed that the effect of strain on filaments rubbed against a different material was negligible below the elastic limit. The more sensitive test of rubbing a strained specimen against an unstrained one of the same material shows that some effect must exist; but no quantitative studies of this phenomenon have been reported.

e. Effect of Size and Shape of Specimens

In experiments in which at least one of the contacting specimens is charged to the limiting field strength, the total charge depends on the geometry of that specimen. Macky[38] has discussed his experiments and related ones, and has shown that the saturation charge is proportional to the capacitance of the system. This result would be expected of course only if the geometries at various capacitances were sufficiently alike that the limiting field strengths at each portion of the system were similar.

Few quantitative studies have been made in experiments in which saturation by breakdown does not occur. Cunningham and Montgomery[15] made a preliminary attempt to establish the dependence of charge generated between filaments of different diameters. Their results show that the charge varies inversely as the filament diameter, a finding difficult to interpret on most theoretical pictures. Further work is necessary before these results can be considered established.

Many experiments have been made with powdered materials. The situation has been reviewed recently by Harper,[67] with particular attention to the work of Knoblauch[68] and of Rudge;[69–71] and by Loeb,[72] with emphasis on the work of Debeau,[11] Hansen,[73] Kunkel,[74–77] and Peterson.[78]

[67] W. R. Harper, *Advances in Phys.* **6,** 365 (1957).
[68] O. Knoblauch, *Z. physik. Chem. (Leipzig)* **39,** 225 (1902).
[69] W. A. D. Rudge, *Phil. Mag.* [6] **23,** 852 (1912).
[70] W. A. D. Rudge, *Phil. Mag.* [6] **25,** 481 (1913).
[71] W. A. D. Rudge, *Proc. Roy. Soc.* **A90,** 256 (1914).
[72] L. B. Loeb, "Static Electrification." J. Springer Verlag, Berlin, 1958.
[73] J. W. Hansen, *Phys. Rev.* **72,** 741 (1947).
[74] W. B. Kunkel, *J. Appl. Phys.* **19,** 1053 (1948).
[75] W. B. Kunkel, *J. Appl. Phys.* **19,** 1056 (1948).
[76] W. B. Kunkel, *J. Appl. Phys.* **21,** 820 (1950).
[77] W. B. Kunkel, *J. Appl. Phys.* **21,** 833 (1950).
[78] J. W. Peterson, *J. Appl. Phys.* **25,** 501 (1954).

In view of the difficulty of describing the results succinctly, the complexity of the phenomena, and the availability of the treatments cited, this topic will not be considered further here.

f. Effect of Charge State

A charged insulator normally will take on additional charge of the same sign to a smaller extent than a neutral body. This action is demonstrated whenever a specimen is rubbed repeatedly without discharging.

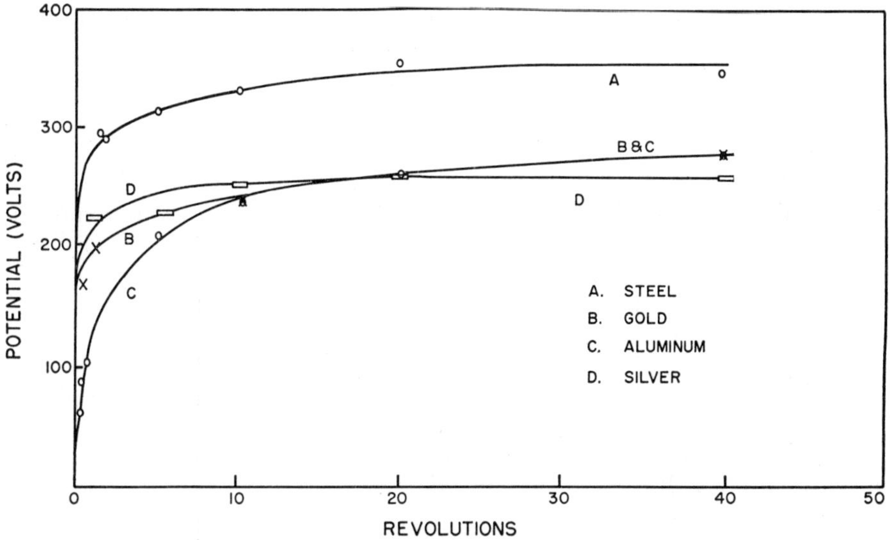

FIG. 9. Variation of charge (expressed as potential) with distance of rub (expressed in number of revolutions) for flat disks of various metals pressed against flat glass disks and rotated about their common axis perpendicular to the plane of contact. [W. A. Macky, *Proc. Roy. Soc.* **A119,** 107 (1928).]

Figure 9 shows the results of Macky[38] in experiments in which a polished disk is revolved against a flat plate. These experiments are quantitative, but the charges at the upper part of the curve are undoubtedly limited by breakdown. Consequently, relations capable of simple interpretation cannot be established between the charge state and amount of charge transfer. Experiments designed to obtain this relation would be desirable.

g. Effect of Temperature

Systematic studies of the effect of ambient temperature on static electrification—with perhaps the exception of certain studies on ice—have been carried out over an extremely limited temperature range centered close to room temperature. Whatever effects have been observed have been interpreted primarily as the result of changes in conductivity. The effects of *local* heating at the region of contact are undoubtedly important in changing the character of the materials at the point of interaction. This heating must be controlled in part by the ambient temperature. Hence it is likely that experiments over a wide range of temperatures would show effects other than those due merely to changed electrical conductivity. The experimental difficulties encountered at low temperatures do not appear insuperable. At high temperatures one encounters the problem that many insulators become sufficiently conducting to lose most of their electrostatic properties.

h. Effect of Atmosphere

(1) *Composition.* Péclet[79] in 1839 studied the possible chemical origin of static electrification between materials rubbed in carbon dioxide as well as air. He found that any differences between the gases were removed

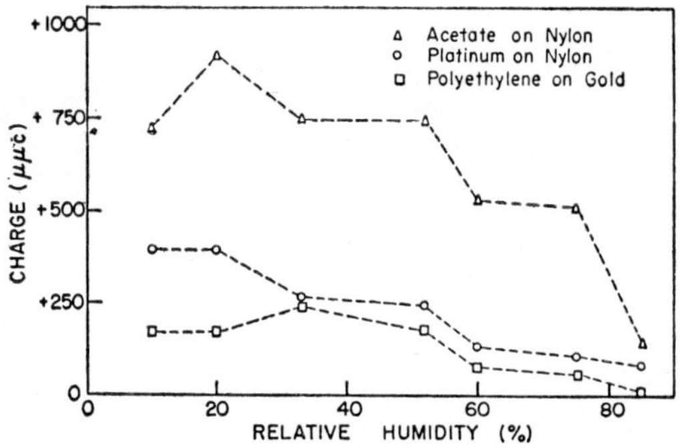

FIG. 10. Net charge transferred as a function of relative humidity for unlubricated filaments. Charge positive on second-named material. Rub type, 45°–45°; temperature, 30°C. [S. P. Hersh and D. J. Montgomery, *Textile Research J.* **25**, 279 (1955).]

by drying them sufficiently. Macky[38] worked with atmospheres of sulfur dioxide, carbon dioxide, coal gas, hydrogen, and oxygen, in each case drying the gas carefully. The differences between atmospheres were very slight. On the other hand, wet atmospheres cause large effects with most materials. Although it is difficult to establish conclusively that the results represent nothing beyond mere change in electrical conductivity, most investigators have assumed that this change is the only matter of consequence. Figure 10 shows representative results by Hersh and Montgomery.[18] Data more susceptible to quantitative analysis are presented in a later section.

[79] E. Péclet, *Ann. chim. (Paris)* [2] **71,** 83 (1839).

(2) Pressure. Several investigators have studied the effect of pressure on static electrification. In all the cases the authors have attributed the effect to the variation in breakdown strength of the atmosphere with pressure, and there seems to be no reason to question this result. To test this point specifically, Cunningham[39] rubbed filaments at varying atmospheric pressure, using the normal force between filaments as a parameter. The charge at a fixed normal force was nearly constant with decreasing pressure until a critical value was reached, when the charge began to fall. Cunningham interpreted this on the principle that the breakdown field intensity is lowered as the pressure decreases, until the atmosphere begins to conduct.

In summary we may state that the primary effects of atmosphere are to limit (1) the charge at saturation by controlling the dielectric strength of the atmosphere; and (2) the charge below saturation by controlling the resistivity of the rubbed materials.

i. Effect of Electromagnetic Fields

No workers have considered the possibility that a static magnetic field could have sufficient effect to justify its study. Several workers have studied the effects of electrostatic fields. Gill and Alfrey[11] carefully discharged small cubes of ebonite or a measured volume of sand, and slid the samples down an inclined plane. The particles were caught in a Faraday cage connected to an electrometer. Another plate, insulated from the slide, was placed parallel to it and about a centimeter above the slide. A potential applied between the two plates created an electric field normal to the surface of the slide. The electrical field had a considerable effect on the static electrification. It was possible to increase, neutralize, or reverse the charge observed in the absence of the field. Peterson[78] refined and extended these experiments, rolling spheres of borosilicate glass down an inclined nickel sheet in a controlled atmosphere. The results confirmed the findings of Gill and Alfrey, who attributed the observed effects to leakage of charge over the dielectric surface, superimposed on a type of contact potential between insulator and plate. Peterson's interpretation, however, was that the additional charge transfer took place by conduction through the absorbed gaseous layer. Harper,[80] citing the findings of Medley,[81] suggested that gas discharges are the source of the modification of the charging by the electric fields. In a later discussion he mentions that leakage might occur either through the atmosphere or through a conducting layer on the particles. This conclusion seems sufficiently safe, but the experiments of Gill and Alfrey

[80] W. R. Harper, *Nature* **167**, 400 (1951).
[81] J. A. Medley, *Nature* **166**, 524 (1950).

and of Peterson need to be re-examined critically to see whether the gaseous discharge alone is not sufficient to explain the results actually reported. In any event, there now seems to be little disagreement on the principle that the effect of applied electric fields is to modify the secondary transfer of charge following the primary transfer associated with direct contact.

The effect of electromagnetic radiation has been studied very little. Of course x-rays have been used to discharge samples, but irradiation during rubbing apparently has not been studied systematically. In view of such correlations as those found by Fukada and Fowler[40] between the distributions of electron traps with depths of the order of 0.1 ev ($\sim$10 μ wavelength), and such concepts as those put forth by Gonsalves[82] and Vick[83] and others, it is surprising that no studies on the effect of visible and infrared radiation have been reported yet.

j. Effect of Path Length

When continually renewed surfaces are presented to one another, as in the case of two filaments rubbing together with axes neither perpendicular nor parallel, the charge is found[18] to be proportional to the path

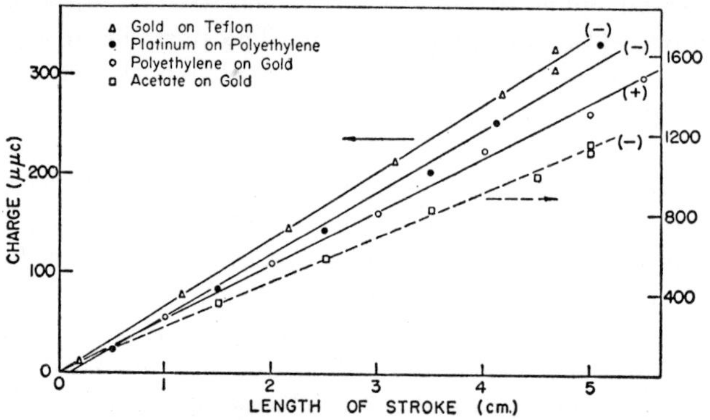

FIG. 11. Net charge transferred as a function of length of stroke for metals rubbed on insulators. Sign of charge on second-named material as indicated. Rub type, 45°–45°; temperature, 30°C; relative humidity, 33%. [S. P. Hersh and D. J. Montgomery, *Textile Research J.* **25,** 279 (1955).]

length if a good insulator rubs on either a good or a poor insulator (see Fig. 11), and is found to be independent of path length for a good conductor on a good conductor. These findings have a natural explanation

[82] V. E. Gonsalves, *Textile Research J.* **23,** 711 (1953).
[83] F. A. Vick, *Brit. J. Appl. Phys.* Suppl. 2, S1 (1953).

in terms of the time required for charge to redistribute itself along the length of the body, compared with time required for the region of contact to move along either body.

When surfaces which are carrying charge are presented to one another, as in the case of a sphere rolling in a cylinder, the charge approaches a limiting value. An analysis of this case is given by Peterson[12] and by Wagner.[19]

k. Effect of Duration of Contact

Most experiments have dealt either with contacts maintained for a time long enough that equilibrium was attained, or with experiments in which the duration was controlled by the velocity and the mechanical parameters of the experimental arrangement. Consequently there is little information directly available on the duration of contact. Experiments are also needed in a related field, namely the measurement of the "contact potential" between pairs of materials of which one at least is an insulator.

l. Effect of Rub Velocity

The trend that has been observed generally is that the charge increases with the rub velocity until a limiting value is reached. In many investigations, the effect of velocity has been studied over only a small range; if this range is in the vicinity of the saturation region, no dependence of charge on velocity is noted. To illustrate the behavior, Fig. 12 shows some results of Cunningham and Montgomery[15] who rubbed clean single-filament insulators against insulating and conducting filaments; Fig. 13 shows the results of Medley[17] who rubbed keratin filaments coated with conducting oil against a metal cylinder. The general course of the charge variation is seen to be the same. It is unlikely that the flattening of the curves is the result of electrical breakdown, for the charges observed by Medley were small. Moreover, Cunningham and Montgomery would have detected sparks by a photomultiplier placed near the filaments. Medley was able to get reasonable agreement with a theoretical relation on the assumption that the flow of charge through the oil controls the charge retained. Cunningham and Montgomery, who worked with uncoated filaments, could only find a more complicated semiempirical relation. These results will be discussed later.

To study the secular effects of velocity, Levy and Dillon[22,84] made thousands of consecutive rubs on a given pair of filaments at fixed velocity, discharging the filaments only at the end of a rubbing period, and then measuring the charge produced by a single rub at the rubbing

[84] J. B. Levy and J. H. Dillon, *Textile Research J.* **26**, 953 (1956).

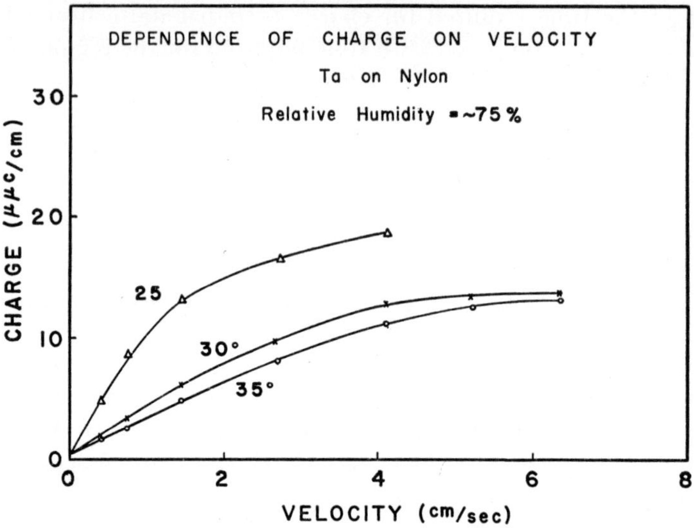

FIG. 12. Net charge transferred as a function of rub velocity for an 8-mil tantalum wire rubbed on an unlubricated 8-mil nylon monofil at various temperatures as shown. Charge positive on nylon. Rub type, 45°–45°; relative humidity, 75%. [R. G. Cunningham and D. J. Montgomery, *Textile Research J.* **28,** 971 (1958).]

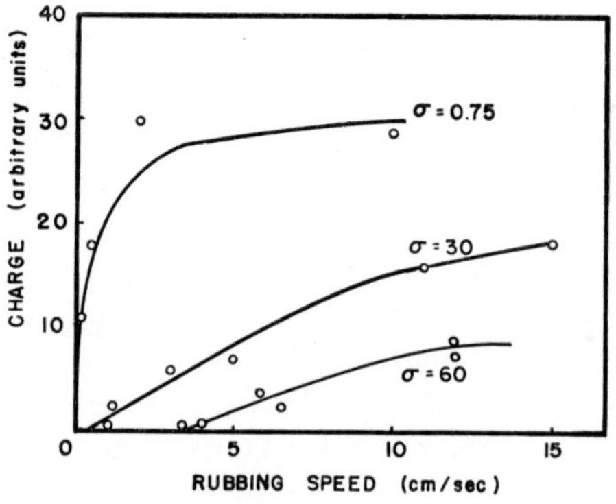

FIG. 13. Net charge transferred as a function of rub velocity for lubricated 0.14-mm keratin fiber (cow's tail) rubbed against a 1.23-mm metal wire (platinum). The fibers have been coated with a solution of antistatic agent in mineral oil. The measured values of the volume conductivity of the solution in μμmho/cm are given on the curves. Charge positive on fiber. Rub type, 0°–90° (one spot on wire rubbed by full length of fiber); temperature, presumably about 20°C; relative humidity, 0%. [After J. A. Medley, *J. Textile Inst.* **45,** T123 (1954).]

velocity. The charge was found to decrease secularly with rubbing velocity. Their explanation is that the increase in heating at higher velocities produces greater transfer of material from one filament to the other, with consequent reduction in the net charge transfer.

m. Effect of Force between Bodies

Quantitative studies of the effect of the normal force between contacting bodies on the amount of charge generated show that the charge increases with increasing force, but with decreasing rate of rise. This has been observed for example by Medley[17] (Fig. 14). In many of the experiments reported by various authors, a saturation value is reached that is in all likelihood attributable to electrical breakdown. Macky,[38] in particular, has demonstrated this effect. The saturation value is nearly

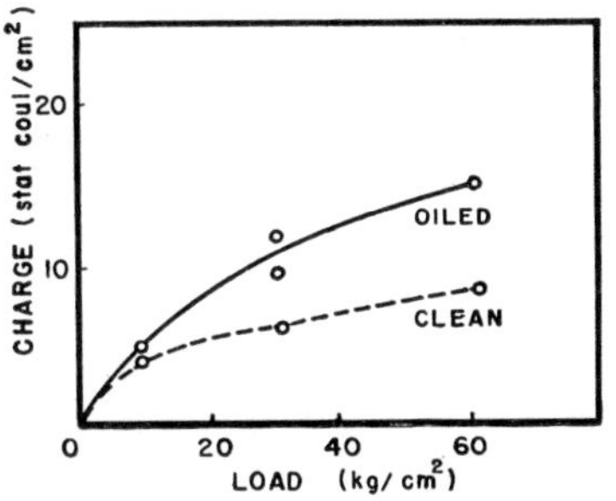

FIG. 14. Net charge transferred as a function of normal force between keratin fiber (both clean and coated with pure paraffin oil) and platinum wire. Charge positive on fiber. Rub type, 0°–90° (one spot on wire rubbed by full length of fiber); temperature, presumably 20°C; relative humidity, 0%. [J. A. Medley, *J. Textile Inst.* **45**, T 123 (1954).]

independent of normal force between the bodies (as well as the velocity of rubbing) under these conditions. Before saturation occurs, the rise in charge can be represented quantitatively by a power law with an exponent of about $\frac{1}{2}$, as shown by the results of Cunningham and Montgomery (Fig. 15). There are theoretical reasons for considering this form of law, but much more experimental work will be required to elucidate this dependence.

The tangential force between given materials at given conditions is controlled by the normal force. Hence studies at varying tangential force cannot be made apart from varying normal force. Nevertheless, it is possible that the correlation between charge and tangential force might be simpler than that between charge and normal force. To examine this possibility, Levy[22] measured the work done against friction in single-

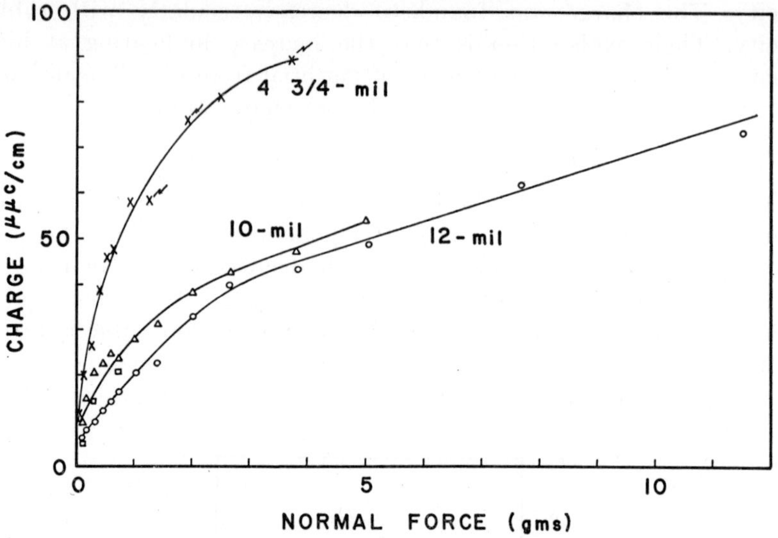

FIG. 15. Net charge transferred as a function of normal force between an 8-mil nylon monofil and polyethylene monofils of the diameters indicated. Charge positive on nylon. Rub type, 45°–45°; temperature, 30°C; relative humidity, 75%. [R. G. Cunningham and D. J. Montgomery, *Textile Research J.* **28**, 971 (1958).]

filament studies. It was established empirically that the charge is linear with the square root of the frictional work. For certain of their data, the linearity becomes a proportionality. In this case the results are in agreement with the observations of Cunningham and Montgomery, under the assumption that the frictional force is proportional to the normal force.

IV. Theoretical Interpretations

It is too much to hope for a complete elucidation of static phenomena at the present stage of knowledge. On the experimental side, the microscopic geometry of the contacting surfaces is not known, the chemical nature of the surfaces is uncertain, and transient temperatures and stresses at the region of contact are unknown. On the theoretical side, it is not clear that sufficient periodicity occurs in the materials normally of interest to allow application of much of the modern theory of solids. Moreover, it appears certain that very accurate quantitative treatments cannot be made in view of the prohibitive labor of the calculations and the unavailability of data on energy levels. Nevertheless, we believe that there is value in attempting to formulate a theoretical scheme which will correlate the present findings to some extent, and which will suggest future experiments. Drastic restrictions and simplifications will have to be made, but this procedure is hardly unusual in the formative stages of a branch of physical science.

An adequate theory of static electrification at the atomic level could not have been formulated before a satisfactory understanding of the nature of metals and insulators existed. The chief theoretical ideas advanced before the advent of quantum mechanics are due to Volta and to Helmholtz. Volta in 1789 postulated (according to Perucca,[85] who does not give the original source) that the source of the frictional electrification of two bodies lies in the *contact* between them, rather than in heating, transfer of material, or other effects due to friction. Rubbing serves merely to increase the area of contact, and thereby yield a larger charge. Helmholtz[86] in 1879 accepted this picture, in addition postulating that a double layer of charge only a few atoms thick is formed at the region of contact, much like the double layer accounting for the contact potential difference between dissimilar metals. The potential difference is small while the objects are in contact, but becomes large upon separation as the capacitance decreases. This *Volta-Helmholtz hypothesis*, as it is frequently called, has a substantial history of acceptance and rejection by various workers, but its recounting is of dubious value here (cf. Hersh[14] for a guide to the relevant papers). Today, workers for the most part believe that contact alone is sufficient to produce appreciable charge transfer with some materials, but that actual rubbing produces additional effects of various kinds. For the portion of the effects arising from contact alone, the existence of a double layer is generally accepted, implicitly or explicitly. The thickness of this layer on the atomic scale, however, may exceed atomic spacings by several magnitudes.

The first attempt to apply the quantum-mechanical theory of solids to static electrification was made by Frenkel[87] in 1941. His note was an outgrowth of a paper[88] presented in 1917 introducing the idea of an "intrinsic potential" for an insulator corresponding to the work function for metal. Frenkel assumed that electrons alone were transferred, these electrons becoming mobile at the high temperatures produced locally through rubbing in the case of insulators. This explanation is successful in interpreting some phenomena, but it has not been followed up to give a unified scheme.

The next attempts were made independently in 1953 by Vick[83] and by Gonsalves.[82] Both authors base their work on the band picture of solids, Vick resorting to surface states to some extent, Gonsalves depend-

[85] E. Perucca, *Z. Physik* **51,** 268 (1928).
[86] H. Helmholtz, *Ann. Physik* **7,** 337 (1879).
[87] J. Frenkel, *J. Phys.* (*U.S.S.R.*) **5,** 25 (1941).
[88] J. Frenkel, *Phil. Mag.* [6] **33,** 297 (1917).

ing on them completely, in making applications to static electrification. Their schemes were improved somewhat by Arthur,[89] in a review appearing at the end of 1955, but he did not carry his work farther.

A theoretical scheme proposed by Peterson[12] in 1954 to explain results with glass spheres rolling against nickel deserves mention. No detailed quantum-mechanical picture for the charge transfer was provided, but the relevant factors were discussed. More importantly, the reduction of initial charge by leakage upon separation was emphasized. A full discussion of these and related papers is given by Loeb.[72]

In 1956 Hersh and Montgomery[90] set forth their assumptions for a theory differing in two important respects from the ideas of Vick and of Gonsalves: first, surface states were not *necessarily* invoked; second, the requirement of uniformity of the Fermi level throughout both objects was emphasized as controlling the directions of charge flow. The portion of their theory involving the transfer of charge during contact is the same as the original theory of contacts set forth by Mott (see p. 168 in ref. 91) and by Fan,[92] with extension to some other dispositions of energy bands. As a matter of fact, this form of treatment, which involves little but the solution of a form of Poisson's equation in which the charge density is determined by quantum-statistical mechanics, is applicable when surface states are included. This theoretical picture has been extended and modified on the basis of further analysis by VanOstenburg and Montgomery,[93] and new experiments by Cunningham and Montgomery.[15] The outlines of this scheme will be presented, not with the conviction that it gives a complete or correct explanation, but with the hope that it will allow the reader to keep the complexity of the problems in perspective by providing a framework.

In common with most other theoretical pictures, this scheme considers static electrification to take place in two steps, the transfer of charge during contact, and the return of a portion of the charge by leakage upon separation. The leakage can take place by quantum-mechanical tunneling across very small gaps, by conduction through the atmosphere, or by conduction on or within the contacting bodies. We shall restrict our considerations to the last-mentioned type of leakage. To formalize the foregoing viewpoint, let us write q, the net charge observed upon separation to large distances of two objects initially in

[89] D. F. Arthur, *J. Textile Inst.* **46**, T721 (1955).
[90] S. P. Hersh and D. J. Montgomery, *Textile Research J.* **26**, 903 (1956).
[91] N. F. Mott and R. W. Gurney, "Electronic Processes in Ionic Crystals," p. 86. Oxford University Press, London and New York, 1st ed., 1940, 2nd ed., 1948.
[92] H. Y. Fan, *Phys. Rev.* **61**, 365 (1941); **62**, 388 (1942).
[93] D. O. VanOstenburg and D. J. Montgomery, *Textile Research J.* **28**, 22 (1958).

contact, as equal to the product of q_0, the charge transferred initially upon contact, and the fraction f of the charge retained:

$$q = q_0 f.$$

The initial charge q_0, which can be of either sign, depends upon the details of the energy-band structure of the materials in contact, and upon their geometry. Let us write q_0 as the product of a band-structure factor b, dependent only on the material and its state of strain, and a mechanical factor g, intrinsically positive and dependent only on the dimensions and form of the objects, and the normal and tangential forces between them:

$$q_0 = bg.$$

The fraction retained, f, which lies between zero and unity, should be expressible as a function of the ratio of a time interval T characterizing the rubbing process, to a time interval τ characterizing the redistribution of charge on the better insulator. The relevant variables would appear to be velocity, dielectric constant, specific resistivity, and a length or thickness characteristic of the better insulator. The equations given above are purely formal, of course. Their value will lie in part in the success encountered in predicting them independently, and in the extent to which they may be combined to give the observed charge q.

The assumptions of the present theoretical scheme are then the following.

A. It is assumed that *electrons alone are transferred.* We believe that the high mobility and the relative ease of detachment of electrons, compared with ions, make it gratuitous to postulate significant ion transfer. Ion transfer must occur at times, but we believe it is seldom controlling.[94]

B. It is assumed that the *direction of charge transfer is determined by the relative position of the Fermi levels in the two contacting materials, the transfer taking place so as to tend to equalize the levels.* The reasons for this assumption are straightforward: in equilibrium, statistical mechanics shows that the levels must be equal; before equilibrium, kinetic theory indicates that electron flow occurs in such a direction as to make the levels equal. This assumption permits determination of the sign of the band-structure factor b, and hence of q_0 and q.

C. It is assumed that *the amount of charge transferred during contact, that is, q_0, is determined by the structure of the energy bands of the bulk of*

[94] For presentation of the opposite point of view, see, e.g., Harper,[67] Loeb,[72] Böning,[95] Medley.[96]

[95] P. Böning, *Z. angew. Physik* **8,** 516 (1956).

[96] J. A. Medley, *Nature* **171,** 1077 (1953).

the contacting materials. This assumption, which is drastic and controversial, must be examined at some length. It is assumed first that enough periodicity or near-periodicity exists within the materials that the conventional concepts and language of the theory of the nearly perfect crystal can be used. Next it is assumed that, at least for a large class of experiments in which rubbing occurs with appreciable severity, the charge transfer is controlled by the properties of the interior of the crystal, rather than by the properties of the surface. In particular it is maintained that absorbed gases and other contaminants usually have little effect (unless of course a hard oxide layer or similar coating is formed), for the forces are so large in most rubbing contacts that the regions of atomic interaction are primarily beyond the original surface layers of the contacting objects. One consequence of this view is that surface states, that is, those additional states arising from the disturbance of periodicity at the interface between a finite crystal and vacuum, are not relevant as far as the initial charge transfer is concerned.[97] (They might of course be important in controlling the charge redistribution upon separation.) A set of "interface states" should exist, however, in place of the surface states. Although it is an oversimplification to neglect the effect of such states, we shall do so nevertheless. Our reasons are twofold. First, the essential processes will be the same for interior states and for interface states. The actual calculations and values of the transferred charge naturally differ, since they depend on the density of states and the depths of the levels. Second, far less is known, from both experiment and theory, about the interface states than the volume states. Indeed, studies in static electrification may prove to provide a fruitful approach to understanding these states.

Calculations based on the assumptions given above have been carried out only for the one-dimensional case. In equilibrium the laws of classical electrodynamics apply. More specifically we may use Poisson's equation in one dimension: $d^2V/dx^2 = -(4\pi\rho/\epsilon)$. Here V is the electrostatic potential, x the spatial coordinate normal to the interface, ϵ the dielectric con-

[97] For presentation of the opposite point of view see Gonsalves.[82] To his presentation we might add a conjecture by N. F. Mott in *Proc. Cambridge Phil. Soc.* **34,** 570 (1938), a paper on which "Electronic Processes in Ionic Crystals"[91] is based: "It is known from the facts about frictional electricity that charge can reside on the surface of a dielectric, and levels such as those postulated by Tamm [*Z. Physik* **76,** 849 (1932)] will probably be found necessary to account for these facts." Mott then shows that only a small fraction of the transferred charge resides on the surface levels in the case of insulators in contact with metals when the work function of the metal is greater than the depth of surface states by at least several times kT. On the other hand, an appreciable fraction of the charge may be held in the surface levels when the work function is less than this depth.

stant, and ρ the volume charge density. The charge density is a function of the potential, the form of the function being determined by the details of the structure. This function can be written as the product of a density of states times the Fermi-Dirac statistical factor giving the probability of occupation. The resulting differential equation, which has the form $V'' \sim f(V)$, always can be solved by an elementary technique, although the solution can seldom be given in closed form. The constants of integration are fixed by the requirements that the field strength vanish at infinity, since the system remains neutral, and that the field strength be continuous across the interface, since the charge density is finite there. One other requirement must be fulfilled in accordance with the requirements of thermodynamics: in equilibrium, the Fermi level is uniform throughout the system. Once the solution for V is obtained, the charge density ρ and its integral, the total charge, follow immediately.

In the case of metal-insulator contacts, four qualitatively different cases arise, whereas there are two for insulator-insulator contacts. All these cases are treated quantitatively by VanOstenburg and Montgomery.[93] The nature of their results can be gleaned from the insulator-insulator contact. Figures 16a and 16b show schematic energy-band diagrams for the two cases. Before contact the levels are constant up to the edge of the crystal. After contact, the quantity $-eV(x)$, the potential energy of an electron within the crystal, will change as a result of the transferred charge, the levels shifting accordingly. The total shift between the objects is $e(\Delta V_1 - \Delta V_2)$, where ΔV_1 is the shift in the electrostatic potential in the substance on the left and ΔV_2 is the corresponding quantity for the substance on the right. Whereas a single parameter, the work function ϕ_0, is relevant in determining the direction of electron transfer in metals, we must, in insulators, be concerned with both the energy χ_0 released when an electron at rest outside the crystal is taken into the lowest level in the lowermost nearly empty band, and the energy υ_0 necessary to remove an electron from the highest level of the uppermost nearly filled band.[98] In the figures, the energy parameters of the substance on the left have the subscript 1, those of the substance on the right the subscript 2. The subscript 0 indicates a value for a parameter before contact. The parameters μ_1, μ_2, γ_1, γ_2 have been introduced for the quantitative treatment.

Figure 16a represents the contact between two insulators. In this case

[98] Just as the work function ϕ_0 for a metal is, in a sense, the solid-state analog of either the electron affinity A or the ionization potential I of a free molecule, χ_0 is the analog of A and υ_0 is the analog of I. We hesitate to give these solid state quantities the same name as their molecular counterparts, and shall refer to them with symbols only.

the energy-band structure is such that electrons can escape from the left-hand insulator to the right-hand one only by thermal agitation from a filled band into an empty band lying above it. Hence, only a small amount of charge is lost from a large spatial region by the left-hand insulator; moreover, an equal charge is gained by the right-hand insulator within a large spatial region. If the energy gap between the top of the nearly filled band on the left and the bottom of the nearly empty one on the

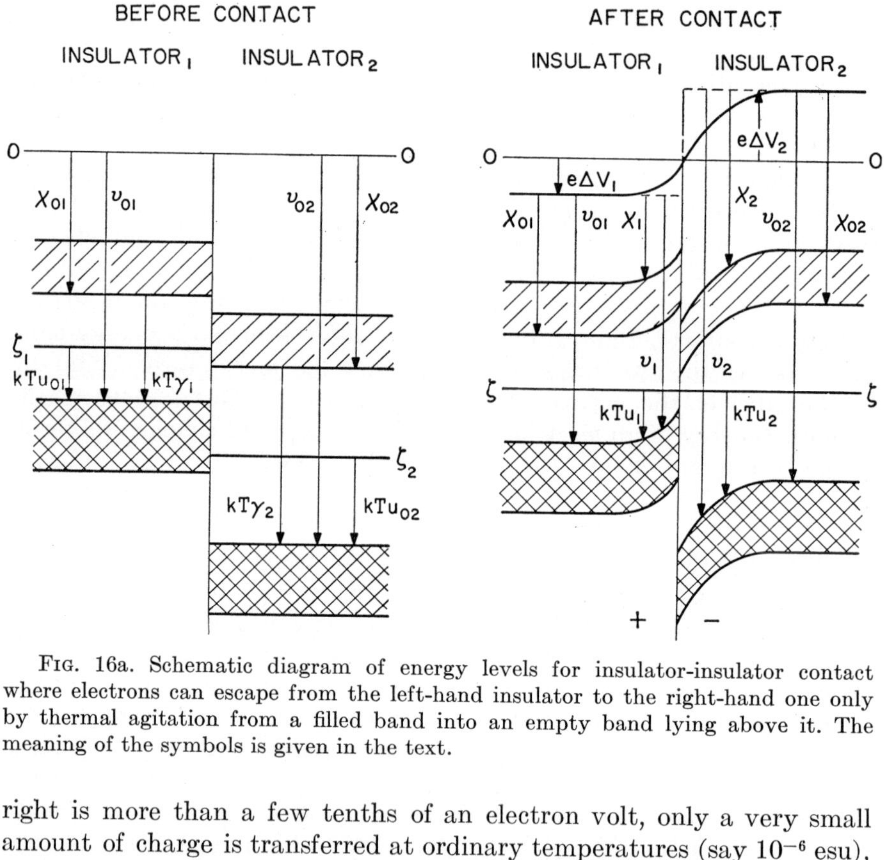

FIG. 16a. Schematic diagram of energy levels for insulator-insulator contact where electrons can escape from the left-hand insulator to the right-hand one only by thermal agitation from a filled band into an empty band lying above it. The meaning of the symbols is given in the text.

right is more than a few tenths of an electron volt, only a very small amount of charge is transferred at ordinary temperatures (say 10^{-6} esu), and the thickness of the layer is almost macroscopic (say 10^5 A).

Figure 16b represents the contact between two insulators in which the energy-band structure is such that electrons can spill from the filled band of the left-hand material into the empty band of the right-hand material. A large amount of charge is lost from a small spatial region by the left-hand insulator and an equal charge is gained by the right-hand material in a small spatial region. If the energy gap between the top of the nearly filled band on the left and the bottom of the nearly empty one on the right is more than a few tenths of an electron volt, charge densities comparable with those obtained in metal-metal contacts are obtained ($\sim 10^4$ esu) in a region of comparable thickness (~ 1 A).

Until additional experimental data are available, the main conclusion to be drawn from this section is that even at ordinary temperatures different band structures permit equilibrium-charge transfers from vanishingly small values to exceedingly large amounts. The transferred

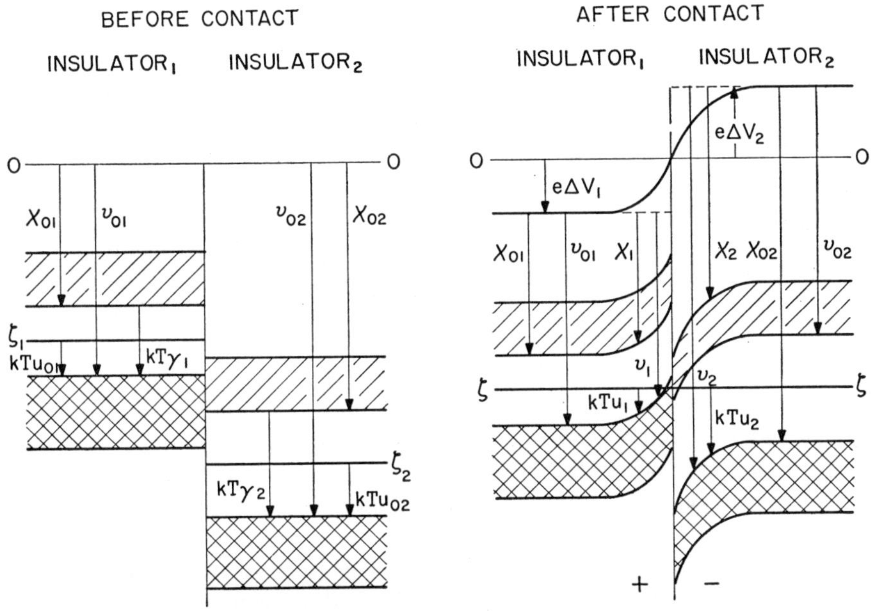

FIG. 16b. Schematic diagram of energy levels for insulator-insulator contact where electrons can spill freely from the filled band of the left-hand material into the empty band of the right-hand material. The meaning of the symbols is given in the text. [D. O. VanOstenburg and D. J. Montgomery, *Textile Research J.* **28,** 22 (1958).]

charge may be of either sign. For certain structures, there is no need to invoke surface states or rubbing. As we shall see later, however, it is likely that local heating (or ionic transfer, perhaps) must be invoked to explain the observed charges in cases in which the energy gaps must be so large that negligible charge would be transferred by contact alone.

D. It is assumed that *the initial charge q_0 is proportional to the area of "true contact," that is, the area of interpenetration of atomic fields.* Almost always, this area must be inferred from the knowledge of other factors, such as the normal force, the length of stroke, and the elastic or plastic properties of the medium. The area of true contact may be nearly equal to the area of apparent contact, or only a small fraction of it, depending on the nature of the materials.

E. It is assumed that *the transferred charge remains localized at the region of contact in insulators after separation, although it is not localized in metals.* This assumption is meaningless unless a time scale is specified. Except during a very short period following contact, the charge in insulators is assumed to remain localized near the place of contact for periods long compared with the time required for charge transfer. This assumption is *ad hoc* as far as the band picture discussed above is concerned, of course. However, experiments carried out long ago established that charge can be localized on, and probably in, insulators. According to the conventional band picture, however, electrons inserted in an empty band, or holes left in a filled band, are free to move throughout the crystal, once the attractive forces from the charge of opposite sign on the other material are removed. Gonsalves[82] has used this argument to formulate a theory based on surface states. Clearly some mechanism of localization must be postulated, and it is natural to assume it is associated with imperfections on or near the surface. Several mechanisms for trapping charge are recognized (cf. discussion in Mott and Gurney,[91] pp. 86–88, 124–131; Seitz,[99] pp. 325–326). We do not wish at this point to suggest any definite mechanisms, but we do assert that charge can be immobilized in insulators by some means.

F. It is assumed that, *when the surfaces are separated after contact, the charge initially transferred is reduced by tunneling through the gap formed by the separating surfaces, or by flow along the materials for a very short time in the immediate neighborhood of the contact.* With respect to the tunneling, quantitative discussions are available for metal spheres which are placed in contact and then separated. Harper,[10] in a careful study, has found excellent agreement between theory and experiment. The insulator-metal case has not yet been analyzed completely. It seems clear that the fraction of charge remaining will be reduced from the metal-metal case, because the inability of the charge which remains on the insulator to redistribute itself will maintain the driving field during the separation.

Considerable discussion of the reverse flow is necessary. Let us restrict ourselves to cases in which appreciable breakdown of the atmosphere does not occur. Return flow then will occur only within or on the bodies. It is well known that the solution of Maxwell's equations *within the volume* of an isotropic homogeneous medium of electrical conductivity σ and dielectric constant ϵ leads to the following expression for the volume

[99] F. Seitz, "The Modern Theory of Solids." McGraw-Hill, New York, 1940.

charge density ρ as a function of time t in the quasi-static case:

$$\rho(t) = \rho(0)e^{-t/\tau}.$$

Here τ, defined as ϵ/σ and frequently called the *relaxation time*, is the time required for the charge to fall to $1/e$ times its initial value. In problems of static electrification, however, we usually deal with charges placed *on the surface* of an insulator, not within the interior. Hence a complicated boundary-value problem arises when we try to find how the surface charges decay in time. In general, the charge decay cannot be characterized by a unique relaxation time. In certain cases, nevertheless, there exists an average relaxation time which is not radically different from that characterizing the redistribution of charge within the interior of a dielectric medium. Appendix B presents the analysis for the distribution of charge which is placed uniformly in length along an infinite cylinder initially. In this special case, the charge relaxes from its initial value to its final value with a single relaxation time. This time, however, is somewhat different from that for the interior of the medium. The time dependence of the charge cannot be described simply for other geometries and less simple initial distributions.

In the cases under consideration in static-electrification experiments, the initial distribution is complicated in both space and time. Despite this fact, we have tried to understand the gross effects by assuming that the behavior can be analyzed in terms of the ratio of a time interval characteristic of the rubbing process and an average relaxation time characteristic of the material. The former interval is given by the time required for the traversal of a characteristic distance l at a velocity v, namely $T = l/v$. The latter interval will be approximated by the relaxation time $\tau = \epsilon/\sigma$ for a homogeneous medium. Hence the characteristic parameter is $T/\tau = (l/v)/(\epsilon/\sigma)$. The same result can be obtained immediately by the standard methods of dimensional analysis if the relevant variables are taken to be ϵ, σ, l, and v. There is ambiguity in this approach in the sense that no indication is given of the way in which the dielectric constant of the surrounding medium enters into combination with that of the charged material. The analytical solution for the special case of the infinite cylinder given in Appendix B shows that the proper combination is the sum of the dielectric constants for the medium and for the dielectric material. The dielectric constant may often be neglected when air is the medium. For other geometries, the proper combination appears to be the sum of the dielectric constant of the medium and an integer greater than one times the dielectric constant of the material. *A fortiori*, the dielectric constant of air may often be neglected.

All that the foregoing argument establishes is that the parameter

T/τ is likely to be useful and that the fraction retained should be examined to see if it can be expressed as a function of this parameter. We have been unable to obtain theoretical insight concerning the form for this function. The primitive assumption that $f \sim e^{-T/\tau}$ fails for unlubricated contacts, even though it has been found adequate for the action of an antistatic coating (Medley[17]).

G. Our final assumption is that *the charge remaining after separation is not neutralized significantly by ions resulting from atmospheric breakdown, in a substantial portion of static-electrification processes.* From one point of view, this statement is not an assumption, but rather a restriction on the range of applicability of the theoretical scheme. The study of static electrification when breakdown occurs requires the study of gaseous discharges as well as the solid state, and we should like to restrict ourselves to the less complex situation as long as possible. It may be maintained, however, that breakdown essentially always occurs. Assumption F then becomes something more than a restriction. The only direct evidence that breakdown does not occur appears to be that found in the experiments of Cunningham,[15] who placed a photomultiplier just below the region of contact and separation. The lower limit of sensitivity was not determined, however, and it is an open question as to whether *very small* discharges did not occur. There is also the question whether an observed pulse could not occur as the result of a process other than a gaseous discharge.

V. Status of the Problem

5. Comparison of Theory and Experiment

a. Qualitative Predictions

According to the theory outlined in the previous section, the direction of charge transfer depends only on the relative positions of the Fermi level. Hence a unique triboelectric series should exist for all materials. This deduction appears to be valid if the conditions of rubbing and the state of the material are made specific enough. First of all, the contacting region for each specimen must be sufficiently homogeneous for it to be considered as constituting one material. We believe that many more materials satisfy this condition than are commonly thought to do so in the case of contacts severe enough to cause surface penetration. It must be realized, however, that the position of the Fermi level depends not merely on chemical composition, but also on temperature and on molecular structure. As a result of possible inhomogeneities in chemical composition and molecular structure, we should expect consistent behavior in direction of charge transfer from one rub to another, or from one sample to another, only if the composition and the structure are such that the Fermi levels have a reasonably wide separation in isolation. In point of fact, except for rather special cases such as the cyclic relation found by Shaw and Jex,[60] investigators agree that a triboelectric series exists. There is a rather general agreement concerning the order of items in the series, although there are many exceptions in detail. The exceptions cannot be taken too seriously at this time, however, for the methods of rubbing are different from one experimenter to the next, and many of the materials have been of indefinite or unknown composition, e.g., rubber, glass, paper. Moreover, temperature may cause a change in position in the series. Similarly, a change in relative humidity may move a material up or down in the series.

One would like to be able to predict the position of a given substance in the series. Estimates of the position of energy levels, based on either theory or experiment, are scarce even for simple substances. In the case of certain fiber-forming polymers we have speculated along the following lines. The Fermi energy ζ of an insulator is about half the sum of χ_0 and υ_0. In the absence of knowledge concerning the energy levels in the polymer, we should like to estimate χ_0 and υ_0 from the electron affinity A and the ionization potential I, respectively, of the constituent molecules. The justification of this procedure is that a long chain of atoms is connected by primary valence bonds having large penetration of electron charge densities in the direction of orientation of a fiber, that is, along the axis of the fiber. Transverse to the axis of the fiber, the cohesive forces consist of secondary valence bonds between atoms of different chains, and there is only slight interpenetration of electron charge clouds. Thus the properties of the constituent atoms of a chain are much affected by being built into a chain (primary valence-bond formation), whereas the properties of the chains themselves are affected little by being built into an aggregate (secondary valence-bond formation). Consequently there is hope that the ease of losing or gaining electrons is controlled largely by the properties of the individual constituent chains, and not by the manner in which they are built into the aggregate. Accordingly the value of χ_0 for a fiber, which is an aggregate of long-chain molecules, would nearly equal A, the electron affinity, for one of the long chain molecules, according to this picture, and υ_0 would nearly equal I, the ionization potential of one of the molecules. In other words, the filament may be thought of as a molecular crystal, and the loss or gain of electrons can be regarded as controlled by the molecular properties. This behavior, is of course, completely different from that of metals, ionic crystals, and covalent crystals.

The dearth of knowledge of ionization potentials and electron affinities prevents direct application of these ideas to polymers. By extrapolating the ionization potentials of homologous series, based on the monomers of addition polymers, Hersh and Montgomery[18] were able to obtain an approximate correlation for insulators. This procedure is not highly reliable for addition polymers, and is likely to be completely invalid for condensation polymers. Further data will test the usefulness of this approach.

Concerning metals, Hersh and Montgomery[18] made preliminary experiments with four metals and three insulators. They identified the thermionic or photoelectric work function with the Fermi energy of the metals. They found that a two-parameter set of levels (χ_0 and υ_0) which permitted a correct description of the experimental results could be ascribed to each of the insulators. Only one of the insulators, polyethylene, was an addition polymer. The value of υ_0 assigned in this scheme could be chosen to be the same as that for the insulator-insulator experiments.

The experimental and the theoretical basis for the foregoing discussion are so tenuous that they cannot be taken seriously at the present time. The arguments do suggest, nevertheless, that it may be possible to characterize the static-electrification behavior of metals semiquantitatively by one parameter, and that of insulators by two.

b. Quantitative Predictions

As far as quantitative predictions are concerned, the theoretical basis is too shaky and the experimental data are too scanty to make an extensive comparison of theory and experiment worth while. We shall examine only two quantitative predictions, restricting ourselves in each case to the simple case of two cylinders which cross at right angles, and are rubbed in the so-called 45°–45° manner (cf. Fig. 8), over distances large compared with the diameters of the cylinders. The examination will be confined to the steady-state values of charge, that is, to the nearly steady, limiting value attained after a few to several score rubs. The justification for studying this value lies partly in its good reproducibility, and partly in the results of studies on contacts by others. When actual surfaces are in contact at light and moderate loads, the load is supported more frequently by many small contacts than by a few large[100] areas. It appears from experiments on wear that the multiple contacts are the result of elastic rather than plastic deformation, and that each protuberance may be rubbed many times without damage.[101,102] It is especially significant for the present study that a protuberance may be

[100] J. Dyson and W. Hirst, *Proc. Phys. Soc.* **B67,** 309 (1954).
[101] J. T. Burwell and C. D. Strang, *J. Appl. Phys.* **23,** 18 (1952).
[102] J. F. Archard and W. Hirst, *Proc. Roy. Soc.* **A238,** 515 (1957).

deformed plastically at its first encounter when materials of comparable hardness are rubbed against each other, and then may relax elastically. At subsequent encounters, the protuberance would bear the same load by elastic deformation.[103] Hence a steady state which will vary only secularly as wear sets in will be reached.

At heavier loads, the number and the size of the regions of contact increase. In the case of materials having a low elastic modulus, such as organic polymers, moderate loads may suffice to press the small local asperities into the general surface of contact. Less is known about the damage suffered under these conditions, but it seems reasonable to suppose that the regions of contact deform elastically after a brief running-in period, with only minor wear effects on each rub.

We wish to examine the dependence of q, the charge generated, on S, the true area of contact swept out. The relation of the latter to A_{true}, the true area of contact, depends on the mechanism of contact. This area is controlled by the normal force F, and may range from a small fraction to nearly the whole of the macroscopic apparent area of contact A_{apparent}.

Consider first the case of loads large enough that the local asperities are pressed flat, so that the true area of contact approaches the apparent area. The deformation becomes elastic in the steady state, and hence $A_{\text{true}} \sim F^{\frac{2}{3}}$, according to the classical work of Hertz (see, e.g., Timoshenko[104]). Experimentally, Pascoe and Tabor[105] found the value of 0.74 for the exponent in the case of polished polymethylmethacrylate cylinders. In similar work Archard[106] obtained a value 0.72. The apparent area was taken as $\frac{1}{4}\pi d^2$ in these experiments. Here d is the width of the track. The frictional force showed the same dependence on the normal force as did the apparent area, whence Archard concludes that $A_{\text{apparent}} = A_{\text{true}}$ for this case. (With rougher surfaces, this relation was observed only at high normal forces.) Then the width of the area of the contact varies as the square root of the apparent area, whence S, the area swept out, is proportional to L, the length of path, times $F^{\frac{1}{3}}$, that is,

$$S \sim LF^{\frac{1}{3}}.$$

On the other hand, the contact will occur only at protuberances, at lighter loads or with rough surfaces. Archard[106] asserts that as the complexity of the model increases, that is, as the actual surface is represented by local irregularities superimposed on other irregularities, and so on, the

[103] J. F. Archard and W. Hirst, *Proc. Roy. Soc.* **A236**, 397 (1956).

[104] S. P. Timoshenko, "Strength of Materials," Chapter VII. D. Van Nostrand Co., New York, 1st ed., 1930, 2nd ed., 1941.

[105] M. W. Pascoe and D. Tabor, *Proc. Roy. Soc.* **A235**, 210 (1956).

[106] J. F. Archard, *Proc. Roy. Soc.* **A243**, 190 (1957).

number of individual contacts becomes more nearly proportional to the load, and their size depends less on it. Under such conditions, $A_{true} \sim F$, the width of each contact region remaining nearly constant under load. Then S, the area swept out, is nearly proportional to L times the number of contacts, which is proportional to F in turn. We then have

$$S \sim LF.$$

In cases of intermediate loading and roughness, both the number and the size increase with load. Hence S will vary at a rate intermediate between those of the cases given. We may summarize the results by writing

$$S \sim LF^n$$

with n decreasing from unity to one-third with increasing F, the smaller limiting value being reached earlier as the elastic modulus and the roughness decrease.

According to the theoretical scheme under discussion, the charge transferred is proportional to the area swept out. As stated earlier, Hersh and Montgomery[18] found that the charge is proportional to the length L. Only graphical results of the normal force F were presented. No attempt was made to get an analytical relation between F and q. Examination of their data shows that the charge q rises with increasing F, reaching a limiting value in some cases. It is difficult to state that these exploratory results concerning the variation of q with F either substantiate or nullify the assumption that q is proportional to area swept out. To decide this question more definitely, Cunningham and Montgomery[15] carried out experiments with an improved apparatus, obtaining the results shown earlier in Fig. 15. A logarithmic plot of their data is shown in Fig. 17. Here it has been attempted to remove the dependence on diameter by dividing the square of the filament diameter into the normal force, as will be discussed subsequently. The data do not lie far from a straight line for a given diameter except at light loads, in which case the contact is irregular as a result of bouncing of the yoke. Perhaps all the data may be fitted by a single line as shown. This line has been drawn with a slope equal to 0.5. A less steep line, having slope 0.3 or 0.4, would represent some of the data for a given diameter better. Indeed, additional experiments gave sets of data which could be fitted with lines of slope 0.3–0.5. The theoretical prediction for the slope at moderate and heavy loads is 0.33, or a little higher, in accordance with the relation $A \sim F^{0.72}$ found by Archard.[106] The agreement then is acceptable tentatively for the dependence of charge on length of stroke and on normal force. The slope should go to unity at light loads.

The observed dependence of charge on filament diameter, on the other hand, cannot be reconciled with this hypothesis. The theoretical expression for the apparent area of contact between the two filaments is difficult to obtain, but it is clear at least that the area cannot decrease with increasing diameter. The experiments, however, show the reverse trend. One might speculate that the increase in average pressure with decreasing diameter at fixed load causes an increase in temperature, thereby increasing the ease of charge transfer. We prefer an alternative explanation, namely, that the chemical nature of the nylon filaments differed from samples of one diameter to those of another, since the

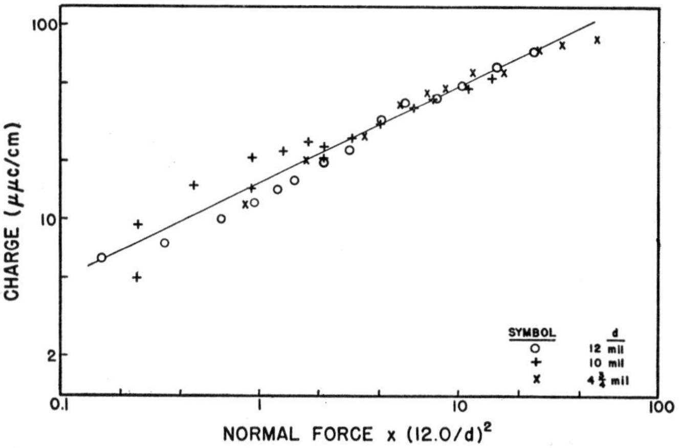

FIG. 17. Logarithmic plot of data of Fig. 15. To remove the dependence of the charge on filament diameter d, the force has been multiplied by the normalizing factor (12.0 mils/d mils)^2. [R. G. Cunningham and D. J. Montgomery, *Textile Research J.* **28,** 971 (1958).]

samples were not made from the same melt. In any event, the experiments should be repeated with filaments from the same melt and of same draw ratio, but having diameters which vary over a wide range.

The second prediction we wish to examine is the effect of velocity on the charge generated. According to the theory presented previously, q can be written as $q_0 f$, with f equal to some function of $\epsilon v/\sigma l$. Here ϵ and σ refer to the material of higher resistivity. A more sensible assumption (see Appendix B) is to replace ϵ by a function of the dielectric constants of the two materials and of air; however, it is unlikely that this refinement is essential at this stage. Cunningham and Montgomery[15] measured the charge as a function of velocity at various temperatures and relative humidities (cf. Fig. 12), with the idea that the principal effect of the change in ambient conditions would be a change in volume resistivity. As shown in Fig. 18, they were able to fit their data by taking $f = [1 - \exp(-\epsilon v/\sigma l)]$, with some uncertainty in the independent evaluation of ϵ and σ. A theoretical explanation of this form of the function was not obtained. The characteristic length happened to be about the same as the filament diameter, but this presumably is fortuitous. Additional experimental data are needed to ascertain if this relation has any generality. As the equation stands, its utility in applications

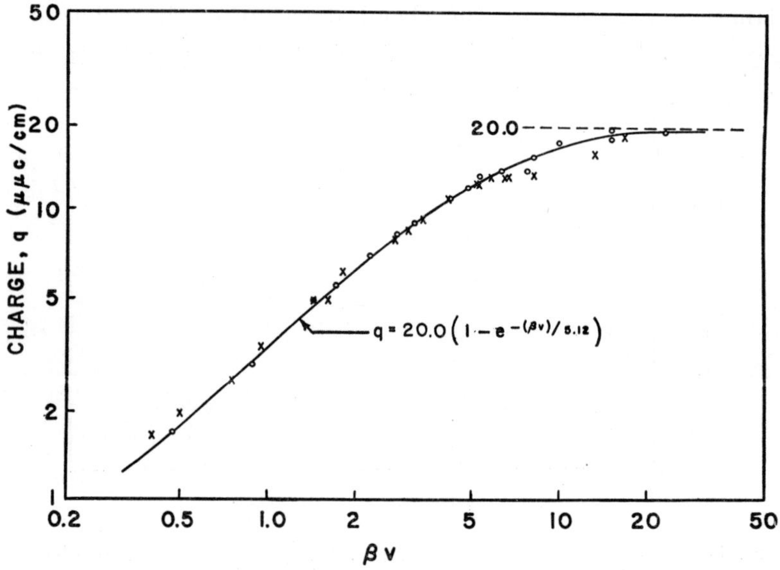

FIG. 18. Logarithmic plot of data of Fig. 12 (crosses) together with data for a similar pair of filaments (circles). The solid curve is calculated according to the empirical relation $q = q_0[1 - \exp(-\beta v/\beta_0)]$. A different value of β has been chosen for each curve of Fig. 12 and of the corresponding figure for the other pair of filaments. The values $q_0 = 20.0$ μμcoul/cm and $\beta_0 = 5.12$ cm/sec have been selected to give the best fit. [R. G. Cunningham and D. J. Montgomery, *Textile Research J.* **28,** 971 (1958).]

rests on the possibility that it can be used to show how changes in resistivity may be compensated by changes in velocity. It is useful in fundamental studies because it permits us to separate the q_0 and f if we work at sufficiently high resistivity or velocity that f becomes unity.

6. Scientific Consequences of the Theory

(a) The merits of the scheme proposed are that it permits a partial separation of the effects of the relevant variables, and that it suggests additional experiments capable of testing its validity and generality. Although the theory is simplified drastically, it does not lead one into the misconception that the processes are inherently simple.

(b) The modification and expansion of the scheme which results from attempts to verify the theory may permit it to repay a part of the debt to other fields of investigation. Static electrification may become an extremely sensitive probe for studies in surfaces and solids. For example, studies in this field eventually may give information concerning energy states on the surface and in volume. The true area of contact may be estimated eventually from charge generated upon rubbing. Slight effects of chemical or mechanical treatments may become detectible readily.

7. Technological Consequences of the Theory

An understanding of the fundamentals of charge production and dissipation undoubtedly would advance technology in several industries. In other regions of technical activity, increases in understanding have led not only to improvements in present processes, but to the introduction of new processes as well. Static electrification should prove to be no exception. Its exploitation may lead to developments in related phenomena, such as pyroelectricity and piezoelectricity.

Currently the methods used for producing desirable charge on insulating solids are almost completely empirical.[107] Electrostatic generators, particle sorters, electrostatic copying devices, paint sprayers, insecticide dusters, might be developed and improved by a better understanding of the mechanism of charge transfer. The present methods of avoiding the formation of unwanted charge[108] on conductors which are rubbed together are also empirical, depending almost inevitably on the reduction of resistivity or speed. There are many processes in which the reduction of neither is desirable (for example in processing of photographic film or synthetic fibers). Methods for dissipating unwanted charge[108] depend on increasing the conductivity or on supplying neutralizing ions from the surrounding atmosphere. The former may be impossible or inconvenient, and the latter may be deleterious to the material (e.g., in the case of photographic emulsion on films) or may endanger personnel (e.g., by action of high voltage, directly on personnel or on inflammable substances; or by biological effects of ionizing radiation). Neither method is

[107] For illustrative cases, see papers in *Brit. J. Appl. Phys.* Suppl. 2, Static Electrification (1953).

[108] For surveys of static elimination, see Loeb,[72] Chapter 5, Section 3; Henry[109]; and other papers in ref. 107.

[109] P. S. H. Henry, *Brit. J. Appl. Phys.* Suppl. 2, S6 (1953).

likely to be efficient. A knowledge of the mechanism of charge immobilization in the solids involved might suggest more effective and attractive methods of dissipating charge. For example, visible or infrared radiation, possibly in conjunction with weak electromagnetic sweeping fields, might release bound charge readily.

8. Directions of Future Work

Many possible lines of activity suggest themselves. In the absence of a definitive theory, it is difficult to pick out the topics likely to be of greatest significance. Subjects for investigation, together with comments concerning their suitability for attack at this time, are listed below.

(*1*) *Completion and confirmation of results already discussed.* Of the variables listed earlier, perhaps the ones most deserving of additional attention are the geometric ones, namely shape and size, and the chemical composition.

(*2*) *Extension to new kinds of materials.* New classes of substance, some better understood than good insulators, might provide good tests of current theory, and suggest a basis for future theory. Semiconductors and ionic crystals are examples of materials that could be investigated after higher-speed apparatus has been constructed.

(*3*) *Extension to very high vacua and to very low temperatures.* Some simplification in interpretation should result from exclusion of atmosphere: adsorption of atmospheric gases would not influence the results; higher field strengths could be obtained before breakdown and allow a wider range of charge values for testing the theory. There is almost no guide concerning new phenomena to be expected at very low temperatures. Static-electrification experiments at either high-vacuum or low-temperature are difficult; substantial investment in time and equipment will be required for them.

(*4*) *Effect of type of contact.* The number of types of contact studied has been small, even with given geometry, and the results are incomplete. For example only the extreme cases of complete symmetry (45°–45°) and high asymmetry (90°–0°, 0°–90°) have been investigated with the cylinder-cylinder geometry. Only rudimentary studies exist in the case of sphere-sphere, sphere-cylinder, sphere-plane, cylinder-plane, and plane-plane geometries.

(*5*) *Initiatory and long-term behavior.* For scientific as well as technological purposes, the initiatory behavior is important. Experiments can be designed to study the initiatory behavior if agreement can be reached on what constitutes the virgin state for a material. Filaments would make excellent specimens, a new portion being unwound from a spool for each test. Such experiments should lead to fundamental advances, if it happens that reproducibility from sample to sample is high. The long-term behavior on the other hand, seems to involve many secondary effects associated with wear phenomena. The study of this behavior is important in itself, of course, but its greatest use for the immediate future would appear to be in delimiting the area of steady-state behavior.

(6) *Effect of coatings.* The effect of deliberate introduction of inhomogeneity by coating a material with a lubricant or an antistatic agent has been the object of many technological studies and of very few scientific ones. It seems reasonable to give this topic only minor scientific attention until a better insight has been obtained into simpler cases.

(7) *Related subjects.* It will be surprising if help in elucidating static phenomena does not come from the related fields of friction, wear, electrical contacts, gaseous discharges, and contact potentials.

(8) *Theoretical considerations.* If one can maintain the conception that the net charge transferred is the product of two factors—an initial equilibrium charge and the fraction retained—the way is open for a partial separation of the effects of the numerous variables. The method of calculating the initial charge appears clear in principle if the collective electron scheme is applicable. It is possible, on the other hand, that a calculation which starts from atomic or molecular orbitals may be fruitful, at least in the case of highly-insulating molecular crystals. With the collective electron scheme, the charge distribution between a metal and a vacuum has been estimated theoretically in only a few cases, and under severely restrictive assumptions. The distribution between two semi-infinite periodic structures in contact along a plane has not been treated satisfactorily. The problem grows appalling when the geometry departs from this simplest case. Until further strides are made in solid state theory, it appears that simplifying devices of a radical nature will have to be sought.

With respect to the problem of determining the fraction of charge retained, we hardly know where to begin. Results from classical electrostatics are valuable, but in only a small part of the problem. The fundamental difficulty is that we do not understand in any detail either the mechanical or the electrical processes involved in the contact and separation of two solid surfaces.

Such is the state of static electrification today. The prospects for the future are bright, for we possess experimental techniques capable of yielding laws, and theoretical concepts capable of unifying the laws. All that is needed is the proper attention.

Appendix A. Chemical Nature of Certain Polymers

Name	Composition
Acetate	Cellulose acetate
Cordura rayon	Regenerated cellulose, viscose process high-tenacity (DuPont)
Dacron	Copolyester of ethylene glycol and terephthalic acid (DuPont)
Dynel	Vinyl copolymer of acrylonitrile (40%) and vinyl chloride (60%) (Union Carbide and Carbon)
Fiberglas	Glass (Owens-Corning)
Lucite	Vinyl polymer of methyl methacrylate (DuPont)
Nylon	Copolyamide of hexamethylene diamine and adipic acid (6,6)
Orlon	Vinyl polymer of acrylonitrile (DuPont)
Polyethylene	Vinyl polymer of ethylene
Polyvinyl alcohol	Vinyl polymer of vinyl alcohol (made by saponification of polyvinyl acetate)
Polyvinyl chloride	Vinyl polymer of vinyl chloride
Saran	Vinyl copolymer of vinylidene chloride and vinyl chloride
Teflon	Vinyl polymer of tetrafluoroethylene (DuPont)
Velon	Vinyl copolymer of vinylidene chloride and vinyl chloride (predominantly vinylidene chloride) (Dow)
Viscose	Regenerated cellulose (viscose process)

Appendix B. Redistribution of Charge on the Surface of a Dielectric[110]

As is well known, the volume charge density in the *interior* of a homogeneous isotropic medium decays exponentially in time. This relation follows from the Maxwell equations

$$\nabla \cdot \mathbf{D} = \rho \qquad \text{and} \qquad \nabla \times \mathbf{H} - \partial \mathbf{D}/\partial t = \mathbf{j}$$

and the constitutive relations

$$\mathbf{D} = \epsilon \mathbf{E} \qquad \text{and} \qquad \mathbf{j} = \sigma \mathbf{E}$$

in the quasi-stationary case. For then

$$\nabla \cdot \mathbf{j} = -\nabla \cdot \partial \mathbf{D}/\partial t.$$

[110] The treatment in this Appendix is based on an analysis by Dr. W. G. Hammerle.

But

$$\nabla \cdot \mathbf{j} = \nabla \cdot \sigma \mathbf{E} = \sigma \nabla \cdot \mathbf{D}/\epsilon = (\sigma/\epsilon)\rho.$$

Further,

$$\nabla \cdot \partial \mathbf{D}/\partial t = \partial \nabla \cdot \mathbf{D}/\partial t = \partial \rho/\partial t,$$

which has the solution

$$\rho(x,y,z,t) = \rho(x,y,z,0)e^{-t/\tau}$$

where

$$\tau \equiv \epsilon/\sigma.$$

In particular, we see that if the charge is zero at any point initially it remains zero for all t.

On the *surface* of a dielectric medium, however, this conclusion does not hold. To demonstrate the course of the redistribution of charge quantitatively, we shall work out the case of a long circular cylinder with a surface charge distribution that is independent of the distance along the cylinder. This problem does not illustrate the full complexity of the redistribution process, but it has the advantage of yielding the result in closed form.

Consider an infinitely long circular cylinder of radius $r = a$, conductivity σ, and dielectric constant ϵ, imbedded in an infinite medium of conductivity $\sigma_0 = 0$ and dielectric constant ϵ_0. Suppose that the surface charge is distributed independently of distance z along the axis. For convenience, suppose that the surface charge density $\omega(\theta)$ is even in the parameter θ, the azimuthal angle about the axis of the cylinder, and, in view of the differential equation to be solved, that ω is expanded in the Fourier series $\omega(\theta) = C_0 + \Sigma C_n \cos n\theta$. All sums are to be taken from $n = 1$ to $n = \infty$. Actually ω depends on time t as well as on angle θ; this dependence will be indicated explicitly where it is essential.

Application of Maxwell's equations yields Laplace's equation for the potential ϕ, defined through the relation $\mathbf{E} = -\nabla\phi$, for all points not on the boundary. By standard techniques, the solution for ϕ may be written as

$$\phi(r,\theta) = \begin{cases} \Sigma A_n r^n \cos n\theta & (r < a) \\ B_0 \ln (r/a) + \Sigma B_n r^{-n} \cos n\theta & (r \geq a). \end{cases}$$

The constants B_0, A_n, and B_n are to be determined by the boundary conditions and by the charge distribution. The requirement that ϕ be continuous at $r = a$, that is, $\phi(r)\Big|_{a+0} = \phi(r)\Big|_{a-0}$, shows that $A_n = B_n a^{-2n}$.

The requirement that the discontinuity in the normal component of the electric displacement at the surface be equal to the surface charge density, that is,

$$\epsilon_0(-\partial\phi/\partial r)\Big|_{a+0} - \epsilon(-\partial\phi/\partial r)\Big|_{a-0} = \omega(\theta),$$

leads to the relation

$$B_n = C_n a^{n+1}/n(\epsilon_0 + \epsilon).$$

Finally, as $r \to \infty$ the dominant term in ϕ is $B_0 \ln (r/a)$; but by direct application of Gauss's law,

$$\phi \to -(q/2\pi\epsilon_0) \ln (r/a) = -(2\pi a C_0/2\pi\epsilon_0) \ln (r/a)$$

where q is the total charge per unit length of the cylinder. The expression for the potential now is

$$\phi(r,\theta) = \begin{cases} \Sigma a C_n[(r/a)^n/n(\epsilon_0 + \epsilon)] \cos n\theta & (r < a) \\ -a(C_0/\epsilon_0) \ln (r/a) + \Sigma a C_n[(a/r)^n/n(\epsilon_0 + \epsilon)] \cos n\theta & (r \geq a). \end{cases}$$

Since the charge distribution changes in time, the coefficients C_0 and C_n change in a way determined by the application of Ohm's law $\mathbf{j} = \sigma\mathbf{E}$. Consider a small pillbox-shaped volume of height Δr containing the surface $\Delta\mathbf{S}$, with ends parallel to the surface and hence normal to $\Delta\mathbf{S}$. By conservation of charge,

$$\begin{aligned} \int(\nabla \cdot \mathbf{j} + \partial\rho/\partial t)\, dV &= \int \mathbf{j} \cdot d\mathbf{S} + \partial/\partial t \cdot \int \rho\, dV \\ &= -\bar{\jmath}_n \Delta S + \text{current through sides} + \partial/\partial t \cdot \bar{\rho}\Delta r \Delta S = 0. \end{aligned}$$

In the limit of vanishing Δr, with $\rho\Delta r \to \omega$, followed by $\Delta S \to 0$, we obtain $j_n = \partial\omega/\partial t = \sigma E_n = -\sigma\partial\phi/\partial r\Big|_{a-0}$. Now

$$\partial\omega/\partial t = \dot{C}_0 + \Sigma\dot{C}_n \cos n\theta$$

and

$$-\sigma\partial\phi/\partial r\Big|_{a-0} = -\Sigma C_n[\sigma/(\epsilon_0 + \epsilon)] \cos n\theta.$$

Since the functions $\cos n\theta$ are linearly independent, their coefficients may be set equal separately. We then get

$$\begin{aligned} \dot{C}_0 &= 0, & C_0 &= \text{constant} \\ \dot{C}_n &= -\tau C_n, & C_n(t) &= C_n(0)e^{-t/\tau}, \end{aligned}$$

where we have defined

$$\tau \equiv (\epsilon_0 + \epsilon)/\sigma.$$

Hence

$$\omega(\theta,t) = C_0 + \Sigma C_n(0)e^{-t/\tau} \cos n\theta = \omega(\theta,0)e^{-t/\tau} + C_0(1 - e^{-t/\tau}).$$

But

$$C_0 = \omega(\theta, \infty) = \int\omega(\theta,0)\, d\theta = q/2\pi a.$$

Therefore the solution may be written as

$$\omega(\theta,t) = [\omega(\theta,0) - \omega(\theta,\infty)]e^{-t/\tau} + \omega(\theta,\infty),$$

that is, the *difference* between the initial charge density and the ultimate charge density (uniformly distributed and equal to $q/2\pi a$) decays exponentially with a time constant equal to the conductivity of the dielectric divided by the sum of the dielectric constants for the dielectric and for the medium.

In the special case of a line concentration of charge q per unit length, that is, $\omega(\theta,0) = (q/2\pi a)\delta(\theta)$, it is seen that the line charge at $\theta = 0$ decays exponentially, but that the surface charge density at every other place is independent of position, increasing asymptotically in time to its ultimate value.

The redistribution of charge cannot be characterized by a unique relaxation time for other geometries. Nevertheless for certain initial distributions, and over a limited time range, the redistribution takes place more or less exponentially, and an approximate relaxation time can be assigned. The details are complicated.

The Interdependence of Solid State Physics and Angular Distribution of Nuclear Radiations

ERNST HEER AND THEODORE B. NOVEY

University of Rochester, Rochester, New York, and Argonne National Laboratory, Lemont, Illinois

I. Introduction

The study of the angular correlation of radiation emitted from nuclei is a very important tool in physics. It provides a general method for the determination of properties of the nuclear levels involved, of the radiations emitted, and of the interactions responsible for the emission.

For practical reasons the experiments usually are carried out on nuclei in condensed matter. As a consequence of this experimental convenience, electric and magnetic fields (originating from the surrounding matter) are present at the position of the nuclei. These fields interact with the electric and magnetic moments of the nuclei and may alter the angular distribution pattern of the emitted radiation through this interaction. There is thus a natural interdependence of nuclear and solid state physics.

Angular correlation has been treated rather extensively from the point of view of the nuclear physicist in several recent review articles.[1–4] We would like to combine the points of view of solid state and nuclear physics in this article and indicate the information concerning the structure of the solid state which may be obtained from this type of experiment.

The terms angular distribution and angular correlation are used for a wide class of phenomena in which the probability $W(\theta)$ for emission of a nuclear radiation is described as a function of the angle θ between the emitted radiation and some fixed direction. The nuclear radiation may be gamma radiation, leptons (β,μ,ν), mesons (π,K), or baryons (nucleons and hyperons). If the nuclei emitting the radiation have their spins distributed at random in space, there is no preferred direction for the emission of the radiation, that is, the angular distribution of radiation emitted from unoriented nuclei will always be isotropic. Therefore, one needs to produce or to select nuclei which have a preferred direction in space (oriented nuclei) in order to obtain an anisotropic pattern of emission. There are a number of ways of doing this.

a. Nuclear orientation produced by low-temperature methods.[4–7] This type of orientation is achieved by cooling the nuclear spin system

[1] H. Frauenfelder, *in* "Beta- and Gamma-Ray Spectroscopy" (K. Siegbahn, ed.), Chapter XIX (I). North Holland, Amsterdam, 1955.

[2] R. M. Steffen, *Adv. Phys. (Phil. Mag. Suppl.)* **4,** 293 (1955).

[3] S. Devons and L. J. B. Goldfarb, *Handbuch d. Phys.* **42** (1957).

[4] R. J. Blin-Stoyle and M. A. Grace, *Handbuch d. Phys.* **42** (1957).

[5] R. J. Blin-Stoyle, M. A. Grace, and H. Halban, *in* "Beta- and Gamma-Ray Spectroscopy" (K. Siegbahn, ed.), Chapter XIX (II). North Holland, Amsterdam, 1955.

[6] S. R. deGroot and H. A. Tolhoek, *in* "Beta- and Gamma-Ray Spectroscopy" (K. Siegbahn, ed.), Chapter XIX (III). North Holland, Amsterdam, 1955.

[7] R. J. Blin-Stoyle, M. A. Grace, and H. Halban, *Progr. in Nuclear Phys.* **3,** 63 (1953).

(which is assumed to be coupled to a fixed spatial direction by electric or magnetic fields) to a very low temperature. At thermal equilibrium the probability of different spin orientations then varies according to the Boltzmann function and a net orientation of the spin system occurs.

b. Nuclear orientation produced by nuclear interactions. This may be achieved by the absorption of a polarized particle into a nucleus or by the absorption or scattering of an unpolarized particle through a channel involving nonvanishing angular momentum. Certain types of interactions and decays in which parity is not conserved may lead to oriented reaction or decay products.

c. Selection of nuclei with aligned spins. Consider a nucleus which emits two radiations successively. Choosing only those nuclei which emit the first radiation into a given direction is equivalent to selecting nuclei whose spins are partially aligned, provided that the first radiation is emitted with nonvanishing angular momentum. This situation is usually referred to under the designation "angular correlation of successive nuclear radiation."

All of the experiments have the following in common: they measure the angular distribution of radiations from nuclei whose spins are pointing preferentially in a certain direction. The interpretation of the results of the measurements requires a knowledge of the degree of orientation at the moment the radiation is emitted. Suppose the degree of orientation is known at some initial time. The emission of the radiation may occur at an appreciably later time to which time the degree of orientation has to be traced. It is during this extrapolation that the structure of the solid state comes into play. The nuclei which are investigated are usually not free but are incorporated in a target or a source, so that large electric and/or magnetic fields may be present. These fields interact with the electric or magnetic moments of the oriented nucleus under study and may change the degree of orientation in the time interval between the orientation of the nucleus and the emission of the particle. This change has two implications: it may hinder the nuclear measurements which one would like to make but it may on the other hand allow one to study the interaction of the nucleus with its surroundings and therefore, may be used as a tool in solid state research.

The information gained from such observations is often similar to that obtained in nuclear and paramagnetic resonance studies but, in some cases, is not obtainable in any other way. For example, in contrast to nuclear resonance measurements, one obtains the information without having to impose a radio-frequency field. This situation is of interest, for example, in the investigation of the fields within a superconductor. One also has a much higher sensitivity in the sense that it is possible to measure the field acting on impurities of the order of 10^8 atoms compared to the required 10^{17} atoms for nuclear resonance experiments.

In the following Sections 1 to 4 we will summarize the necessary formalism connected with the field-free angular correlation problem. We will then examine the effects of various types of fields which are likely to be encountered (Sections 5 to 17).

II. The Unperturbed Angular Correlation of Nuclear Radiation

1. Angular Correlation of Successive Nuclear Radiations

Since the reader of this article presumably will be interested more in the solid state than in the nuclear physics aspect of angular correlation, we will treat the theory only as far as is necessary to understand the influence of external fields on angular correlation.

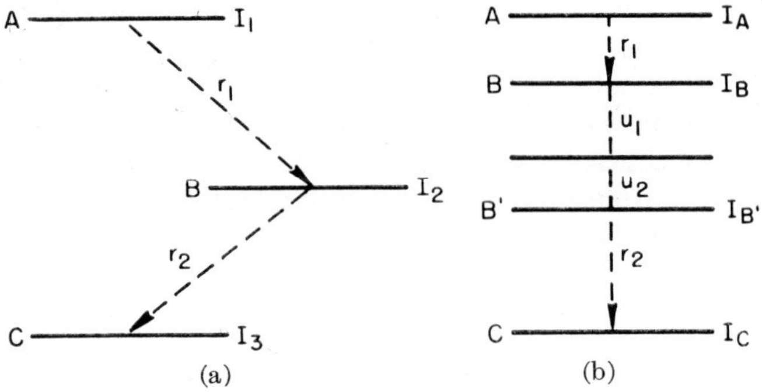

FIG. 1. (a) Nuclear decay scheme of a level A which decays by the emission of a radiation r_1 into a level B and then by the emission of a radiation r_2 into a level C. In agreement with the usual notation of nuclear physics, the vertical distance between the levels on the drawing corresponds to the energy difference of the levels and the lateral displacement indicates the difference in charge (i.e., in proton number). (b) Same for the special, but important case of a gamma cascade. Here it is assumed that the level A decays by consecutive emission of 4 gamma rays into the level C. Energy-sensitive counters then make it possible to measure the angular correlation between r_1 and r_2, leaving u_1 and u_2 unobserved.

Let us assume we have an assembly of randomly oriented nuclei in state A which decays by successive emission of two radiations r_1 and r_2 to the levels B and then C (see Fig. 1a). We then measure the angular distribution of the second radiation r_2 with respect to the first radiation r_1. This is done with an arrangement shown schematically in Fig. 2 consisting of two detectors which subtend a variable angle θ at the source. In the simplest case, the detector 1 accepts only the radiation r_1 and the detector 2 is sensitive only to radiation r_2. The detectors may count all the radiations which fall in their solid angles; however, the coincidence analyzer selects principally only pairs of radiations r_1 and r_2 which are genetically related to each other. It accomplishes this by accepting a signal from detector 1 only if a signal from detector 2 arrives at the same time. More precisely, only those radiations which are emitted within the resolving time τ_R of the coincidence circuit are accepted. τ_R is typically of the order of 10^{-6} to 10^{-9} second. The possibility of chance coincidence between unrelated radiations may be reduced to a tolerable magnitude by proper selection of resolving time and source strength. The expression for the number of coincidences per unit time as a function of angle is usually called the angular correlation function $W(\theta)$.

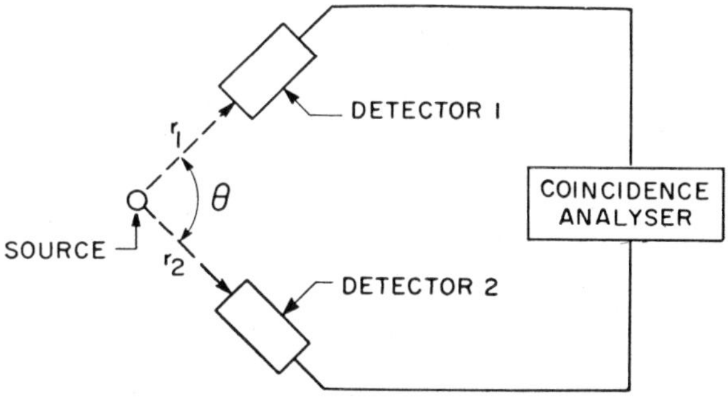

FIG. 2. Schematic diagram of the experimental arrangement used for the measurement of an angular correlation.

The radiation r_2 is not emitted simultaneously with r_1, but is emitted according to the law of exponential decay of the level B with its lifetime τ_N. Lifetimes of nuclear levels vary over a very wide range extending from 10^{-16} sec to millions of years. The fact that the angular correlation can be measured only if one establishes a genetic relationship between the radiations r_1 and r_2 sets a practical upper limit for the lifetime τ_N which is suitable for an angular correlation measurement. In general, $\tau_N \leqslant 10^{-5}$ sec. Because the effect of external fields plays a role during the lifetime τ_N of the intermediate state and because the strength of these fields is limited, we are mainly interested from the solid state point of view in those nuclei for which $\tau_N \geqslant 10^{-9}$ sec.

The formulation of this section will assume that during this intermediate lifetime τ_N nothing happens to the nucleus, i.e., the orientation of its spin does not change. This holds for an isolated nucleus in the vacuum. The corresponding angular correlation is termed the "undisturbed correlation" or the "true correlation."

The angular correlation can be written in the form

$$W(\theta) = \sum_k A_k P_k(\cos\theta) \tag{1.1}$$

where the $P_k(\cos\theta)$ are Legendre polynomials. In all cases in which parity is conserved—and we will deal only with those in this section—only even subscripts k occur. Furthermore, in all cases of practical importance only A_0, A_2, and A_4 are different from 0 so that we have

$$W(\theta) = A_0 + A_2P_2(\cos\theta) + A_4P_4(\cos\theta) \tag{1.2}$$

which for convenience is sometime written as

$$W(\theta) = a_0 + a_2\cos^2\theta + a_4\cos^4\theta \tag{1.3}$$

or

$$W(\theta) = b_0 + b_2\cos 2\theta + b_4\cos 4\theta. \tag{1.4}$$

The coefficients A_k can be written as

$$A_k = f_{1k}f_{2k} \tag{1.5}$$

where f_{1k} is a factor depending only on the properties of the first transition from level A to level B and similarly f_{2k} depends only on the properties of the transition from level B to level C.

If more than two radiations are emitted in the radioactive decay and if a number of unobserved radiations u_i occur between the observed radiations r_1 and r_2 as in Fig. 1b then Eq. (1.5) must be modified to

$$A_k = f_{1k}\cdot U_k\cdot f_{2k} \tag{1.6}$$

where

$$U_k = \prod_i U_{ki}. \tag{1.7}$$

The coefficients U_{ki} depend only on the unobserved transitions u_i. Explicit formulas for the U_{ki} are given in Sections 16 and 17 of Devons and Goldfarb.[3]

If only one angular momentum is involved in the emission of the radiation r_1 (in other words in the case of pure multipole radiation), the f_{ik} can be written as

$$f_{ik} = b_{ik}F_{ik}. \tag{1.8}$$

Here b_{ik} is the so-called "particle parameter" and is normalized to unity for photons. Formulas and tables for other particles (α,β, and e^-) can be found in Devons and Goldfarb[3] and Biedenharn and Rose.[8] $F_{ik} = F_{ik}(L_iL_iI_iI)$ is a coefficient which depends on the angular momentum L_i of the emitted radiation r_1, on the spins I_i of the level A or C, and on the

[8] L. C. Biedenharn and M. E. Rose, *Revs. Modern Phys.* **25**, 729 (1953).

spin I of the level B. If there are unobserved radiations between r_1 and r_2, then, in F_{1k}, I is the final spin of the first transition and, in F_{2k}, I is the initial spin of the second transition (see Fig. 1b). The coefficients F have been calculated numerically and tabulated[9] for a large range of values of L_i, I_i, and I. Less extensive tables are given by Biedenharn and Rose.[8,10]

If the radiation r_i is not a pure multipole radiation, i.e., if selection rules allow the presence of more than one angular momentum, then Eq. (1.8) becomes more complicated. Usually only the lowest possible angular momentum $L = L_{\min}$ and the next higher one $L' = L_{\min} + 1$ are of importance. In this case, (1.8) becomes

$$f_{ik} = 1/(1 + \delta^2) \cdot (b_{ikL}F_{ikL} + \delta^2 b_{ikL'}F_{ikL} + 2\delta b_{ikLL'}F_{ikLL'}) \qquad (1.9)$$

where δ is the ratio of amplitudes and δ^2 is the ratio of intensities of the multipoles of orders L' and L; $b_{ikLL'}$ and $F_{ikLL'}$ are obvious generalizations of the particle parameters b_{ik} and coefficients F_{ik} introduced earlier. They are also defined and tabulated in the papers referred to above.

In concluding this section we may say that, if a level A decays by a radiation r_1 into level B and then subsequently by a radiation r_2 into a level C, we can calculate the angular correlation $W(\theta)$ between the two radiations if we know the types of radiations (which determine the particle parameters b_{ik}) and the spins of the states involved. For what follows, it will not be of importance to know how this correlation is calculated. We will use the result, however, that the correlation $W(\theta)$ can be written in the form given by Eq. (1.1), i.e., as a sum of Legendre polynomials. It is also important to note that the coefficients of these polynomials, the so-called "angular correlation" coefficients, can be written as a product of two factors, describing the first and the second radiations, respectively. Thus the first factor describes the degree of orientation of the nuclei selected by the condition that r_1 is emitted in the direction of detector 1. The second factor describes the angular distribution of r_2 with respect to the axis of orientation of the nuclei.

One further point needs to be mentioned. In the case of a correlation in which both radiations involve only one angular momentum, the coefficients A_k can be calculated quite easily. If the experiment gives values which differ from the theoretical ones, one has established that the correlation is not the undisturbed one and that the influence of the external fields must be taken into account. On the other hand, if one or

[9] M. Ferentz and N. Rosenzweig, Argonne National Laboratory Report ANL-5234 (1953).

[10] M. E. Rose, *in* "Beta- and Gamma-Ray Spectroscopy" (K. Siegbahn, ed.), Appendix IV. North Holland, Amsterdam, 1955.

both of the two radiations are mixtures of two multipoles, then the theoretical angular correlation coefficients also depend on the mixing ratios δ_1 (and δ_2). Since these mixing ratios usually cannot be predicted with sufficient accuracy by nuclear theory, and since they are hard to measure by any method other than those based on angular correlation, it is not possible to calculate the undisturbed angular correlation and then, by comparison with the experiment, establish whether or not the correlation is disturbed. This emphasizes the importance of understanding the influence of external fields on angular correlation from the point of view of nuclear physics.

As an example of a measured angular-correlation function, we describe measurements on the gamma-gamma cascade in the decay of Co^{60} which decays by beta emission into the second excited state of Ni^{60} and then decays to the ground state of Ni^{60} by the emission of two successive gamma rays. Since the spins of the three levels are known to be 4, 2, and 0, it is possible to calculate the angular correlation function:

$$W(\theta) = 1 + 0.102P_2(\cos\theta) + 0.009P_4(\cos\theta).$$

The experimental results of many investigators are in excellent agreement with the predicted function. (See Fig. 3.)

In addition to this angular correlation, a great number of other gamma-gamma angular correlations have been measured. The number of measured alpha-gamma and beta-gamma correlations and correlations involving conversion electrons is appreciably smaller, mostly because of the larger experimental difficulties encountered and the scarcity of nuclear systems with suitable properties. Thus the latter types of correlations are less important for our purpose.

2. Oriented Nuclei

Let us consider an assembly of nuclei each with spin I. The projection m of I on an (arbitrary) axis of quantization can then have one of the $(2I + 1)$ integral (or half-integral) values $-I, -I + 1, \ldots I - 1, I$. We denote by $W(m)$ the probability of finding a nucleus in a state with magnetic quantum number m. An assembly of nuclei is unoriented if $W(m)$ is independent of m. For our purposes we call an assembly of nuclei oriented if the $W(m)$ differ by at least a few per cent. There are two special cases of orientation, namely, polarization and alignment. An assembly of nuclei is said to be polarized with a degree of polarization P_1 if

$$P_1 = \frac{1}{I} \sum_m mW(m) \tag{2.1}$$

is different from zero. In this case the assembly has a net moment. It is said to be aligned with a degree of alignment P_2 if

$$P_2 = \frac{3 \sum_m m^2 W(m) - I(I+1)}{I(2I-1)} \tag{2.2}$$

is different from zero. The alignment is a measure of the difference in the total population of states with different values of $|m|$.

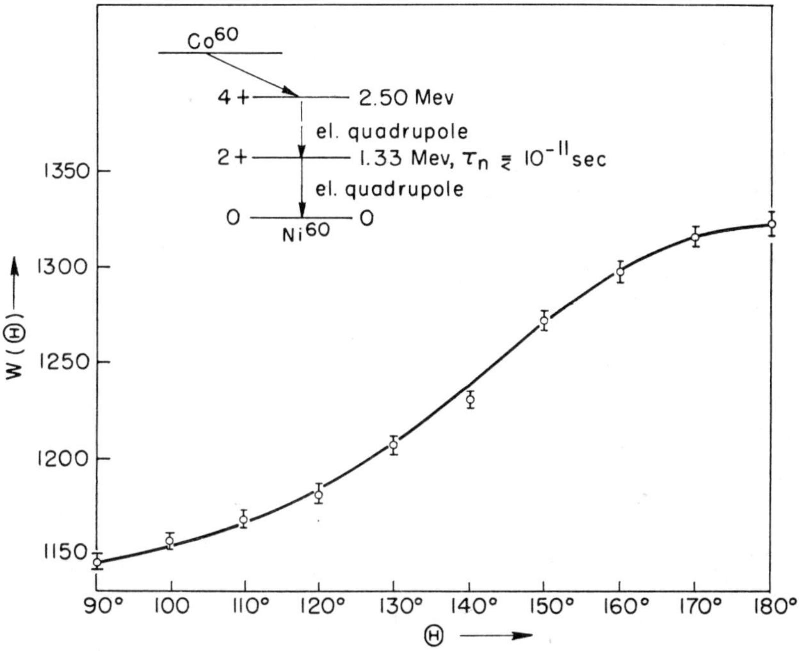

FIG. 3. The decay scheme of Co^{60} and the angular correlation of the gamma rays in the cascade in Ni^{60} [R. M. Steffen, *Adv. Phys.* (*Phil. Mag. Suppl.*) **4,** 293 (1955)].

Several methods have been proposed and some of them have been used to produce oriented nuclei. Since we are not interested here in these methods themselves, but only in their application in connection with angular distribution studies, we refer the reader to the pertinent literature[4,5,7] for a more detailed treatment.

(a) Direct interaction of the nuclear magnetic dipole moment $\mu = g\mu_k I$ with a large external magnetic field H causes a splitting of the nuclear magnetic states, $E(m) = g\mu_k H \cdot m$. The population $W(m)$ of individual states is then given by the Boltzmann function, $W(m) \sim e^{-m\beta}$, where $\beta = g\mu_k H/kT$. A sizable polarization occurs if $\beta \sim 1$. To reach

$\beta = 1$, temperatures of 10^{-3}°K are required with a field of 5×10^4 gauss, or $H/T \sim 5 \times 10^7$ gauss/°K. The extremely low temperatures or high fields required makes this method rather difficult.

(b) Large fields of the order of 10^5 gauss or more, are present at the nucleus in a paramagnetic atom. At low temperatures each nucleus will be oriented with respect to the direction of the electronic magnetic moment. Since at these temperatures the electronic magnetic moments may be appreciably oriented in a small magnetic field, nuclear orientation follows immediately.

(c) The interaction of the nuclear quadrupole moment Q with the electric field gradient dE/dz in a single crystal causes a hyperfine splitting, $\Delta\nu_Q$, of the nuclear states and therefore a nuclear alignment at low temperatures similar to (a). The condition for sizable alignment is that $\Delta\nu_Q/T \sim 2 \times 10^4$ Mc/°K.

(d) A separation of the nuclear sublevels which will cause nuclear alignment at low temperatures also occurs through the anisotropic hyperfine splitting in suitable paramagnetic single crystals.

(e) The absorption of circularly polarized resonance radiation by atoms leads to a polarization of the atoms (optical pumping).[11] A polarization of the nuclei will then result through the hyperfine interaction. In dilute gases where depolarization through collision is small, and for sufficient light intensities, an appreciable degree of polarization may be obtained.

(f) If one saturates the paramagnetic resonance of the free electrons in metals or other suitable media then the nuclear spins I will be oriented[12] because of the coupling $I \cdot S$ with the electron spins S.

If the radioactive nuclei are aligned by any of the foregoing methods, this will result in general in an anisotropic angular distribution of the radiations emitted in the decay of these radioactive nuclei.

In the context of this paper—which deals with extranuclear effects—the only important type of experiment is a measurement of the angular distribution of the gamma rays following alpha or beta decay of aligned nuclei. The angular distribution of the gamma rays with respect to the direction of alignment is then given by

$$W(\theta) = \sum_k A_k P_k(\cos \theta), \qquad k \text{ even} \tag{1.1}$$

with

$$A_k = B_k U_k f_{2k}. \tag{2.3}$$

Here U_k, as in Eq. (1.6) describes the unobserved radiation (which is either beta or alpha radiation) and f_{2k} is still given by Eq. (1.8) and

[11] A. Kastler, *Proc. Phys. Soc.* **A67**, 853 (1954).

[12] A. W. Overhauser, *Phys. Rev.* **92**, 411 (1953).

describes the observed (gamma) radiation. To make the comparison of Eqs. (2.3) and (1.6) complete we see that the newly introduced coefficient B_k corresponds to f_{1k}. In the case of the angular correlation f_{1k} described the degree of nuclear orientation is achieved by the selection of those nuclei which emit the first radiation in the direction of the detector No. 1. In the present case the coefficient B_k describes the alignment achieved by the nuclear orientation process. We have

$$B_k = \sum_m (2k + 1)^{\frac{1}{2}} C(IkI,m0) W(m). \tag{2.4}$$

These coefficients have been calculated for the most important methods of nuclear orientation.[4]

In the case where there is polarization ($P_1 \neq 0$) but no alignment ($P_2 = 0$), there is no anisotropic spatial distribution of the emitted γ radiation. There will be, however, a circular polarization of the radiation which can be used as a measure of the polarization. Equation (1.1) still applies and $W(\theta)$ now is the probability of observing circularly polarized radiation at an angle θ with respect to the axis of polarization. In this case only terms with odd k appear.

We have assumed in this section that the orientation of the nuclear spin does not change during the time the nucleus remains in the level which emits the observed γ ray. If the lifetime of the intermediate level is larger than 10^{-9} sec, this assumption may not be valid and extranuclear effects must be taken into account as in the angular correlation case.

3. Nuclear Reactions

If a target nucleus interacts with incident radiation it is possible that after the reaction the remaining, often radioactive, nucleus is oriented to a certain degree. Such orientation can occur either if the incoming particle is polarized itself or if the reaction proceeds through a channel with nonvanishing angular momentum. The remaining nucleus then decays by emission of nuclear radiation whose angular distribution with respect to the direction of the incoming particle reflects the nuclear orientation created by the reaction. If the reaction mechanism is known in detail, then the angular distribution may be calculated.[3] It can in general be written in the form

$$W(\theta) = \sum_k A_k P_k(\cos \theta). \tag{1.1}$$

The coefficients A_k are again calculated under the assumption that the degree of orientation created by the nuclear reaction remains unchanged until the excited nucleus emits its radiations. If the lifetime involved is longer than 10^{-9} sec, we must as in the case of angular correlation of successive nuclear radiation, take into account the depolarizing effects of the electric and magnetic fields present in the target material.

As a specific and important example we mention the process of Coulomb excitation in which the nuclei are excited by the electromagnetic interaction with the bombarding nucleus and not by any strong nuclear interaction. The angular distribution of the gamma rays emitted in the subsequent de-excitation is closely related to the angular correlation of successive nuclear radiation (see Fig. 4). We still have Eqs. (1.1) to (1.8).

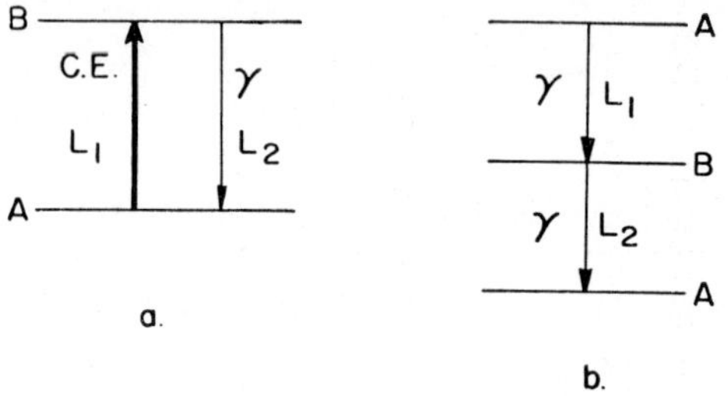

FIG. 4. In the Coulomb excitation (a) the nucleus is excited from its ground state (level A) to an excited state (level B). This level decays after a certain time back into the ground state. The angular distribution of the gamma radiation emitted in the de-excitation of level B is similar to the angular correlation of two successive gamma radiations from the level A to B and then A (b).

The only difference is that the particle parameter for the gamma ray is unity, whereas it is dependent on the target nucleus, the type of incident particle, and its energy in the case of the Coulomb excitation. The particle parameters for the Coulomb excitation are calculated and tabulated by Alder *et al.*[13]

4. Decay of Elementary Particles

As a consequence of the nonconservation of parity and charge conjugation in the weak interactions, large longitudinal polarizations are produced in the decay of fundamental particles, such as π mesons, and in the beta decay of complex nuclei. If another radiation is emitted after this weak-interaction decay, the spatial distribution or polarization of this radiation reflects the original polarization. Two examples are of special interest for us.

a. *The π-μ-e Decay*[14]

Here the π decay produces a completely polarized μ meson. The direction of polarization is parallel to the momentum of the μ meson.

[13] K. Alder, A. Bohr, T. Huus, B. Mottelson, and A. Winther, *Revs. Modern Phys.* **28**, 432 (1956).

[14] T. Kinoshita and A. Sirlin, *Phys. Rev.* **108**, 844 (1957).

Usually the μ meson is stopped in some material and decays with its lifetime of 2.2×10^{-6} sec into an electron, a neutrino, and an antineutrino. The angular distribution of the decay electron is given by

$$W(\theta) = 1 + (v/c)A \cos \theta \tag{4.1}$$

v is the velocity of the electron, c the velocity of light. The coefficient A can be calculated for the various types of weak interactions. Again Eq. (4.1) is based on the assumption that the direction of polarization is not affected between the time of its creation (i.e., when the π meson decays) and the time of the emission of the electron. Since the lifetime of the μ meson is rather long, extranuclear effects in general have to be taken into account.

b. Beta Decay of Complex Nuclei

If a radioactive nucleus undergoes β decay, and if the interaction has a contribution of the Gamow-Teller type (i.e., if the spins of the emitted electron and neutrino are parallel), the remaining nucleus is polarized. Using the notation of Section 2 we have $P_1 \neq 0$, but $P_2 = 0$. This leads to a spatially isotropic emission of subsequent gamma rays just as for completely unoriented nuclei; however, the radiation emitted in a given direction will be circularly polarized. A measurement of the degree of circular polarization can be used in order to determine the degree of initial polarization. The angular distribution $W(\theta,\tau)$ for the emission of circularly polarized gamma rays is again given by (1.1) with

$$A_k = (-\tau)^k f_{1k} U_k f_{2k}. \tag{4.2}$$

Here f_{1k} refers to the beta decay and f_{2k} to the subsequently observed gamma radiation, and τ is equal to $+1$ (-1) if the gamma rays detected are right (left) circularly polarized.

Again the effect of extranuclear fields will enter into the picture if there is a sufficiently long elapsed time between the creation of polarization (the moment of the β decay) and its observation (the emission of the gamma ray).

III. Extranuclear Effects Involving Nonparamagnetic Atoms

5. Introduction

In the preceding sections we have described how one can predict the usually anisotropic angular distribution of radiation emitted from an oriented state. We assumed that the orientation is unchanged from the moment it is produced to the moment the radiation is emitted. From this point on, however, we wish to discuss the possible sources of change in this orientation and the resulting effect on the angular distribution.

Observation of these so-called "disturbed" angular correlations can provide us with information concerning the interaction and fields by which they are produced.

The angular correlation is expressed in principle as a product of two factors. The first (called f_{1k} or B_k) describes the initial orientation and the second (f_{2k}) describes the angular behavior of the emitted radiation. As is expected, the influence of an external field which produces a change in orientation can be described by an additional factor in the product. Thus we now have

$$W(\theta,t) = \sum_k f_{1k} G_k(\tau) f_{2k} P_k(\cos\theta). \tag{5.1}$$

Here G is the "attenuation factor," which carries the information concerning the way in which magnetic population has been changed by the interaction. It will depend on the type of interaction, the strength of interaction ω, and the orientation of the field with respect to the initial preferred direction. This G factor will be stated explicitly in the following sections for cases of special interest and simplicity.

The factor G is a function of time, being unity for $t = 0$, the time at which the initial orientation was produced. The time dependence of G and therefore the angular correlation may be periodic or it may decrease monotonically. In the latter case the angular distribution finally may become completely isotropic, i.e., $G_k \to 0$.

For the cases treated in Section 1 the angular correlation was not time-dependent and therefore the resolving time of the coincidence analyzer could not affect the measurements. Since in the presence of external fields the angular correlation is time-dependent, it is necessary to specify the timing of the measurements in more detail. Coincidence circuits have a resolving time τ_R. If we begin the measurement at a time t_D after the initial orientation has been created ($t = 0$), then the angular correlation function takes the form:

$$W(\theta) = \int_{t_D-\tau_R}^{t_D+\tau_R} W(\theta,t) e^{-t/\tau_N}\, dt \Big/ \int_{t_D-\tau_R}^{t_D+\tau_R} e^{-t/\tau_N}\, dt \tag{5.2}$$

where τ_N is the mean life of the oriented state. Two special cases are of importance. The first one is the so-called integral angular correlation. This can be characterized by the conditions $\tau_R \gg \tau_N$ and $t_D = 0$ so that one averages over all radiations independently of the time at which they are emitted. The integral correlation is the simplest to measure experimentally but some information concerning the interaction mechanism is usually lost in the averaging process.

The second case is that of the differential angular correlation where $\tau_R < \tau_N$. If the angular correlation is measured as a function of t_D the experiment usually is called delayed angular correlation. In the limit of $\tau_R \ll \tau_N$, $1/\omega$ the measurement yields the attenuation factor $G_k(t)$ directly. Therefore it is obvious that delayed correlation measurements are of great importance for the study of extranuclear effects.

In Sections 6 to 17 which follow we will deal mostly with the extranuclear effects on angular correlation of successive nuclear radiation. Application to angular distribution of radiations emitted from oriented nuclei and to the angular distribution of nuclear reaction products is straightforward. The decay of elementary particles will be treated separately in Section 18.

6. Static External Magnetic Fields

The interaction of the magnetic field H with the magnetic dipole moment of the nucleus, $\mu = gI\mu_n$, induces the Larmor precession frequency

$$\omega_L = \frac{g\mu_n H}{\hbar} \tag{6.1}$$

in which μ_n is the nuclear magneton and g is the nuclear gyromagnetic ratio; ω_L is independent of the magnetic quantum number m. In the simple vector model, ω_L corresponds to the angular frequency with which the nuclear spin precesses about the field direction.

If the field is parallel to the initial direction of orientation, then the Larmor precession does not change the projection of the spin along this direction. In other words, there will be no change of magnetic sublevel population relative to the axis of orientation and no change of the angular distribution of radiation, i.e., $G_k = 1$.

If the field has a component perpendicular to the initial direction of orientation, then the sublevel population will change with time. If we limit ourselves to measuring the outgoing radiation in directions perpendicular to the field, the differential angular distribution is given by

$$W(\theta,t) = \sum_k A_k P_k[\cos(\theta - \omega_L t)]. \tag{6.2}$$

That is, the angular correlation pattern precesses around the field direction, with a frequency ω_L.

It is assumed in this formulation that the detector in the direction $\theta = 0$ accepts the first radiation and the detector at the azimuthal direction θ accepts the following radiation.

The integral correlation obtained from Eqs. (5.2) and (6.2) is

$$W(\theta) = \frac{\int_0^\infty \sum A_k P_k \cos(\theta - \omega_L t) e^{-t/\tau_N}\, dt}{\int_0^\infty e^{-t/\tau_N}\, dt} \tag{6.3}$$

$$W(\theta) = \sum_k \frac{b_k}{\sqrt{1 + (k\omega_L \tau_N)^2}} \cos k(\theta - \Delta\theta) \tag{6.4}$$

where

$$\Delta\theta = \frac{1}{k} \operatorname{arc\,tan} (k\omega_L \tau_N). \tag{6.5}$$

In (6.4) b_k are the coefficients of the undisturbed correlation function, [Eq. (1.4)]. From Eq. (6.4) it follows that the magnetic field leads to an attenuation of the anisotropy by a factor $1/\sqrt{1 + (k\omega_L\tau_N)^2}$ and a rotation of the correlation function by an angle given by (6.5), which for small rotation is $\Delta\theta \sim \omega_L\tau_N$.

For detectors equally sensitive to both radiations (6.2) must be replaced by

$$W(\theta,t) = \sum_k \frac{A_k}{2} [P_k \cos(\theta - \omega_L t) + P_k \cos(\theta + \omega_L t)] \tag{6.6}$$

and (6.4) by

$$W(\theta) = \sum_k \frac{b_k}{1 + (k\omega_L\tau_N)^2} \cos k\theta = \sum_k b_k{}^* \cos k\theta \tag{6.7}$$

i.e., rotation of the radiation pattern is not observed but an attenuation is.

If the field has an arbitrary direction with respect to the initial preferred direction, then the G factor is a complicated function of the angle involved and should be taken from Alder *et al.*[15]

The measurement of the integral angular correlation as a function of the magnetic field or of the delayed angular correlation at constant magnetic field as a function of delay time gives the Larmor precession frequency ω_L and, if the magnetic field is known, the nuclear magnetic moment. Such a measurement therefore is related closely to a nuclear magnetic resonance experiment which, in its simplest form, namely for free nuclei in a magnetic field, measures exactly the same quantity. Two differences, however, should be pointed out. The first is the much smaller precision of the angular correlation method. This comes from the fact that in the angular correlation experiment one measures the rotation of the angular correlation pattern by the angle $\Delta\theta \cong \omega_L t$. With detectors

[15] K. Alder, H. Albers-Schönberg, E. Heer, and T. B. Novey, *Helv. Phys. Acta* **26**, 761 (1953).

of practical size such a rotation may be measured to a precision of one degree, which for a typical rotation of $\Delta\theta = 100°$ gives a limiting accuracy of 1%. In the nuclear resonance case one measures ω_L directly, the precision in this case may be as good as 10^{-5}%, being limited only by the line width. It should be pointed out, however, that the angular correlation experiment can be improved appreciably by the use of the stroboscopic method,[16] provided that one can fulfill the condition $\omega_L\tau_N \gg 1$. In this case one measures

$$\frac{\sum_n \int_{nT}^{a+nT} W(\theta,t)e^{-t/\tau_N}\,dt}{\sum_n \int_{nT}^{a+nT} e^{-t/\tau_N}\,dt} = X(\theta,T). \tag{6.8}$$

A measurement of the quantity $X(T)$ for a fixed angle θ has the general property of a resonance measurement in which the line width is of the order of $\Delta\omega = 1/\tau_N$. This method is an improvement over the more direct measurements only for very long-lived nuclei or extremely high fields.

The second difference between nuclear resonance and angular correlation resides in the fact that angular correlation measurements can be used to determine the magnetic moments of short-lived nuclear levels. The available field strengths of ordinary ferromagnets set the lower limit for the lifetimes in the region of 10^{-9} sec. Pulsed fields may allow the reduction of this by one or two orders of magnitude. The upper limit is partly determined by the technical problem of accidental coincidences. The ratio of true to accidental coincidences is proportional to the source strength and to the lifetime. For a given ratio of true to accidental coincidences, the longer the lifetime the smaller must be the source strength. The measuring time will then be correspondingly increased. In addition to this, for lifetimes much longer than 10^{-6} sec it becomes increasingly difficult, although not impossible, to avoid depolarizing effects or relaxation processes (see Sections 9, 19).

The measurements of the magnetic moments of short-lived excited states that have been made by the above described method are listed in Table I.

Figure 5 shows the result of a measurement[17] of the magnetic moment of the 247-kev excited state of Cd^{111}. In this case, the integral angular correlation has been measured as a function of the magnetic field. The

[16] R. A. Lundy, J. C. Sens, R. A. Swanson, V. L. Telegdi, and D. D. Yovanovitch, *Phys. Rev. Letters* **1**, 38 (1958).

[17] H. Albers-Schönberg, E. Heer, T. B. Novey, and P. Scherrer, *Helv. Phys. Acta* **27**, 547 (1954).

quantity plotted is the coefficient b_2^* of the expansion

$$W(\theta) = 1 + b_2^* \cos 2\theta.$$

The radioactive material for this experiment was in the form of an aqueous solution of indium chloride. The liquid source may not be completely free of disturbing effects. This topic is discussed in detail in Section 10. We may mention that several authors have treated the prob-

TABLE I. SUMMARY OF THE MAGNETIC MOMENTS OF SHORT-LIVED EXCITED NUCLEAR LEVELS MEASURED BY ANGULAR CORRELATION METHODS

Properties of the investigated nuclear level							
Nucleus	Energy kev	Spin	Mean life mμsec	Magn. moment	Type of source	Method of measure-ment	Ref.
$_9F^{19}$	197	5/2	100	3.70 ±0.45	liquid, KF in H_2O	integral a.c.	*a*
				>3.0 ±0.7	solid CaF_2	integral a.c.	*b*
$_{48}Cd^{111}$	247	5/2	125	−0.70 ±0.12	solid, In met. in Ag	integral a.c.	*c*
				−0.783 ±0.028	liq., $InCl_3$ in H_2O	diff. a.c. rotation	*d*
				−0 725 ±0.047	liq. In metal	integral a.c.	*e*
$_{60}Nd^{150}$	130	2	~2	0.44 ±0.08	aq. sol. of $Nd(NO_3)_3$	integral a.c.	*f*
$_{61}Pm^{147}$	92	7/2	3.5	⩽3.5	aq. sol.	integral a.c.	*g*
$_{62}Sm^{152}$	125	2	2.0	0.42 ±0.08	aq. sol. of $Sm(NO_3)_3$	integral a.c.	*f*
$_{62}Sm^{154}$	84	2	~4	0.42 ±0.08	aq. sol. of $Sm(NO_3)_3$	integral a.c.	*f*
$_{66}Dy^{160}$	84	2	2.6	0.36 ±0.16	liq. and solid	integral a.c.	*h*
$_{73}Ta^{181}$	482	5/2	15.3	+3.0	liq., HfF_4 in HF	integral a.c. rotation	*i*
				+3.25 ±0.17	liq., HfF_4 in Hf	integral a.c.	*j*
				+2.70 ±0.17	liq., HfF_4 in Hf	integral a.c. rotation	*g*

TABLE I. SUMMARY OF THE MAGNETIC MOMENTS OF SHORT-LIVED EXCITED NUCLEAR LEVELS MEASURED BY ANGULAR CORRELATION METHODS (*Continued*)

Properties of the investigated nuclear level							
Nucleus	Energy kev	Spin	Mean life mμsec	Magn. moment	Type of source	Method of measurement	Ref.
$_{82}Pb^{204}$	1273	4	374	+0.28 ±0.12	met. Tl solid and liquid	integral a.c. rotation	*k*
				+0.22 ±0.02	liq., $TlNO_3$ in HNO_3	rotation	*l*
$_{93}Np^{237}$	60	5/2	91	2 ±0.5	liq., $AmClO_4$ in $HClO_4$	integral a.c.	*m*

[a] P. Lehmann, A. Leveque, and R. Pick, *Phys. Rev.* **104,** 411 (1956).
[b] K. Sugimoto and A. Mizobuchi, *Phys. Rev.* **103,** 739 (1956).
[c] H. Aeppli, H. Albers-Schönberg, H. Frauenfelder, and P. Scherrer, *Helv. Phys. Acta* **25,** 339 (1952).
[d] R. M. Steffen and W. Zobel, *Phys. Rev.* **97,** 1188 (1955).
[e] H. Albers-Schönberg, E. Heer, T. B. Novey, and P. Scherrer, *Helv. Phys. Acta* **27,** 547 (1954).
[f] G. Goldring and R. P. Scharenberg, *Phys. Rev.* **110,** 701 (1958).
[g] T. Lindquist and E. Karlson, *Arkiv Fysik* **12,** 519 (1957).
[h] P. Debrunner, W. Kündig, J. Sunier, and P. Scherrer, *Helv. Phys. Acta* **31,** 326 (1958); and private communication.
[i] S. Raboy and V. E. Krohn, *Phys. Rev.* **95,** 1689 (1954).
[j] P. Debrunner, E. Heer, W. Kündig, and R. Rüetschi, *Helv. Phys. Acta* **29,** 463 (1956).
[k] H. Frauenfelder, J. S. Lawson, and W. Jentschke, *Phys. Rev.* **93,** 1126 (1954).
[l] V. E. Krohn and S. Raboy, *Phys. Rev.* **97,** 1017 (1955).
[m] V. E. Krohn, T. B. Novey, and S. Raboy, *Phys. Rev.* **105,** 234 (1957).

lem of obtaining the correct value of the magnetic moment in the case of the (usually) small disturbance in the liquid source.[2,18]

7. Static Electric Quadrupole Interaction

We now consider the interaction of the electric quadrupole moment of the nucleus with the gradient of the electric field present at the site of the nucleus. Electric field gradients sufficiently strong to create appreciable effects cannot be produced artificially but are present in crystal lattices. The quadrupole interaction for axially symmetric fields is described by the interaction energy:

$$E_Q(m) = \frac{3m^2 - I(I+1)}{4I(2I-1)} eQ \frac{\partial E_z}{\partial z} = \frac{3m^2 - I(I+1)}{4I(2I-1)} \Delta\nu_Q h \quad (7.1)$$

[18] P. Debrunner, E. Heer, W. Kündig, and R. Rüetschi, *Helv. Phys. Acta* **29,** 463 (1956).

$$\Delta\nu_Q = eQ\frac{\partial E_z}{\partial z} = \text{“quadrupole interaction strength”}$$

$\partial E_z/\partial z$ = electric field gradient referred to the symmetry axis
Q = quadrupole moment of the polarized state
I = nuclear spin
m = component of nuclear spin along the axis of orientation.

Cubic crystals, or more accurately crystals which have a field with cubic symmetry at the site of the nuclei being investigated, have a vanishing field gradient and therefore do not influence the angular correlation.

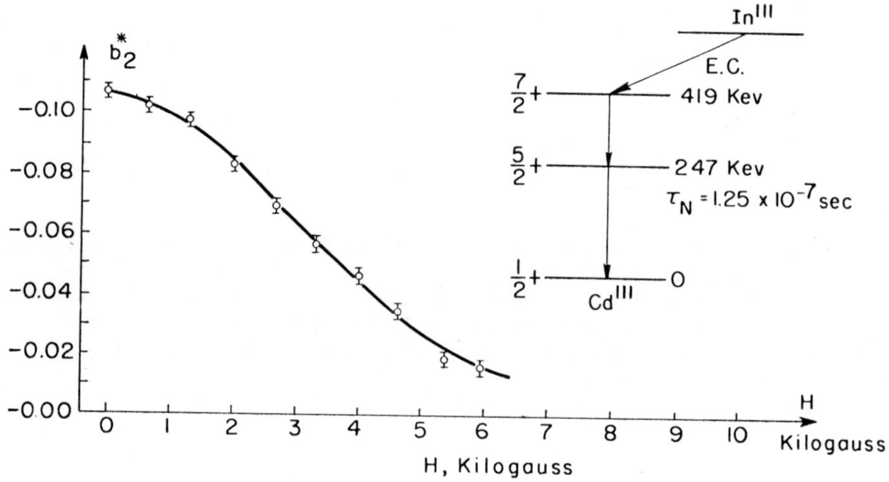

FIG. 5. Measurement of the magnetic moment of 247-kev excited state of Cd^{111}. The integral angular correlation is measured as a function of magnetic field strength. b_2^* is the coefficient in the expansion $W(\theta) = 1 + b_2^* \cos 2\theta$; see Eq. (6.7) [H. Albers-Schönberg, E. Heer, T. B. Novey, and P. Scherrer, *Helv. Phys. Acta* **27**, 547 (1954)].

In noncubic crystals the quadrupole interaction strength $\Delta\nu_Q$ is typically of the order of one hundred megacycles and may be as high as several thousand megacycles. Thus one expects an effect on the angular correlation if the lifetime involved is longer than 10^{-9} second.

For convenience, we distinguish between the following possibilities.

a. *Single Crystals with Axial Symmetry*

This is similar to the magnetic field case treated in Section 6. There is no influence, i.e., $G = 1$, if the axis of the crystalline field is parallel to the original direction of orientation. If the field axis is perpendicular to this direction there results a rotation of the angular correlation pattern, which is, however, somewhat more complicated because of the fact that different precession frequencies occur for substates of different m, and precession frequencies with opposite sign occur with equal probability.

These precession frequencies correspond to the energy differences between neighboring levels

$$\omega_Q(m,m') = \frac{E_Q(m) - E_Q(m')}{\hbar}. \tag{7.2}$$

All these frequencies can be expressed as multiples of the fundamental frequencies. For integral I,

$$\omega_Q = \frac{3}{4I(2I-1)} \frac{eQ}{\hbar} \cdot \frac{\partial E_z}{\partial z} = \frac{3 \cdot 2\pi}{4I(2I-1)} \Delta\nu_Q \tag{7.3}$$

and for half-integral I,

$$\omega_Q = \frac{3}{2I(2I-1)} \frac{eQ}{\hbar} \frac{\partial E_z}{\partial z} = \frac{3 \cdot 2\pi}{2I(2I-1)} \Delta\nu_Q. \tag{7.4}$$

The attenuation factor G has the same general form as in Section 6, but since different precession frequencies are present, a sum over the m quantum number is required. G is given in detail by Alder *et al.*,[15] in which the case of arbitrary field directions is also discussed.

The electric interaction strength may be deduced from a measurement of the differential angular correlation as a function of the delay time. It can also be obtained from the integral correlation by measuring the angular correlation for several directions of the crystalline field. Any measurement however, will give only the product of field strength and quadrupole moment. Since field gradients are difficult to calculate, one can only measure ratios of Q of nuclei; however, as it is possible to obtain estimated values of Q from nuclear models, the measurement may be used successfully to study the field gradient magnitudes and the symmetry properties of crystalline material.

The information obtained from these measurements is similar to that obtained from pure quadrupole resonance experiments in single crystals.[19] The same differences described in Section 6 also occur here.

Only one experiment has been carried out so far with single crystals. The angular correlation was measured[17] for the two successive gamma rays emitted in the decay of In^{111} where the radioactive indium atoms were in a tetragonal lattice of indium metal. The angular correlation was measured as a function of the orientation of the crystalline axis **C** with respect to the direction of the two emitted gamma rays $\mathbf{k}_1$, and $\mathbf{k}_2$.

[19] M. H. Cohen and F. Reif, *Solid State Phys.* **5**, 321 (1957).

Figure 6 shows part of these results. The curve in this figure was found by a least square fit to the experimental points. The results yielded a quadrupole interaction $\Delta\nu_Q = 13.4$ Mc.

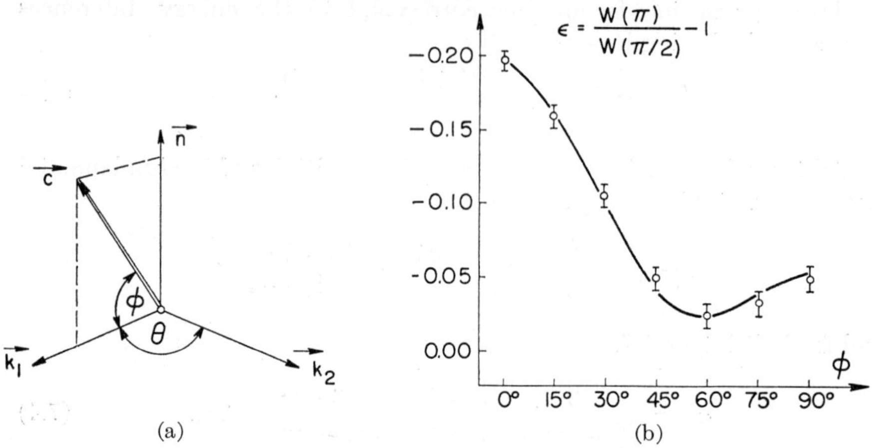

FIG. 6. Measurement of the electric quadrupole interaction in the 247-kev excited state of Cd^{111} with an axially symmetric single crystal source [H. Albers-Schönberg, E. Heer, T. B. Novey, and P. Scherrer, *Helv. Phys. Acta* **27,** 547 (1954)]. (a) Experimental arrangement. **C** is the symmetry axis of the crystal. $\mathbf{k}_1$ and $\mathbf{k}_2$ are the directions of detection of the radiations and **n** is the normal to the plane of the detectors. (b) Comparison of experimental results with theory. The solid curve is a least-square fit from which was derived an interaction energy, $\Delta\nu_Q = 13.4$ Mc.

b. Polycrystalline Sources of Crystals with Axial Symmetry

In this case the field acting on an individual nucleus is the same as in the case of a single crystal. The single crystal formulas apply but for the polycrystalline source must be averaged over all directions. This averaging leads to a simple formula for the attenuation factor of the delayed angular correlation.

$$G_k(t) = \sum_m S_m^{kk} \cos m\omega_Q t \tag{7.5}$$

where ω_Q is given by Eq. (7.3) or (7.4). The S_m^{kk} depend on the spin and are tabulated by Alder *et al.*[15] and Abragam and Pound.[22]

The differential angular correlation is a periodic function. Thus the static quadrupole interaction does not destroy the angular correlation. It merely produces a rotation, provided, of course, that the field strength is constant in time. In practice the last condition is not completely fulfilled, because of lattice vibrations and crystal distortions, for example. Hence the correlation is destroyed in a time given by the relaxation time familiar from nuclear resonance investigations.

The integral correlation

$$G_k = \sum_m \frac{S_m^{kk}}{(1 + m^2\omega_Q^2\tau_N^2)} \tag{7.6}$$

behaves in such a way that the attenuation factor does not go to zero even for infinitely strong interaction but to the minimum value

$$G_{k\,\min} = \frac{1}{2k + 1} \tag{7.7}$$

if we assume the foregoing definition of static field.

The influence of the quadrupole interaction in polycrystalline sources on the angular correlation is closely related to the pure quadrupole nuclear resonance measurements on polycrystalline samples.

Much experimental work has been carried out using a polycrystalline material as a vehicle for the radioactive sources. As a matter of fact, most of the early angular correlation measurements were made with polycrystalline sources at a time when the quadrupole interaction was not known to be present. It is not possible from a measurement of the integral correlation alone to obtain the coefficients of the undisturbed correlation or even to realize that the correlation is disturbed. In the case of a single crystal, we are able to orient the crystalline axis with respect to the direction of emission of the radiations. We then are able to compare the measured to the theoretical angular correlation for all orientations and, in this way, to find both the undisturbed correlation and the interaction strength.

The analogous experiment with a polycrystalline source involves a measurement of the differential, or delayed, angular correlation, since this measurement also allows one to find the undisturbed correlation and the interaction strength. Figure 7 shows a measurement made with a metallic In^{111} source.[20] The solid curve has been calculated according to Eq. (7.5) assuming a quadrupole interaction $\Delta\nu_Q = 17.7$ Mc. The fact that the experimental points lie closely on the theoretical curve is evidence that, to a good approximation, only static electric quadrupole interaction is present in this indium crystal. A similar measurement has been carried out on the lead isotope Pb^{204} present in metallic polycrystalline thallium sources.[21]

Besides the method based on the use of a single crystal and the delayed angular correlation in polycrystalline sources, there exists another, rather insensitive, way to study the static electric quadrupole interaction. This

[20] P. Lehmann and J. Miller, *J. phys. radium* **17,** 526 (1956).
[21] G. K. Wertheim and R. V. Pound, *Phys. Rev.* **102,** 185 (1956).

method consists in a measurement of both A_2G_2 and A_4G_4 in a polycrystalline source. If A_2 and A_4 are known, G_2 and G_4 may be calculated and compared to the theoretical values as functions of the quadrupole interaction strength. If G_2 and G_4 can be fitted with the same interaction strength, it may be concluded that only a pure electric quadrupole interaction is present. Since the error of the measurements is quite large, this method is usually unsatisfactory in practice.

A few remarks on the experimental situation concerning the so-called minimum correlation remain to be made. As pointed out in connection

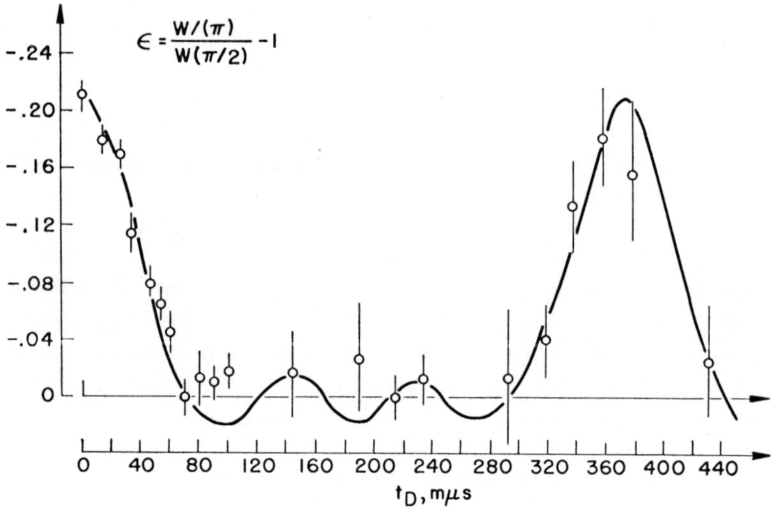

FIG. 7. Measurement of the delayed angular correlation in In^{111} with a polycrystalline indium metal source. From a comparison with Eq. (7.5) an interaction strength $\Delta\nu_Q = 17.7$ Mc can be deduced [P. Lehmann and J. Miller, *J. phys. radium* **17,** 526 (1956)].

with Eq. (7.7), the attenuation factor G never becomes zero in a polycrystalline source, but always stays above a minimum value regardless of the strength of the interaction. A number of measurements of the angular correlation in polycrystalline sources have been reported (mainly measurements of the decay of In^{111}) where the attenuation factor is below its minimum value. Measurements[2] of the angular correlation of successive gamma rays emitted in the decay of In^{111} in a source of $InCl_3$ yielded $G = 0.04$. It has been pointed out by Abragam and Pound[22] that the calculation of the limiting value of G according to Eq. (7.7) is correct only for axial symmetric fields. On the other hand the same authors also have shown that the assumption of nonaxial symmetric fields cannot

[22] A. Abragam and R. V. Pound, *Phys. Rev.* **92,** 943 (1953).

explain attenuation factors smaller than those expected according to Eq. (7.7). Therefore, the conclusion is that we do not have pure static electric quadrupole interaction in this case. The effects which may have to be taken into account are (a) time-dependent electric quadrupole interaction, and (b) effects arising from the excited electron shell after the K capture. Both effects will be discussed later.

c. *Fields without Axial Symmetry*

Very little work has been carried out on this very complicated case. The general formula is given by Devons and Goldfarb.[3] Quantitative calculations have been completed only for the minimum correlation in the case of rhombic symmetry for several values of the spin of the oriented state.

For a numerical treatment of the angular correlation in a nonaxial single crystal the results of calculations made in connection with pure quadrupole nuclear resonance may be utilized.[19]

Table II shows a summary of the information on static electric quadrupole interaction obtained by angular correlation methods. As mentioned before, all the measurements determine the interaction strength, $\Delta\nu_Q$, the product of the quadrupole moment and the electric field gradients. Some attempts have been made to calculate the electric field gradient for the indium lattice in order to obtain an estimate for the electric quadrupole moment of the excited state under investigation (the 247-kev excited state of Cd^{111}). Such a calculation is very difficult to do with a precision that would be of interest to the nuclear physicist. One of the main reasons for this is as follows. The initial radioactive atom incorporated in the tetragonal indium crystal is In^{111} which supposedly occupies a regular lattice position and is under the influence of the regular quadrupole interaction. The gamma cascade, which is used to measure the angular correlation, however, takes place only after the radioactive In^{111} nucleus has decayed by orbital electron capture to an excited Cd^{111} nucleus. The quadrupole interaction, which may be measured in the experiment, therefore, is the interaction of the nucleus of the Cd^{111} atom, which now is an impurity atom in an In lattice. An additional difficulty comes from the fact that the electron shells of the Cd^{111} atom are disturbed because of the preceding capture of an electron from the K shell.

However, the following applications of this method may prove to be useful.

(1) If the radioactive decay goes through more than one level that has a nonvanishing quadrupole moment and sufficiently long lifetime,

TABLE II. LIST OF THE QUANTITATIVE INFORMATION ON STATIC ELECTRIC QUADRUPOLE INTERACTION OBTAINED FROM ANGULAR CORRELATION MEASUREMENTS

Properties of the investigated nuclear level							
Nucleus	Energy kev	Spin	Mean life mμsec	Type of source	Method of measurement	$\Delta\nu_Q$ 10^6 sec^{-1}	Ref.
$_{48}Cd^{111}$	247	5/2	125	In metal polycryst	del. a.c.	17.7	*a*
				In metal monocryst	*	13.4	*b*
				In metal polycryst	†	14.5	*c*
				In_2O_3 polycryst	†	14	*c*
				$CdSO_4$ polycryst	del. a.c.	200	*d*
$_{73}Ta^{181}$	482	5/2	15.3	HfO_2 polycryst	†	>350	*e*
				$(NH_4)_3HF_7$ polycryst	†	59	*e*
				Rb_2HfCl_6 polycryst	†	16.5	*e*
$_{80}Hg^{199}$	158	5/2	3.45	Hg met. polycryst	†	593	*f*
				$HgCl_2$ polycryst	†	1100	*f*
$_{82}Pb^{204}$	1273	4	375	Tl met. polycryst	del. a.c.	12	*d*

* Measurement of the integral angular correlation as a function of the orientation of the single crystal with respect to the detectors.

† Measurement of the integral angular correlation and comparison with the undisturbed angular correlation as established from measurements in liquid sources.

[a] P. Lehmann and J. Miller, *J. phys. radium* **17,** 526 (1956).

[b] H. Albers-Schönberg, E. Heer, T. B. Novey, and P. Scherrer, *Helv. Phys. Acta* **27,** 547 (1954).

[c] R. M. Steffen, *Phys. Rev.* **103,** 116 (1956).

[d] P. Lehmann and J. Miller, *Compt. rend.* **240,** 298 (1954).

[e] P. Debrunner, E. Heer, W. Kündig, and R. Rüetschi, *Helv. Phys. Acta* **29,** 463 (1956).

[f] R. V. Pound and G. K. Wertheim, *Phys. Rev.* **102,** 396 (1956).

one can assume that the electric field gradient is the same for all the levels and can obtain ratios of quadrupole moments with good precision. This is only true, however, if time-dependent effects caused by excited electron shells originating from *K* capture or internal conversion are of no importance.

(2) If the radioactive nucleus is an isomer (i.e., if it decays by gamma emission only), it does not change its electronic configuration during the decay, so that calculations of the electric field gradient are more likely to be applicable. In this case it should be possible to measure the ratio of the quadrupole moment of the excited state to that of a stable isotope of the same element by measuring the quadrupole interaction strength of the stable isotope in the same lattice by nuclear resonance techniques. The ratio of the interaction strength to that of the excited state determined by angular correlation methods then gives the ratio of quadrupole moments.

(3) Investigations of lattice imperfections also might utilize this type of measurement. This topic is discussed in Part VIII.

8. Combined Static Magnetic and Axially Symmetric Electric Interaction

In the case of a static quadrupole interaction, the nuclear level splitting in an external magnetic field is given by

$$E_m = eQ\frac{\partial E_z}{\partial z}\frac{3m^2 - I(I+1)}{4I(2I-1)} + g\mu_N Hm = \frac{3m^2 - I(I+1)}{4I(2I-1)}\Delta\nu_Q h + g\mu_N Hm. \quad (8.1)$$

The influence of this combination can be calculated easily if the fields are parallel. This requires the use of a single crystal having axial symmetry. The results take a simple form if we assume the detectors lie in the plane perpendicular to the field direction and if only A_0 and A_2 are different from 0. If we write the correlation in the form

$$W(\theta) = 1 + b_2 g_2 \cos 2\theta = 1 + b_2^* \cos 2\theta \quad (8.2)$$

the attenuation factor g_2 for the delayed angular correlation is given by

$$g_2(t,\omega_Q,\omega_L) = \sum_m S_{m2}{}^{22} \cos(m\omega_Q t)\cos(2\omega_L t). \quad (8.3)$$

The attenuation factor for the integral angular correlation is given by

$$g_2(\omega_Q,\omega_L) = \tfrac{1}{2}\sum_m S_{m2}{}^{22}\left[\frac{1}{1+(m\omega_Q + 2\omega_L)^2} + \frac{1}{1+(m\omega_Q - 2\omega_L)^2}\right]. \quad (8.4)$$

The coefficient $S_{m\mu}{}^{k_1k_2}$ is tabulated by Alder *et al.*;[15] ω_Q given by Eq. (7.3) or (7.4); ω_L by Eq. (6.1). If the correlation involves higher terms than $P_2(\cos\theta)$ or if the detectors are not in the plane specified, appropriate formulas should be taken from Alder.[15] The behavior of $g_2(\omega_Q,\omega_L)$ as a function of the magnetic field shows the interesting feature that it first increases with increasing ω_L. This can be understood with the help of the vector model. The electric interaction causes the nuclei to precess around the crystalline axis, both senses of rotation being equally probable. The magnetic interaction, on the other hand, causes a precession in only one direction. When both interactions are present they will interfere in such a way that the precession of half of the nuclei is reduced. This is equivalent to a reduced influence and gives rise to a maximum in the g factor. Then by measuring this g factor as a function of the magnetic field it is possible

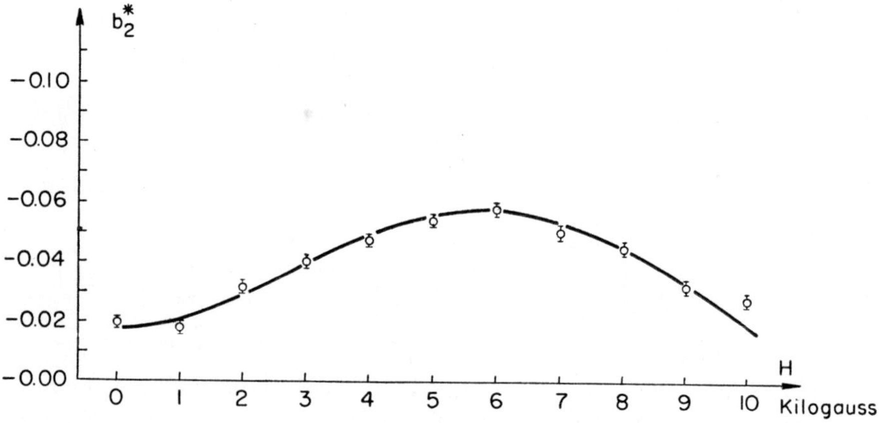

FIG. 8. Effect of combined electric and magnetic interaction on the angular correlation in Cd^{111}. The coefficient b_2^* of the expansion $W(\theta) = 1 + b_2^* \cos 2\theta$ is plotted as a function of the externally applied magnetic field. The solid line is a least-square fit to the experimental points [H. Albers-Schönberg, E. Heer, T. B. Novey, and P. Scherrer, *Helv. Phys. Acta* **27,** 547 (1954)].

to obtain both the electric and magnetic interaction strengths. A measurement showing the effect of combined electric and magnetic interaction on angular correlation has been carried through for the excited state of Cd^{111}. Figure 8 shows the experimental result. The solid curve was obtained by a least square fit using Eq. (8.4). Both ω_Q and ω_L were obtained. For nonparallel fields the theory has not been worked out. This case is of interest for angular correlation measurements involving polycrystalline sources in a magnetic field. Here, as in the single crystal case, both the electric and magnetic interaction strengths can be obtained from a measurement of the g factor as a function of the magnetic field. The information gotten is closely related to that obtained from nuclear magnetic resonance in polycrystalline sources. Some calculations have been carried out for magnetic field interactions small compared to electric interactions and vice versa.[22]

9. Time-Dependent Electric Quadrupole Interaction

Time-dependent quadrupole interactions may occur in crystalline sources because of thermal lattice vibrations; however, the most important case is that occurring in liquids. The liquids may be characterized by $\langle(\partial E/\partial z)^2\rangle$, which is the average over time and space of the square of the electric field gradient, and by the correlation or relaxation time τ_c. This time τ_c was introduced by Debye.[23] It characterizes the time required for a liquid to change its configuration and thus gives the time that a quadrupole field acts on a given nucleus in a given direction.

Under the condition $\langle\omega_Q^2\rangle^{\frac{1}{2}}\tau_c \ll 1$, the influence of such a time-dependent quadrupole interaction can be calculated easily and yields for the differential correlation: $W(\theta,\tau) = \Sigma A_k G_k P_k(\cos\theta)$ with

$$G_k(t) = e^{-\lambda_k t} \tag{9.1}$$

where

$$\lambda_k = \langle\omega_Q^2\rangle \cdot \tau_c$$

or

$$\lambda_k = \frac{3}{80}\frac{k(k+1)[4I(I+1) - k(k+1) - 1]}{I^2(2I-1)^2} 4\pi^2\langle\Delta\nu_Q^2\rangle\tau_c \tag{9.2}$$

$$\langle\omega_Q^2\rangle = \frac{3}{80}\left(\frac{eQ}{\hbar}\right)^2\left\langle\left(\frac{\partial E}{\partial z}\right)^2\right\rangle_{\text{Av}} \frac{k(k+1)[4I(I+1) - k(k+1) - 1]}{I^2(2I-1)^2}$$

$$\langle\Delta\nu_Q^2\rangle = \left(\frac{eQ}{\hbar}\right)^2\left\langle\left(\frac{\partial E}{\partial t}\right)^2\right\rangle_{\text{Av}}. \tag{9.3}$$

The integral correlation is obtained as before by the integration

$$G_k = \frac{1}{\tau_N}\int_0^\infty e^{-t/\tau_n} G_k(t)\,dt = \frac{1}{1+\lambda_k\tau_n} = \frac{1}{1+\langle\omega_Q^2\rangle\tau_c\tau_N}. \tag{9.4}$$

The differential correlation (9.1) in contrast to the case of the static quadrupole interaction is not a periodic function but decreases monotonically. In other words, after sufficiently long time, the original orientation of the assembly of nuclei is completely lost. This is also reflected in the integral correlation (9.4), which can become zero for large field strengths or long lifetimes.

If we compare the integral correlation for the cases of static and time-dependent quadrupole interactions, we see that there is appreciable attentuation ($G_k \sim \frac{1}{2}$), if in the first case $\omega_Q^2\tau_n^2$, and in the second case $\langle\omega_Q^2\rangle\tau_c\tau_n$ are comparable to one. If $\langle\Delta\nu_Q^2\rangle$ in the time-dependent case is about the same as $\Delta\nu_Q^2$ in the static case, then there can be a much longer lifetime τ_n in a liquid than in a solid before an attenuation is

[23] P. Debye, "Polar molecules." Dover, New York, 1945.

observed; this follows because in a liquid τ_c is very small. For example, if $\Delta\nu_Q$ is 1000 Mc, the nuclear lifetime in a solid must be short compared to 10^{-9} sec. However, in a liquid whose correlation time is 10^{-11} sec, the lifetime must only be short compared to 10^{-7} sec.

From these considerations it is clear why liquid sources are preferred for the study of unperturbed correlations. Liquid metals or other non-polar liquids are especially well suited for, in addition to having short correlation times, they have smaller average interaction strengths principally because atomic groupings possessing dipole moments are absent.

When the integral correlation is studied as a function of the correlation time, the attenuation factor decreases with increasing correlation time as can be seen from (9.4), and may reach a very small value at times long compared to $(1/\tau_n)\langle\omega_Q{}^2\rangle$. On the other hand, as $\langle\omega_Q{}^2\rangle^{\frac{1}{2}}\tau_c$ becomes comparable to one, Eq. (9.4) no longer applies. The attenuation factor begins to increase and approaches the value for the static field case for large interactions.

Measurements of delayed angular correlations in liquid sources have been reported by several experimenters.[24–27] The measurements in which an aqueous solution of indium salts was employed showed a slow decrease of the attenuation factor with increasing delay time. The decrease is compatible with an interaction strength $\langle\Delta\nu_Q{}^2\rangle^{\frac{1}{2}} = (450 \pm 100)$ Mc assuming $\tau_c \sim 10^{-11}$ sec. Measurements also have been made[25] with indium salts dissolved in a mixture of glycerine and water. According to simple assumptions τ_c is given by

$$\tau_c = (4\pi a^3/3kT)\eta \tag{9.5}$$

where a is a characteristic length (effective molecular diameter) and η is the viscosity.[23,28] Adding glycerine to the water changes the viscosity and therefore the relaxation time. Figure 9 shows the experimental results. Here $A_2{}^*(t) = A_2G_2(t)$ is measured for sources of different viscosity. The coefficient $A_2{}^*$ shows the expected exponential behavior [Eq. (9.1)] if we ignore the initial slope which has not been corrected for the finite resolving time of the coincidence analyzer.

Figure 10 shows the interaction parameter λ_2 determined from this experiment as a function of the viscosity. According to Eq. (9.5), a linear dependence is expected. There are marked deviations which may arise from the following facts. First, the change in glycerine concentra-

[24] V. E. Krohn, T. B. Novey, and S. Raboy, *Phys. Rev.* **105,** 234 (1957).
[25] R. M. Steffen, *Phys. Rev.* **103,** 116 (1956).
[26] H. Albers-Schönberg, E. Heer, and P. Scherrer, *Helv. Phys. Acta* **27,** 637 (1954).
[27] J. S. Fraser and J. C. D. Milton, *Phys. Rev.* **94,** 795 (1954).
[28] N. Bloembergen, E. M. Purcell, and R. V. Pound, *Phys. Rev.* **73,** 679 (1948).

tion not only changes τ_c but $\langle \Delta\nu_Q{}^2 \rangle$ and a as well. Second, there may be after-effects present from the preceding electron capture. This topic will be discussed in Section 17.

Delayed angular correlation measurements have been reported[24] for the alpha-gamma correlation in the decay of Am^{241}, where again liquid

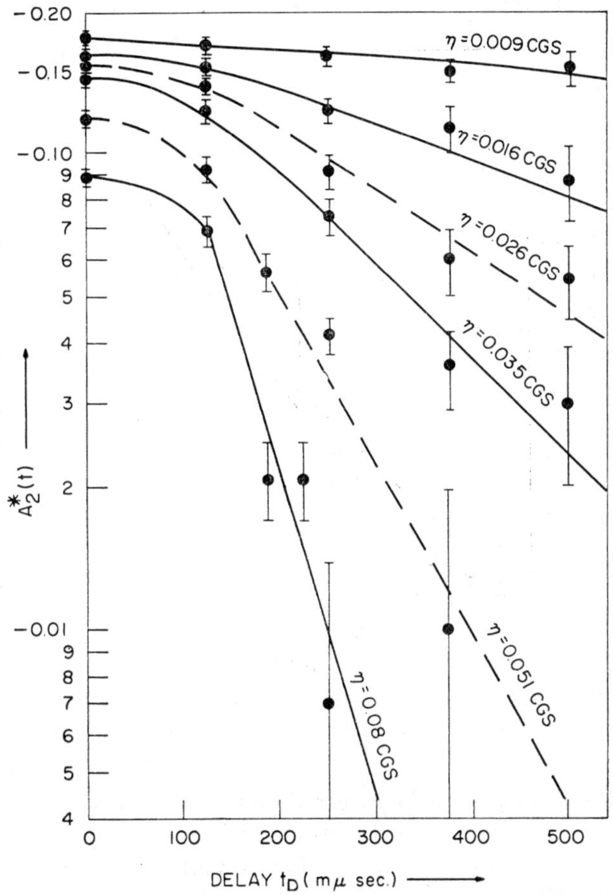

FIG. 9. The delayed correlation $A_2{}^*(\tau)$ observed in liquid In^{111} sources as a function of the delay time t_D. The source is a dilute aqueous solution in $InCl_3$ whose viscosity was varied by adding glycerine [R. M. Steffen, *Phys. Rev.* **103**, 116 (1956)].

sources are used. The authors find that for some liquids (for example dilute perchloric acid) the attenuation factor behaves according to Eq. (9.1). An interaction strength $\langle \Delta\nu_Q{}^2 \rangle^{\frac{1}{2}} \sim 700 \pm 100$ Mc can be calculated from the measurements. (See Fig. 11.) Some other liquids (such as acetic acid solutions) do not give an exponential behavior for G_2. This may be a consequence of the fact that this liquid, showing a strong interaction, does not fulfill the condition $\langle \omega_Q{}^2 \rangle^{\frac{1}{2}} \tau_c \ll 1$.

An explanation of the source of the much larger interaction strength found in this case lies in the configuration of the americium ion in solution. In the strong acids the ion is surrounded with water molecules whose dipole moments tend to be symmetrically distributed and hence do not

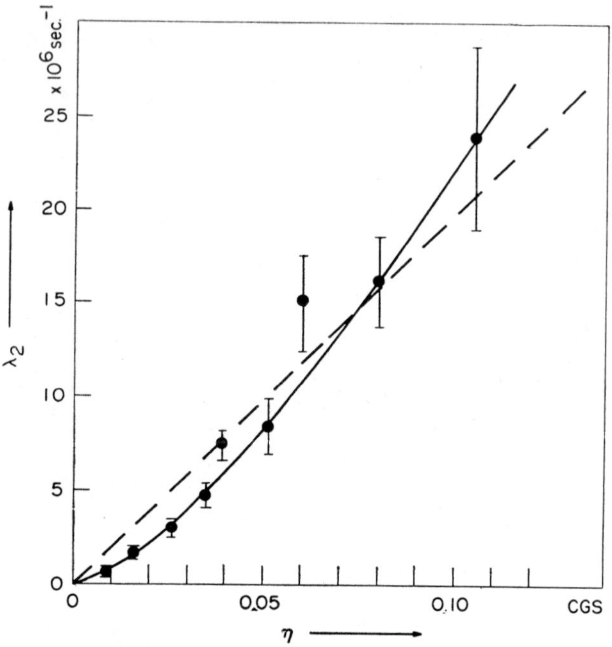

FIG. 10. The interaction parameter λ_2 as determined from the measurements in Fig. 9 *versus* the viscosity. The dashed line represents the expected theoretical behavior [R. M. Steffen, *Phys. Rev.* **103,** 116 (1956)].

produce abnormally large field gradients at the nucleus. In acetic acid solution, however, one acetate group is known to be firmly attached (complexed) to the americium ion. This asymmetrical grouping produces a very large field gradient at the nucleus and hence a larger interaction strength.

Delayed angular correlation measurements not only prove useful because they reveal the type of influence present, but they are also important because they are able to show whether or not an influence is actually present. It has been suggested by many authors that liquid metallic sources are especially well suited for measurements of the undisturbed angular correlation. Three reasons can be given. (a) The correlation times are very short ($\tau_c \sim 10^{-12}$ sec). (b) The average quadrupole interaction frequency is usually rather small; in solid indium metal $\omega_Q \sim 10$ Mc, in liquid indium metal, we expect something of the same order of magnitude, or less. (c) The influence of preceding K capture is small (see Section 17).

Figure 12 shows a delayed angular correlation measurement made in a liquid metallic indium source. The result indicates that $\langle \Delta\nu_Q^2 \rangle \tau_c <$

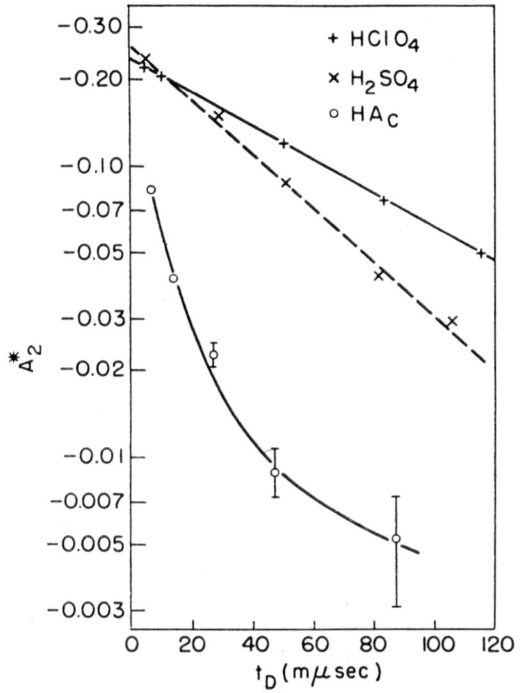

FIG. 11. The attenuation factor $A_2^*(\tau)$ for the delayed angular correlation observed in various liquid Am^{241} sources *versus* the delay time [V. E. Krohn, T. B. Novey, and S. Raboy, *Phys. Rev.* **105**, 234 (1957)].

5×10^4 sec^{-1}. Since the condition for negligible influence on the correlation is $\langle \omega_Q^2 \rangle \tau_c \tau_N \ll 1$, it follows that this condition holds in these liquid sources for nuclear lifetime shorter than 10^{-6} sec.

Integral correlations in liquid sources have been measured in great numbers. As in the case of the polycrystalline source, not much can be learned from these measurements, since they usually cannot show whether an interaction is present or of what nature and strength it is. There have been, however, some attempts to disentangle the situation by more elaborate experiments.[25,26] The following ones may be worth mentioning.

(a) The measurement of the integral correlation in a liquid source as a function of the viscosity of the source can give λ_k by comparison with Eq. (9.4) and can give the undisturbed correlation by extrapolation to zero viscosity. Comparison of the measured λ_k with the one calculated from the viscosity, η, can test the validity of the basic assumptions of the liquid interactions.

Figure 13 shows the result of such a measurement.[25] The source again is In^{111} dissolved in glycerine-water mixtures. Again there are deviations from the expected theoretical curve.

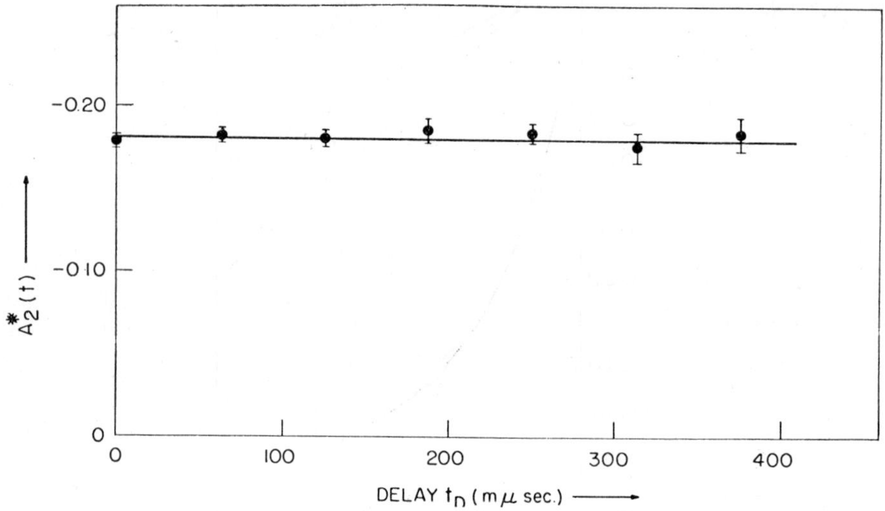

FIG. 12. The coefficient $A_2^*(\tau)$ of the delayed angular correlation observed with a liquid indium metal source as a function of delay time [R. M. Steffen, *Phys. Rev.* **103,** 116 (1956)].

(b) If the correlation function contains two nonvanishing terms (e.g., A_2 and A_4), a measurement[18] of the correlation function for two sources having different but known viscosities (and therefore correlation times) can be used to determine A_2, A_4, and the interaction strength. This method requires a precision measurement of A_4 which is very difficult to achieve.

Table III shows a summary of the information on time-dependent electric quadrupole interaction obtained by angular correlation methods.

10. Time-Dependent Electric Quadrupole Interaction Plus External Magnetic Field

In the last section we have shown that liquid sources usually show only small attenuation and are therefore well suited for measurements of magnetic moments of short-lived excited states with the use of methods described in Section 6. Liquid metal sources show very little attenuation ($\langle\langle\omega_Q^2\rangle\tau_c < 5 \times 10^4$ sec^{-1}). Aqueous solutions show considerably more attenuation ($\langle\langle\omega_Q^2\rangle\tau_c \sim 2 \times 10^7$ sec^{-1}). In aqueous solutions and for lifetimes longer than 10^{-8} sec the attenuation therefore cannot be neglected in a magnetic moment measurement.

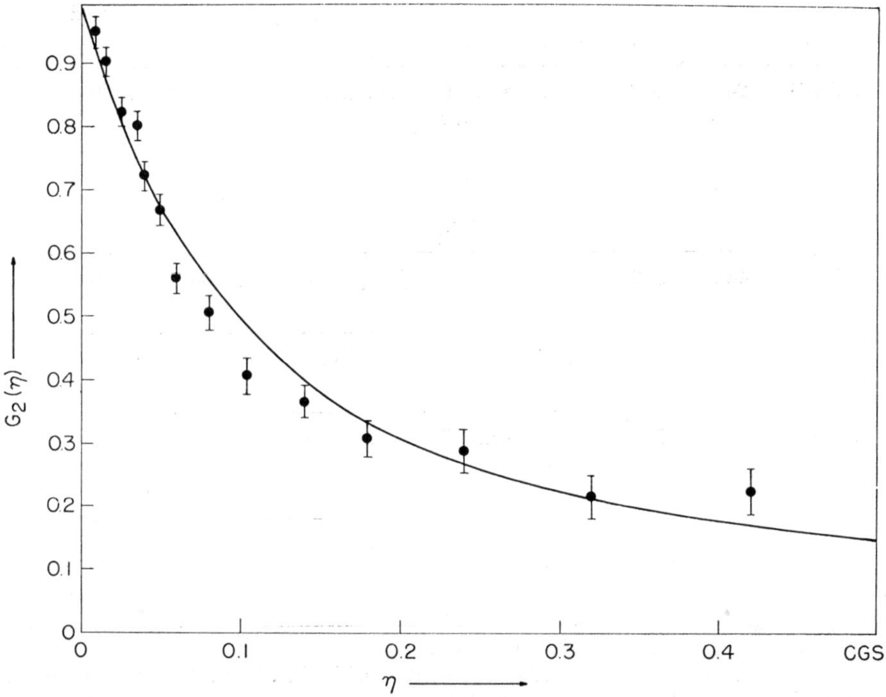

FIG. 13. The attenuation factor G_2 for the integral correlation observed in liquid In^{111} sources as a function of the viscosity. The source is a dilute aqueous solution of indium chloride whose viscosity was varied by adding glycerine [R. M. Steffen, *Phys. Rev.* **103,** 116 (1956)].

With a magnetic field applied in a direction normal to the plane of the detectors, the correlation function is given by[22]

$$W(t) = \sum_k G_k(t) A_k P_k(\cos [\theta - \omega_L t]) \tag{10.1}$$

where $G_k(t)$ is the attenuation factor for time-dependent electric quadrupole interaction, Eq. (9.1). From the foregoing equation explicit expressions for delayed and integral angular correlations have been derived.[18,22,29]

[29] R. M. Steffen and W. Zobel, *Phys. Rev.* **97,** 1188 (1955).

TABLE III. LIST OF THE QUANTITATIVE INFORMATION ON TIME-DEPENDENT ELECTRIC QUADRUPOLE INTERACTION IN LIQUID MEDIA OBTAINED BY ANGULAR CORRELATION METHODS

Properties of the investigated nuclear level								
Nucleus	Energy kev	Spin	Mean life mμsec	Type of source	Method of measurement	$<\Delta\nu_Q^2>$ 10^6 sec^{-1}	Assumed τ_c sec	Ref.
$_{48}Cd^{111}$	247	5/2	125	aqueous sol. of $InCl_3$	del. a.c.	~450	10^{-11}	*a*
				liq. In metal	del. a.c.	<140	10^{-12}	*a*
$_{73}Ta^{181}$	482	5/2	15.3	HfF_4 in HF	integral a.c.	<1000	10^{-11}	*b*
				HfF_4 in Hf	del. a.c.	<1600	10^{-11}	*c*
$_{93}Np^{237}$	60	5/2	91	$AmClO_4$ in $HClO_4$	del. a.c.	73	10^{-11}	*d*
				Am_2SO_4 in H_2SO_4	del. a.c.	93	10^{-11}	*d*

[a] R. M. Steffen, *Phys. Rev.* **103,** 116 (1956).
[b] P. Debrunner, E. Heer, W. Kündig, and R. Rüetschi, *Helv. Phys. Acta* **29,** 463 (1956).
[c] F. K. McGowan, *Phys. Rev.* **93,** 471 (1954).
[d] V. E. Krohn, T. B. Novey, and S. Raboy, *Phys. Rev.* **105,** 234 (1957).

In addition to aiding in the determination of magnetic moments, an external field also may be used to restore the disturbed angular correlation to its field-free value. This may be understood in the following way. In a nonparamagnetic liquid the electric field gradients are averaged out over times of the order of τ_c, the relaxation time of the liquid. In the presence of a magnetic field, the nuclear spin has a Larmor precession frequency ω_L and the effective relaxation time of the liquid is of the order of $1/(1/\tau_c + \omega_L)$. It follows that, in order for an appreciable restoration of the angular correlation to occur, ω_L must be large compared to $1/\tau_c$. This requires fields of the order of $10^6 - 10^7$ gauss for the relaxation times observed for water. Of course the field must be applied parallel to the direction of initial orientation in order to avoid influencing the correlation. (See Section 6.)

11. Time-Dependent Magnetic Interaction

A time-dependent magnetic interaction can be present in a liquid source in which radioactive atoms experience the field of neighboring paramagnetic atoms or neighboring nuclei having a magnetic moment. In this section we consider only measurements on nonparamagnetic radioactive atoms.

The formalism for this case is very similar to that for the time-dependent electric quadrupole cases. Equations (9.1) and (9.4) still hold, but (9.2) must be replaced by

$$\lambda_k = \langle \omega_L{}^2 \rangle \tau_c \sim g^2 \mu_n{}^2 \langle H^2 \rangle_{\text{Av}}. \tag{11.1}$$

The magnetic field which appears in Eq. (11.1) is caused by neighboring atomic and nuclear magnetic moments. Expressions for calculating these fields are given by Bloembergen *et al.*[28] and Van Vleck.[30] The field caused by neighboring nuclear moments is of the order of a fraction of a gauss and therefore can be neglected for lifetimes shorter than 10^{-3} sec, even for a very long correlation time, and for much longer lifetimes in most liquids. This situation is reflected in the very long spin-spin relaxation times, of the order of seconds or longer, for protons in nonparamagnetic media where the correlation times are very short.

The effects of atomic moments are larger by a factor of 1836 but the concentration of paramagnetic atoms usually cannot be made larger than $\approx 10^{21}$/cc. Since the field decreases with the concentration, it usually will be at most of the order of a few hundred times larger than for the nuclear moments. This, however, is still a small effect and has not been observed. However, it has been found in nuclear resonance studies; in these cases the relaxation time usually is determined by the interaction with paramagnetic impurities and is as short as 10^{-5} to 10^{-6} sec.

IV. Extranuclear Effects Involving Paramagnetic Atoms

12. Introduction

In the previous sections we discussed the influence of external magnetic and electric fields on the angular correlation. A basic assumption was that the electron shells of the nuclei under investigation have no electric quadrupole or magnetic moments and therefore do not interact with the nuclear moments.

[30] J. H. Van Vleck, *Phys. Rev.* **74,** 1168 (1948).

In this and the following Sections 13–15, we will drop this assumption. The presence of the electronic moments influences the angular correlation in two ways. First, there is a strong coupling between the electron shell and the nucleus which produces an attenuation of the angular correlation. Second, because of this strong coupling, the influence of externally applied fields is different from that described in the case of "free" nuclei previously treated.

The coupling between electronic and nuclear moments has been treated in detail in connection with hyperfine structure in optical spectra.[31] Its influence on the angular correlation was the first type to be considered.[32] These calculations and later[33] numerical calculations for some special cases were made for free atoms in a stationary state. This restriction is approximately fulfilled for gaseous sources. We will treat this case in the next section. In Section 14 paramagnetic atoms in solutions will be considered, and in Section 15 paramagnetic atoms in solids.

The effects of paramagnetism created by the perturbation of the electron shell produced by preceding decay processes will be discussed in Sections 16 and 17.

13. Free Paramagnetic Atoms

The interaction which we consider here is that between the moments of the nucleus and the electron shell surrounding it. Both the electric and the magnetic interaction are present in general; however, as is well known from optical hyperfine splitting, the electric hyperfine splitting is small compared to the magnetic. Therefore in what follows we consider only the magnetic interaction for which we have the differential correlation:

$$G_k(t) = \sum_{FF'} (2F + 1)(2F' + 1) W(IJkF|F'I)^2 \cos \omega_{FF'} t \qquad (13.1)$$

and the integral correlation:

$$G_k = \sum_{F,F'} \frac{(2F + 1)(2F' + 1) W(IJkF|F'I)^2}{1 + (\omega_{FF'}\tau_N)^2} \qquad (13.2)$$

where J is the total angular momentum of the electron shell, I is the nuclear spin, $F = J + I$ is the total atomic angular momentum, and $\omega_{FF'}$ is the hyperfine splitting. Its value may be taken from the pertinent calculations.[31]

[31] H. Kopfermann, "Nuclear Moments." Academic Press, New York, 1958.
[32] G. Goertzel, *Phys. Rev.* **70**, 897 (1946).
[33] K. Alder, *Helv. Phys. Acta* **25**, 235 (1952).

For $\omega_{FF'}\tau_N \gg 1$ we obtain the minimum correlation (hard core)

$$G_{k\ \min} = \sum_F (2F + 1)^2 W(IJKF|FI)^2. \tag{13.3}$$

In order to see more clearly the way the correlation is influenced we consider the special case of $J = \frac{1}{2}$. One then obtains

$$G_k = 1 - \frac{k(k+1)}{(2I+1)^2} \frac{(\omega\tau_N)^2}{1 + (\omega\tau_N)^2} \tag{13.4}$$

$$G_{k\ \min} = 1 - \frac{k(k+1)}{(2I+1)^2}. \tag{13.5}$$

The differential correlation and the integral correlation show a behavior similar to that in the case of a polycrystalline source. The differential correlation is periodic in time. The integral correlation does not approach isotropy even for very strong interactions. To obtain an estimate of the size of the effect, we may assume a hyperfine splitting of 10^{-2} cm^{-1}, corresponding to a magnetic field of the order of 10^6 gauss at the nucleus. The interaction then is $\sim$500 Mc and it results in a noticeable effect on the angular correlation if $\tau_n > 10^{-9}$ sec.

If the angular correlation is influenced by such a paramagnetic interaction, it can be restored, in principle, by applying a sufficiently strong magnetic field parallel to the initial direction of orientation. The field strength must be sufficiently large so that $(HJ) \gg (IJ)$. This requires of the order of 10 kilogauss. This decoupling can be understood as follows. The strong external magnetic field causes the electronic moment J to precess rapidly about the external field direction and in this way averages out the component due to J perpendicular to the external field. The remaining component of the electronic moment parallel to the preferred direction does not affect the angular correlation as we have seen earlier (Section 6).

The equations of this section are valid for free paramagnetic atoms only. In practice they might be applied to the case of dilute paramagnetic gases if the collision time between paramagnetic atoms or between the paramagnetic atoms and the walls of their container were long compared to the nuclear lifetime.

No experimental work related to the effects discussed in this section has come to our attention to date.

14. Paramagnetic Atoms in Solutions

In this section we consider solutions of paramagnetic ions and those gases in which the time between collisions of the paramagnetic atoms is short compared to the nuclear lifetime. In a solution the electronic moment is coupled to the lattice vibrations. This coupling produces a random motion of the electronic moment S which can be characterized by an electronic relaxation time τ_S. The effect on the angular correlation is very similar to the case of time-dependent interaction already treated. Equation (9.1) still holds and (9.2) has to be replaced by

$$\lambda_k = \tfrac{2}{3}\tau_S\omega_S{}^2I(I+1)S(S+1)[1-(2I+1)W(I1kI|II)] \quad (14.1)$$

where S is the electronic spin angular momentum and $\hbar\omega_S\mathbf{I}\cdot\mathbf{S}$ is the interaction energy. As before, this treatment is valid only for $\omega_S\tau_S \ll 1$.

Experimental evidence for this type of interaction has been reported in one case.[34] The angular correlation of the gamma rays from Coulomb excitation of several rare earth isotopes in aqueous solution has been measured and compared to theoretical calculations.[13] The nuclides Nd^{150}, Sm^{152}, and Sm^{154} showed no disturbance within the accuracy of the measurement. On the other hand Gd^{154}, Gd^{156}, and Gd^{160} showed attenuation factors of about one-half. All of these nuclides have approximately equal lifetimes of 2×10^{-9} sec. On first sight this behavior is somewhat unexpected since the Gd isotopes have a rather small hyperfine splitting compared to those of Nd and Sm because the Gd ground state is a $S_{\frac{7}{2}}$ level whereas that of Sm is $H_{\frac{5}{2}}$. However, because Gd is an S state, its spin-lattice coupling is much weaker and therefore the relaxation time τ_S is much longer and the attenuation accordingly larger. From the fact that Nd and Sm show no disturbance we can calculate that the relaxation time τ_S must be of the order of 10^{-12} sec assuming a hyperfine splitting of 6×10^{-3} cm^{-1}.

In the case of Gd we can calculate a relaxation time of 10^{-8} sec, assuming a hyperfine splitting of 5×10^{-5} cm^{-1} which is only a hundredth of the splitting for the case of the other rare earths. The different Gd isotopes show the same attenuation. Thus we can conclude that the perturbation is not caused by electric quadrupole interaction, since the calculated quadrupole moments vary by a factor of three.

To verify that the attenuation results from the long spin lattice relaxation time in the case of Gd, one could measure the attenuation factor as a function of temperature. In general τ_S shortens as the temperature increases and therefore the attenuation factor should increase correspondingly.

Another way to restore the angular correlation would be to apply a sufficiently strong magnetic field along the direction of initial polarization. When $g\mu_kH/\hbar$ is large compared to the larger of ω_S and $1/\tau_S$ the

[34] G. Goldring and R. P. Scharenberg, *Phys. Rev.* **110,** 701 (1958).

angular correlation would be restored completely. For the above-mentioned correlation of Gd a field of 10 kilogauss would completely restore the correlation.

The measurement of a nuclear magnetic moment is somewhat complicated by the presence of the paramagnetic electron shell. In a paramagnetic atom, a field of the order of 10^6 gauss is produced at the nucleus. In the absence of an external magnetic field, this field is averaged out over the lifetime of the oriented state if $\tau_S \ll \tau_N$. When an external field is applied, the atomic levels are split and the sublevels are populated unequally according to the usual Boltzmann distribution. This results in a net field at the nucleus in addition to the external field.

The above statements can be understood more easily if we consider the example of a spin $\frac{1}{2}$ electron which can be described as having orientations "up and down" along an arbitrary axis of quantization which we take to be the direction of the magnetic field if present. During the time the spin is "up" the nuclear moment will precess in one direction, and when the spin is "down" it will precess in the other direction. If no external field is present, then the probability for the spin to be up and down is the same. If $\tau_S \ll \tau_N$ the spin is up as often as down for each individual atom; therefore, there is no net precession and no attenuation of the angular correlation.

In an external field, on the other hand, the probability for the spin to be one way is larger than for the other. If again $\tau_S \ll \tau_N$ this results in a net precession of the nuclear moment, corresponding to a net magnetic field which can be calculated from the difference in population. This precession is superimposed on the usual Larmor precession of the nucleus around the external magnetic field.

Thus influence on the angular correlation can be calculated by using the formulation of Section 6 where in all equations the magnetic field is replaced by an effective magnetic field.[34]

$$H_{\text{eff}} = \left(1 + \frac{\mu_0}{\mu_n} \frac{A}{3kT} \frac{g_{\text{eff}}^2}{g_e g_n}\right) H \tag{14.2}$$

where μ_0 = Bohr magneton

μ_n = nuclear magneton

A = is the constant in the hyperfine splitting interaction $A(\mathbf{J} \cdot \mathbf{I})$

$g_{\text{eff}}^2 = \dfrac{3kTX_0}{N\mu_0^2}$, where X_0 is the atomic susceptibility and N is the number of atoms per cubic centimeter

$g_e = 1 + \dfrac{S(S+1) + J(J+1) - L(L+1)}{2J(J+1)}$

g_n = nuclear g factor.

Measurements of nuclear magnetic moments in paramagnetic atoms have been made in Nd^{150}, Sm^{152}, and Sm^{154};[34] H_{eff} has been calculated as $(2.3 \pm 0.03)H$ for Nd and $(1.7 \pm 0.2)H$ for Sm, both at room temperature. Similar measurements have been made in Dy^{160} and Eu^{152}.[35]

15. Paramagnetic Atoms in Solids

In solids the electronic moment is not free as in the previously considered cases of paramagnetic liquids and gases, but is coupled through the electronic orbits with the inhomogeneous electric field of the crystal. The influence on the angular correlation has not been calculated for the most general type of anisotropic magnetic hyperfine structure. In the special case in which the spin part of the Hamiltonian

$$K = AI_ZS_Z + BI_XS_X + CI_YS_Y$$

can be approximated by $K = AI_ZS_Z$ and for $S = \frac{1}{2}$ the attenuation factor is given[22] by

$$G_k = \frac{1}{2k+1} \sum_{\mu} \frac{1}{1 + \left(\frac{\mu A \tau_N}{2K}\right)^2}. \tag{15.1}$$

For strong interaction this goes to the minimum correlation

$$G_{k\ \min} = \frac{1}{2k+1}. \tag{15.2}$$

The application of an additional external magnetic field has not been treated in detail; however, the restoration of the undisturbed angular correlation by such a field has been discussed.

In the case of free paramagnetic atoms we have seen that a magnetic field applied in the direction of initial orientation and fulfilling the condition $g\mu_0H/\hbar \gg \omega_S$ completely restores the angular correlation attenuated by the paramagnetic interaction. This is so because the electron spin precesses with a frequency $g\mu_0H/\hbar$ around the field direction. Components of the magnetic field perpendicular to the external field are averaged out in this way in times comparable to the precession time of the nuclear moment about the electron spin direction.

In the case of a paramagnetic atom in a solid the effective electron spin is coupled tightly to the lattice[36] and therefore is not free to rotate. Thus components of the magnetic field perpendicular to the direction of initial orientation cannot be averaged out with available magnetic fields.

[35] P. Debrunner, W. Kündig, J. Sunier, and P. Scherrer, *Helv. Phys. Acta* **31,** 326 (1958); and private communication.

[36] B. Bleaney, *Phil. Mag.* [7] **42,** 441 (1951).

No experimental results concerning the topic of this section are available.

V. Extranuclear Fields Created by the Nuclear Process (Radiation Damage)

16. The Effect of the Nuclear Recoil

In the process of polarization the nucleus may receive an appreciable recoil energy. This recoil energy is given in Table IV for a number of cases. A numerical value of the recoil energy for a typical case is also given.

Table IV

Process	Recoil energy in ev	Typical case: Energy	A	Recoil energy
Gamma emission	$533E_\gamma^2/A$ ev E_γ in Mev A, atomic weight of nucleus	0.5 Mev	100	1.33 ev
Neutrino emission (electron capture)	$533E_\gamma^2/A$ ev E_γ in Mev (total transition energy)	0.25 Mev	150	0.22 ev
Beta-particle emission	$\frac{139}{A}\left[\left(\frac{E_\beta}{.511}+1\right)^2-1\right]$ ev E_β in Mev	1.0 Mev	100	11 ev
Electron emission (internal conversion)	$\frac{139}{A}\left[\left(\frac{E_e+1}{.511}\right)^2-1\right]$ ev E_e in Mev	0.5 Mev	100	4.1 ev
Alpha-particle emission	$4 \times 10^6 \frac{E_\alpha}{A}$ ev E_α in Mev	4 Mev	100	160,000 ev

If the energy is of the order of a few electron volts or less, the recoiling nucleus does not leave the original site in the lattice and effects arising from such displacement are not expected or observed. For example, in the case of the angular correlation of the radiations following the K capture in In^{111}, measurements in single crystals gave the correlation expected for a nucleus at the correct lattice position which is subjected to an electric field having the symmetry of the crystal.[17]

Of greatest importance is the recoil in alpha decay for which recoil energies of the order of a hundred kilovolts occur. It is expected that the nucleus will come to rest in a time of the order of 10^{-11} sec. In general, the recoil will influence the angular correlation in two ways. First the nucleus may travel through regions of large electric field gradients during the flight time resulting in a time-dependent electric quadrupole interaction. Second, the recoiling atom may be ionized. This produces a time-dependent paramagnetic interaction.

Since the effect of the recoil acts only over a time of 10^{-11} sec, it is hard to prove that the effect exists by a delayed angular correlation method. However, an indication is given by the following experimental facts. Several cases of alpha-gamma angular correlations in even-even nuclei (Ra^{224}, Ra^{226}, Pu^{238}, Th^{228}, Th^{230}) have been measured.[37] Since the spin sequence is 0–2–0, the theoretical angular correlation can be predicted unambiguously. The experimentally measured correlations show a marked attenuation, the attenuation factor G_2 being of the order of 0.7 to 0.4. The attenuation factor decreases with increasing Z. The lifetimes of the excited states are all of the order of $6\text{–}8 \times 10^{-10}$ sec. The observed attenuation requires an interaction strength of $\omega_Q = 2000$ Mc. It was not possible to conclude from this result alone whether the effect arises from a static quadrupole interaction or an effect of the recoil. One piece of evidence favors the latter possibility. In the case of the alpha-gamma decay of Am^{241}, which has an appreciably longer lifetime of 90 mμsec, the delayed angular correlation has been measured.[27] An influence is found but is associated with an interaction strength of only about 200 Mc. Since the quadrupole moments and the electric field gradients probably are of the same magnitude as in the foregoing even-even cases, there is a marked discrepancy which can be explained assuming that the 2000-Mc interaction originates in the effect of the nuclear recoil. Since it acts only over a very short time it is not revealed in the delayed angular correlation measurement in Am^{241} and since mixture of several alpha multipoles occur in this case, the theoretical angular correlation cannot be calculated.

17. The Effect of Excitation of the Electron Shell

During radioactive decay or during the process which produces the preferred orientation of the nuclei along a direction, there is a possibility that a paramagnetic state will be produced by excitation of the electronic shell. This is most likely to occur in electron capture or in internal conversion since a hole is produced in one of the inner electron shells. The Auger processes which follow may lead to an excited electronic state which is paramagnetic. This also may occur in other nuclear processes ($\beta^{\pm}$ decay, α decay, nuclear reactions) but with less probability. The effect of the paramagnetic state then depends on its lifetime in the surrounding medium. This lifetime is known to be very short (about 10^{-12} sec) in metals. On the other hand, it may be as long as seconds in insulators. The interaction strength is again of the order of the hyperfine

[37] J. C. D. Milton and J. S. Fraser, *Phys. Rev.* **95**, 628 (1954).

splitting, i.e., 500 Mc, and therefore we expect an influence on the angular correlation if $\tau_N \sim 10^{-9}$ sec or longer. No detailed theory has been worked out for this case. It is possible to decouple the nuclear and electronic moments and restore the full correlation by applying fields which fulfill the condition $(HJ) \gg (JI)$ in free atoms or $(HJ) \gg (JI)$, $1/\tau$ in liquids. Decoupling is not practical in solids.

There is some experimental information concerning the influence of the excited electron shell on angular correlation. It may be noted first that, when the influence of external fields on angular correlations was discovered in 1951,[38] the experimenters found a strong influence in non-metallic sources and a weak (or vanishing influence) in metallic sources. At the time, this was believed to result from the fact that in insulators the electron shell excitation arising from K capture remains for an appreciable time, whereas in metals it decays very rapidly. Although this hypothesis could not be proved at the time, it turned out to be correct in part.

More quantitative evidence for the influence of the excited electron shell was obtained from the following experiments.

(a) In the study of the angular correlation of the gamma rays which follow K capture in In^{111}, measurements with polycrystalline salts (e.g., $InCl_3$) showed attenuation factors lower than the minimum allowed value [Eq. (7.7)]. On the other hand, similar measurements[39] made on the gamma rays from the decay of the 48-min isomer, Cd^{111*}, yielded no attenuation coefficients below the minimum value. Since in the decay of the isomer Cd^{111*} electronic excitation can not occur, it is possible that the electronic excitation following K capture is responsible for the extremely low attenuation factors observed.

(b) Measurements have been made, using the same radioactive source where the gamma-gamma as well as the conversion electron-gamma correlation have been measured. In the latter case, instead of the first gamma ray a conversion electron is ejected and its correlation with the subsequent second gamma ray is measured. This comparison, which utilizes the same source, should give a very good indication of the importance of the excited electron shell since the surroundings and therefore the quadrupole interaction are the same. Measurements have been made[40–42] with the radioactive isotopes Hf^{181} and Hg^{196}. At present

[38] H. Aeppli, A. S. Bishop, H. Frauenfelder, M. Walter, and W. Zunti, *Phys. Rev.* **82,** 550 (1951).
[39] J. J. Kraushaar and R. V. Pound, *Phys. Rev.* **92,** 522 (1953).
[40] F. Gimmi, E. Heer, and P. Scherrer, *Helv. Phys. Acta* **29,** 147 (1956).
[41] E. Snyder and S. Frankel, *Phys. Rev.* **106,** 755 (1957).
[42] H. C. Coburn, J. V. Kane, and S. Frankel, *Phys. Rev.* **105,** 1293 (1957).

the results of the different investigators disagree as to whether the attenuation factors for the gamma-gamma and conversion electron-gamma cases are the same.

(c) The only clear-cut information concerning the question of the influence of the excited electron shells comes from measurements[20] of the delayed angular correlation in various sources using the radioactive isotope In^{111} which decays by K capture into Cd^{111}. Figure 14 shows the result for a source of In_2O_3. Similar results were obtained with $In(OH)_3$.

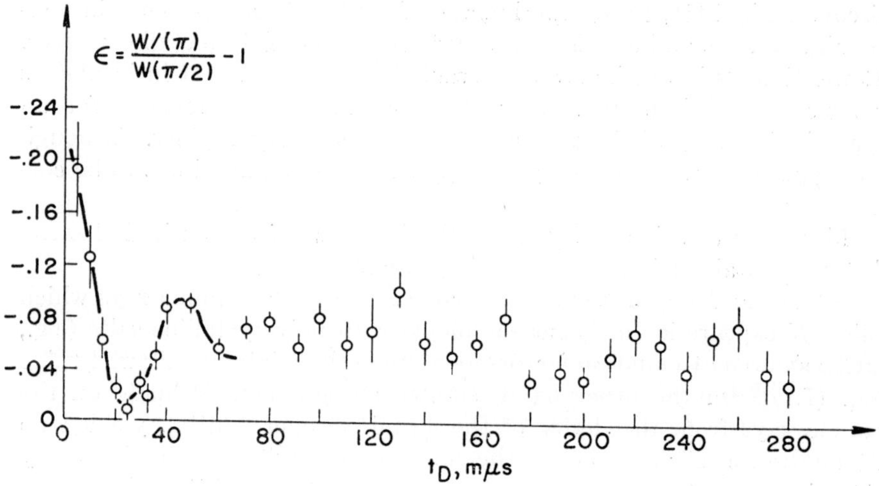

FIG. 14. Delayed angular correlation measurement of the gamma-gamma cascade in the decay of In^{111} in a source of polycrystalline In_2O_3. The measured anisotropy $\epsilon = \frac{W(\pi)}{W(\pi/2)} - 1$ is plotted *versus* delay time of the second gamma [P. Lehmann and J. Miller, *J. phys. radium* **17,** 526 (1956)].

If one compares Fig. 14 with Fig. 7 which contains results of measurements by the same authors with a source of In metal, one realizes the striking difference. Metallic sources show the periodic behavior characteristic of static quadrupole interaction, whereas nonmetallic sources show first a rapid decay of the correlation within about 10^{-8} sec and then possibly some fluctuation around the low value.

Further evidence for the influence of the excited electronic shell following *K* capture has been found[43] in investigations of the angular correlation of the gamma rays from oriented Co^{57}.

[43] G. R. Bishop, M. A. Grace, C. E. Johnson, A. C. Knipper, H. R. Lemmer, J. Perez, Y. Jorba, and R. G. Scurlock, *Phil. Mag.* [7] **46,** 951 (1955).

We might conclude that there is considerable experimental evidence for the presence of the influence of the excited electron shell but that further investigation is needed.

VI. Special Topics

18. The Decay of Elementary Particles

As discussed in Section 4, polarized particles are produced in the decay of several elementary particles and this polarization can be determined by a measurement of the angular correlation or polarization of subsequently emitted particles.

The case of the μ meson is of special interest to us because its lifetime of 2.2×10^{-6} sec is sufficiently long that the reaction mechanisms we have discussed may depolarize the mesons before they decay. Polarized μ mesons have been stopped in a great many types of substance and the polarization has been determined by a measurement of the asymmetry of electron emission relative to the initial direction of motion of the meson (i.e., the direction of polarization). This asymmetry varies widely, depending on the material and the external magnetic fields applied. The interactions of positive and negatively charged μ mesons with atoms are considerably different as the negative μ meson may be captured into an atomic orbit and cascade by x-ray emission down to the K shell from where it will either decay or be captured into the nucleus with neutrino emission.

Since the μ meson very likely has spin $\frac{1}{2}$ it cannot have a quadrupole moment and so in the unbound state can only be influenced by magnetic fields. However, it can form a bound state with an electron (muonium) which may have a quadrupole moment in the triplet state and hence may interact with crystalline fields.

The magnitude of the asymmetry of electron emission from μ mesons stopped in various materials varies from zero to the full expected correlation.[43a] Some of the anisotropies obtained [see Eq. (4.1)] are shown in Table V. The asymmetry is generally high in metals, somewhat smaller in liquids, and low in nonconductors although there are exceptions to this rule. For example, bromoform gives a very high asymmetry. These effects are not understood and indeed the existence of muonium is still in question. Evidence for it resides in the fact that one can retain the polarization by applying a strong magnetic field in the direction of the incoming meson, that is, along its axis of polarization. This field decouples the spins of the electron and μ meson and thereby prevents the μ meson from precessing about the electron moment and becoming depolarized.

[43a] R. A. Swanson, *Phys. Rev.* **112,** 580 (1958).

TABLE V. DECAY ASYMMETRIES FOR POSITIVE MUONS

Target	Asymmetry
Graphite	0.24
Diamond	0.04
Aluminum	0.21
Silicon	0.25
Fused quartz	0.04
Crystalline quartz	0.01
Water	0.14
Polystyrene	0.07
Benzene	0.05
Chloroform	0.18
Chloroform + DPH	0.23
Sulfur	0.01
Nuclear emulsions	0.07–0.14

There also is evidence for the production of μ-mesic atoms. This will produce appreciable magnetic fields at the μ meson and hence depolarization.

19. Studies of Very Long-Lived States

The experiments described up to this point give evidence concerning depolarization effects which occur in a time shorter than 10^{-5} sec. Thus very small depolarization effects, that is those associated with very long relaxation times, cannot be studied this way. The experiments require that coincidences be detected; because of the problem of random coincidences, resolving times longer than 10^{-5} sec are seldom used.

If the nuclear orientation can be measured by the detection of one radiation only, then this limitation is removed and long relaxation times can be studied. As an example we will discuss in detail the polarization of Li^8 nuclei produced by capture of polarized thermal neutrons by Li^7.[44,45]

The Li^8 decays by beta emission with a half-life of 0.8 sec to an excited state of Be^8 which then dissociates into two alpha particles. The Li^8 polarization was detected by the preferential emission of beta particles opposite to the direction of the nuclear spin of the incident neutron. This is a case of a very long spin relaxation time and offers a new and sensitive method for studying spin-spin relaxation in solids as well as a general procedure for producing polarized nuclei.

Polarized neutrons are captured by unpolarized Li^7 in a polycrystal-

[44] M. T. Burgy, W. C. Davidon, T. B. Novey, G. T. Perlow, and R. R. Ringo, *Bull. Am. Phys. Soc.* [2] **2,** 206 (1957).

[45] D. Kurath, *Bull. Am. Phys. Soc.* [2] **2,** 206 (1957).

line source of LiF or Li_2CO_3. The compound nucleus has either spin -1 or -2. This state decays by photon emission to the ground state of Li^8 which then beta decays with a 0.8-sec half-life. The initial polarization of the ground state of Li^8 can be calculated as can the size of the anisotropy expected in the beta emission. The experimental anisotropy is then a measure of the amount of nuclear relaxation which has occurred before beta decay. The full anisotropy expected was obtained. This implies a relaxation time longer than the lifetime. Under continued irradiation the effect was observed to decrease in time probably due to the production of paramagnetic centers in the crystal lattice. This effect could be reduced by heating to a few hundred degrees centigrade or inhibited by the application of an increased magnetic field along the direction of polarization. The thermal cycling presumably removes the F centers by allowing the trapped electrons to diffuse to their normal sites. The effect of the magnetic field is a very interesting one and is applicable in cases of this type where essentially isolated nuclei having a given g factor are embedded in a lattice of nuclei having another g factor. One can decouple these unlike nuclei with the aid of an external field because of their differing precession frequencies in this field. Thus one maintains the initial polarization along the direction of the applied field. This technique cannot be used in nuclear resonance work because the concentration of nuclei required to give a detectable signal is sufficiently large that there are near neighbors with identical g factors and they precess about each others moment independent of an external field. In the case of radioactive decay, however, one can remove the spin-spin relaxation and maintain a nuclear polarization even at room temperature, providing that the spin lattice relaxation time is sufficiently long. The spin lattice relaxation in the case of LiF has been measured to be two weeks.[46]

VII. Remarks on Experimental Techniques

20. Introduction

In all of the experiments described in the preceding sections, the angular distribution of nuclear radiations emitted from oriented nuclei are measured. One of the main experimental problems is that of orienting the nuclei. This may be done in many different ways such as by angular correlation of successive nuclear radiation, low-temperature methods, optical and microwave methods, nuclear reactions, and Coulomb excitation. Each of these different methods requires a completely different technique and it is far outside the scope of this article to present them in detail. We will describe, however, the necessary equipment for the

[46] A. Abragam and W. G. Proctor, *Phys. Rev.* **109**, 1441 (1958).

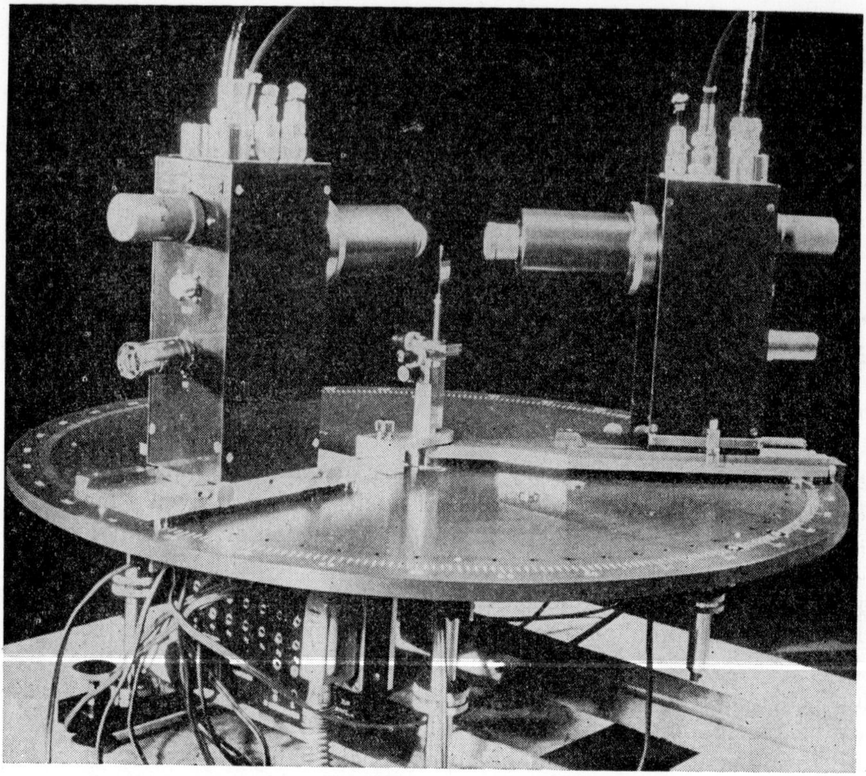

FIG. 15. Gamma-gamma angular correlation apparatus.

angular correlation measurements, since this method has found the widest application for the solution of the problems in which we are interested. Figures 15 and 16 show a typical angular correlation apparatus.

The next important part of the experimental technique is concerned with the measurement of the angular distribution of the radiation. This requires detectors and electronic equipment which will be described in the following two sections.

Third, since we are concerned here mainly with the influence of external fields on angular distributions, another experimental problem centers about the source of the target, i.e., the material containing the radioactive nuclei under study (Section 23).

In Section 24 we will discuss some problems connected with the evaluation of the data and in the final Section 25 we will discuss the requirements for suitable nuclei.

FIG. 16. Electron-gamma, alpha-gamma angular correlation apparatus.

21. DETECTORS

In most cases particle detectors are scintillation detectors because of their high efficiency for all radiations and the possibility of good energy resolution. They also provide faster pulses with much less uncertainty in time than Geiger or proportional counters. This, of course, is of great importance in measurements requiring coincidence detection. The type of scintillator used depends upon the radiation to be detected and the speed of response desired.

Thallium-activated sodium iodide is very commonly used for the detection of gamma rays. The high Z of the iodine provides a large photoelectric cross section, which allows one to choose the gamma ray desired by pulse-height selection. The relatively long phosphor decay time of 0.25 μsec may appear to be too long for fast coincidence work; however, by operating a fast coincidence circuit on the initial part of the pulse, resolving times of one-tenth to one-hundredth of the decay time may be achieved for gamma rays of 100 kev or more.

Electrons are conveniently detected with an organic scintillator. Anthracene provides the highest light output in cases in which maximum energy resolution is required. Stilbene, plastic scintillators, and other organic scintillators provide somewhat less light but have faster decay times, which are desirable when the ultimate in time resolution is to be obtained (10^{-10}–10^{-11} sec). Sodium iodide is not so desirable because it is sensitive to moisture. This makes it necessary to use a thin window which may distort the electron energy by absorption. In addition, the sensitivity to gamma rays makes it more difficult to discriminate between electrons and photons.

The organic scintillators also are used often for alpha-particle detection, although there is a severe loss in light efficiency. Where good energy resolution is required, sodium iodide has been used in spite of its sensitivity to humidity.

The light pulses from the scintillators are detected with photomultiplier tubes which provide amplification factors up to 10^8. These tubes provide electrical pulses sufficiently large to operate electronic detection equipment directly. Where high fields are present of course, one must protect photomultipliers with magnetic shielding.

In experiments involving liquid-helium temperatures and therefore a cryostat one has special problems with electron and alpha-particle detection. Photons, which are able to penetrate a considerable amount of material, can be detected outside the cryostat, but for the other particles one is forced to locate the detector within the cryostat. This can be done by placing the scintillator in the low-temperature region and transmitting the light to the photomultiplier by means of a light pipe. This is necessary because the photosensitive surface of the multiplier loses its efficiency at very low temperatures.

22. Electronics

a. Angular Correlation Measurements

As explained earlier the orientation of the nuclei is obtained in angular correlation measurements by selecting only those nuclei that radiate in a given direction. The angular distribution of radiation emitted subsequently is then measured relative to this direction. The physical apparatus must involve at least two detectors, one of which can be moved so that the angle subtended at the source by the detectors may be varied. One then requires a coincidence analyzer which selects those radiations falling on the detector that have been emitted in the same nuclear decay.

The radiations may be selected further with respect to energy and type of particle. A typical electronic system for performing these measurements is shown in Fig. 17. The electrical pulses from the detectors are branched into two lines, "fast" and "slow" (slow is of the order of 10^{-6} sec, fast is 10^{-7} to 10^{-9} sec pulse width). This division is made because linear pulse-height analyzers that will make possible the use of sufficiently short coincidence resolving times are not available. Fast coincidence circuitry is essential in order to discriminate against chance coincidences at the high counting rates encountered in these experiments. In the fast line the pulses may, if necessary, be amplified and applied to a fast coincidence circuit. Recent development of high-gain photomultiplier tubes make it possible to eliminate the need for amplification in some cases.

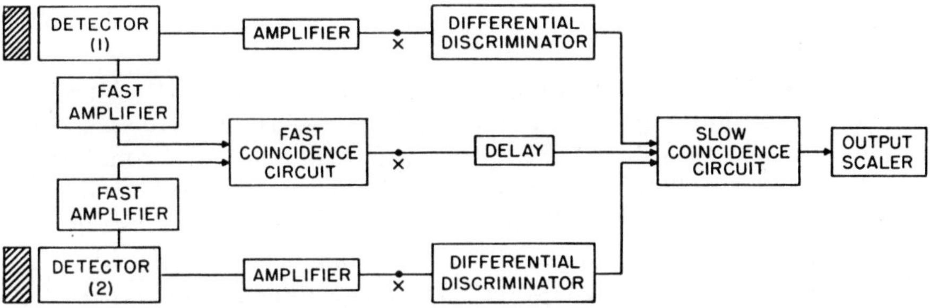

FIG. 17. A typical fast-slow coincidence analyzer system for angular correlation measurements [S. Devons and L. J. B. Goldfarb, *Handbuch d. Phys.* **42,** 403 (1957)].

At the same time the pulses are applied in sequence to slow linear amplifiers and to pulse-height analyzers which allow energy selection, if this is desired. In the detectors used most commonly, namely scintillation detectors, the pulse height is proportional to the energy of the incident particle.

The outputs of the analyzers together with the output of the fast coincidence circuit then are applied to a slow multiple coincidence circuit which selects the desired event.

The two main types of angular correlation techniques involve the use either of coincidence resolving time which is long compared to the lifetime of the oriented nuclear state (integral correlation) or resolving time short compared to the lifetime of the oriented state (differential correlation). In the second case one usually studies the angular distribution as a function of the time between the formation of the oriented state and its decay, that is, one selects coincidences in which the second pulse is delayed with respect to the first. This is done by introducing a variable delay line or electronic delay in the line of the first detector.

b. Polarized and Aligned Nuclei and Particles

When the axis of orientation of the nucleus or particle is determined by an external field or process, a reference direction is known and, in principle at least, a single particle detector alone is needed. Thus the electronic equipment can be a single-channel system. In practice, however, a coincidence system may be needed to indicate the arrival of a polarized particle as for example in mu-meson studies.

23. Sources

Aside from problems of determining the fields present in the matter containing the oriented nuclei, one must consider the effects of the mass of the source material in scattering or degrading the energy of the particles being detected. Such effects can distort the angular correlation being measured.

Because of the low absorption coefficients of photons, the mass of a source for gamma-ray measurement is not so critical. Glass capsules a few millimeters in diameter which contain the radioactive material in the form of crystalline powder or in solution often are used. Small single crystals also have been used and usually are mounted in an arrangement which makes it possible to orient the crystalline axis as desired.[17]

Sources for electron counting[40] are more difficult to prepare. The multiple scattering of the electrons which can distort the angular distribution begins at thicknesses of the order of 1% of the range. The sources can be prepared by evaporation from solution or, more preferably, by vacuum evaporation from a filament or crucible onto a thin supporting foil or film.

Liquid sources for electron counting have been prepared in spite of the requirements of very low surface density by the evaporation of active material into a thin liquid metal film.[40]

Alpha-particle source preparations are similar to electron source preparations but scattering is less important although the ranges of alpha particles are smaller than for electrons of the same energy. Because small-angle scattering predominates, the alpha-particle tracks are nearly straight. Thus it is possible to make studies with thin liquid-film alpha sources with thicknesses of the order of one milligram per square centimeter.[24]

For the counting of photons, the path length of air between the source and detector are not of concern because very little scattering or absorption occurs at the distances normally encountered. Experiments may even be made with an appreciable amount of material because of Dewar vessels, liquid helium, etc., surrounding the source. For electrons, however, since scattering is so important, measurements usually are made in a vacuum or at least in a hydrogen or helium atmosphere.

Since scattering in air is not important for alpha particles, it is only necessary to keep the absorbing material below the particle range.

24. Corrections

In addition to being concerned with the perturbing effects of external fields in making measurements of angular distribution of radiations there are a number of experimental effects of which one must be aware.

If coincidence systems are used, it is necessary to correct the data for chance coincidences. In general one adjusts the source disintegration rate and/or the coincidence resolving time to reduce the chance contribution to a small fraction of the total so that relatively large errors in the correction introduce only small errors in the true coincidence rate. The chance coincidence contribution can be determined experimentally by two different procedures which eliminate the true coincidences. The first involves the use of a delay in one side of the system, and the second utilizes two separate sources each of which is seen by one detector only.

There also may be important corrections for scattering of radiations because of source thickness and solid angle corrections dependent upon source and detector shapes and sizes. These corrections have been considered in detail by several workers and the procedures well summarized by Frauenfelder.[1]

Backscattering of radiation, particularly electrons and photons, must be considered in apparatus design. Vacuum chambers of several inch or larger radius built of low Z material are usually satisfactory. If photon backscattering is a problem, as for example when one is selecting a photon having an energy of a few hundred kilovolts in the presence of photons of higher energy, the photon detector usually can not be shielded against backscattered photons, since the backscattering can occur in the shielding. The best approach is provided by employing large separations and low masses in designing the parts of the apparatus.

25. Criteria for Suitable Radioactive Nuclides

The most important requirement, of course, is that the lifetime of the oriented state be such that changes in the orientation can be observed. As was discussed previously, the range between 10^{-6} and 10^{-9} sec is the one most generally suitable. If, however, very long spin relaxation times are present or are being studied, this range may be extended to many hours or longer.

In angular correlation measurements the lifetime of the parent state from which the orienting radiation is observed cannot be so short that the source activity decays away before sufficient data are recorded unless one uses sources that are continuously renewed or prepared. This consideration divides the experimental measurements into two groups: those where one prepares a source of activity and makes measurements on it for some hours or longer, and those where the apparatus must be closely connected to an accelerator or reactor producing the activity in a more or less continuous fashion. The critical value of the half-life dividing the two groups is in the range of an hour.

VIII. Angular Distributions of Nuclear Radiations as a Tool in Solid State Physics

In this article, we have attempted to present the pertinent theory and the experimental evidence concerning the interrelation of the angular correlation of nuclear radiations and the structure of condensed matter.

Most of the work described has been undertaken with the point of view of the nuclear physicist who is attempting to obtain information regarding the properties of nuclei without becoming too enmeshed in the intricacies of the many-body problems of the solid state.

The question remains to be answered whether or not one could obtain useful information if one would undertake such investigations primarily from the vantage point of the solid state scientist. It is the intent of this paper to emphasize that this approach is useful. We hope those aware of the problems of the solid state will concur.

In conclusion we would like to make a few brief suggestions regarding problems which might be attacked advantageously with these methods.

Generally speaking, the methods of angular correlation allow the determination of the magnetic fields and electric field gradients at the position of the radioactive nucleus whose radiation is studied. Most of the information obtained is related closely to that obtained from studies of nuclear magnetic and electron paramagnetic resonance. Although the resonance methods are more precise in general, the techniques of angular correlation offer two advantages. First, the number of atoms required for a measurement is much smaller. This allows one, for example, to work with cases in which a very small number of impurity centers are present in the lattice and to measure the fields present at such impurity centers.

Second, the fields in a solid can be measured without the need of applying an external radio-frequency field. This should enable one to measure the fields in a superconductor. One could study the displacement of the field from a superconductor when the transition temperature is passed. Actually an unsuccessful search for a change in the electric field gradient at the transition temperature has already been made.[47] More-

[47] H. Albers-Schönberg and E. Heer, *Helv. Phys. Acta* **28**, 389 (1955).

over, the penetration of the magnetic field into the superconductor can be measured.[48]

Measurements of the type considered here also may find useful applications in the study of the rates of chemical reactions in such a way as to include complex formation and exchange. They may also be useful in the investigation of the rearrangement of excited electron shells.

[48] H. Frauenfelder, private communication; H. R. Lewis, Jr., Thesis, University of Illinois, 1958 (unpublished).

Oscillatory Behavior of Magnetic Susceptibility and Electronic Conductivity

A. H. KAHN AND H. P. R. FREDERIKSE
Solid State Physics Section
National Bureau of Standards, Washington, D.C.

I. Introduction

Oscillations in the magnetoresistance of bismuth crystals were first observed by Shubnikov and de Haas[1] in 1930. This observation led to the discovery of the oscillatory behavior of the diamagnetic susceptibility known as the de Haas-van Alphen effect.[2] The susceptibility oscillations were particularly perplexing at that time because they violated the principle that free electrons should not exhibit diamagnetism, predicted by classical theory. The explanation of the behavior of the susceptibility was provided by Landau[3] and Peierls[4] in their discussions of the quantum theory of electrons in solids in the presence of a magnetic field. The major experimental work at high fields was concerned with the observation of the susceptibility, until 1950. The transport phenomena have received more attention recently. The Naval Research Laboratory with its 100-kilogauss magnet has contributed significantly in this area.

[1] L. Shubnikov and W. J. de Haas, *Leiden Comm.* 207a, 207c, 207d, 210a (1930).
[2] W. J. de Haas and P. M. van Alphen, *Leiden Comm.* 208d, 212a (1930), and 220d (1933).
[3] L. D. Landau, *Z. Physik* **64,** 629 (1930).
[4] R. Peierls, *Z. Physik* **80,** 763 (1933).

The understanding of the transport effects has developed slowly. Titeica[5] made progress in the theory of conductivity of metals in strong magnetic fields but did not treat the oscillations. Explanations of the magnetoresistance of bismuth were put forward by Davydov and Pomeranchuk.[6] More detailed analyses have appeared in the last few years in articles by Zilberman,[7] Lifshits,[8,9] Argyres,[10,11,12] Kubo *et al.*,[13] and Adams and Holstein.[14] An understanding of the susceptibility requires a knowledge of the density of states of the electron gas when the magnetic field is present; treatment of conductivity requires, in addition, a knowledge of the scattering mechanisms and their dependence upon the field.

The information obtained from the susceptibility measurements on metals has been described extensively by Shoenberg,[15] whereas galvanomagnetic and thermomagnetic effects at low fields have been reviewed by J.-P. Jan in Volume 5 of this series.[16] In this paper, we shall discuss high-field quantization effects, placing emphasis on transport effects. The first half is concerned mainly with the free electron gas. The second part contains a review of information derived from magnetic and conductivity data, major attention being given to the properties of bismuth.

II. Theory

The oscillatory effects exhibited by free carriers at high magnetic fields can be approached best by means of a free electron model. The result of much of the theoretical study of electrons in a periodic potential in the presence of a magnetic field has been to provide justification for the use of an effective mass approximation, i.e., the use of free electron wave functions with an effective mass which may depend upon direction and energy. Accordingly we shall first discuss the properties of the free electron gas in a magnetic field.

[5] S. Titeica, *Ann. Physik* [5] **22,** 129 (1935).

[6] B. Davydov and I. Pomeranchuk, *J. Phys.* (*U.S.S.R.*) **2,** 147 (1940).

[7] G. E. Zilberman, *J. Exptl. Theoret. Phys.* (*U.S.S.R.*) **29,** 762 (1955), transl., *Soviet Phys. JETP* **2,** 650 (1956).

[8] I. M. Lifshits and A. M. Kosevich, *J. Phys. Chem. Solids* **4,** 1 (1958).

[9] I. M. Lifshits, *J. Phys. Chem. Solids* **4,** 11 (1958).

[10] P. N. Argyres and E. N. Adams, *Phys. Rev.* **104,** 900 (1956).

[11] P. N. Argyres, *J. Phys. Chem. Solids* **4,** 19 (1958).

[12] P. N. Argyres, *Phys. Rev.* **109,** 1115 (1958).

[13] R. Kubo, H. Hasegawa, and N. Hashitsume, *Phys. Rev. Letters* **1,** 279 (1958).

[14] E. N. Adams and T. D. Holstein, to appear.

[15] D. Shoenberg, "Progress in Low Temperature Physics," Vol. 2. North Holland, Amsterdam, 1957.

[16] J.-P. Jan, *Solid State Phys.* **5,** 1 (1957).

The effect of electron spin will be neglected in most of the following discussion. In almost all cases in which the oscillatory effects have been observed, the splitting arising from spin paramagnetism, $e\hbar H/2m_0c$, is small in comparison with the diamagnetic splitting, $e\hbar H/m^*c$, because of the small effective mass.

1. De Haas-van Alphen Effect

a. Electron Gas in a Strong Magnetic Field[5]

The Hamiltonian for an electron in a magnetic field is

$$\mathcal{H} = \frac{1}{2m}\left(\mathbf{p} + \frac{e}{c}\mathbf{A}\right)^2 \tag{1.1}$$

in which m is the effective mass, $\mathbf{p}$ is the canonical electron momentum, e the numerical value of the electronic charge, and $\mathbf{A}$ the vector potential. The kinetic momentum is related to the canonical momentum by the equation

$$m\mathbf{v} = \mathbf{p} + \frac{e}{c}\mathbf{A}. \tag{1.2}$$

In performing transport calculations, the most convenient choice of gauge for a uniform magnetic field is given by the relation $\mathbf{A} = (0,Hx,0)$, for a field H in the Z direction. For this case the Hamiltonian becomes

$$\mathcal{H} = \frac{1}{2m}\left[p_x^2 + (p_y + m\omega_0 x)^2 + p_z^2\right] \tag{1.3}$$

where $\omega_0 = eH/mc$. With this choice of gauge, the eigenfunctions of $\mathcal{H}$ are given by

$$\psi_{nk_yk_z} = \frac{1}{\sqrt{L_yL_z}}\,\phi_n\left(x + \frac{\hbar k_y}{m\omega_0}\right) e^{ik_yy}e^{ik_zz} \tag{1.4}$$

where k_y and k_z are constants which may assume any real values, n is an integer, and ϕ_n is a one-dimensional harmonic oscillator wave function. The x part of the wave function is centered at position $x_0 = -\hbar k_y/m\omega_0$. The relation (1.4) is normalized to unity in the region $-\infty < x < \infty$, $0 \leqslant y \leqslant L_y$, $0 \leqslant z \leqslant L_z$.

The energy levels associated with the wave functions $\psi_{nk_yk_z}$ are

$$\epsilon_{nk_yk_z} = (n + \tfrac{1}{2})\hbar\omega_0 + \frac{\hbar^2 k_z^2}{2m}. \tag{1.5}$$

Thus we look upon the motion of the electron as if it were free in the z direction and trapped in the transverse directions with the energy levels of the harmonic oscillator. The current operator $-emv$ has an average value $-e\hbar k_z/m$ for an electron in state $\psi_{nk_yk_z}$.

The density of states is drastically different in the presence of a magnetic field. Applying periodic boundary conditions and going to a continuous **k** space, we find that the number of electronic states in range $dk_y\,dk_z$, for fixed n, is given by the quantity

$$\frac{L_yL_z}{(2\pi)^2}\,dk_y\,dk_z \tag{1.6}$$

if we neglect electron spin. Integrating (1.6) over k_y, summing over the discrete levels n, and changing to a summation over $\epsilon_{nk_yk_z}$, we find that the number of states in energy range $d\epsilon$ is

$$N(\epsilon)\,d\epsilon = \frac{V}{(2\pi)^2}\,\frac{1}{2}\,\hbar\omega_0\left(\frac{2m}{\hbar^2}\right)^{\frac{3}{2}}\sum_{n=0}^{n_{\max}}\frac{1}{\sqrt{\epsilon-(n+\frac{1}{2})\hbar\omega_0}}\,d\epsilon \tag{1.7}$$

where V is the volume $L_xL_yL_z$, and the summation involving n extends over all positive integers for which the radical is real. In performing the integration over k_y, we use the limitation that $0 \leqslant x_0 = (-\hbar k_y/m\omega_0) \leqslant L_x$. This neglects the effects of the spill-out of $\phi_n(x - x_0)$ beyond the edges of the volume considered. Thus our present treatment is satisfactory for bulk effects, but neglects all surface contributions. We show a plot of the energy levels given by Eq. (1.5) and the density of states associated with Eq. (1.7) in Fig. 1. It is seen that the density of states per unit energy range diverges for energies at the bottom of each oscillator level. This divergence has a profound effect on the partition function and the scattering rates and is responsible for the oscillatory behavior of the magnetic susceptibility and transport effects.

As in the Sommerfeld free electron theory, we employ Fermi-Dirac statistics for the determination of the occupation probabilities of the one-electron states. The probability $f(nk_yk_z)$ that state $\psi_{nk_yk_z}$ is occupied, under conditions of thermal equilibrium, is given by the Fermi function

$$f(nk_yk_z) = f(\epsilon_{nk_yk_z}) = \frac{1}{e^{(\epsilon_{nk_yk_z}-\epsilon_F)/kT}+1} \tag{1.8}$$

where ϵ_F is the Fermi energy. The Fermi energy is determined through normalization of the distribution, i.e., by setting

$$N = 2\int_{\frac{1}{2}\hbar\omega_0}^{\infty} N(\epsilon)f(\epsilon)\,d\epsilon \tag{1.9}$$

where the factor 2 accounts for the effect of spin, and N is the total num-

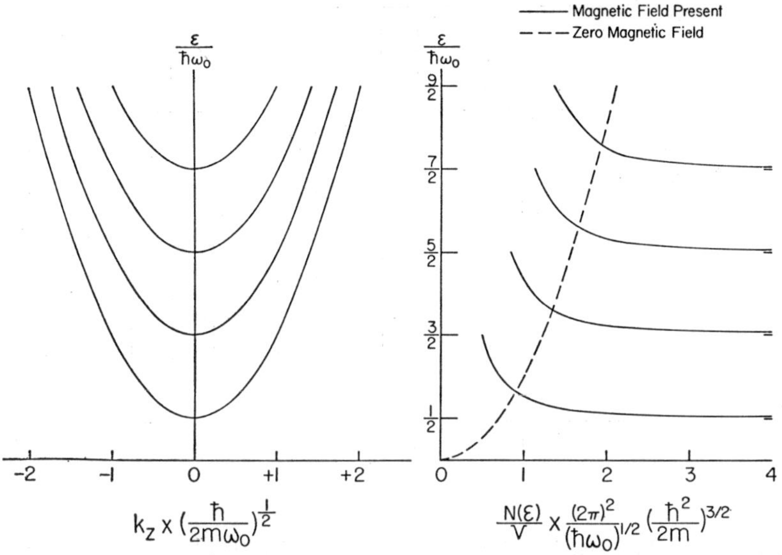

FIG. 1. Energy levels and density of states in a magnetic field. Energy levels in (a) are from Eq. (1.5); density of states in (b) is from Eq. (1.7).

ber of electrons. In the case of extreme degeneracy, i.e., for $kT \ll \epsilon_F$, and for an otherwise arbitrary magnetic field, evaluation of Eq. (1.9) yields

$$\frac{N}{V} = \frac{\hbar\omega_0}{2\pi^2}\left(\frac{2m}{\hbar^2}\right)^{\frac{3}{2}} \sum_n \sqrt{\epsilon_F - (n + \tfrac{1}{2})\hbar\omega_0}. \tag{1.10}$$

If we introduce the quantity $\epsilon_F{}^0 = (\hbar^2/2m)(3\pi^2 N/V)^{\frac{2}{3}}$, the Fermi energy for $H = 0$ when the degeneracy is complete, we obtain

$$\tfrac{2}{3}\left(\frac{\epsilon_F{}^0}{\hbar\omega_0}\right)^{\frac{3}{2}} = \sum_{n=0}^{n_{\max}} \sqrt{\frac{\epsilon_F}{\hbar\omega} - (n + \tfrac{1}{2})}. \tag{1.11}$$

In Fig. 2 a plot of $\epsilon_F/\hbar\omega_0$ *vs* $\epsilon_F{}^0/\hbar\omega_0$, based on Eq. (1.11), is shown. It is seen that, as the magnetic field decreases, the Fermi energy approaches that of the zero field case; however, it exhibits a deviation almost periodic in $1/H$. At very high fields, the Fermi level may come so close to the zero-point energy that the condition $\epsilon_F \gg kT$ is no longer satisfied. Then there is a transition to the nondegenerate case.

In the case of magnetic fields for which the condition $\epsilon_F/\hbar\omega_0 \gg 1$ is satisfied, Eq. (1.9) can be integrated approximately by application of

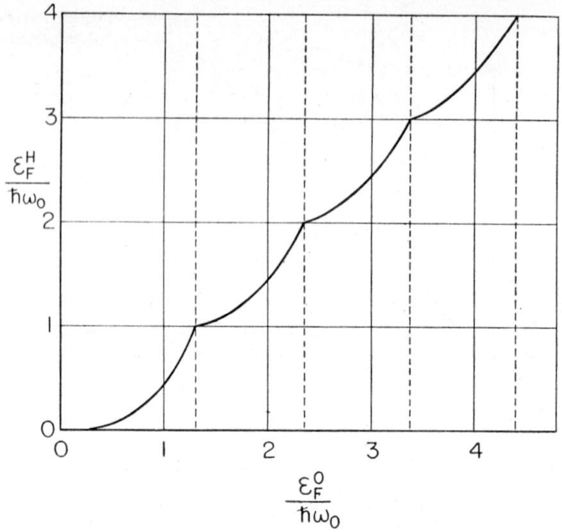

FIG. 2. Relation between Fermi levels with ($\epsilon_F{}^H$) and without ($\epsilon_F{}^0$) magnetic field.

Poisson's summation formula.[17] The result is

$$\frac{N}{V} = \frac{(2m\epsilon_F)^{\frac{3}{2}}}{3\pi^2\hbar^3}\left[1 - \frac{3\pi\sqrt{2}}{4}\frac{kT}{\hbar\omega_0}\left(\frac{\hbar\omega_0}{\epsilon_F}\right)^{\frac{3}{2}}\sum_{r=1}^{\infty}\frac{(-1)^{r+1}}{r^{\frac{1}{2}}}\frac{\cos\left(2\pi r\frac{\epsilon_F}{\hbar\omega_0} + \frac{\pi}{4}\right)}{\sinh\frac{2\pi^2 rkT}{\hbar\omega_0}}\right]. \tag{1.12}$$

Equation (1.12) applies for all temperatures. The oscillatory terms are very small at temperatures such that $kT \gg \hbar\omega_0$. Quantization effects then disappear, and ϵ_F approaches its zero-field value.

b. *Magnetic Susceptibility*

The magnetic susceptibility of the electron gas is obtained from the Helmholtz free energy F by the relation

$$x = -\frac{1}{H}\frac{\partial F}{\partial H}. \tag{1.13}$$

The free energy, for Fermi statistics, is given by the formula[18]

$$F = \frac{N}{V}\epsilon_F - 2kT\sum_i \log\left[1 + e^{(\epsilon_F - \epsilon_i)/kT}\right]. \tag{1.14}$$

[17] R. Courant and D. Hilbert, "Methods of Mathematical Physics," Vol. 1, p. 76. Interscience, New York, 1953.

The free energy and susceptibility have been calculated by Landau.[3] An elegant derivation of Landau's results, based on a density matrix formulation, has been given by Wilson and Sondheimer.[18,19] Landau's formula for the magnetic susceptibility, including the paramagnetic part, is

$$\chi = \frac{1}{12\pi^2}\frac{e^2}{\hbar c}\left(\frac{2\epsilon_F}{m^*c^2}\right)^{\frac{1}{2}}\left[3\left(\frac{m^*}{m_0}\right)^2 - 1 - \frac{3\pi kT}{\hbar\omega_0}\left(\frac{2\epsilon_F}{\hbar\omega_0}\right)^{\frac{1}{2}}\sum_{r=1}^{\infty}\frac{(-1)^r\cos\dfrac{r\pi m^*}{m_0}\sin\left(\dfrac{2\pi r\epsilon_F}{\hbar\omega_0}-\frac{1}{2}\right)}{r^{\frac{1}{2}}\sinh\dfrac{2\pi^2 rkT}{\hbar\omega_0}}\right]. \tag{1.15}$$

Usually only the first oscillatory term, $r = 1$, is important. Equation (1.15) demonstrates clearly that the oscillatory effects will be observable most easily at low temperatures and in materials in which the electrons have small effective masses and low Fermi levels, and hence have low concentrations of free carriers.

Dingle[20] has studied the influence of electron collisions on the susceptibility. He takes collisions into account by considering the energy levels to be broadened. The broadened levels are given a Lorentzian (damped oscillator) shape. He finds that the rth oscillatory terms of Eq. (1.15) for the susceptibility is multiplied by a factor $\exp(-2\pi r/\omega_0\tau)$, where τ is the mean time between collisions. This produces a diminished amplitude of oscillation. In the temperature range for which $2\pi^2 kT/\hbar\omega_0 \ll 1$, the first and largest oscillatory term of the susceptibility becomes

$$\frac{1}{\pi}\frac{e^2}{\hbar c}\left(\frac{\epsilon_F}{m^*c^2}\right)^{\frac{1}{2}}\left(\frac{\epsilon_F}{\hbar\omega_0}\right)^{\frac{1}{2}}\left(\frac{kT}{\hbar\omega_0}\right)\exp\left[-\frac{2\pi^2k(T+T')}{\hbar\omega_0}\right]\cdot\sin\left(\frac{2\pi\epsilon_F}{\hbar\omega_0}-\frac{\pi}{4}\right)\cos\frac{\pi m^*}{m} \tag{1.16}$$

in which we have used $T' = 2\hbar/\pi k\tau$. Thus the collisions produce an effect which is very similar to an increase in the temperature. The parameter T' is sometimes called the Dingle temperature. The mean collision time τ has been introduced in a phenomenological way; its relation to the mobility relaxation time is not clearly understood.[21]

[18] A. H. Wilson, "The Theory of Metals," 2nd ed. Cambridge University Press, London and New York, 1954.

[19] E. H. Sondheimer and A. H. Wilson, *Proc. Roy. Soc.* **A210,** 173 (1951).

[20] R. B. Dingle, *Proc. Roy. Soc.* **A211,** 517 (1952).

[21] R. J. Sladek, *Phys. Rev.* **108,** 590 (1957).

c. Effect of Electron Spin

The inclusion of electron spin requires an additional term in the Hamiltonian of Eq. (1.1), namely $g\mu_B\mathbf{H}\cdot\mathbf{S}$, where g is the numerical value of the spectroscopic splitting factor, μ_B is the Bohr magneton, and S is the electron-spin angular momentum operator in units of $\hbar$. All spin-orbit effects are neglected here. The energy eigenvalues, Eq. (1.5), then are augmented by the term $\pm g\mu_B H/2$, corresponding to the two possible orientations of spin. Now we shall be interested primarily in oscillatory effects in materials having electrons with effective masses of the order of hundredths of the free electron mass. Thus the splitting arising from spin may be ignored relative to the splitting between oscillator levels, the latter being $\hbar\omega_0 = 2\mu_B H \times (m/m^*)$. The density of states per unit energy range is altered only slightly under these conditions.

2. Transport Effects

Before proceeding with the quantum-mechanical discussion of electrical conductivity it may be worth examining the classical picture of electronic motion in electric and magnetic fields. In the case of parallel electric

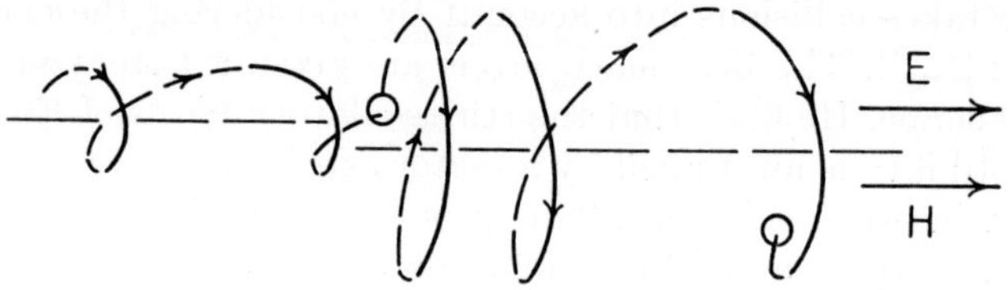

Fig. 3. Classical motion of a charged particle in parallel electric and magnetic fields. The circles indicate collisions. In the absence of collisions the current parallel to the fields would diverge.

and magnetic fields the classical electronic paths are helices. The effect of the electric field is to accelerate the electron in the direction of the fields. Collisions with crystalline imperfections are viewed as intermittently stopping the motion, thus limiting the current. This is illustrated in Fig. 3. In the situation of crossed electric and magnetic fields, the path of an electron is cycloidal. With collisions absent, the net current would be perpendicular to both fields. Figure 4 shows the path of an electron in crossed fields, intermittently brought to rest by collisions. It is to be noted that, in the transverse case, i.e., crossed fields, the component of current along the electric field is produced and enhanced by the collisions, if the scattering is weak. In view of these simplified considerations, it is not surprising that the calculations of longitudinal and transverse conductivities require different techniques.

a. Longitudinal Conductivity

In the study of longitudinal conductivity the Boltzmann technique may be used provided one generalizes the distribution function to describe populations of electrons over the magnetic quantum states. This generalization of the zero field method seems reasonable since the factor in the electronic wave function associated with motion along the electric field, $e^{ik_z z}$, is the same, with or without magnetic field. Thus the description of the motion of wave packets[18] moving in the direction of the electric field remains unchanged. A rigorous proof of this generalization has been given by Argyres.[12] We present a discussion of longitudinal conductivity in the domain of scattering by impurity ions.

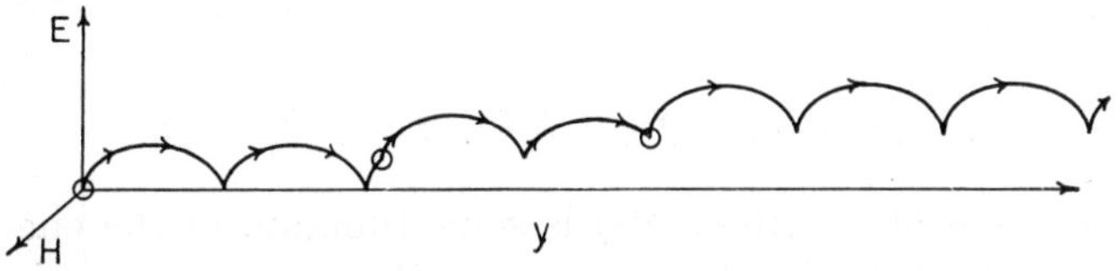

FIG. 4. Classical motion of a charged particle in crossed electric and magnetic fields. The collisions, indicated by circles, are shown as if bringing the particle to rest. The current parallel to the electric field would vanish if collisions were absent.

Let $f(n,k_y,k_z)$ be the probability of occupation of state $\psi_{nk_yk_z}$ of Eq. (1.4). Then the rate of change of f caused by drift in the electric field is given by

$$\frac{-eE_z}{\hbar}\frac{\partial f}{\partial k_z} \tag{2.1}$$

just as in the case of zero magnetic field. The effect of collisions with impurity ions on the distribution function is given by

$$\left(\frac{\partial f}{\partial t}\right)_{\text{coll}} = -\sum_{nk_yk_z}\{f(nk_yk_z)[1 - f(n'k_y'k_z')]W_{nk_yk_z;\, n'k_y'k_z'} - f(n'k_y'k_z')[1 - f(nk_yk_z)]W_{n'k_y'k_z';\, nk_yk_z}\} \tag{2.2}$$

where $W_{nk_yk_z;\, n'k_y'k_z'}$ is the probability per second that an electron will be scattered from state $\psi_{nk_yk_z}$ to state $\psi_{n'k_y'k_z'}$, and the factors f and $1 - f$ account for the exclusion principle. The transition probability per unit time, valid to the second order of perturbation theory, is given by

$$W_{nk_yk_z;\, n'k_y'k_z'} = \frac{2\pi}{\hbar}|\langle nk_yk_z|V|n'k_y'k_z'\rangle|^2\delta(\epsilon_{nk_yk_z} - \epsilon_{n'k_y'k_z'}) \tag{2.3}$$

in which V is the potential of the random array of scattering centers. We note that for a static V, the transition rates are symmetric in the primed and unprimed states. The transitions conserve energy. Thus we obtain from Eq. (2.2) the more simplified form:

$$\left(\frac{\partial f}{\partial t}\right)_{\text{coll}} = -\sum_{n'k_y'k_z'} [f(nk_yk_z) - f(n'k_y'k_z')]W_{nk_yk_z;\, n'k_y'k_z'}. \tag{2.4}$$

In the steady state we write the Boltzmann equation

$$\left(\frac{\partial f}{\partial t}\right)_{\text{coll}} + \left(\frac{\partial f}{\partial t}\right)_{\text{drift}} = 0. \tag{2.5}$$

We now assume that the first-order effect of the electric field is to shift the equilibrium distribution in the direction of the electric field. This may be written in the form

$$f(nk_yk_z) = f_0(nk_yk_z) + E_z\phi(\epsilon)k_z \tag{2.6}$$

where f_0 is the Fermi function, $\phi(\epsilon)$ is some function of the energy, and E_z is the electric field. Under this assumption the effect of scattering can be described by a relaxation time. We have

$$\begin{aligned}\left(\frac{\partial f}{\partial t}\right)_{\text{coll}} &= -\sum_{n'k_y'k_z'} E_z[\phi(\epsilon)k_z - \phi(\epsilon')k_z']W_{nk_yk_z;\, n'k_y'k_z'} \\ &= -(f - f_0) \sum_{n'k_y'k_z'} \frac{k_z - k_z'}{k_z} W_{nk_yk_z;\, n'k_y'k_z'}\end{aligned} \tag{2.7}$$

where we have used the energy conservation condition in the second step. We define the relaxation time by

$$\frac{1}{\tau_{nk_yk_z}} = \sum_{nk_yk_z} \frac{k_z - k_z'}{k_z} W_{nk_yk_z;\, n'k_y'k_z'} \tag{2.8}$$

and write the Boltzmann equation in the form

$$\frac{eE_z}{\hbar}\frac{\partial f}{\partial k_z} = \frac{f - f_0}{\tau}. \tag{2.9}$$

To first order in the electric field, the solution of the Boltzmann equation is given by

$$f = f_0 + \frac{eE_z\tau}{\hbar}\frac{\partial f_0}{\partial k_z}. \tag{2.10}$$

The current is found by averaging the z component of current over the distribution as follows:

$$J_z = -\frac{e\hbar}{m} \sum_{nk_yk_z} k_z f(nk_yk_z). \tag{2.11}$$

At low temperatures the derivative of the Fermi function may be approximated by $-\delta(\epsilon - \epsilon_F)$. Then Eq. (2.11) takes the form

$$J_z = \frac{E_z e^2}{m} \sum_n N_n \tau_n(\epsilon_F) \tag{2.12}$$

where N_n is the number of electrons in oscillator level n, and $\tau_n(\epsilon_F)$ is the relaxation time for electrons in level n at the Fermi level. The relaxation times will depend upon the value of the Fermi energy relative to the bottom of the oscillator levels. As the magnetic field is varied, the Fermi level will cross the oscillator levels, causing oscillatory behavior of the conductivity.

The longitudinal relaxation time has been calculated for lattice scattering and for ionized impurity scattering (screened Coulomb potential) by Argyres and Adams[10] in the quantum limit, i.e., the case in which only the levels for which $n = 0$ are occupied, and by Argyres[11] for lattice scattering when higher levels are occupied. The results for impurity scattering have been extended by the present authors to the situation in which levels with the quantum numbers $n = 0$ and $n = 1$ are occupied. Figure 5 shows the calculated resistivity at $T = 0$ as a function of magnetic field for ionized impurity scattering in a degenerate semiconductor having a donor concentration 2×10^{15} cm^{-3} and an effective mass $0.013m_0$, values appropriate to InSb. It is assumed that all carriers are ionized. The discontinuity at 4×10^3 oersteds is associated with a very short relaxation time for electrons at the bottom of the $n = 1$ oscillator level, when the Fermi level is just at $\frac{3}{2}\hbar\omega_0$. Actually, the time-dependent perturbation theory would no longer be valid when the relaxation time is very short. Also we note that spatial inhomogeneity of impurity concentration would tend to wash out the sharpness of the discontinuity. The oscillation is of the correct order of magnitude for InSb.

Argyres[11] finds that the conductivity in the acoustical phonon scattering range is given by

$$\left(\frac{\sigma_H}{\sigma_0}\right)_{\text{phonon scatt}} = 1 - \pi^2 \sqrt{2}\, \frac{kT}{\hbar\omega_0} \left(\frac{\hbar\omega_0}{\epsilon_F}\right)^{\frac{1}{2}} \sum_{r=1}^{\infty} \frac{(-1)^r r^{\frac{1}{2}} \cos\left(2\pi r \dfrac{\epsilon_F}{\hbar\omega_0} - \dfrac{\pi}{4}\right)}{\sinh\left(2\pi^2 r kT/\hbar\omega_0\right)}. \tag{2.13}$$

This expression is subject only to the condition $\epsilon_F \gg \hbar\omega_0$. Barrie[22] finds similar results. It is seen that increasing temperature diminishes the amplitude of the oscillations, causing the conductivity to approach its zero field value.

At very high magnetic fields, the Fermi level can become depressed so that the condition $\epsilon_F \gg kT$ no longer applies, i.e., the electron gas becomes nondegenerate. In this limit and under the condition that $\epsilon_F < \hbar\omega$,

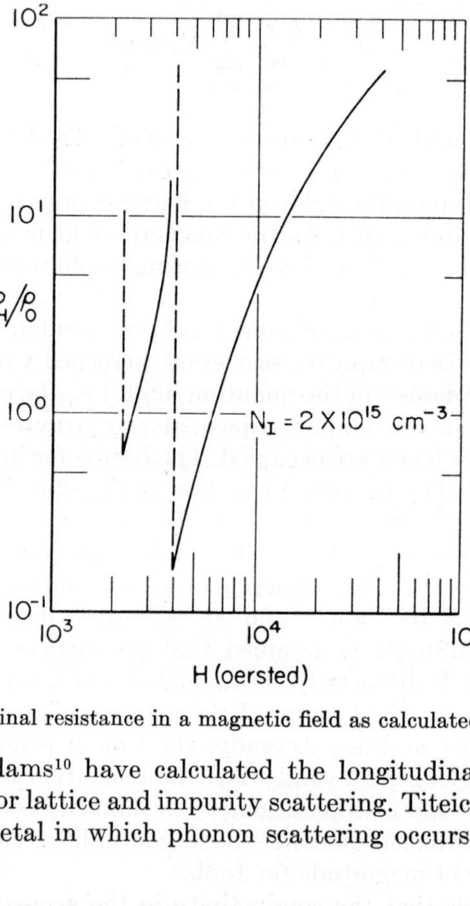

FIG. 5. Longitudinal resistance in a magnetic field as calculated from Eq. (2.12).

Argyres and Adams[10] have calculated the longitudinal resistivity of a semiconductor for lattice and impurity scattering. Titeica[5] has studied the behavior of a metal in which phonon scattering occurs in this high-field case.

b. Transverse Conductivity, Hall Effect, and Magnetoresistance

For the study of the motion of electrons in crossed electric and magnetic fields, the distribution functional analysis fails. It can be shown that to first order in the electric field, the distribution function does not deviate from its thermal equilibrium value. There is the additional diffi-

[22] R. Barrie, *Proc. Phys. Soc.* **B70**, 1008 (1957).

culty that the diagonal elements of the velocity operators v_x and v_y vanish in the eigenfunction scheme of Eq. (1.4). Thus one may not use a distribution over states appropriate for the case of zero electric field when calculating transverse currents. The best method of handling this situation is probably that based on the use of the density matrix. Lifshits[9] and Argyres[12] have used density matrix approaches, but they did not include the electric field in the quantization. More recent work of Kubo, Hasegawa, and Hashitsume[13] and Adams and Holstein[14] indicates that the former approach is incorrect, although the actual point of error has not yet been located. We shall use the method of Adams and Holstein as applied to the conductivity of a free electron gas which is subject to scattering by impurity ions, a case which may serve as a model for describing the situations met in studies of metals and semiconductors at low temperatures. The discussion will be carried out as if the exclusion principle could be ignored. However, we shall employ an argument by Kohn and Luttinger,[23] according to which the results obtained in this way may be applied to an electron gas by using the Fermi-Dirac distribution function for the diagonal elements of the thermal equilibrium density matrix.

We use a one-electron approximation for the description of the electron gas. Let the time-dependent wave function for the ith electron be

$$\psi(r,t) = \sum_n a_n(t)\psi_n(r)$$

in which the ψ_n are a complete set of orthonormal functions and the a_n are time-dependent coefficients. The density matrix ρ[24,25] is defined by the relation

$$\rho_{mn} = \langle a_m(t)a_n{}^*(t)\rangle$$

where the brackets indicate an average taken over an ensemble of systems. The time dependence of the density matrix follows from the Shroedinger equation and is given by

$$\frac{dp}{dt} = \frac{i}{h}[\rho,\mathcal{H}] \tag{2.14}$$

where $\mathcal{H}$ is the Hamiltonian for an electron and the square brackets indicate the commutator of the operators within. The average, over the ensemble, of any variable Q is calculated by taking the trace of its product with the density matrix:

$$\langle Q\rangle = \mathrm{Tr}(Q\rho). \tag{2.15}$$

[23] W. Kohn and J. M. Luttinger, *Phys. Rev.* **84,** 814 (1951).
[24] J. von Neumann, *Göttinger Nachr.* pp. 245 and 273 (1927).
[25] U. Fano, *Revs. Modern Phys.* **29,** 74 (1957).

The diagonal elements of the density matrix are interpreted as giving the probability of occupation of the various states, i.e., they play the role of a distribution function.

We shall discuss the time-development of the density matrix for the electron gas in electric and magnetic fields, in a reference system which includes the electric field in the basic wave functions. Let the magnetic field H be along the Z axis and the electric field E_x be in the direction of the X axis. Using the gauge $A = (0,Hx,0)$, the Hamiltonian becomes

$$\mathcal{H} = \frac{1}{2m}[p_x^2 + (p_y + m\omega_0 x)^2 + p_z^2] + eE_x x \tag{2.16}$$

where the symbols have the same meaning as in Eq. (1.3). The associated eigenvalue problem is similar to the one in the absence of the electric field and is easily solved.[5] The eigenfunctions are

$$\psi_{nk_yk_z} = \frac{1}{\sqrt{L_yL_z}}\,\phi_n\left(x + \frac{\hbar k_y}{m\omega_0} + \frac{eE_x}{m\omega_0^2}\right)e^{ik_yy}e^{ik_zz} \tag{2.17}$$

and the corresponding eigenvalues are

$$E_{nk_yk_z} = (n + \tfrac{1}{2})\hbar\omega_0 + \frac{\hbar^2k_z^2}{2m} - \frac{eE_x\hbar k_y}{m\omega_0} - \frac{1}{2}\frac{e^2E_x^2}{m\omega_0^2}. \tag{2.18}$$

We use an upper case letter E to represent the energy in the presence of the electric field and an ϵ to represent the part of the energy eigenvalue independent of electric field. Thus, the effect of the electric field is to remove the k_y degeneracy and to shift the center position of the orbits. We shall need matrix elements of the operators for the X coordinate and the velocity components. In the above reference system they are given as follows:

$$\begin{aligned}\langle nk_yk_z|x|n'k_y'k_z'\rangle &= -\left(\frac{\hbar k_y}{m\omega_0} - \frac{eE_x}{m\omega_0}\right)\delta_{nn'}\delta_{kk'} \\ &\quad + \sqrt{\frac{\hbar}{2m\omega_0}}\,(\sqrt{n+1}\,\delta_{n',n+1} + \sqrt{n}\,\delta_{n',n-1})\delta_{kk'}\end{aligned} \tag{2.19}$$

$$\langle nk_yk_z|v_x|n'k_y'k_z'\rangle = i\sqrt{\frac{\hbar\omega_0}{2m}}\,(-\sqrt{n+1}\,\delta_{n',n+1} + \sqrt{n}\,\delta_{n',n-1})\delta_{kk'} \tag{2.20}$$

$$\begin{aligned}\langle nk_yk_z|v_y|n'k_y'k_z'\rangle &= \sqrt{\frac{\hbar\omega_0}{2m}}\,(\sqrt{n+1}\,\delta_{n',n+1} + \sqrt{n}\,\delta_{n',n-1})\delta_{kk'} \\ &\quad - \frac{eE_x}{m\omega_0}\,\delta_{n,n'}\delta_{kk'}\end{aligned} \tag{2.21}$$

$$\langle nk_yk_z|v_z|m'k_y'k_z'\rangle = \frac{\hbar k_z}{m}\,\delta_{kk'}\delta_{nn'}. \tag{2.22}$$

Here, and elsewhere, whenever indices on k are not specified, it will be understood that k stands for both components.

In the presence of the electric and magnetic fields and the impurity potential V, the equation of motion of the density matrix is

$$-i\hbar \frac{\partial \rho}{\partial t} = [\rho, H_{\mathrm{kin}} + eE_x\chi + V] \tag{2.23}$$

where H_{kin} is the kinetic part of the Hamiltonian, i.e., the Hamiltonian of Eq. (2.16) without the electric field term.

We now employ a device used by Kohn and Luttinger,[23] that of introducing a slow time variation into the electric field term. We replace E_x by $E_x e^{st}$, where s is a small positive number which will later be allowed to approach zero. This has the effect of disconnecting the electric field at time $t = -\infty$. At that time we assume the system to be in thermal equilibrium. The density matrix will then have its thermal equilibrium value, ρ_0, where

$$(\rho_0)_{nk_yk_z;\, n'k_y'k_z'} = \frac{1}{e^{(\epsilon_{nk_yk_z} - E_F)/kT} + 1}\,\delta_{nn'}\delta_{kk'}. \tag{2.24}$$

In the above equation, E_F is the Fermi energy at zero electric field and ϵ the kinetic energy of the state α. Now let the electric field increase at a rate so slow that the electrons in the states $\psi_{nk_yk_z}$ of Eq. (2.17) adiabatically follow the field. In the absence of scattering, the electron density will remain constant and the density matrix will still be given by Eq. (2.24), with nk_yk_z representing the states at the instantaneous value of electric field and with ϵ equal to the kinetic part of the energy eigenvalue. This follows from the fact that any change in the total energy caused by the electric field must be exactly counterbalanced by an equal change in the electrochemical potential if the density of electrons is to remain unchanged.

We now proceed to find the density matrix in the electric field when scattering is present. Let

$$\rho = \rho_0 + f e^{st} \tag{2.25}$$

and substitute this into the equation of motion, Eq. (2.23). Then the correction term f, representing the effect of scattering in the presence of the electric field, satisfies the equation

$$-i\hbar s e^{st} f = [\rho_0 + f e^{st}, H_{\mathrm{kin}} + E_x e\chi e^{st}] + [\rho_0 + f e^{st}, V]. \tag{2.26}$$

We solve for f at $t = 0$. By taking matrix elements of Eq. (2.26) in the reference system of Eq. (2.17) one obtains the relation

$$(E_{nk} - E_{n'k'} - i\hbar s) f_{nk,n'k'} = [\rho_0(\epsilon_{nk}) - \rho_0(\epsilon_{n'k'})] V_{nk,n'k'} + [f, V]_{nk,n'k'}. \tag{2.27}$$

The elements of f diagonal in the wave vector satisfy a simpler equation,

$$[\hbar\omega_0(n - n') - i\hbar s]f_{nk,n'k} = [\rho_0(\epsilon_{nk}) - \rho_0(\epsilon_{n'k})]V_{nk,n'k} + [f_1, V]_{nk,n'k}. \quad (2.28)$$

The latter are the elements which determine the current, as may be seen from Eqs. (2.15) and (2.20). We now seek an approximate solution for $f_{nk,n'k}$ of the lowest order in V which will give rise to a net current. This is obtained by solving Eq. (2.27) for the general off-diagonal element of f to first order in V. This element has the value

$$f_{nk,n'k'} = \frac{[\rho_0(\epsilon_{nk}) - \rho_0(\epsilon_{n'k'})]V_{nk,n'k'}}{E_{nk} - E_{n'k'} - is\hbar}. \quad (2.29)$$

Substituting Eq. (2.29) into the commutator of Eq. (2.28), one obtains the desired solution:

$$f_{nk,n'k'} = \frac{1}{\hbar\omega_0(n - n') - is\hbar} \sum_{n''k''} \left[\frac{\rho_0(\epsilon_{nk}) - \rho_0(\epsilon_{n''k''})}{E_{nk} - E_{n''k''} - is\hbar} - \frac{\rho_0(\epsilon_{n'k}) - \rho_0(\epsilon_{n''k'})}{E_{n''k''} - E_{n'k} - is\hbar} \right] V_{nk,n''k''}V_{n''k'',n'k}. \quad (2.30)$$

We now let s approach zero and use the relation[23]

$$\frac{1}{x - is} = P\left(\frac{1}{x}\right) + i\pi\delta(x) \quad (2.31)$$

in carrying out the summation. The principal part term, P, will not give rise to currents and we shall ignore it. The correction term to the density matrix then becomes

$$f_{nk,n'k} = \frac{i\pi}{\hbar\omega_0(n - n')} \sum_{n''k''} V_{nk,n''k''}V_{n''k'',n'k}\{[\rho_0(\epsilon_{nk}) - \rho_0(\epsilon_{n''k''})]\delta(E_{nk} - E_{n''k''}) + [\rho_0(\epsilon_{n'k}) - \rho_0(\epsilon_{n''k''})]\delta(E_{n'k} - E_{n''k''})\}. \quad (2.32)$$

We note that the electric field enters Eq. (2.32) through the delta functions which represent energy conservation in the scattering transitions. We expand the right-hand side of Eq. (2.32) in powers of E_x and retain the lowest term. After some manipulation we find for off-diagonal element

$$f_{nk,n'k} = \frac{i\pi e E_x}{(n - n')m\omega_0^2} \sum_{n''k''} V_{nk,n''k''}V_{n''k'',n'k}(k_y - k_y'')\rho_0'(\epsilon_{n''k''}) \cdot [\delta(\epsilon_{n''k''} - \epsilon_{nk}) + \delta(\epsilon_{n'k''} - \epsilon_{n'k})] \quad (2.33)$$

where ρ_0' is the derivative of the Fermi function. To this approximation the elements of f diagonal in n and k vanish. Using Eqs. (2.33), (2.20), and (2.21), we find the expectation value of the current components:

$$\begin{aligned} \langle J_x + iJ_y \rangle &= -e\mathrm{Tr}(v_x + iv_y)\rho = \frac{ie^2E_x}{m\omega_0}\sum_{nk}\rho_0(\epsilon_{nk}) \\ &+ \frac{i\pi e^2E_x}{m\omega_0^2}\sqrt{\frac{\hbar\omega_0}{2m}}\sum_{nk,n''k''}\sqrt{n}\,V_{n-1,k;\,n''k''}V_{n''k'';\,nk} \\ &\cdot (k_y - k_y'')\rho_0'(\epsilon_{n''k''})[\delta(\epsilon_{n'',k''} - \epsilon_{n-1,k}) + \delta(\epsilon_{n'',k''} - \epsilon_{n,k})]. \end{aligned} \tag{2.34}$$

Considerable simplification of the above is possible.[14] If the scattering potential is that of a random distribution of impurity centers, the current components reduce to the expressions

$$J_x = -\frac{\pi\hbar e^2E_x}{(m\omega_0)^2}\sum_{\substack{n''k''\\nk}}\rho_0'(\epsilon_{n''k''})(k_y - k_y'')^2|V_{nk,n''k''}|^2\delta(\epsilon_{nk} - \epsilon_{n''k''}) \tag{2.35}$$

$$J_y = \frac{Ne^2E_x}{m\omega_0}. \tag{2.36}$$

Using these results we may identify the elements of the conductivity tensor σ. These are given by

$$\begin{aligned} J_x &= \sigma_1E_x - \sigma_2E_y \\ J_y &= \sigma_2E_x + \sigma_1E_y. \end{aligned} \tag{2.37}$$

In terms of σ_1 and σ_2 the Hall coefficient R and the transverse conductivity ρ_t are given by

$$R = -\frac{1}{H}\frac{\sigma_2}{\sigma_1^2 + \sigma_2^2}$$

and

$$\rho_t = \frac{\sigma_1}{\sigma_1^2 + \sigma_2^2}.$$

Thus to second order in the scattering the Hall constant is unchanged, while the transverse conductivity will depend entirely on the scattering. Higher order calculations would show an effect of the scattering on the Hall constant and cause it to deviate from its value $1/Nec$. The transverse resistivity, to second order in the scattering, is given by the expression

$$\rho_t = \frac{-\pi\hbar}{N^2e^2}\sum_{nkn''k''}\rho_0'(\epsilon_{n''k''})(k_y - k_y'')^2|V_{nkn''k''}|^2\delta(\epsilon_{nk} - \epsilon_{n''k''}). \tag{2.38}$$

At sufficiently low temperatures the derivative of the Fermi function in Eq. (2.38) may be replaced by $-\delta(\epsilon_{n''k''} - \epsilon_F)$. Introduction of V_q, the Fourier components of V, leads to the result

$$\rho_t = \frac{m}{2\pi\hbar N^2 e^2} \sum_{nn''} \frac{1}{\sqrt{\epsilon_F - (n + \frac{1}{2})\hbar\omega_0}} \frac{1}{\sqrt{\epsilon_F - (n'' + \frac{1}{2})\hbar\omega_0}} \sum_{q_x q_y} |V_{\mathbf{q}}|^2 |e_{nk,n''k''}{}^{i\mathbf{q}\cdot\mathbf{r}}|^2 \qquad (2.39)$$

in which the z component of $\mathbf{q}$ is that which conserves energy between primed and unprimed states. This final expression predicts singularities in the resistivity for magnetic field values at which $\epsilon_F = [n + (\frac{1}{2})]\hbar\omega_0$. One can produce significant expressions for the resistivity by selecting a suitable cutoff,[14] or by introducing a damping theory for solving Eq. (2.23).[13] However, Eq. (2.39) does establish the oscillatory behavior of the resistivity as a function of magnetic field. We shall not go into the details of the calculation of the amplitudes of the oscillatory terms. This would involve the specific scattering potential. Various scattering mechanisms will lead to differing phases and field dependences of the amplitudes of the oscillatory terms. Adams and Holstein[14] have shown that the effect of collision broadening of the energy levels introduces an exponential damping term in the conductivity oscillations, just as it does in the susceptibility.[20]

3. Magnetic Energy Levels in a Periodic Lattice

The problem of finding the energy levels of an electron in a magnetic field and a periodic potential has received little attention until recently. In the tight-binding approximation, Peierls[4] demonstrated that the magnetic Hamiltonian is obtained from the zero field electronic energy $E(k)$ by replacing k, the wave vector, by the operator $-i\nabla - e/hcA$. This result has been shown to have greater generality by Luttinger.[26] It follows from the foregoing result that all the results of the previous sections may be applied in the case of electrons or holes which have energies proportional to k^2 if one replaces the electron mass by an effective mass.

Onsager[27] has suggested that the quantization of the transverse motion in a magnetic field for more complicated energy surfaces may be determined by the Bohr-Sommerfeld rule. Using the operator substitution of the above paragraph, it is found that the following communication rela-

[26] J. M. Luttinger, *Phys. Rev.* **84,** 814 (1951).
[27] L. Onsager, *Phil. Mag.* [7] **43,** 1006 (1952).

tions are satisfied if the magnetic field is in the z direction:

$$[k_x,k_z] = [k_y,k_z] = 0$$
$$[k_x,k_y] = \frac{ie}{hc} H_z.$$

Thus k_x and k_y play the role of conjugate variables. The Bohr-Sommerfeld quantization condition becomes

$$\frac{ch}{2\pi eH_z} \oint k_y \, dk_x = n + \gamma \tag{3.1}$$

where n is an integer, γ is a constant between zero and one, and the integral is taken over the closed curve determined by the intersection of a plane perpendicular to H with the energy surface. The integral is equal to the cross-sectional area of the energy surface in $\mathbf{k}$ space. This method of quantization has been derived more rigorously from a WKB approximation by Harper[28] and Brailsford.[29] The latter points out that the energy levels will be continuous rather than discrete if the cross section is not a closed curve. Using Onsager's approach, Lifshits and Kosevich[30] have derived expressions for the oscillatory susceptibility and magnetoresistance when the energy is an arbitrary function of the wave vector. They obtain the expression

$$-\sin\left[\frac{\hbar c}{eH} A_m(\epsilon_F) \mp \frac{\pi}{4} - 2\pi\gamma\right] \tag{3.2}$$

for the first oscillatory term in the susceptibility with an amplitude given by

$$\left(\frac{e\hbar}{c}\right)^{\frac{1}{2}} \frac{kTA_m(\epsilon_F)e^{-\frac{\pi ckT}{e\hbar H}}}{\hbar(2\pi)^{\frac{3}{2}} \left|\frac{\partial^2 A_m}{\partial k_z{}^2}\right|^{\frac{1}{2}} H^{\frac{3}{2}}} \cos\left[\frac{\hbar^2}{2m_0} \frac{dA(\epsilon_F)}{d\epsilon_F}\right]. \tag{3.3}$$

Here A_m is the extremal area of the Fermi surface; the minus or plus signs correspond to the cases in which the area is a maximum or a minimum, respectively. In the case of free electrons, we set $\gamma = \frac{1}{2}$; then the susceptibility agrees with Landau's formula.

In their analysis of the electrical conductivity, Lifshits[9] and Lifshits and Kosevich[8] relate the oscillations in conductivity to the oscillations in magnetic susceptibility. Their work is based on an approximation in which the Fourier amplitudes of the velocity and collision operators are replaced by their classical values, and in which the density of states

[28] P. G. Harper, *Proc. Phys. Soc.* **A68,** 874, 879 (1955).
[29] A. D. Brailsford, *Proc. Phys. Soc.* **A70,** 275 (1957).
[30] I. M. Lifshits and A. M. Kosevich, *Soviet Phys. JETP* **2,** 636 (1956).

remains that appropriate to quantization in a magnetic field. Making this approximation, the authors find that the oscillatory term in the conductivity tensor is proportional to the product of the classical mobility tensor and the oscillatory term of the magnetization. Lifshits[9] has performed a detailed analysis only for the case in which the scattering matrix is set equal to unity, i.e., under the assumption that the effect of collisions is such as to cause all elements of the density matrix to approach their equilibrium values at the same rate. This applies to elastic scattering arising from phonons or from the potential associated with a random array of delta functions. He also indicates the procedure to be used for a more general scattering mechanism. The results for a constant isotropic effective mass agree with those of Argyres.[12] This work also neglects the possibility of quantization in the presence of the electric field.

The oscillatory effects have been used as a probe for determining the geometry of the Fermi surface. The extremal value of the cross-sectional area normal to a given direction can be obtained from the period of oscillation when the magnetic field is along the given direction. Lifshits and Pogorelov[31] have shown the way in which to construct a simply connected centrosymmetric surface from its extremal cross-sectional areas. Thus, for the more simple Fermi surfaces, the periodic value of $1/H$, when given as a function of the direction of H, can be used to map the entire Fermi surfaces. In more complicated cases, the cross-sectional areas do not determine the surface completely, but help an analysis appreciably. This was the case, for example, in the study of aluminum by Heine.[32]

III. Oscillatory Behavior in Bismuth

Bismuth has provided the standard example for the de Haas-van Alphen effect for many years. No material has been studied more extensively than this semimetal since the discovery of the effect. Accurate measurements of the magnetic susceptibility and of many different transport phenomena observed in bismuth can be found in the literature of the last 25 years. Moreover, a large number of theoretical papers have been written in a desire to explain the oscillatory effects observed in a number of the properties which have been studied in this material.

Although a detailed calculation of the band structure seems to be almost prohibitively difficult, a detailed picture of the conduction and valence band can be constructed from a combination of theoretical considerations and experimental data. In fact, a great deal of our knowledge of the band structure has been obtained from the de Haas-van Alphen

[31] I. M. Lifshits and A. V. Pogorelov, *Doklady Akad. Nauk S.S.S.R.* **96,** 1143 (1954).
[32] V. Heine, *Proc. Roy. Soc.* **A240,** 340 (1957).

effect. Other experiments have also contributed to the present model of the energy bands, however. We may mention: (a) galvanomagnetic measurements at low magnetic field strengths, (b) cyclotron resonance, and (c) electronic specific heat. To facilitate an understanding of the anisotropic nature of the oscillatory effects in bismuth, we will first discuss the energy band picture as it exists today.

4. Band Structure of Bismuth

Bismuth has a rhombohedral structure. The fifth Brillouin zone, shown in Fig. 6a, is able to accommodate all five valence electrons; the unusual properties of Bi indicate, however, that a small number of electrons spill

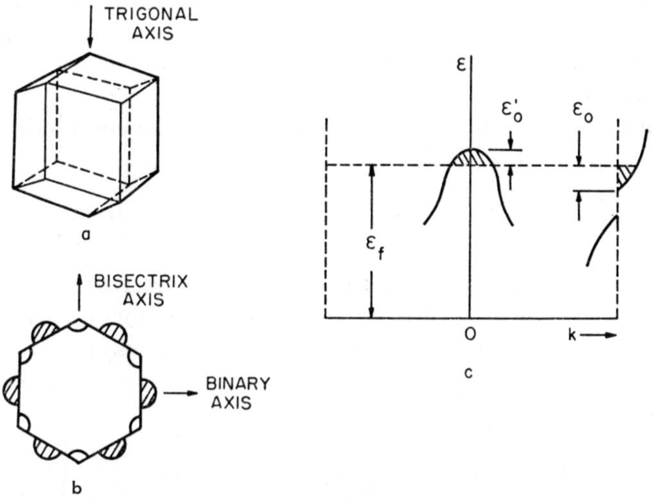

Fig. 6. Energy band structure of bismuth [after B. Lax, *Revs. Modern Phys.* **30,** 122 (1958)].

over into the next zone. A cross section of the Brillouin zone perpendicular to the trigonal axis ("3") contains the 3 binary axes ("1") and the 3 bisectrix axes ("2") (see Fig. 6b).

The first proposal of a model of the band structure was made by Jones.[33] Using susceptibility data, the behavior of dilute Bi alloys and structural considerations, he suggested that the energy surfaces near the extrema are a set of ellipsoids of revolution. Blackman[34] modified the model somewhat by taking the anisotropy into account. This model is essentially the same as that assumed by Abeles and Meiboom[35] in their

[33] H. Jones, *Proc. Roy. Soc.* **A147,** 369 (1934); **A155,** 653 (1936).

[34] M. Blackman, *Proc. Roy. Soc.* **A166,** 1 (1938).

[35] B. Abeles and S. Meiboom, *Phys. Rev.* **101,** 544 (1956).

interpretation of the galvanomagnetic effects at low magnetic field strengths. Shoenberg,[36] however, concluded from his own measurements of the de Haas-van Alphen effect that the ellipsoids are tilted out of the trigonal plane by a small angle.

The present picture of the energy surfaces is therefore as follows. The portion of the band containing the conduction electrons consists of 3 sets of ellipsoids. The three ellipsoids have their major axes nearly but not completely parallel to the trigonal axis, one binary axis and one bisectrix axis of the crystal respectively. Moreover, the ellipsoids are centered on the (221) planes. In the case of the valence band, it is generally assumed that there is one ellipsoid at the center of the zone. It is quite possible, however, that this surface is warped. The two bands overlap as indicated in Fig. 6c; the Fermi level is determined by the curvatures of the bands and the condition that the number of electrons and holes are equal in pure bismuth.

The energy surface of one of the ellipsoids in the conduction band is given by

$$E = \frac{\hbar^2}{2m_0}(\alpha_{11}k_x{}^2 + \alpha_{22}k_y{}^2 + \alpha_{33}k_z{}^2 + 2\alpha_{23}k_yk_z) \tag{4.1}$$

in which the x, y, and z axes are parallel to the binary axis, the bisectrix axis, and the trigonal axis, respectively. The parameters α_{ij} are related to the components of the mass tensor in the following way:

$$\left.\begin{aligned} \frac{m_{11}}{m_0} &= \frac{1}{\alpha_{11}}, \quad \frac{m_{22}}{m_0} = \frac{\alpha_{33}}{\alpha_{22}\alpha_{33} - \alpha_{23}{}^2}, \\ \frac{m_{33}}{m_0} &= \frac{\alpha_{22}}{\alpha_{22}\alpha_{33} - \alpha_{23}{}^2}, \quad \frac{m_{23}}{m_0} = -\frac{\alpha_{23}}{\alpha_{22}\alpha_{33} - \alpha_{23}{}^2}. \end{aligned}\right\} \tag{4.2}$$

From these relations one can deduce the effective masses associated with various orientations of the magnetic field H with respect to the crystal axes:

$$H\|\text{trigonal axis:} \quad m_{\text{trig}}{}^* = m_0\left[\frac{m_{11}(m_{22}m_{33} - m_{23}{}^2)}{m_{33}}\right]^{\frac{1}{2}} \tag{4.3a}$$

$$H\|\text{bisectrix axis:} \quad m_{\text{bis. 1}}{}^* = m_0\left[\frac{m_{11}(m_{22}m_{33} - m_{23}{}^2)}{m_{22}}\right]^{\frac{1}{2}} \tag{4.3b}$$

$$m_{\text{bis. 2}}{}^* = m_0\left[\frac{4m_{11}(m_{22}m_{33} - m_{23}{}^2)}{3m_{11} + m_{22}}\right]^{\frac{1}{2}} \tag{4.3c}$$

$$H\|\text{binary axis:} \quad m_{\text{bin. 1}}{}^* = m_0\,[m_{22}m_{33} - m_{23}{}^2]^{\frac{1}{2}} \tag{4.3d}$$

$$m_{\text{bin. 2}}{}^* = m_0\left[\frac{4m_{11}(m_{22}m_{33} - m_{23}{}^2)}{m_{11} + 3m_{22}}\right]^{\frac{1}{2}}. \tag{4.3e}$$

[36] D. Shoenberg, *Proc. Roy. Soc.* **A170**, 341 (1939).

This means that the dependence of the susceptibility and the transport phenomena on the magnetic field will show oscillations with a single period when H is parallel to the trigonal axis, whereas superimposed oscillations involving two periods will be found in measurements in which H is in the directions of the binary or bisectrix axes.

In recent years, several workers have investigated cyclotron absorption in bismuth. Dexter and Lax[37] and Foner *et al.*[38] used a 23 kMc/sec cavity with the steady magnetic field either perpendicular or parallel to the plane of the specimen. Although the absorption is nonresonant, the derivative of the power absorbed shows several peaks. The values of the effective mass deduced from these experiments are of the same order as those reported by Shoenberg,[39] however the anisotropy of the masses is different.

Galt *et al.*[40] used a similar setup. The microwave radiation was circularly polarized in this case. The interpretation of their results in terms of two types of hole seems to be very questionable.[41]

It has been shown by Azbel and Kaner[42] that it is possible to observe another type of cyclotron resonance in metals in the presence of the anomalous skin effect. The authors predicted that the surface impedance would show oscillations periodic in $1/H$ when the field H is applied parallel to the surface of the metal. Such an experiment has been performed by Aubrey and Chambers[43] on bismuth. Their results are in good agreement with theoretical predictions and confirm the Jones-Shoenberg model of the energy surfaces qualitatively. Their effective masses, however, differ somewhat from the de Haas-van Alphen values. Thus their work reduces the discrepancy between the model of Shoenberg and that of Abeles and Meiboom appreciably. It also removes the difficulty concerning the anisotropy of the collision time τ [44,45] to a large degree. The effective masses quoted by Aubrey and Chambers are: $m_{11} = 6.0 \times 10^{-3}m_0$, $m_{22} = m_0$, $m_{33} = 2 \times 10^{-2}m_0$, $m_{23} = -0.1m_0$.

There still remains the disagreement in the values for ϵ_F, the Fermi

[37] R. N. Dexter and B. Lax, *Phys. Rev.* **100,** 1216 (1955).

[38] S. Foner, H. J. Zeiger, R. L. Powell, W. M. Walsh, and B. Lax, *Bull. Am. Phys. Soc.* [2] **1,** 117 (1956).

[39] D. Shoenberg, *Phil. Trans. Roy. Soc.* **A245,** 1 (1952).

[40] J. K. Galt, W. A. Yager, F. R. Merritt, B. B. Cetlin, and H. W. Dail, Jr., *Phys. Rev.* **100,** 748 (1956).

[41] M. Tinkham, *Phys. Rev.* **101,** 902 (1956).

[42] M. Ya. Azbel and E. A. Kaner, *J. Exptl. Theoret. Phys.* (*U.S.S.R.*) **30,** 811 (1956), transl., *Soviet Phys. JETP* **3,** 772 (1956).

[43] J. E. Aubrey and R. G. Chambers, *J. Phys. Chem. Solids* **3,** 128 (1957).

[44] B. Lax, K. J. Button, H. J. Zeiger, and L. M. Roth, *Phys. Rev.* **102,** 715 (1956).

[45] B. Lax, *Revs. Modern Phys.* **30,** 122 (1958).

energy, in pure bismuth. Reynolds *et al.*[46] find a period of oscillation equal to 1.5×10^{-5} gauss^{-1} associated with $\epsilon_F = 0.012^5$ ev. Results of several other workers agree closely with this value of the period. Combining this with the effective masses reported by Aubrey and Chambers[43] one calculates $\epsilon_F = 0.013^5$. Abeles and Meiboom[35] conclude that the energy of overlap is 0.012 ev, a result which can be reconciled with the foregoing values. Shoenberg, on the other hand, quotes a value of 0.017^7 ev for ϵ_F.

Most investigators[43,47] agree on a model of the form of the energy surface for holes consisting of one ellipsoid of revolution which has its main axis parallel to the trigonal axis of the crystal. The effective masses are of the order of 1 or 2 and the axial ratio is 2.3, as suggested by Abeles and Meiboom.[35]

5. Oscillatory Behavior in the Properties of Bismuth

a. *Susceptibility*

The first measurements which displayed the sinusoidal behavior of the susceptibility of bismuth were made by de Haas and van Alphen.[2] They used the Faraday method and investigated the susceptibility in magnetic fields between 2.4 and 20.4 kilo-oersteds at 20.4 and at 14.2°K. Several maxima and minima were observed but no quantitative interpretation was given. A more detailed study was made a few years later by Shoenberg and Uddin.[48] Measurements at temperatures between 4 and 77°K revealed that the amplitude of the oscillations decreases rapidly with increasing temperature; oscillations are not observable above 40°K. Introduction of small amounts of impurities (Te and Pb) also decreased the magnitude of the effect. In 1939 Shoenberg[36] repeated these experiments using the torque method instead of the Faraday technique. In the torque method one determines the couple acting on a specimen suspended in a homogeneous magnetic field. The results are considerably more accurate than those obtained with the Faraday technique, since one measures the differences of susceptibilities in various directions rather than absolute values. The oscillatory components in the expression for the susceptibility differences provide, however, the same information concerning the energy surfaces. Shoenberg interpreted his results in terms of Landau's expression [Eq. (1.15)] and the energy band model described above. It is clear from Eq. (1.15) that the periodic term vanishes whenever the relation $[(2m^*c\epsilon_F/e\hbar H) - \delta]\pi = n\pi$ is satisfied (n = integer and

[46] J. M. Reynolds, H. W. Hemstreet, T. E. Leinhardt, and D. D. Triantos, *Phys. Rev.* **96**, 1203 (1954).
[47] V. Heine, *Proc. Phys. Soc.* **A69**, 505 (1956).
[48] D. Shoenberg and M. Z. Uddin, *Proc. Cambridge Phil. Soc.* **32**, 499 (1936).

δ = phase). Consequently a plot of the reciprocal of the values of H for which the periodic term is zero as a function of n will yield a straight line having a slope of $e\hbar/2m^*c\epsilon_F$ and an intercept δ. The value of m^* is, of course, dependent on the direction of H with respect to the crystalline axes.

The intercept appeared to yield a value of $\delta = -\pi$, quite different from the theoretically predicted value of $\pi/4$. The question of the phase of the oscillations still is in an unsatisfactory state. Later studies on bismuth and on other materials have given values of 0, $\pi/2$, $-(\frac{3}{4})\pi$, etc. Most workers believe the theoretical value $\pi/4$ is correct and offer several reasons why the experimental results do not agree with this value.

It follows from Eq. (1.16) that the absolute value of m^* can be determined from the temperature and field dependence of the amplitude A of oscillation. In early work, Shoenberg[36] was not aware of Dingle's correction[20] (collision broadening of the energy levels). Furthermore, he replaced the hyperbolic sine by an exponential and plotted ln $[(A/T)H^{\frac{3}{2}}]$ *vs* T/H for given temperatures. The curves obtained were not very good straight lines especially at low temperatures because of the poor approximation for the hyperbolic function. A rough estimate of the slope, which is $2\pi^2km^*c/e\hbar$, could however be made. Another method used was that of determining the slope for given magnetic field strength in the foregoing plot. Again, the slope should be $(2\pi^2km^*c/e\hbar)$; however, the data always indicated values smaller than those presently accepted.

The estimated electronic parameters obtained from this work were corrected later by taking collision broadening into account. Shoenberg lists the following values in an article[37] written in 1952: $m_{11} = 2.4 \times 10^{-3}m_0$, $m_{22} = 2.5m_0$, $m_{33} = 0.05m_0$, $m_{23} = -0.25m_0$, and $\epsilon_F = 0.018$ ev.

Very accurate measurements, carried out with fields extending to 32 kilo-oersteds, have been made recently by Dhillon and Shoenberg.[49] These experiments were made with orientations of crystalline axes and magnetic field such that only a single period was observed. A value of ϵ_F equal to 0.0175 ev, in good agreement with the previous result was deduced from this study. The Dingle temperature was estimated to be 1.5°K.

b. *Transport Phenomena*

Investigations of the oscillatory transport phenomena in bismuth include measurement of the magnetoresistive effect (transverse resistivity ρ_t and longitudinal resistivity ρ_l) and the Hall effect, with coefficient R_H, as well as the influence of a magnetic field on the thermoelectric power Q and the thermal conductivity κ. These experiments extend over a wide range of temperature, involve magnetic fields up to 100,000 oersteds, and

[49] J. S. Dhillon and D. Shoenberg, *Phil. Trans. Roy. Soc.* **A248,** 1 (1955).

cover all the principal crystal directions. In most of the papers describing these experiments, the authors have emphasized the period of oscillation and have compared their results with the periodicity of other transport phenomena and of the susceptibility. Table I contains a compilation of the periods of oscillation found by different workers. The measurements of Gerritsen and de Haas[50] show the oscillatory nature of the Hall coefficient clearly. The oscillations are, however, superimposed on a steadily increasing component which is negative at low fields and positive at high fields. It is quite possible that such behavior is a consequence of the particular geometry of sample and probes. Figure 7 indicates that the oscillations of the Hall effect are in phase with those of the susceptibility.

TABLE I. PERIODS OF OSCILLATION [$\Delta(1/H)$] IN BISMUTH (IN UNITS OF 10^{-5} OERSTED^{-1})

	χ Shoenberg[a]	R_H Gerritsen and de Haas[b]	R_H Reynolds *et al.*[c]	R_H and ρ_t Connell and Marcus[d]	R_H Overton and Berlincourt[e]	ρ_t Alers and Webber[f]	ρ_l Babiskin[g]	Q,κ,ρ Steele and Babiskin[h]
$H\parallel$trig.	1.4	...	1.5	1.6	1.57	...	1.57	...
$H\parallel$bin.	7.4	~7	...	7.5	...	...	7.1	7.1
	0.25	...	...	...	...	...	0.3	...
$H\parallel$bis.	4.3	~5	...	4.2	...	4.0	...	4.1
	8.5	...	...	8.8	...	7.9	...	8.2

[a] D. Shoenberg, *Phil. Trans. Roy. Soc.* **A245**, 1 (1952).
[b] A. N. Gerritsen and W. J. de Haas, *Physica* **7**, 802 (1940).
[c] J. M. Reynolds, H. W. Hemstreet, T. E. Leinhardt, and D. D. Triantos, *Phys. Rev.* **96**, 1203 (1954).
[d] R. A. Connell and J. A. Marcus, *Phys. Rev.* **107**, 940 (1957).
[e] W. C. Overton, Jr., and T. G. Berlincourt, *Phys. Rev.* **99**, 1165 (1955).
[f] P. B. Alers and R. T. Webber, *Phys. Rev.* **91**, 1060 (1953).
[g] J. Babiskin, *Phys. Rev.* **107**, 981 (1957).
[h] M. C. Steele and J. Babiskin, *Phys. Rev.* **98**, 359 (1955).

The influence of hydrostatic pressure on the de Haas-van Alphen oscillations of the Hall effect of Bi has been investigated by Overton and Berlincourt.[51] They find that the period increases by 1.5% for a pressure of 1700 psi. This can be explained qualitatively if we assume that the conduction band moves upward and the valence band downward under pressure, as it has been found in germanium and other semiconductors. In this paper Overton and Berlincourt derive an experimental expression for

[50] A. N. Gerritsen and W. J. de Haas, *Physica* **7**, 802 (1940).
[51] W. C. Overton, Jr., and T. G. Berlincourt, *Phys. Rev.* **99**, 1165 (1955).

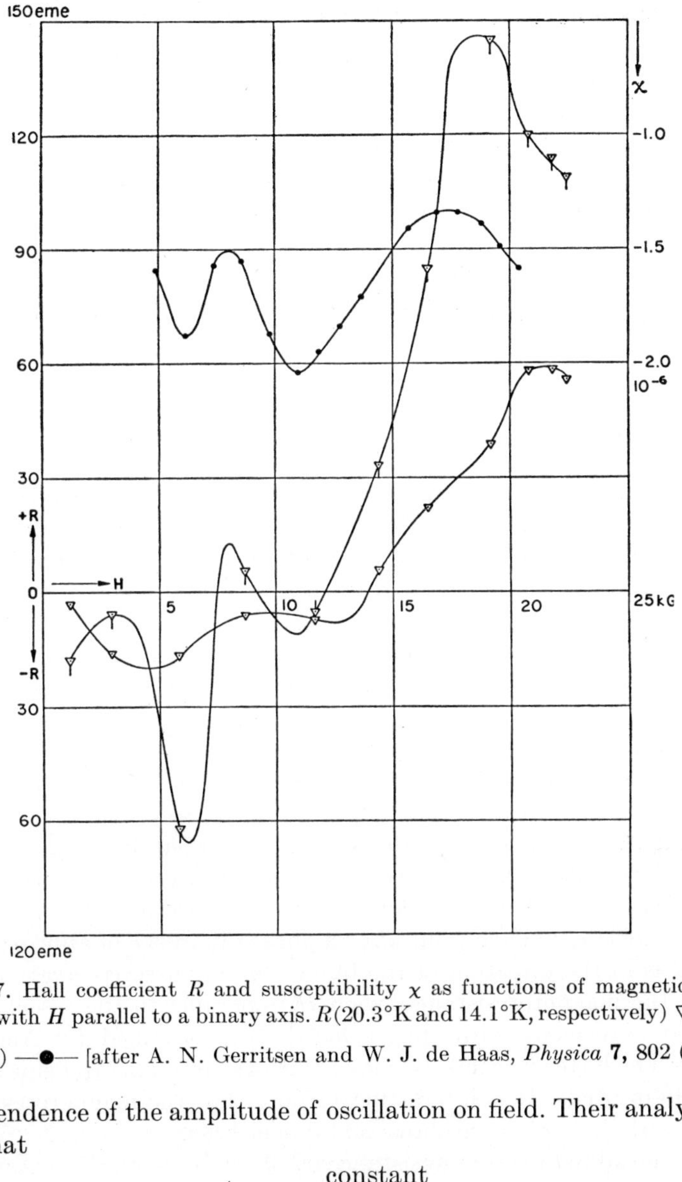

FIG. 7. Hall coefficient R and susceptibility χ as functions of magnetic field for bismuth with H parallel to a binary axis. R(20.3°K and 14.1°K, respectively) ▽ and ▽̩ ; χ(14.1°K) —●— [after A. N. Gerritsen and W. J. de Haas, *Physica* **7**, 802 (1940)].

the dependence of the amplitude of oscillation on field. Their analysis indicates that

$$A = \frac{\text{constant}}{H^{C_2} \sinh (C_3/H)} \tag{5.1}$$

with $C_2 = 2.4 \pm 0.1$ and $C_3 = (67 \pm 1) \times 10^3$ oersteds. It is stated in the paper that one should expect theoretical values of $C_2 = 1.5$ and $C_3 = 2\pi^2 kTm^*c/e\hbar$ by analogy with the de Haas-van Alphen oscillations in the susceptibility. It is clear from more recent theoretical treatments[9,12,14] of the transport phenomena, however, that an analysis of the experimental data in terms of expression (5.1) and with use of a value of $C_2 = 1.5$ has no sound basis.

Reynolds *et al.*[46] have used a different expression for the Hall coefficient. This expression is based on consideration of the number of de Haas-van Alphen electrons[18] and has the following form:

$$R = R_0 + rH^{\frac{3}{2}} \frac{2\pi^2 kT}{\beta^* H} \frac{\sin\left(\dfrac{2\pi\epsilon_F}{\beta^* H} - \dfrac{\pi}{4}\right)}{\sinh \dfrac{2\pi^2 kT}{\beta^* H}} \tag{5.2}$$

where β^* is the effective double Bohr magneton, $e\hbar/m^*c$, R_0 and r are constants, and T includes the "Dingle temperature." The field dependence of the amplitude contains a factor $H^{\frac{1}{2}}$. This form of the amplitude factor has been obtained from the expression for the susceptibility by an analogy argument. Lifshits' treatment[9] of transport phenomena suggests that the oscillatory term of formula (5.2) is essentially right. Unfortunately, it is very difficult to check the power of H, for the variation arising from this dependence is so much smaller than that arising from the hyperbolic function. Reynolds *et al.* find a period of 1.5×10^{-5} gauss^{-1}, in good agreement with results of other workers,[51] but quite a bit larger than the value quoted by Shoenberg.[39]

Recently Connell and Marcus[52] investigated the Hall coefficient and the transverse magnetoresistance of bismuth at low temperatures. Measurements were made with the magnetic field parallel to the trigonal, binary, or bisectrix axis. These experiments indicate that the oscillations in the Hall effect and in the magnetoresistance are in phase when H is parallel to the trigonal axes, but have a phase difference of approximately π when H is in the direction of the binary or the bisectrix axes. The behavior of the magnetoresistance when H is parallel to the bisectrix axis is in good agreement with the earlier results obtained by Alers and Webber.[53] The latter workers extended the measurements to fields as high as 100 kilo-oersteds. It is interesting to note that the magnetoresistance follows a fairly linear course above 30 kilo-oersteds.

The longitudinal magnetoresistance of Bi has been studied extensively by Babiskin.[54] Two orientations were explored: H parallel to the trigonal axis and parallel to the binary axis. In the latter case, Babiskin was able

[52] R. A. Connell and J. A. Marcus, *Phys. Rev.* **107**, 940 (1957).
[53] P. B. Alers and R. T. Webber, *Phys. Rev.* **91**, 1060 (1953).
[54] J. Babiskin, *Phys. Rev.* **107**, 981 (1957).

to observe both the long period of oscillation at low fields and a very short period of oscillation at high fields. The first set of measurements ($H \|$ trigonal axis) shows the cusp-like nature of the oscillations when n is small very clearly. The half-periods of oscillations 3 through 9 are alternately smaller and larger than the median value and approach the true half-period for $n > 9$ (at $T = 3°K$).

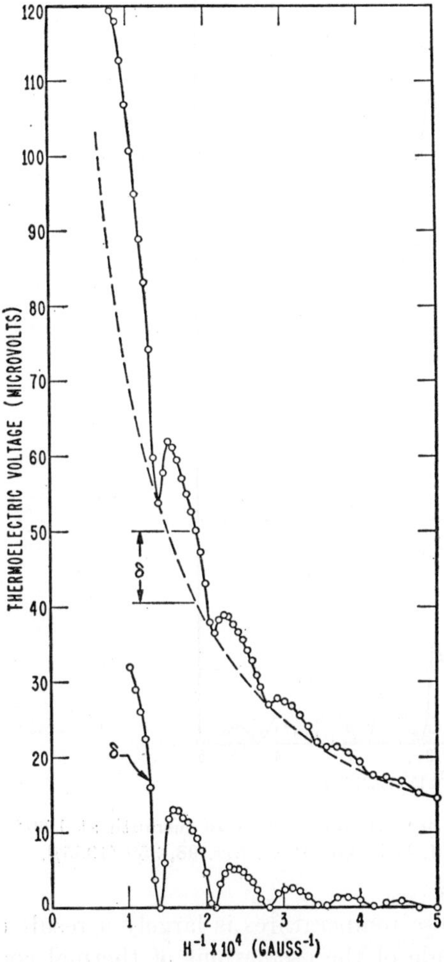

FIG. 8. Oscillatory behavior of the thermoelectric voltage of bismuth at 1.604°K ($H \|$binary axis) [after M. C. Steele and J. Babiskin, *Phys. Rev.* **98,** 359 (1955)].

The oscillatory behavior of the thermoelectric power Q, and the thermal conductivity κ, and of the transverse magnetoresistance ρ_t was studied by Steele and Babiskin.[55] Figures 8 and 9 show their results for the first two quantities. The values of β^*/ϵ_F obtained from Q, κ, and ρ_t are identi-

[55] M. C. Steele and J. Babiskin, *Phys. Rev.* **98,** 359 (1955).

cal within experimental error and agree well with the periods calculated from susceptibility and Hall data. These measurements indicate that the thermoelectric power, the thermal conductivity, and the electrical resistance are very closely in phase. The phase relationship between the latter two is rather surprising. One should not forget, however, that the

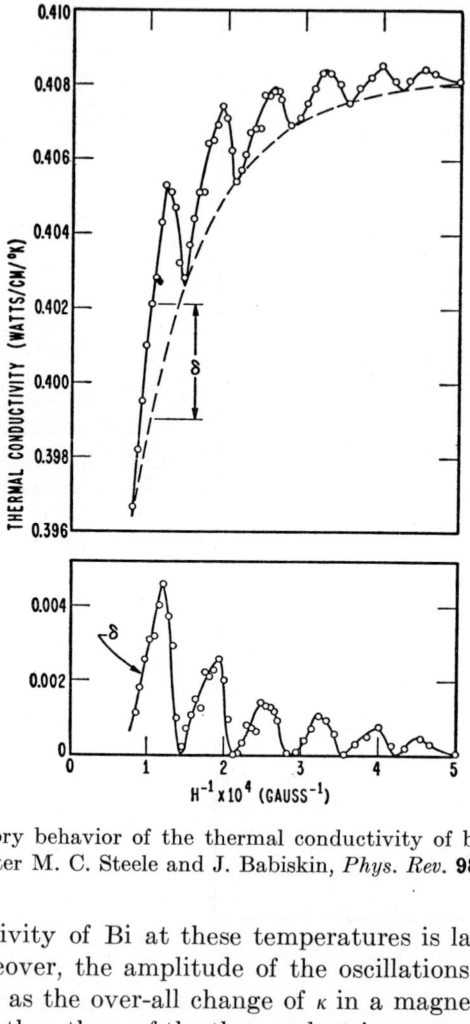

FIG. 9. Oscillatory behavior of the thermal conductivity of bismuth at 1.604°K ($H \parallel$ binary axis) [after M. C. Steele and J. Babiskin, *Phys. Rev.* **98,** 359 (1955)].

thermal conductivity of Bi at these temperatures is largely a result of the lattice. Moreover, the amplitude of the oscillations of thermal conductivity as well as the over-all change of κ in a magnetic field are considerably smaller than those of the thermoelectric power and the electrical resistance. The oscillations in κ might, therefore, be caused by a periodic change of the electronic contribution or by an oscillation in the lattice component arising from interaction with the electrons. It should be noted that these experiments were performed only for the case in which H is parallel to the binary axis; we have seen previously that the phase relations between different transport phenomena might depend on the direction of the field with respect to the crystalline axes.[52] (Measurements of Alers on zinc[56] and tin[57] showed essentially no phase difference between the oscillations of the thermal and electrical magneto*resistance*.)

The amplitude of oscillation is found to depend on the magnetic field through a factor of the form $H^{2.5}$ in the case of the thermoelectric power and through a factor $H^{2.0}$ in the case of the thermal conductivity. This result is surprising, for the theoretical formulas predict something closer to an exponential behavior. The monotonic increase of the thermoelectric power is linear, whereas the over-all thermal conductivity and electrical resistance increase approximately quadratically with H.

IV. Other Metals

Susceptibility oscillations (the de Haas-van Alphen effect) have now been observed in 15 metals, including graphite. This work has been reviewed recently by Shoenberg[15] and by Lax.[45] Both articles contain a table listing typical values of the period, values of Fermi energy, and the number of electrons per atom as well as effective masses.

Oscillations in the transport phenomena have been measured on 6 materials contained in the foregoing list. In addition to observations on bismuth, experiments have been performed on Zn, Sn, Sb, Ga, and graphite. The periods deduced from these measurements are compared with those derived from the susceptibility in Table II. These results have been analyzed in terms of a model of band structure only in the case of graphite.[58] The effective masses so calculated are in good agreement with results of de Haas-van Alphen experiments[39] and with interpretations of observations of cyclotron resonance.[59–61]

We should mention the experimental and theoretical work on aluminum here briefly, because the band structure analysis is one of the most complete for any of the metals. The de Haas-van Alphen effect has been investigated in great detail recently by Gunnersen.[62] From the results he has calculated the size and shape of the energy "pockets" which give rise to the oscillations in susceptibility.

[56] P. B. Alers, *Phys. Rev.* **101,** 41 (1956).
[57] P. B. Alers, *Phys. Rev.* **107,** 959 (1957).
[58] J. W. McClure, *Phys. Rev.* **104,** 666 (1956).
[59] J. K. Galt, W. A. Yager, and H. W. Dail, Jr., *Phys. Rev.* **103,** 1586 (1956).
[60] B. Lax and H. J. Zeiger, *Phys. Rev.* **105,** 1466 (1957).
[61] P. Nozieres, *Bull. Am. Phys. Soc.* [2] **1,** 321 (1956).
[62] E. M. Gunnersen, *Phil. Trans. Roy. Soc.* **A249,** 299 (1956–1957).

TABLE II. COMPARISON OF PERIODS FROM DIFFERENT OSCILLATORY PROPERTIES

	$\frac{m^* c \epsilon_F}{e\hbar}$ (= period)	From	Reference
Zn	6.4×10^{-5}	Susceptibility	Dhillon and Shoenberg[a]
($H\|$hexag. ax.)	6.4×10^{-5}	Susceptibility	Berlincourt and Steele[b]
	$5.3\text{–}5.8 \times 10^{-5}$	Resistance	Alers[c]
	6.5×10^{-5} (6.8×10^{-5})	Hall coefficient	Grenier *et al.*[d]
	$5.7\text{–}6.0 \times 10^{-5}$	Thermal conductivity	Alers[c]
Sn	5.9×10^{-7}	Susceptibility	Shoenberg[e]
($H\|$tetr.)	5.5×10^{-7}	Susceptibility	Berlincourt[f]
	6.2×10^{-7}	Susceptibility	Croft *et al.*[g]
	6.1×10^{-7}	Resistance	Alers[h]
	5.7×10^{-7}	Thermal conductivity	Alers[h]
Ga orthorhomb.	(a)* 2×10^{-6} (b) 3×10^{-6} (c) 5×10^{-6}	Susceptibility	Shoenberg[i]
	5×10^{-6}	Hall coefficient and resistance	Yahia and Marcus[j]
Sb rhomboh.	(a) 14.5×10^{-7} (c) 10.0×10^{-7}	Susceptibility	Shoenberg[i]
	(a) 14.1×10^{-7} (c) 9.8×10^{-7}	Hall coefficient and resistance	Steele[k]
Graphite	(a) 2.2×10^{-5} (c) 1.65×10^{-5}	Susceptibility	Shoenberg[i]
hex.	(a) 2.15×10^{-5} (c) 1.61×10^{-5}	Susceptibility	Berlincourt and Steele[l]
	2.15×10^{-5}	Hall coefficient and resistance	Berlincourt and Steele[l]
	(a) 2.12 (c) 1.59	Resistance	Soule[m]
	(a) 2.08 (c) 1.56	Hall coefficient	Soule[m]

* (a), (b), and (c) refer to the major axes of the crystal.

[a] J. S. Dhillon and D. Shoenberg, *Phil. Trans. Roy. Soc.* **A248,** 1 (1955).

[b] T. G. Berlincourt and M. C. Steele, *Phys. Rev.* **95,** 1421 (1954).

[c] P. B. Alers, *Phys. Rev.* **101,** 41 (1956).

[d] C. G. Grenier, J. M. Reynolds, and Sayed Ali, *Program 5th Intern. Conf. on Low Temp. Phys. and Chem.*, University of Wisconsin, p. 75 (1957).

[e] D. Shoenberg, *Proc. Roy. Soc.* **A170,** 341 (1939).

[f] T. G. Berlincourt, *Phys. Rev.* **88,** 242 (1952).

[g] G. T. Croft, W. F. Love, and F. C. Nix, *Phys. Rev.* **95,** 1403 (1954).

[h] P. B. Alers, *Phys. Rev.* **107,** 959 (1957).

[i] D. Shoenberg, *Phil. Trans. Roy. Soc.* **A245,** 1 (1952).

[j] J. Yahia and J. A. Marcus, *Bull. Am. Phys. Soc.* [2] **2,** 184 (1957).

[k] M. C. Steele, *Phys. Rev.* **99,** 1751 (1955).

[l] T. G. Berlincourt and M. C. Steele, *Phys. Rev.* **98,** 956 (1955).

[m] D. E. Soule, *Bull. Am. Phys. Soc.* [2] **3,** 105 (1958).

A detailed discussion of the total band structure has been given by Heine.[63] His deductions are based on (a) results of de Haas-van Alphen experiments, (b) anomalous skin effect data, (c) low-temperature specific-heat data, and (d) a fairly detailed theoretical calculation of the band structure.

A list of references concerning the oscillatory effects in the metals Bi, Zn, Sn, Ga, Sb, and in graphite is given under Table II.

V. Semiconductors

During the last few years oscillatory effects have also been observed in the semiconductors Ge, InSb, and InAs.

(*1*) *Germanium.* In 1955 Lautz and Ruppel[64] reported briefly the observation of periodic oscillations in the magnetoresistance of fairly pure germanium at 11°K. Recently some Russian workers[65] have measured the electrical resistance of n-type germanium in pulsed magnetic fields up to 110 kilogauss; their data show a strong oscillatory behavior periodic in $1/H$. Van Itterbeek *et al.*[66] have searched for susceptibility oscillations without success.

(*2*) *Indium antimonide.* Measurements of the magnetoresistance of n-type InSb by Frederikse and Hosler[67] have shown an oscillatory behavior in both the transverse and the longitudinal effect. Similar results have been obtained by Busch *et al.* recently.[68] Oscillations were observed in specimens having impurity contents corresponding to $1 \times 10^{15} - 7 \times 10^{15}$ free electrons per cubic centimeter. The Fermi energy ϵ_F depends on the number of electrons and can be calculated from measurements of the Hall coefficient. Consequently the period of oscillation varies with the purity of the sample. The measured periods appeared to be in fair agreement with the values derived from Hall effect data. No directional dependence was observed in accordance with the isotropic nature of the conduction band. An effective mass was calculated from the temperature and field dependence of the amplitude of oscillation. Considerable scatter of the experimental points decreased the accuracy of the results. A value of $(0.009 \pm 0.004)m_0$ was obtained. This figure must be compared with the results of experiments on cyclotron resonance,[69] the

[63] V. Heine, *Proc. Roy. Soc.* **A240,** 340, 354, 362 (1957).

[64] G. Lautz and W. Ruppel, *Physik. Verhandl.* **6,** 193 (1955).

[65] I. G. Fakidov and E. A. Zavadskii, *J. Exptl. Theoret. Phys.* (*U.S.S.R.*) **34,** 1036 (1958), transl., *Soviet Phys. JETP* **4,** 716 (1958).

[66] A. Van Itterbeek, L. de Greve, and W. Duchateau, *Appl. Sci. Research* **B4,** 300 (1955).

[67] H. P. R. Frederikse and W. R. Hosler, *Phys. Rev.* **108,** 1136 (1957).

[68] G. Busch, R. Kern, and B. Lüthi, *Helv. Phys. Acta* **30,** 471 (1957).

[69] G. Dresselhaus, A. F. Kip, C. Kittel, and G. Wagoner, *Phys. Rev.* **98,** 556 (1955).

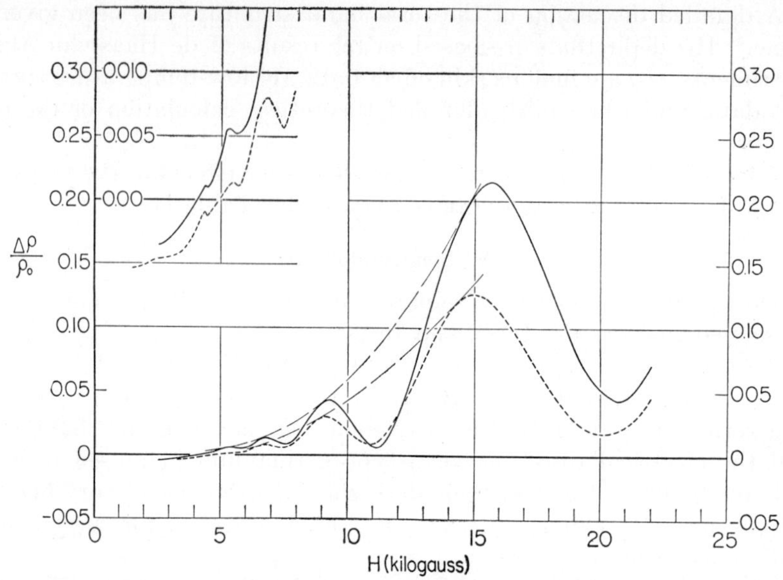

FIG. 10. Transverse magnetoresistance of indium arsenide as a function of magnetic field at 3.05°K (solid line) and 4.2°K (dashed line) [after H. P. R. Frederikse and W. R. Hosler, *Phys. Rev.* **110,** 880 (1958)].

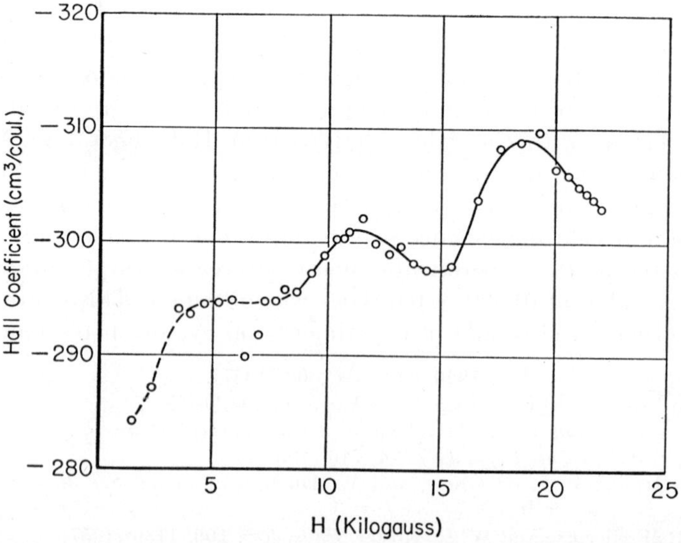

FIG. 11. Hall effect in indium arsenide as a function of magnetic field at 1.7°K [after H. P. R. Frederikse and W. R. Hosler, *Phys. Rev.* **110,** 880 (1958)].

oscillatory magnetoabsorption effect,[70,71] and several other investigations of transport and optical properties.

The Dingle temperature is about 4°K. Using this value, one calculates a mobility from (1.20) which is surprisingly close to the measured Hall mobility.

TABLE III. CHARACTERISTIC VALUES FOR InAs

Sample	T °K	n $\times 10^{16}$ cm^{-3}	Obs. period $\times 10^{-5}$ oersted^{-1} Trans.	Long.	Calc. period $\times 10^{-5}$ oersted^{-1}	T' °K Trans.	Long.	$\frac{m^*}{m_0}$ Trans.	Long.
NOL No. 5[a]	1.7	2.0	4.3	4.5	4.32	15.2	16.8	0.018	0.017
P-1[b]	1.28	2.8	3.50	3.44	3.42	12.6	11.9	0.019	. . .
S-1[b]	1.25	7.6	1.84	1.85	1.77	16.8	16.2	0.021	0.018

[a] H. P. R. Frederikse and W. R. Hosler, *Phys. Rev.* **110,** 880 (1958).
[b] R. J. Sladek, *Phys. Rev.* **110,** 817 (1958).

(3) Indium arsenide. Oscillations have been detected in both the magnetoresistance[72,73] and the Hall effect[73] (see Figs. 10 and 11). The fact that both monocrystalline and polycrystalline samples show the effect confirms the view that the energy surfaces of the conduction band are spherical. Results for three samples are shown in Table III. The agreement between observed and calculated periods, as well as between results of different experimenters, is very satisfactory. The effective masses are somewhat lower than previous experimental[74,75] and theoretical[76] estimates, all of which are in the $(0.02 - 0.03)m_0$ range.

The phase observed by Sladek is in fair agreement with the value of $\pi/4$ predicted by theory. Comparison of Figs. 10 and 11 shows that the phase difference between magnetoresistance and Hall effect is zero.

ACKNOWLEDGMENTS

The authors wish to express their gratitude to Dr. P. N. Argyres and Dr. U. Fano for helpful and illuminating discussions, to Dr. E. N. Adams and Dr. T. D. Holstein for advance copies of their work on this subject. They are also indebted to Dr. F. Seitz for a critical reading of the manuscript.

[70] E. Burstein, G. S. Picus, H. A. Gebbie, and F. J. Blatt, *Phys. Rev.* **103,** 826 (1956).
[71] S. Zwerdling, B. Lax, and L. M. Roth, *Phys. Rev.* **108,** 1402 (1957).
[72] H. P. R. Frederikse and W. R. Hosler, *Phys. Rev.* **110,** 880 (1958).
[73] R. J. Sladek, *Phys. Rev.* **110,** 817 (1958).
[74] W. G. Spitzer and H. Y. Fan, *Phys. Rev.* **106,** 882 (1957).
[75] R. J. Sladek, *Phys. Rev.* **105,** 460 (1957).
[76] F. Stern, *Bull. Am. Phys. Soc.* [2] **2,** 347 (1957).

Heterogeneities in Solid Solutions

ANDRE GUINIER
*University of Paris, France**

* Conservatoire National des Arts et Métiers, 292 rue Saint-Martin, Paris 3°, France.

I. Object of the Article

In this article, we will discuss the structure of metallic solid solutions which, to a first approximation, are *homogeneous, disordered* substitutional solutions. More precisely, the alloy appears as a single phase in classical examinations with microscopy or x-rays. We will consider only the case in which no long-range order exists; thus we will not treat the order-disorder transition, which has been the object of a previous article of this series.[1]

To a first approximation, the structure of the solid solution can be described as a unique crystalline lattice with atoms disposed at random at its mesh points or nodes. The parameters of the lattice are slowly varying functions of the concentration of the dissolved atoms when the latter have a diameter different from that of the solvent atoms.

This model arises directly from the x-ray diffraction patterns of solid solutions.[2] It is known that they are closely analogous to those of a pure metal. The powder diffraction lines, even those with large indices, are narrow and there is no evident anomaly in the variation of their intensity with the indices. However, this simplified model of the structure of a solid solution is not sufficient to explain many pecularities of various physical properties of the alloys. Since such properties are structure-sensitive, one can imagine that the presence of the dissolved atoms is linked with lattice faults which are present in a quantity that is appreciable compared to the quantity of the faults always present in a typical crystal of even a pure metal.

First of all, the fact that different atoms are found at each node of the geometrically perfect lattice corresponds to a defect in the periodicity. This imperfection, inherent in the ideal solid solution, can be exaggerated by several phenomena in the actual solid solution. The atoms may not be distributed perfectly at random; moreover, they can be displaced from the positions of the nodes of a regular lattice. Finally the energies of the valence electrons are different from those of the pure metal. This last

[1] L. Guttman, *Solid State Phys.* **3**, 146 (1956).

[2] C. S. Barrett, "Structure of Metals," p. 198. McGraw-Hill, New York, 1949.

question is considered in an article by Friedel[3] and will not be discussed here.

Our principal goal is the geometrical description of the lattice of a real solid solution. One of the greatest obstacles to the progress of certain aspects of solid state physics is the extreme difficulty of observing in a direct way in a pure metal the lattice faults which are known to play an important role in determining the physical properties. This is due, in part, to the fact that the techniques of x-ray diffraction are insufficient, since the faults are too rare to produce detectible effects. When the concentration of the solute in a solid solution is not too low, however, the conditions are more favorable for observations with x-rays. In this case, one can hope to determine the nature of the faults and to correlate them with changes in the macroscopic properties of an alloy. Thus, theoretical solid state physics may be aided by studies of the real lattice of metallic solid solutions.

We will present the principles of the method of studying lattice faults in Part II and will provide brief indications of the experimental techniques for the application of these principles.

In Parts III and IV we will discuss two categories of solid solutions, namely, solid solutions which are in thermodynamic equilibrium, and nonequilibrium solutions, particularly supersaturated metastable solid solutions, which are in the process of evolving toward the state of stable equilibrium.

II. Method of Studying Faults in the Periodicity by X-Ray Scattering

1. Nature of the Imperfections

It is established beyond doubt that the model commonly associated with the ideal solid solution is a good first approximation. The most significant fact is that diffraction lines of high order (back-reflection lines) are narrow and well-defined. This proves that, in large domains (>1000 A), the atoms are near the nodes of the average periodic lattice which corresponds to the observed diffraction pattern. The crystalline imperfections that occur in such cases are said to be of the first kind.[4] Imperfections that destroy order at distances larger than a few atomic diameters are said to be of the second kind. The diffraction pattern of a lattice possessing such imperfections is characterized by the broadening and by the diminution of intensity of the high-index diffraction lines. These effects are not observed in any solid solution. Thus we can eliminate consideration of this kind of imperfection and will not discuss it further here.

[3] J. Friedel, *Solid State Phys.* to be published.

[4] R. Hosemann, *Ergeb. exakt. Naturw.* **24,** 163 (1951).

In order to illustrate this idea with a simple example in one dimension, let us consider a "linear alloy" formed from equal proportions of two atoms, A and B, having different radii, r_A and r_B, respectively. The ideal model would be composed of a regular structure having a period equal to $r_A + r_B$, i.e., twice the average of r_A and r_B, the distribution of the atoms A and B being rigorously arbitrary (Fig. 1a). In a real alloy, there can be a tendency to order (alternating ABAB . . .) or a tendency to segregation (AAABBB . . .); moreover, the atomic centers can be displaced compared to the nodes of the average structure. However, we can assume that all these displacements are small compared to the lattice parameter (Fig. 1b).

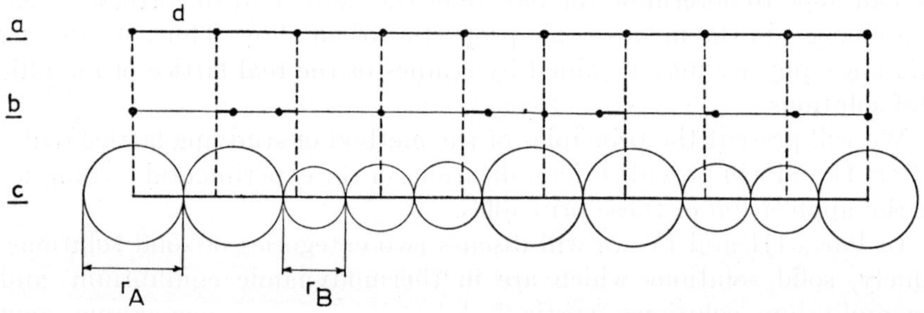

FIG. 1. Structure schemes of a linear alloy of composition AB($r_A > r_B$). a. Regular average lattice. b. Imperfect structure (first kind). c. Close-packed structure (disorder of the second kind).

One can imagine another way of building the lattice of this alloy. Assume the position of a given atom is determined only by its immediate neighbors and not by the nodes of an average lattice. For example, one could pile the atoms against each other along a line, the separation between successive atoms being the sum of their radii (see Fig. 1c). Under these conditions, the difference between the distance of the nth node and the value $n(r_A + r_B)$ can be large compared to $(r_A + r_B)$ and may increase with n. This is an example of an imperfection of the second kind. The associated diffraction pattern, which can be calculated,[5] is so different from the observed one that this model is not useful for a real metallic solid solution.

2. Effect of Imperfections on X-Ray Diffraction

It is convenient first of all to describe the effects of defects in the periodicity of a real crystal on x-ray diffraction and to deduce the possible uses of x-ray methods in the study of the structure of solid solutions.

Let us indicate some of the difficulties which induce serious and

[5] A. Guinier, *Bull. soc. franc. minéral. et crist.* **77**, 680 (1954).

unavoidable limitations. First, we must emphasize, that small deviations from a perfect crystal produce weak effects; the periodic component of the structure gives intense diffraction phenomena which undergo only second-order alterations as a result of nonperiodic faults. We will specify the orders of magnitude later (however, we shall state now that it is hopeless to try to study a disordered structure if the number of atoms which participate in the disorder is not an appreciable fraction of the total number of atoms, let us say, more than 10^{-3} or even 10^{-2}).

Another point to emphasize is that the methods of measurement should be very sensitive and sufficiently precise. It is indispensable to start with correct experimental information if one wishes to use the complex methods of calculations, such as Fourier transforms. Otherwise the results of the calculation have no real significance.

Finally, it is well known that diffraction diagrams do not contain all the items of information necessary for the unambiguous reconstruction of the structure of the diffracting object. In determining crystalline structures in which one has the problem of arranging a given number of known atoms in a unit cell, there often is a rigorous solution. In general, this is not the case in the search of the correct disordered structure. The maximum information that can be drawn from the experiments is of a statistical nature and very often is incomplete.

In contrast to these pessimistic remarks, it is necessary to stress that, in spite of their disadvantages, x-rays have provided the only direct method of approaching real atomic structures up to the present. For example, the electron microscope has given very little information concerning crystalline irregularities in alloys. Fortunately the results obtained recently on pure metals with dislocations[6] give rise to hope for progress in this area.

The principal source of information we will use is the *scattering outside the Bragg reflections.* In addition to producing such scattering, the structural irregularities have two possible effects on the Bragg reflections, namely, *broadening of the reflecting domain* in reciprocal space and modification of the law governing the *decrease in intensity* as a function of the indices of the reflection. We will cite some applications of such measurements, but pay special attention to the diffuse scattering.

For any crystal, this diffuse scattering is the sum of scattering having three origins:

(1) Compton scattering, which originates in an atomic phenomenon independent of the structure;

(2) scattering arising from thermal agitation which exists in all crystals at all temperatures;

[6] P. B. Hirsch, R. W. Horne, and M. J. Whelan, *Phil. Mag.* [8] **1**, 677 (1956).

(3) the scattering of interest to us, which is characteristic of the disorder of the crystal.

Thus in a very precise study, it is necessary to correct the measured intensity of the scattering for the first two effects. It is possible to make the correction for Compton scattering by calculation, taking the sum of the Compton scattering factors for all the atoms.[7] This is only an approximation, however, for it has been shown that the Compton scattering of an atom inside a crystal is less than the Compton scattering of the isolated atom.[8,9,10]

In the case of thermal scattering, the correction can be made by assuming that the scattering is equal to that of the same crystal, in a state known to be free of structural faults,[11] or, for alloys of weak concentration, to that of the pure metal at the same temperature. It is also possible to make a series of measurements at low temperatures and assuming a linear dependence of the temperature scattering evaluate the effect by extrapolation.[12]

In any case, the two corrections cannot be made very precisely. Therefore it is necessary that the scattering arising from the disorder surpass the Compton and thermal effects considerably. Both have a minimum intensity for small angles of diffraction;[7,13] this explains why it is advantageous to explore the central region of reciprocal space.

3. General Results of the Theory

Without giving proofs, we shall recall the principal results of the theory that we shall use.[14] Since the disorder is presumed to be weak, there corresponds to each cell of the real crystal a cell of the average perfect crystal from which the former is derived without ambiguity. In order to simplify the notation, let us take the example of a crystal possessing a single atom per unit cell. The calculations can be extended to the crystal with several atoms per unit cell; the basic results remain

[7] A. H. Compton and S. K. Allison, "X-rays in Theory and Experiment," p. 199. Macmillan, London, 1935.

[8] J. Laval, *Compt. rend.* **215**, 359 (1942).

[9] H. Curien, *Bull. soc. franc. minéral. et crist.* **75**, 197, 343 (1952).

[10] C. B. Walker, *Phys. Rev.* **103**, 547 (1956).

[11] M. Lambert, Thesis, University of Paris, 1958.

[12] J. M. Cowley *J. Appl. Phys.* **21**, 24 (1950).

[13] R. W. James, "The Crystalline State," Vol. II. Bell and Sons, London, 1948.

[14] A detailed account of the theory may be found in "Théorie et Technique de la Radiocristallographie," Part 5. Dunod, Paris, 1956. See also: W. H. Zachariasen, "Theory of X-ray Diffraction by Crystals." Wiley, New York, 1945; R. W. James, "The Crystalline State," Vol. II. Bell and Sons, London, 1948; A. J. C. Wilson, "X-ray Optics." Methuen, London, 1949; J. Hoerni and W. A. Wooster, *Acta Cryst.* **5**, 626 (1952).

the same. The nth atom is displaced from the nth node by the vector $\mathbf{\Delta x}_n$ and has an atomic scattering factor f_n (Fig. 2). The structure factor of the nth cell is:

$$F_n = f_n \exp(-2\pi i(\mathbf{s} \cdot \mathbf{\Delta x}_n)). \tag{3.1}$$

Here $\mathbf{s}$ represents the vector $(\mathbf{S} - \mathbf{S}_0)/\lambda$, which plays a fundamental role in any theory of x-ray diffraction; $\mathbf{S}_0$ and $\mathbf{S}$ are the unit vectors in the direction of the incident and diffracted rays respectively; and λ is the wavelength of the incident monochromatic radiation.

The disorder results in the fact that the structure factor F_n is a function of n, the numerical index of the cell, instead of being the same for all the cells, as in a perfect crystal. The scattering factor f_n can take one of the values f_A, f_B, which correspond to the different atoms A, B, . . . capable of occupying a given site. In addition, $\mathbf{\Delta x}_n$ may vary from one cell to another. The "disorder factor" of the cell φ_n is represented by the difference $F_n - \bar{F}$, where $\bar{F}$ is the average value of F_n for all the cells of the lattice.

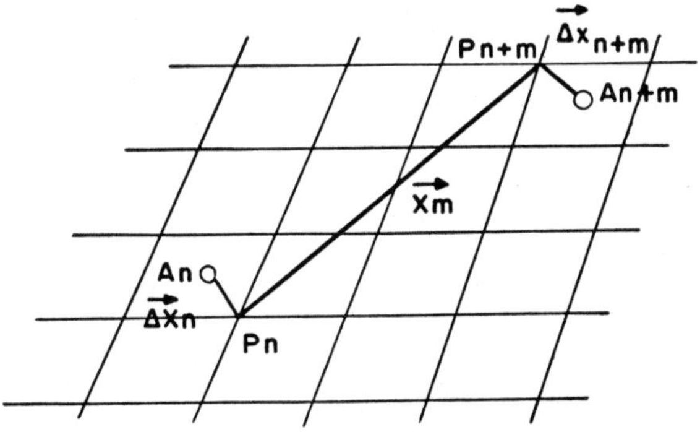

FIG. 2. Position of atoms in disordered crystals.

The diffuse scattering outside the Bragg reflections is obtained in the following manner. A function Φ_m is introduced, which is defined as follows:

$$\Phi_m = \overline{\varphi_n \varphi_{n+m}{}^*}. \tag{3.2}$$

Here Φ_m is the average of the product of the disorder factors in two cells separated by the vector $\mathbf{x}_m$, one of the translation vectors of the average lattice.[15] The diffuse scattering then is the Fourier transform of Φ_m:

$$I(\mathbf{s}) = \sum_m \Phi_m \exp(2\pi i \mathbf{s} \cdot \mathbf{x}_m). \tag{3.3}$$

There is a striking analogy between Φ_m and a Patterson function.

[15] m stands for a triplet of three integers $m_1m_2m_3$, where 1, 2, 3 correspond to the three axes of the crystal.

The most important conclusion to be drawn from Eq. (3.3) is that the experimental measurement of the scattered intensity does not lead directly to a determination of the disorder factors but to the function Φ_m, which can be obtained by a Fourier inversion of $I(\mathbf{s})$. *This function contains all the information that can be obtained a priori from the x-ray experiments without any other information or supplementary hypothesis.*

Theoretically, it always is possible to obtain the function Φ_m; however, there is no general solution to the problem of finding a model of disorder in the crystal being studied. We will see, through examples treated in the following Parts, the varied success of this technique.

For example, in the case of pure substitutional disorder (partial order or segregation in solutions), the degrees of order for the various pairs of sites in the lattice, which define the statistics of the structure of the alloy completely, are obtained directly from Φ_m. In this particular case, the statistical information is very useful, for it provides a good description of the state of the crystal.

In contrast, if the disorder is concentrated in small zones imbedded in an intact crystal and if simultaneously the atoms in these zones are of a different nature and are displaced, the function Φ_m is difficult to interpret, because it gives an average between regions of structures which definitely are different (intact areas, edges of the zones, center of the zones, etc. . . .). In such a case, the method consists of deducing from the pattern or, more precisely, from the function Φ_m, the conditions which an acceptable model of the structure of the perturbed zone ought to satisfy. We will place emphasis on these general characteristics in the following Sections. The scattered intensity expected from a series of possible models will be calculated; the one that conforms best to the experiment will be chosen. It must be noted that quite different models often give diffraction patterns which are difficult to distinguish because of the experimental uncertainties encountered in the techniques now available.

4. General Information on the Nature of Crystalline Disorder Given by X-Ray Scattering

The characteristic feature of Eq. (3.3) is that the scattered intensity depends, through the function Φ_m, on the *correlation* between the disorder of neighboring cells and not alone on the degree of disorder in each cell.

When there is no correlation between two cells separated by the vector $\mathbf{x}_m$, the function φ_n and φ_{n+m} are completely independent if m is different from zero; thus the average product Φ_m, except Φ_0, is zero. If the correlation between cells decreases with increasing m and is practically zero for $|x_m| > r$, r being of the order of a few atomic distances, the function Φ_m

decreases rapidly with m and is different from zero only for $m < 4$ or 5. In contrast, the transform of I in reciprocal space decreases very slowly. This means that the scattered intensity between the nodes of the reciprocal lattice of the average crystal is noticeable. Experimentally, very broad diffuse spots are observed (Fig. 3b).

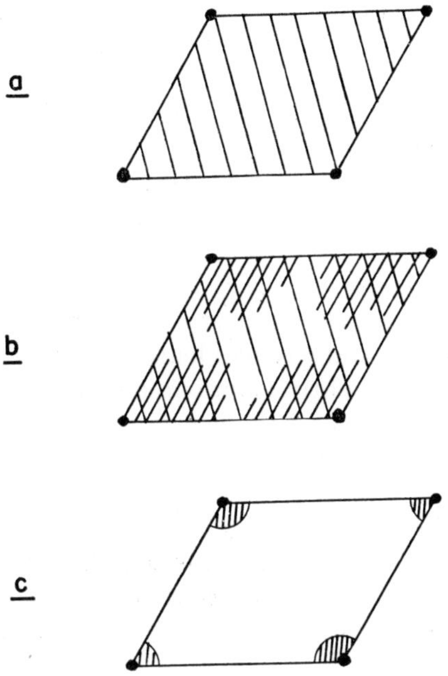

FIG. 3. Repartition of the scattered intensity in one cell of the reciprocal space. a. Disorder without any correlation between neighboring cells: uniform intensity. b. Disorder with correlation at small distances: extended scattering. c. Disorder with correlation at large distances: scattering restricted around the nodes.

The extreme case is that of the *perfectly disordered* solid solution in which the atoms remain rigorously at the nodes of the average lattice. For a binary alloy AB, the structure factor of the unit cell is f_A or f_B with the probabilities C_A and $C_B = 1 - C_A$. The average structure factor is $\bar{F} = C_A f_A + C_B f_B$ and the disorder parameter φ is $\varphi_A = C_B(f_A - f_B)$ or $\varphi_B = -C_A(f_A - f_B)$, for the sites occupied by an A or B atom respectively. The parameter Φ_m is zero for all values of m with the exception of $m = 0$:

$$\Phi_0 = C_A \varphi_A{}^2 + C_B \varphi_B{}^2 = C_A C_B (f_A - f_B)^2. \tag{4.1}$$

Thus the function Φ can be represented by a Dirac function at the origin.

The scattered intensity is the Fourier transform of the Dirac function. It has the uniform value:

$$I = C_A C_B (f_A - f_B)^2. \tag{4.2}$$

This is the scattering first calculated by Laue and known as "*Laue monotonic scattering*"[16] (Fig. 3a).

If, on the contrary, the disorder is correlated at large distances in the crystal, the function Φ_m decreases slowly with m. Conversely, its transform, the intensity I, decreases rapidly in reciprocal space and the scattering is concentrated in the immediate vicinity of the nodes of the reciprocal lattice. The spots in a single crystal pattern, or the lines in a powder pattern, are broadened. Between these broadened spots or lines, however, the scattering is very weak or even practically zero, as in the perfect crystal.

As a typical example of this case, let us consider a crystal composed of crystallites several hundred angstroms in diameter, the individual crystallites being perfect and having identical structures, but possessing a parameter which varies slightly from one to the other. Here the correlation is perfect in the volume of the crystallite and non-existent between two points situated in different crystallites. It is easy to see that the Debye-Scherrer lines of these crystals are broadened by the fluctuation of the lattice parameters. However, there is no more general scattering between the lines than in the pattern of a perfectly crystallized powder (Fig. 3c). A structure which approaches this schematic arrangement is that of a metal having heterogeneous internal strains.[17]

5. Planar and Linear Faults

An interesting case is that in which the correlation between neighboring cells in the crystal is *anisotropic*. In particular, two phenomena are met frequently in imperfect crystals, namely, planar and linear faults.

Suppose that, in a sufficiently large domain, the reticular planes of a certain family keep their lattice structure rigorously in two dimensions without disturbance, and that successive planes remain parallel. The disorder of the lattice is produced by one or several of the following causes.

(1) The interreticular distances are irregular.

(2) The successive planes are translated parallel to themselves (stacking faults).

(3) Successive planes are occupied by different atoms without regularity.

[16] M. von Laue, *Ann. Physik* [4] **56**, 497 (1918).

[17] B. E. Warren and B. L. Averbach, *in* "Imperfections in Nearly Perfect Crystals" (W. Shockley, ed.), p. 152. Wiley, New York, 1952.

The correlation between neighboring cells is limited to a short distance normal to this plane but extends to large distances parallel to it. This kind of disorder produces a scattering in reciprocal space which is concentrated on the rows of the reciprocal lattice normal to the planes considered (Fig. 4). Wooster[18] uses the abbreviated word "relrod." On the x-ray diagram the relrods correspond to *streaks* passing through the normal

a

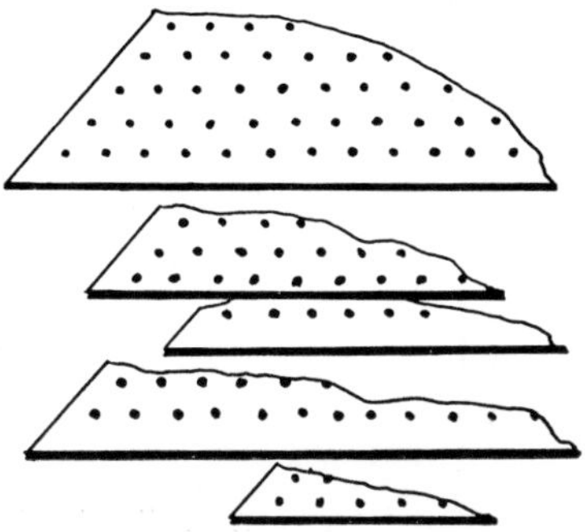

b

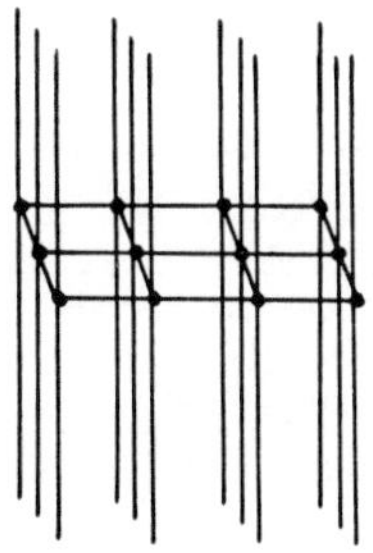

FIG. 4. a. Planar disorder: irregularly displaced reticular planes. b. Corresponding reciprocal space; scattering along "relrods."

nodes of the average lattice. A typical example is provided by the age-hardened Al-Cu alloy in which the faulted planes are the (100) planes. This case will be studied in detail in Part IV.

The *linear fault* is that in which a family of rows of a lattice stay unchanged, being rigorously periodic; however, the rows are displaced irregularly even though they always remain parallel to each other. It can be shown that the regions of scattering in reciprocal space are localized in the reticular planes ("relplanes" according to the terminology of

[18] G. N. Ramachandran and W. A. Wooster, *Acta Cryst.* **4,** 335 (1951).

Wooster) normal to the rows considered in the crystalline lattice (Fig. 5). This type of fault is also found in some age-hardened alloys.

Such faults, in which the periodicity is partially conserved, are important because they give rise to the most intense type of anomalous x-ray scattering. Being concentrated in narrow regions of the reciprocal space, the scattering is relatively intense and much easier to detect than a general background which has slowly varying modulations.

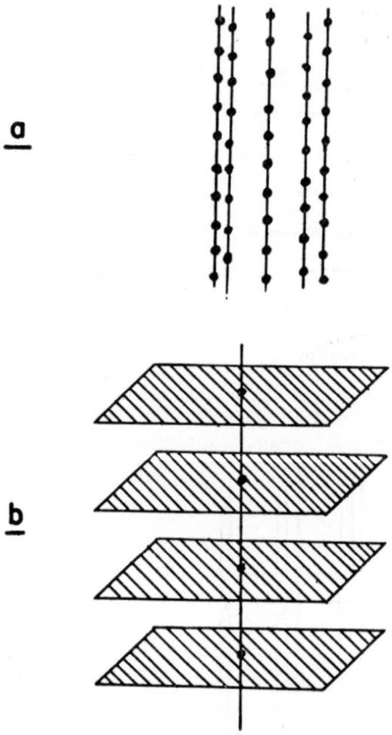

FIG. 5. a. Linear disorder: irregularly displaced rows. b. Scattering concentrated in a family of "relplanes."

From a simple qualitative study of the x-ray patterns, it is possible to make a clear distinction between faults of periodicity arising on the one hand, from the substitution of different atoms in homologous sites of the lattice and, on the other, from the displacement of atoms outside the nodes of the average regular lattice. These disorders are called respectively *substitution* and *displacement disorders*. Naturally the two types of faults can be present simultaneously.

The essential results of the theoretical study are the following.

(1) The substitutional disorder produces a scattering which is repeated periodically around the different nodes of the reciprocal lattice.

(2) For a pure displacement disorder, the scattered intensity is roughly proportional to the square of the distance from the center of reciprocal space. In particular, the scattered intensity is nil in the neigh-

borhood of the center. In this case, *no small angle scattering is observed. Such scattering is characteristic of substitutional disorder.*[19]

Sometimes the substitutional disorder is correlated directly to a disorder of displacement. Let us suppose, for example, that there is segregation of atoms which are, at the same time, larger and possess a greater scattering power than the average atom in the crystal. A correlation between the nature of the atoms and their displacements produces an *asymmetry* of the scattering around each node of the reciprocal lattice. In this particular example the scattering is more intense on the small-angle than on the large-angle side.[20] We will see applications of this general property in the structures of some solid solutions (Part III) and of some aging alloys (Part IV).

6. Experimental Methods

Without going into the details of the actual techniques, it is convenient to present an idea of the equipment necessary for the study of crystalline imperfections.[21] The principles involved in the design of equipment are the result of the theoretical considerations reviewed in the preceding Sections.

There is a choice between two x-ray detectors, namely, the G-M counter and photographic film. The advantages of the first are well-known; its use in laboratories, especially for routine powder patterns, is so developed that it is useful to point out the particular advantages of the photographic method for the type of problems with which we are interested. The important feature one is looking for in the diffraction diagram is any kind of anomalous scattering. Experience shows that such scattering can appear under very varied forms (points, narrow lines, extended and blurry spots, etc. . . .). The recording of the scattered intensity during the rotation of the crystal around an arbitrary axis would be very hard to interpret. Thus the photographic pattern is nearly indispensable in discovering phenomena. Once the scattering is located, it is easy to find the best arrangement of the crystal and counter in order to make a quantitative study of the phenomenon. If, however, a counter is used without preliminary photographs, there is only a small chance of being able to interpret the phenomena correctly.

An essential point that directs the choice of equipment originates in

[19] A. Guinier and G. Fournet, "Small Angle Scattering of X-Rays," p. 201. Wiley, New York, 1955.

[20] G. D. Preston, *Proc. Roy. Soc.* **A167**, 526 (1938).

[21] A. Guinier, "Théorie et Technique de la Radiocristallographie," p. 623. Dunod, Paris, 1956.

the fact that the intensity of the diffuse scattering is *very weak*. Therefore, it is necessary to do the following.

(1) Use intense sources. Tubes with a fine focus and a rotating anode are extremely useful.

(2) Eliminate effects of parasitic scattering that can mask the principal phenomenon.

In general, *strictly monochromatic* radiation obtained by crystalline reflection ought to be used.[22,23] It is necessary to remember, however, that a crystal reflects, besides the wanted radiation, harmonic radiations other than that desired, $\lambda/2$, $\lambda/3$, . . . etc.[24] These radiations can be discriminated by means of appropriate absorbing screens. In some cases,

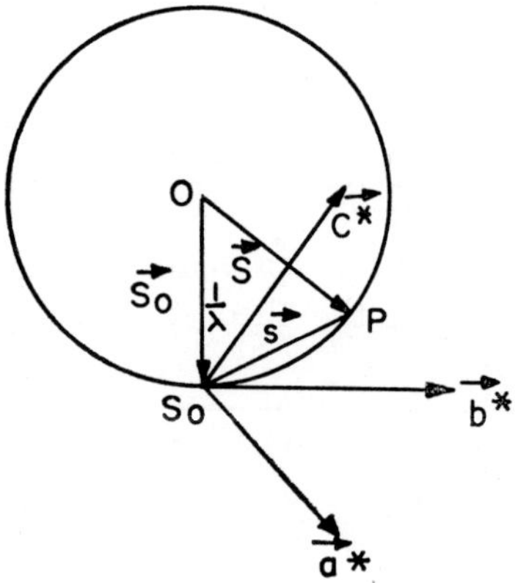

FIG. 6. Interpretation of a diagram given by a single crystal with a monochromatic radiation (λ, direction $\mathbf{S}_0$). Section of the reciprocal space by the Ewald sphere. In the direction **S**, the observed scattered intensity corresponds to the scattering power at the point *P* in the reciprocal lattice (**a***, **b***, **c***).

however, they may be so troublesome that one is obliged to suppress them. Lowering the tube voltage below the excitation of $\lambda/2$ reduces the intensity of the principal wavelength too much. The use of a scintillation counter followed by a discriminator which isolates the desired radiation constitutes great progress in simplifying the problem.[25]

The purpose of the experiment is to explore reciprocal space to localize the regions of scattering. The surest method is to operate with a stationary single crystal. Each diagram corresponds to a section of reciprocal space intersected by the Ewald sphere (Fig. 6). Successive diagrams for various

[22] H. K. Hardy, *Progr. in Metal Phys.* **5,** 185 (1954).

[23] A. J. Geisler and J. K. Hill, *Acta Cryst.* **1,** 238 (1948).

[24] N. Norman and B. E. Warren, *J. Appl. Phys.* **22,** 483 (1951).

[25] P. H. Dowling, C. F. Hendee, T. R. Kohlin, and W. Parrish, *Philips Tech. Rev.* **18,** 262 (1956).

positions of the crystals give a representation of the whole reciprocal space. This involves very long experiments,[26] especially when a counter is used.

In particular cases, it is possible to accelerate the research; for example, by oscillating the crystal. This gives on one diagram the integrated effects of a series of diagrams with a stationary sample. It is also possible to use goniometric cameras; these provide diagrams which are especially simple to interpret, since the portion of the reciprocal space intersected by a plane chosen for its particular interest is seen directly.[27]

Finally, in very special cases, it is possible to use powders which facilitate sample preparation considerably. However, this is rather a rare exception (see Section 17).

III. Solid Solutions in Equilibrium

In this Part we will examine the structure of solid solutions that are *in equilibrium*. The typical case consists of an alloy kept at a temperature in the range of a stable single phase, or a disordered solid solution that is maintained above the critical temperature of ordering. We will examine the experimental information that is now available concerning real structures and the state of the theories which concern them.

7. Perfect and Imperfect Disorder

The first problem to be discussed is that of the distribution of atoms at the nodes of the lattice of the solid solution. For the moment, we will assume that the atoms, regardless of their nature, are situated exactly at these nodes. Thus we disregard any displacement disorder.

Within a solid solution at any temperature, there are continual changes in the configuration through the mechanism of self-diffusion. Even when the solid solution is disordered, i.e., when all the possible geometric configurations are equally probable, large fluctuations in composition may arise by pure chance in a given small volume of the crystal.

Probability theory permits us to calculate easily the chance of having a given composition in a domain comprising a given number of sites. In an equiatomic alloy AB, an aggregate of n neighboring B atoms on n sites is for example much less frequent as n increases, because this state corresponds to a single configuration, whereas the mixture of $n/2$ A atoms and $n/2$ B atoms on the same set of sites can be realized by a very large number of possible configurations. Smoluchowski[28] has shown that it is necessary to consider these fluctuations in composition in the calcula-

[26] J. M. Cowley, *J. Appl. Phys.* **21**, 24 (1950).
[27] R. Glocker, W. Köster, J. Scherb, and G. Ziegler, *Z. Metallk.* **43**, 208 (1952).
[28] R. Smoluchowski, *Phys. Rev.* **84**, 511 (1951).

tion of certain physical properties. The effect can be appreciable. Moreover, these fluctuations do not give rise to a particular x-ray scattering effect. The calculation that leads to the value of the Laue monotonic scattering [formula (4.2)] is based on the assumption of perfect randomness of the solid solutions. Thus it takes account of every possible fluctuation with the right probability.

It is natural to assume that the interaction between neighbors in the solid solution depends on their nature. Consequently the free energies of the different configurations are not the same and the equilibrium state is not that of the perfectly disordered distribution. Thus the Laue calculation is not valid and the x-ray scattering is different from the Laue monotonic scattering. It can be foreseen that the study of x-ray scattering provides a very direct method of studying the atomic distribution.

In fact, the Laue scattering has not been observed in most of the measurements of scattering that have been made with homogeneous alloys. It seems then that most of the solid solutions depart from perfect irregularity, at least for concentrations of solute atoms that are not very small.

Two conditions are necessary if the x-ray experiments are to be feasible and easy to interpret.

(1) It is necessary that there be a large difference between the scattering factors of the two atoms (we shall consider only binary alloys).

(2) It is necessary that the atomic volume of the two atoms be nearly the same. This makes the hypothesis that there is no displacement of the atoms away from the nodes of a regular lattice more reasonable.

8. Short-Range Order; Characteristics of the Scattering Produced

When the disorder of the solid solution is not perfect and when there is no long-range order for the distribution of the different atoms, it is said that there is a short-range order (SRO).[29] The proportion of the A or B atoms is different from the relative amounts, C_A or C_B, of the two component atoms, in the few first shells of neighbors of a given atom, because this proportion is influenced by the nature of the atom at the origin.

Let us take the simple schematic example of a linear lattice formed by atoms A and B present in equal quantity. If the alloy has a tendency towards order, ABAB . . . , the probability that an A atom has a B neighbor is greater than one-half. Conversely, when the two kinds of atoms have a tendency to segregate, this probability will be smaller than one-half. The degree of short-range order for the first neighbors is

[29] L. Guttman, *Solid State Phys.* **3**, 152 (1956).

taken equal to $\alpha = 1 - 2n_{AB}$, where n_{AB} is the relative number of pairs of neighbors of nature AB. Thus α is negative for the alloy having a tendency to order, positive for the alloy with a tendency to segregate. It is equal to 1 for perfect order and to zero for the perfectly disordered alloy.

It is easy to calculate[30] the diffraction pattern of a linear alloy defined by a given SRO coefficient α. One finds the following.

(1) For $\alpha < 0$ (tendency toward order), there are broad maxima of scattering which coincide with the positions of the superlattice diffraction lines of the ordered alloy ABAB . . . . These maxima are more intense and sharper as α increases (Fig. 7a).

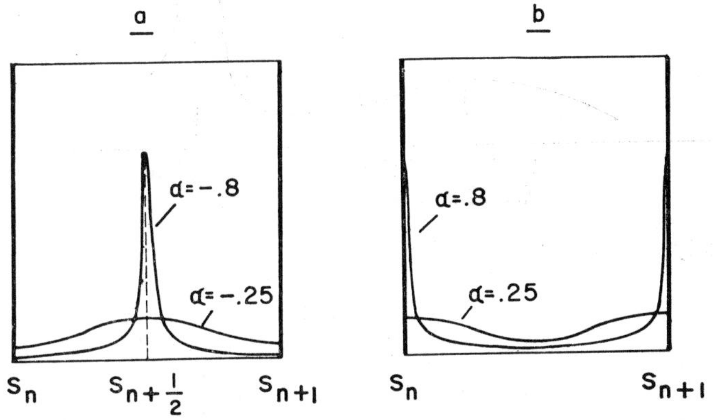

FIG. 7. Scattering by a linear equiatomic alloy for different degrees of order. a. Tendency toward order. b. Tendency toward segregation.

(2) For $\alpha > 0$ (tendency toward segregation), the maxima of scattering coincide with the center of the reciprocal space and the normal nodes of the lattice. The normal diffraction lines are superimposed on these maxima (Fig. 7b).

Similar results are valid for a three-dimensional lattice. The values of the SRO coefficient for each pair of atomic sites in the lattice define the distribution of the scattered intensity in the reciprocal lattice. Maxima around superlattice nodes indicate a tendency to order and scattering around the normal nodes a tendency to segregate.

9. Experimental Evidence of Short-Range Order

First of all, a qualitative description of the observed phenomena will be given for two typical examples, Au-Ag and Al-Ag.

At any temperature, gold and silver form a continuous series of solid

[30] A. Guinier, *Acta Cryst.* **1**, 188 (1948).

solutions whose parameters vary only slightly, from 4.08 A to 4.07 A. Thus the atomic diameters are very nearly equal. Furthermore there is a large difference between their atomic scattering factors [$(f\text{Au}/f\text{Ag}) \sim (79/47)$]. According to formula (4.2), the intensity of the Laue scattering is a maximum for the equiatomic alloy ($C_A = C_B$). In a preliminary study,

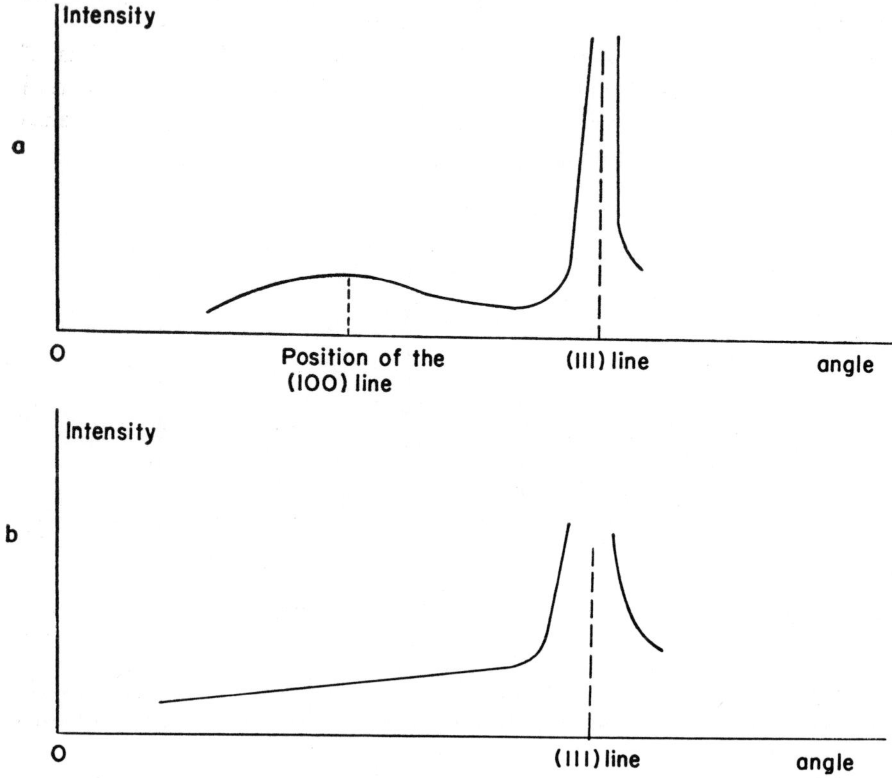

FIG. 8. Scattering by a, a 50–50 Au-Ag alloy; b, a mixture of Ag and Au powder in the same proportion.

Guinier[31] used a polycrystalline sample and found a very broad and very weak maximum of the scattering in the powder pattern at a position corresponding closely to the absent (100) line (Fig. 8).

A detailed quantitative study has been made with a single crystal by Norman and Warren;[24] weak maxima which disappear in the face-centered cubic lattice are found around the nodes of the simple cubic lattice, 100, 110, etc. . . . . It can be concluded from these results that the scattering definitely is different from the Laue monotonic scattering.

[31] A. Guinier, *Proc. Phys. Soc.* **57**, 310 (1945).

The positions of the maxima of the scattered intensities indicate that the function Φ_m (see Section 3) has important Fourier components whose periodicity is that of the simple lattice.

Thus it is likely that the distribution of Au and Ag is governed by this lattice: A atoms tend to surround themselves with B atoms in preference to A atoms. According to a general result of the theory, this

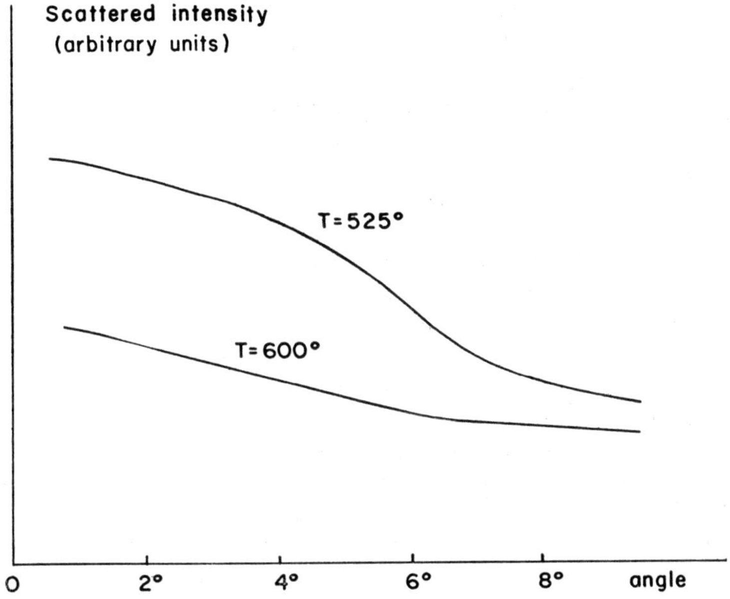

FIG. 9. Scattering by a homogeneous solid solution in equilibrium at 525°C and 600°C [C. B. Walker, J. Blin, A. Guinier, *Compt. rend.* **235**, 254 (1952)] (Radiation Cu $K\alpha$).

tendency is characterized by minima of scattering situated at the center and at the nodes of the reciprocal lattice associated with maxima at intermediate points, the superstructure nodes of the perfectly ordered lattice.[32]

The alloy Al-Ag containing 20 weight per cent of Ag forms a single disordered phase above 450°C according to the equilibrium diagram. The small-angle scattering has been studied by Walker[33,34] at 525°C and 600°C (Fig. 9). A maximum in the scattered intensity has been found at the center of the reciprocal lattice. This maximum is less intense and

[32] A. Guinier, "Théorie et Technique de la Radiocristallographie," p. 579. Dunod, Paris, 1956.

[33] C. B. Walker and A. Guinier, *Acta Met.* **1,** 568 (1953).

[34] C. B. Walker, J. Blin, and A. Guinier, *Compt. rend.* **235,** 254 (1952).

broader at the higher temperature. The scattering from a polycrystalline sample of 10 weight per cent Ag was measured in a larger angular region by Rudman and Averbach[35] who found, in addition to a maximum at an angle of 0°, an increase in the scattering in the neighborhood of the Debye-Scherrer diffraction lines.

This type of scattering is characteristic of a tendency for segregation.[32] It indicates that the atoms of silver have a tendency to form clusters more frequently than is predicted by the statistics of a perfectly disordered alloy. The majority of the clusters of silver atoms are very small. The order of magnitude of the size of these groups can be deduced simply from the law of the decrease in scattered intensity with angle. On a very approximate manner, this can be placed in the form

$$I_0 \exp\left(-\frac{4\pi^2\epsilon^2 R^2}{3\lambda^2}\right)$$

R being the radius of gyration of one group of silver atoms and ϵ the scattering angle;[36] R is the order of 7 A. At higher temperatures, the number of clusters and their size decrease since the scattered intensity diminishes.

These experiments demonstrate the existence of different types of departures from perfect disorder in solid solutions and show how it is possible to draw qualitative ideas in a simple way concerning the atomic arrangement from the scattering patterns.

10. Quantitative Study of Short-Range Order in Solid Solutions

If good measurements of the scattering intensity are available, x-rays allow one to obtain a quantitative description of the binary solid solution AB, for the application of the general theory is quite simple in this particular case.

In effect, the parameter Φ_m (defined in Section 3), which is deduced from the measured intensity for each translation vector $\mathbf{x}_m$ of the crystalline lattice by Fourier transformation, is proportional to the degree of short-range order α_m for pairs of atoms A and B separated by the vector $\mathbf{x}_m$. Let C_A and C_B be the proportion of the atoms A and B with scattering factors f_A and f_B. Let n_{BA} be the relative proportion of pairs whose origin is at a B atom and whose extremity is at an A atom. The coefficient α_m is defined as:

$$\alpha_m = 1 - \frac{n_{BA}}{C_A} \tag{10.1}$$

[35] P. S. Rudman and B. L. Averbach, *Acta Met.* **2**, 576 (1954).

[36] A. Guinier and G. Fournet, "Small Angle Scattering of X-rays," p. 25. Wiley, New York, 1954.

so that $\alpha_m = 0$ in the case of complete disorder since $n_{BA} = C_A$. When the degree of order increases progressively, α_m tends toward the value corresponding to the perfect superlattice.

Let us recall that (3.2)

$$\Phi_m = \overline{\varphi_n \varphi_{n+m}}. \tag{10.2}$$

A simple calculation shows that:

$$\Phi_m = C_A C_B (f_A - f_B)^2 \alpha_m. \tag{10.3}$$

Therefore, from (3.3)

$$I(s) = C_A C_B (f_A - f_B)^2 \Sigma \alpha_m \exp(-2\pi i \mathbf{s} \cdot \mathbf{x}_m). \tag{10.4}$$

Since $\alpha_m = \alpha_{-m}$ and $\alpha_0 = 1$

$$I(s) = C_A C_B (f_A - f_B)^2 [1 + 2\Sigma_m \alpha_m \cos(2\pi \mathbf{s} \cdot \mathbf{x}_m)]. \tag{10.5}$$

The scattered intensity appears as the sum of a term independent of **s**, which describes the Laue scattering, and terms, which are modulated functions of **s** and have the average value zero. These modulated terms have the periodicity of the reciprocal lattice: if $\mathbf{r}_p$ is one of the reciprocal lattice vectors,

$$I(\mathbf{s} + \mathbf{r}_p) = I(\mathbf{s}) \tag{10.6}$$

for all values of p. This is verified easily since

$$\mathbf{r}_p \cdot \mathbf{x}_m = \text{integer}.$$

Thus $\cos(2\pi(\mathbf{s} + \mathbf{r}_p) \cdot \mathbf{x}_m) = \cos(2\pi \mathbf{s} \cdot \mathbf{x}_m)$.

We see that it is sufficient to determine the distribution of the scattered intensity in only one cell of the reciprocal lattice; it is advantageous experimentally to choose a cell near the origin for the following reasons.

(1) The scattering factors f_A and f_B are larger.

(2) The magnitude of scattering arising from the Compton effect and thermal agitation are smallest around the origin of reciprocal space.

(3) The perturbing effects originating in the displacements of the atoms are as small as possible in the first cell. This justifies the approximation in which they are neglected.

Just as the knowledge of all the parameters of order, α_m, determines the scattering pattern, it also follows that all the parameters of short-range order can be deduced from the scattering pattern (Cowley[26]). If the complete map of the values of the scattered intensity in one cell of the reciprocal lattice has been established, the parameter α_m is given, by inversion of the sum (10.4), in the forms:

$$\alpha_m = \frac{1}{C_A C_B (f_A - f_B)^2} \int_{\text{cell}} I(\mathbf{s}) \exp(+2\pi i \mathbf{s} \cdot \mathbf{x}_m)\, dv_s. \tag{10.7}$$

The symmetries of the lattice permit abbreviation of the experimental work, since only a fraction of the unit cell of the reciprocal lattice must be determined to know the entire cell. Measurements of the relative value of the scattered intensity are sufficient, since $\alpha_0 = 1$.

11. Results of Observations on Short-Range Order

Results for the following solid solutions are now available: Cu_3-Au,[26,37,38] Cu-Au,[38] Ag-Au,[31,24] Co-Pt,[39] Li-Mg,[40] Cu-Pt,[41] Al-Zn,[42] Al-Ag,[33,35] Ag-Al,[42] and Au-Ni.[42a] One example of a determination which has been carried out as far as possible is given by the work of Cowley on Cu_3Au.[26] In Table I, the SRO coefficients are given for the first seven shells of neighbors. The coefficient is significantly different from zero for the six nearest shells; the third happens to be equal to zero. The coefficient is not larger than the possible error for the shells of neighbors beyond the seventh. In most of the other experiments, only one coefficient, namely α_1 for the immediate neighbors, is given because of the limited precision of the measurements in the experiments.

TABLE I. DEGREE OF SHORT-RANGE ORDER PARAMETER IN Au-Cu_3 AT 405°C (AFTER COWLEY[a])

	Coordinates of the site	Perfect order	Experimental
1	110	−0.33	−0.152
2	200	1.00	0.186
3	211	−0.33	0.009
4	220	1.00	0.095
5	310	−0.33	−0.053
6	222	1.00	0.025
7	321	−0.33	−0.016
	...	...	...

[a] J. M. Cowley, *J. Appl. Phys.* **21,** 24 (1950).

Table II gives the result of measurements of α_1 for several alloys by Averbach and his colleagues. Note that among the alloys with $\alpha < 0$ (a tendency to order), some have an ordered phase in equilibrium at low

[37] R. Griffoul and A. Guinier, *Rev. mét.* **45,** 387 (1948).
[38] B. W. Roberts, *Acta Met.* **2,** 597 (1954).
[39] P. S. Rudman and B. L. Averbach, *Acta Met.* **5,** 65 (1957).
[40] P. H. Herbstein and B. L. Averbach, *Acta Met.* **4,** 414 (1956).
[41] C. B. Walker, *J. Appl. Phys.* **23,** 118 (1952).
[42] P. S. Rudman, Sc.D. thesis, Mass. Inst. Technol. (1955).
[42a] P. A. Flinn, B. L. Averbach, and M. Cohen, *Acta Met.* **1,** 664 (1953).

TABLE II. NEAREST NEIGHBOR SHORT-RANGE ORDER COEFFICIENTS α_1 COMPILED BY AVERBACH[a]

Alloy (atomic fractions)	Temperature (°C)	α_1
0.75 Cu, 0.25 Au	405	−0.15
0.50 Cu, 0.50 Au	425	−0.13
0.75 Ag, 0.25 Au	250	−0.05
0.50 Ag, 0.50 Au	250	−0.08
0.50 Ni, 0.50 Au	900	−0.03
0.50 Co, 0.50 Pt	860	−0.14
0.50 Li, 0.50 Mg	25	−0.08
0.50 Cu, 0.50 Pt	890	$\alpha_1 = 0.00$, $\alpha_2 = 0.20$
0.50 Al, 0.50 Zn	400	0.16
0.90 Al, 0.10 Ag	540	0.15
0.815 Ag, 0.185 Al	450	−0.09

[a] B. L. Averbach, "Dictionary of Physics" (Solid Solutions). Pergamon, New York, to be published.

temperatures (Cu-Au), whereas none are known for others (Au-Ag). A peculiar case is given by the Au-Ni system which separates into two phases at low temperatures, one rich in Ni and the other rich in Au (Fig. 10). In contrast, in the equilibrium phase at high temperatures the atoms have a tendency to *order* themselves. Thus it is not possible to say in this case that the fluctuations of order are embryos which will serve as nuclei for the phases forming at low temperature. In fact, it is observed that the precipitation always occurs in a discontinuous manner, nucleating at the grain boundaries without particular orientations relative to the lattice. Furthermore the rate of precipitation is slow, indicating that thermal fluctuations must overcome the local order.

On the other hand, for the solid solution of Ag or Zn in Al, the clusters observed at high temperature may be considered as the nuclei of the precipitates which grow in the two-phase region. Precipitation takes place throughout the crystal and the precipitate is oriented in relation to the solid solution (Widmanstätten structure).

Quantitatively, the clustering of the dissolved atoms also can be described by degrees of order as a function of the distance between two sites which are given by formula (10.7) (Averbach and Rudman[39]). Since the clusters are spherically symmetrical, it is also possible to determine a distribution function $P(r)$, which represents the probability of finding another B atom at the distance r from a B atom in the alloy AB.

Münster and Sagel[43] have studied Al-Zn alloys richer in Zn, in the

[43] A. Münster and K. Sagel, *Molecular Phys.* **1,** 23 (1958).

domain of concentration (30 to 70% Zn) where the solid solution segregates into two solutions of different compositions below a critical temperature (Fig. 11). They have found a very strong small-angle scattering just above this critical temperature, in an interval of a few degrees.

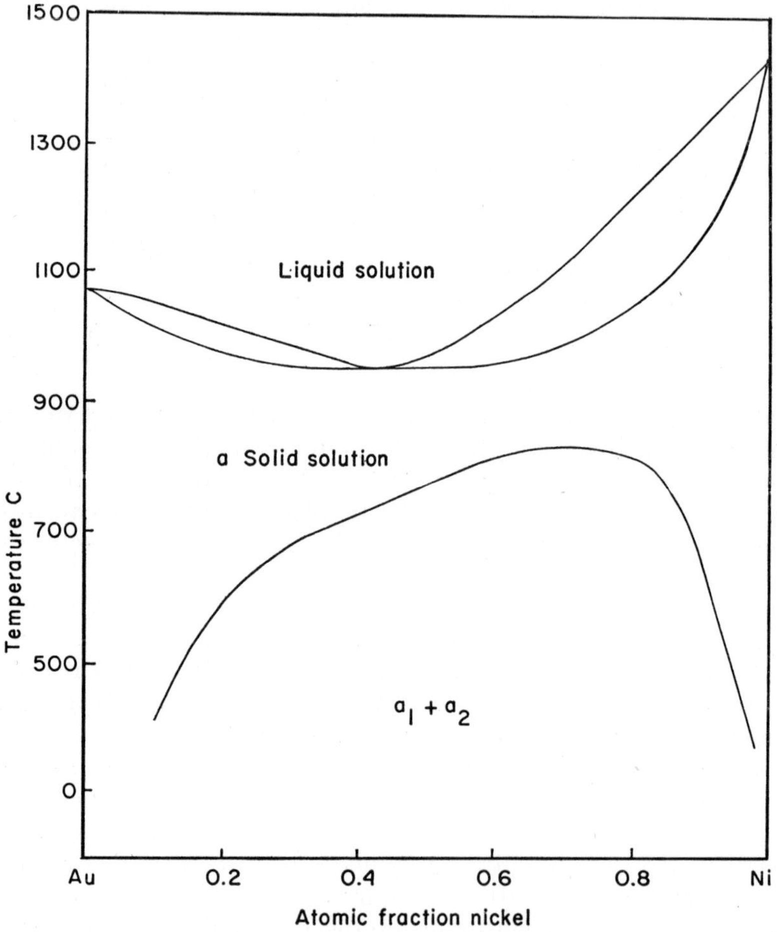

FIG. 10. Equilibrium diagram of the Au-Ni system.

This phenomenon is comparable to the critical scattering of light (opalescence) in a liquid mixture in the immediate vicinity of the temperature of segregation. At 0.5°C above the critical temperature, the order of magnitude of the size of the domains causing this fluctuation is 31 A for the alloy with 39.5% atomic Zn.

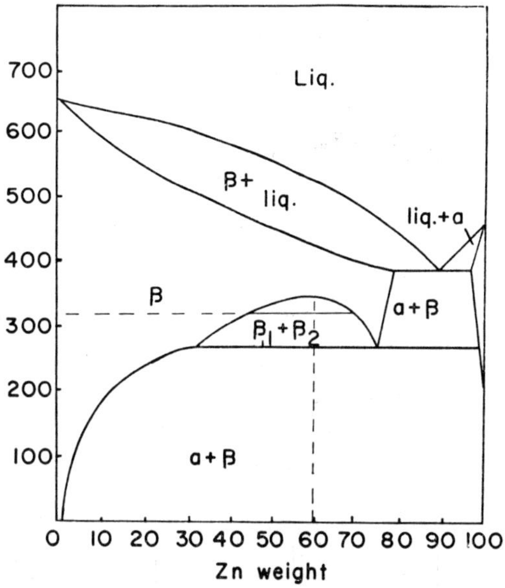

FIG. 11. Equilibrium diagram of the Al-Zn system.

12. Displacement Disorder in Solid Solutions

Except in cases in which the atoms have very similar dimensions (Au-Ag, Al-Ag, Al-Zn), it is very likely that an irregular distribution of the two species of atoms causes displacements outside the nodes of the lattice. However, such a well-defined average lattice exists: so that it can be assumed that only the distances between nearest neighbors are modified, whereas the separation between more distant atoms remains nearly equal to the corresponding distance of the average lattice; there is a sufficient number of intermediate atoms whose individual effects compensate.

The simplest method of interpreting the diffraction patterns has been given by Warren and colleagues.[44] It is assumed that the distances between atoms is a function of their nature. Thus, there would be three possible distances for the pairs AA, BB, and AB for the first neighbors, equal respectively to $2r_A$, $2r_B$ and $r_A + r_B$. The effect is certainly more complex but probably less important for more distant pairs and Warren neglects them. With these conditions, he has been able to calculate the scattered intensity as a function of the short-range order coefficient and a parameter β_1 which includes the differences between r_A and r_B and the average atomic radius calculated from the parameter of the average lattice.

[44] B. E. Warren, B. L. Averbach, and B. W. Roberts, Jr., *J. Appl. Phys.* **22,** 1493 (1951).

The formula of Warren for the intensity scattered by a single crystal is

$$\dot{I} = C_A C_B (f_B - f_A)^2 \Big[\sum_{m_1} \sum_{m_2} \sum_{m_3} \alpha_{m_1 m_2 m_3} \cos 2\pi(m_1 s_1 + m_2 s_2 + m_3 s_3)$$
$$- \sum \sum \sum 2\pi\beta_{m_1 m_2 m_3}(m_1 s_1 + m_2 s_2 + m_3 s_3)$$
$$\sin 2\pi(m_1 s_1 + m_2 s_2 + m_3 s_3) \quad (12.1)$$

where $\alpha_{m_1 m_2 m_3}$ is the degree of order and $\beta_{m_1 m_2 m_3}$ the "disorder coefficient" for the site of coordinates m_1, m_2, m_3.

The formula (12.1) explains an observation made on a disordered single crystal of Cu_3Au. The diffuse background along an [100] axis in reciprocal space is alternatively weaker and stronger between the points 000–100, 100–200, 200–300, etc. (Fig. 12). If the 12 nearest neighbors are considered alone, the second term of Eq. (12.1) has the value $-16\pi\beta_1 s_1 \sin 2\pi s_1$ for any point on the [100] axes. This function is sinusoidal with a

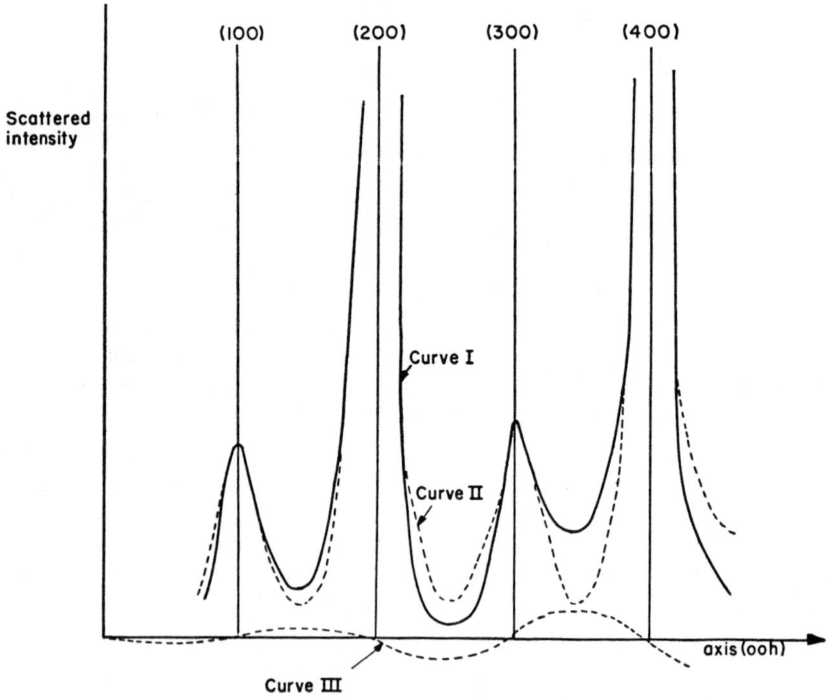

FIG. 12. Scattering by a disordered single crystal of Cu_3Au along the [001] axis [B. W. Warren, B. L. Averbach, B. W. Roberts, Jr., *J. Appl. Phys.* **22,** 1493 (1951)]. Curve I: experimental; Curve II: short-range order effect; Curve III: nearest neighbor size effect. The measured intensity is corrected for the polarization factor and the decrease of the atomic scattering factor.

progressively increasing amplitude. It is verified easily that the low background corresponds to the domains where the term is negative.

From a rather rough comparison between the experimental curve and the theory Warren[44] finds an empirical value for β_1 and the value of the distance between two gold atoms in the alloy, namely 2.78 A. This result should be compared with the Au-Au distance in pure gold namely 2.89 A. The agreement is not very good but the difference is understandable: there is a diminution of the Au-Au distance, because of the influence of the smaller Cu-Cu distance (2.54). Warren also has shown how it is possible to determine β_1 even from a powder diagram.

A more precise theory has been given by Borie.[45] It not only predicts the effect of modulation of the Laue scattering described above but two supplementary effects as well:

(1) A decrease in the intensity of the diffraction lines as their indices increase, resembling the effect produced by thermal agitation. It is possible from the intensity measurements of the lines to deduce a value for the root-mean-square separation of an atom from the average node. This effect has also been studied by Russian workers on a series of alloys and mixed crystals.[46–49] The intensity which is lost in the diffraction peaks appears in the diffuse scattering just as in the case of scattering associated with thermal agitation when the vibration of the atoms are supposed to be independent.

(2) A special scattering in the neighborhood of the nodes of the average lattice. This effect has also been predicted by Huang[50] and Cochran.[51] It is weak and difficult to observe because the thermal scattering around the nodes is more important. Although Borie has verified this effect qualitatively, it is not possible to use it to deduce quantitative conclusions. A similar but more intense effect will be described in one type of supersaturated alloys (Section 21).

Borie has shown in addition that the SRO peaks are broadened a little by the size effect when the alloy has a certain degree of short-range

[45] B. Borie, *Acta Cryst.* **10,** 89 (1957).

[46] G. V. Kurdjumov, V. A. Ilina, V. K. Kritskaya, and L. I. Lysak, *Izvest. Akad. Nauk S.S.S.R., Ser. Fiz.* **17,** 297 (1953).

[47] V. I. Iveronova and A. A. Katsnelson, *Izvest. Akad. Nauk S.S.S.R., Ser. Fiz.* **15,** 44 (1951).

[48] V. I. Iveronova, I. Kuzmina, S. I. Futergenstler, and E. I. Stetaf, *Izvest. Akad. Nauk S.S.S.R., Ser. Fiz.* **15,** 44 (1951).

[49] V. I. Iveronova and A. P. Zvjagina, *Izvest. Akad. Nauk S.S.S.R., Ser. Fiz.* **20,** 729 (1956).

[50] K. Huang, *Proc. Roy. Soc.* **A190,** 102 (1947).

[51] W. Cochran, *Acta Cryst.* **9,** 259 (1956); W. Cochran and G. Kartha, *Acta Cryst.* **9,** 941, 944 (1956).

order. He has also found experimentally that they are shifted slightly. The theory, however, does not give an adequate account of this shift.

Warren's theory clearly is very approximate, that of Borie more elaborate. It is especially important, however, to see the type of information provided by these theories. One way of presenting the results is to calculate the radius of the atoms in the solid solution from the disorder parameters. From the measurement of intensity of the normal diffraction peaks and from the modulated scattering, Borie has found that the copper and gold atoms in Cu_3Au seem to maintain the size they have in the pure elements.

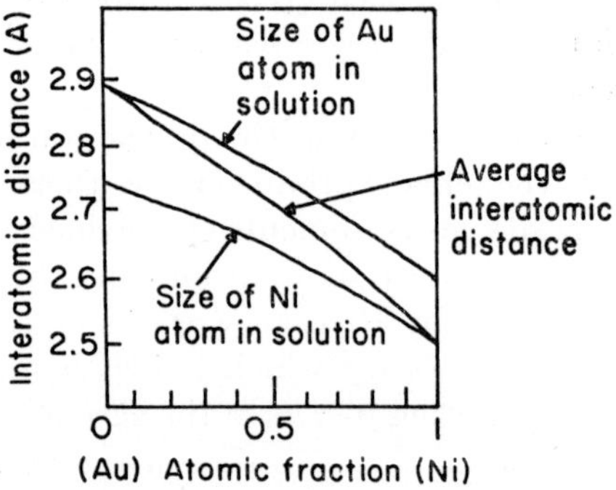

FIG. 13. Radii of the Au and Ni atoms in the solid solution calculated from the scattered intensity [P. S. Rudman and B. L. Averbach, *Acta Met.* **5**, 65 (1957)].

This conclusion is far from general. The calculated diameters of the atoms in the solid solutions AuNi and CoPt[52] are not the diameters of the atoms in pure metals (Fig. 13). There is a contraction of the larger atom and a dilation of the smaller. Thus the distortions of the lattice are reduced but not canceled because the two atoms do not have the same volume.

The other result is that the root-mean-square displacement arising from disorder is of the order of half of that arising from thermal agitation at room temperature. For example, the r.m.s. static displacement is 0.08 A for the atoms of Cu and Au in the Cu_3Au alloy; the r.m.s. thermal displacement is 0.14 A at 295°K and 0.09 at 90°K. It is interesting to note that the displacements are *not small* compared to the difference of the radii of the two atoms in pure metals, 0.14 A.

In a certain range of concentration, the solid solutions Li-Mg have a lattice parameter which is not intermediate between the parameters of the pure metals as expected, but is smaller than both of them (Fig. 14).

[52] B. L. Averbach, *in* "Theory of Alloy Phases," p. 306. Am. Soc. Testing Materials, Philadelphia, 1955.

Herbstein and Averbach[40] suppose that lithium and magnesium keep their pure metallic radii in the pairs Li-Li and Mg-Mg. It follows from the data that the pairs Li-Mg ought to undergo appreciable contraction; this indicates a modification of the atoms in dissimilar pairs. It is conceivable that this is an effect arising from slight ionization since the ions are smaller than the atoms in the metallic state.

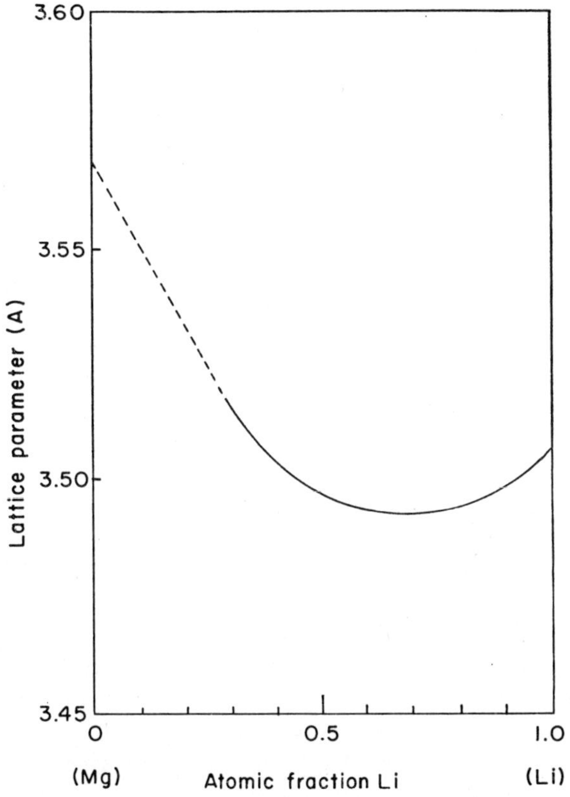

FIG. 14. Parameter of the Li-Mg solid solution at 20°C [P. H. Herbstein and B. L. Averbach, *Acta Met.* **4**, 414 (1956)].

13. Comparison of the Theories of Solid Solutions with the Experimental Data

It is interesting to consider the way in which the information on the structure of solid solutions obtained recently has permitted the theory to advance.

The most important issue is to correlate the macroscopic thermodynamic properties with the real structures of solid solutions. The chief thermodynamic properties are the enthalpy and the entropy of mixing which are determined by calorimetric measurements or by measurements of the electromotive forces or vapor pressure.[53]

The simplest theory is the quasi-chemical theory,[54,55] which is based on the fundamental hypothesis that the energy of the crystal can be considered to be the sum of the energies of all pairs. As a first approximation, this sum is often taken uniquely over nearest neighbors. The term

$$V = E_{AB} - \frac{E_{AA} + E_{BB}}{2}$$

appears in the enthalpy of formation H of a binary solution AB.

Here E_{ij} represents the energy of the pair ij. An ideal solution ($H = 0$) corresponds to $V = 0$. If $V \neq 0$, $H = P_{AB}V$, where P_{AB} is the number of pairs of unlike nearest neighbors.

There is a very simple relation between the degree of short-range order and V or H. If $V < 0$, there is a tendency for order at equilibrium and if $V > 0$, a tendency for segregation.

The quasi-chemical theory has been successful for the solid solutions of the Au-Cu system. There is good agreement between the degrees of order calculated from thermodynamic and x-ray measurements;[56] the theory has even been improved by consideration of second neighbors.[57] In some systems, however, the disagreement is complete. For example Au-Ni shows a tendency to order (negative SRO coefficient) even though the enthalpy of mixing is found to be positive ($V > 0$).

The failure of the quasi-chemical theory has been attributed to the strain energy which results from the positional disorder in the solid solution. Calculations of this strain energy have been made with the aid of elasticity theory (Friedel,[58] Rudman[59]). The enthalpy of formation of Au-Ni alloy is accounted for very satisfactorily in this manner. According to this point of view, the tendency of Au-Ni to order at high temperatures does not arise from the difference in energies of like and unlike pairs, but from the strain energy which is less in a compact arrangement when the large and small atoms alternate than when they are segregated.

[53] J. Lumsden, "Thermodynamics of Alloys," p. 154. Institute of Metals, London, 1952.
[54] E. A. Guggenheim, "Mixtures." Oxford University Press, London and New York, 1952.
[55] R. A. Oriani, *Phys. and Chem. Solids* **2,** 327 (1957).
[56] R. A. Oriani, *Acta Met.* **2,** 608 (1954).
[57] G. Fournet, *J. phys. radium* **13,** 14A (1952); **14,** 226 (1953).
[58] J. Friedel, *Phil. Mag.* Suppl. 3, 887 (1954).
[59] P. S. Rudman, Sc.D. thesis, Mass. Inst. Technol. (1955).

The use of the concepts of elasticity in the theory of alloys has been criticized by Oriani.[55] He points out that there is no reason why the elastic model should not be applied to the Au-Cu system as well to Au-Ni. Nevertheless it is found experimentally that the strain energy is only a minor part of the enthalpy of formation of Au-Cu. On the other hand, the enthalpy of formation of liquid alloys is of the same order of magnitude as that for solid alloys. According to Oriani, the existing data show a correlation between the enthalpy of mixing in the solid and liquid states. This point of view implies that both enthalpies depend on a common phenomenon which obviously cannot be elastic strains, for they are not present in the liquids.

Another difficulty is encountered in the experimental measurements of the entropy of mixing: the measured entropy is sometimes greater than the contribution originating from the randomness of the distribution of atoms, which can be deduced from the measured degree of order. The origin of this excess has been attributed partly to elastic deformations and partly to a change in the spectrum of elastic vibrations in the lattice.[60] However, quantitative results concerning these effects are not available.

The theories we have reviewed in the preceding paragraphs do not consider the nature of the interaction between atoms in solid solutions or the possible changes in the electronic structure which occur when a foreign atom is substituted in a crystal. It is certain that this effect is important. For example, it is well known that the number of valence electrons as well as the atomic size factor has an influence upon the stability of a metallic phase. However, as we have indicated in the Introduction, this question is outside the scope of our paper and will be treated in an article by Friedel in this series.

To summarize, it is now possible to describe the statistical distributions of atoms in a solid solution accurately from experimental information. Although it is possible to demonstrate the presence of lattice distortions, it is still difficult to describe these distortions quantitatively without making very rough approximations.

From the theoretical point of view, it is known that there are many cases in which the quasi-chemical approximation is insufficient. As yet, however, we do not have complete theories which can place what is known of the structure in harmony with the macroscopic thermodynamic properties.

IV. Nonequilibrium Solid Solutions

The principal object of this chapter will be the discussion of solid solutions which, to a first approximation, form a single disordered phase

[60] B. L. Averbach, P. A. Flinn, and M. Cohen, *Acta Met.* **2**, 92 (1954).

even though the homogeneous phase is not in equilibrium. The most important case to be considered is the supersaturated solid solution when it is still a single phase. The corresponding equilibrium state consists of a mixture of two phases: the *depleted matrix*, in which the concentration of dissolved atoms has reached the normal value corresponding to the limit of solubility, and the *precipitate*, in which the excess dissolved atoms have been collected.

Another case that we will do no more than mention here is that of the disordered solid solution whose equilibrium state is perfectly ordered. Metastable states intermediate between ordered and disordered and having peculiar atomic arrangements have been observed. These phenomena evidently are closely linked to the order-disorder transformation and hence lie outside of the limited scope of this paper.[61–63]

14. Continuous and Discontinuous Precipitation

Microscopic examination of a supersaturated solution during its transition towards the equilibrium state shows that there are two possible types of phenomena.[64,65] In the first, intact regions of the matrix and regions decomposed by precipitation are juxtaposed; the volume of the two-phase region increases with time. This is called "discontinuous precipitation." Generally the precipitated regions grow from the edges of grains and consume the crystal progressively. The crystals of the original matrix become a mixture of precipitate and new crystals of an impoverished matrix. Generally there is no orientation relation between the new and the initial crystals. Decomposition is governed by the nucleation and the nuclei of the precipitate grow by the diffusion of atoms dissolved in the matrix. It is necessary to imagine that there is disorganization of the matrix crystal at first and, then, complete reorganization of the atoms around the nuclei.

The sharp change in structure occurs without intermediate stages, the disorder being limited to narrow regions along the interfaces between the matrix and the precipitated solutions. The disordered domains are too small to be detected.

In the second type of decomposition, which constitutes the most interesting case from our point of view, the precipitate appears simul-

[61] S. Ogawa, D. Watanabe, and T. Komoda, *Acta Cryst.* **11,** 872 (1958).

[62] A. J. C. Wilson, *Proc. Roy. Soc.* **A181,** 360 (1943).

[63] R. Griffoul and A. Guinier, *Rev. mét.* **45,** 387 (1948).

[64] R. F. Mehl and L. K. Jetter, *in* "Age-Hardening of Metals" Am. Soc. Metals Symposium (R. Smoluchowski, ed.), p. 342. Cleveland, Ohio, 1940.

[65] A. H. Geisler, *in* "Phase Transformation in Solids" (R. Smoluchowski, ed.), p. 426. Wiley, New York, 1951.

taneously throughout the mass of the matrix crystal, and, as a general rule, there is a rigorous orientation relation between precipitate and matrix (Widmanstätten figures[66]). The matrix crystal remains unchanged in its exterior shape and orientation. There is only a slight variation of the parameters as a consequence of the variation of composition. Thus it is possible to conclude that the transformations take place without upsetting the initial crystal and by atomic movements within the framework of the lattice of the matrix. Occasionally the two types of precipitates—continuous and discontinuous—may occur simultaneously at different places in the alloy.[67]

There are structural changes in the solid solutions before any precipitation appears. These structural changes involve an important fraction of the total volume so that they produce detectable x-ray diffraction effects. In addition, very marked variations in the physical and mechanical properties occur.

We shall study this stage of evolution in detail; it can be called the *pre-precipitation.* It is characteristic of the behavior of age-hardening alloys.[68]

15. The Pre-precipitation

Let us recall the general conditions for age-hardening. The alloy is in the state of a homogeneous solid solution above a certain temperature T_c, and the limit of solubility of the dissolved atoms must decrease strongly with decreasing temperature (Fig. 15).

By quenching, from a temperature above T_c, a supersaturated solid solution that does not decompose immediately is obtained at room temperature. However, the physical properties of the quenched alloy change with time, either at ordinary temperatures (natural aging) or during an anneal at a temperature much lower than T_c (artificial aging).

During these anneals the optical microscope reveals no second phase in the bulk of crystal, at least at the beginning. Occasionally there are a few grains of precipitate along the boundaries.[67] This localized precipitation is far too rare, however, to account for the important changes of the properties.

[66] R. F. Mehl and C. S. Barrett, *Trans. AIME* **93,** 78 (1931).

[67] A. H. Geisler, C. S. Barrett, and R. F. Mehl, *AIME, Tech. Publ.* No. 1557 (1943).

[68] The age-hardening of alloys has been reviewed in the following general papers: R. F. Mehl and L. K. Jetter, *in* "Age-Hardening of Metals," Am. Soc. Metals Symposium (R. Smoluchowski, ed.), p. 342. Cleveland, Ohio, 1940; A. H. Geisler, *in* "Phase Transformations in Solids" (R. Smoluchowski, ed.), p. 387. Wiley, New York, 1951 (with a thorough bibliography up to 1951); G. S. Smith, *Progr. in Metal Phys.* **1,** 163 (1949); H. K. Hardy, *ibid.* **5,** 279 (1954); *Light Metals* **7,** 328, 383 (1944).

Observations with the electron microscope using ordinary replica techniques also do not reveal a second phase clearly. There only are indications of a heterogeneity in the matrix.[69,70] Thus, in the replicas taken with an Al-Cu alloy aged at room temperature, white spots which are irregularly dispersed through the crystal are visible. These spots are the result of irregularities in oxide formation but they do not prove the existence of a well-defined phase. Recently, however, examinations[71,72] by transmission through thin layers have given very interesting results in some cases, especially Al-Cu, which confirm the structures revealed by x-rays and described in Section 18.

There are two possibilities for explaining the changes in physical properties during aging. One may assume either that the alloy contains

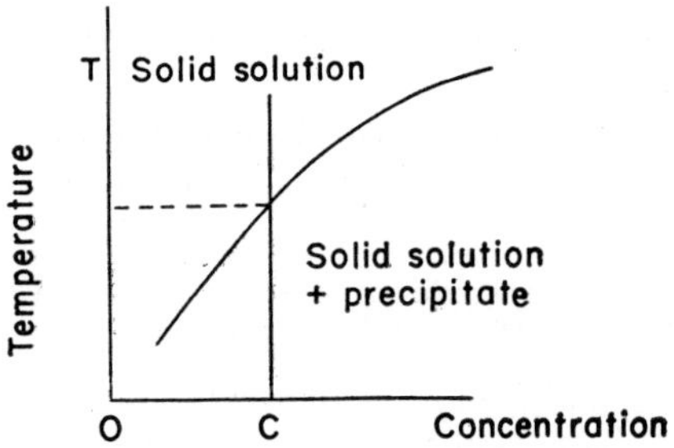

FIG. 15. Equilibrium diagram of a solid solution; solubility increasing with T.

precipitates so fine that they escape observation, or that the solid solution passes through a state of pre-precipitation characterized by a different atomic structure.

The first idea is the basis for the first theory of age-hardening (Merica,[73] Geisler[74]). While Merica did not give direct experimental proofs of his hypothesis, Geisler based his concepts upon an interpretation of the observed anomalous x-ray scattering. He assumed that the diffuse spots observed in the diagrams of the aged alloys really were diffraction spots arising from very small precipitated particles, broadened by the well-known size effect. The variation of the shape during aging would correspond to changes in the shape and dimensions of the growing precipitate. From the observations of the diffuse scattering in several

[69] G. Thomas and J. Nutting, *Inst. Metals Monogr. and Rept. Ser.* No. 18, 57 (1955).
[70] A. Saulnier and R. Syre, *Rev. mét.* **49,** 81 (1951).
[71] R. Castaing, *Rev. mét.* **52,** 669 (1955).
[72] R. B. Nicholson and J. Nutting, *Phil. Mag.* [8] **3,** 531 (1958).
[73] P. D. Merica, R. G. Waltenberg, and A. Scott, *Trans. AIME* **64,** 41 (1920).
[74] A. H. Geisler and J. K. Hill, *Acta Cryst.* **1,** 238 (1958).

alloys, especially Al-Ag, and Al-Mg-Si, Geisler[74] concluded that there is a *sequence* of stages during the evolution of the alloy corresponding to the progressive growth of the precipitated particles.

The first nucleus is formed by the clustering of a few dissolved atoms; this nucleus then grows into a needle directed along an important crystallographic axis. The needle is subsequently transformed into a thin platelet parallel to a reticular plane. Finally the platelet becomes a grain, which is large enough in three dimensions to give normal diffraction spots.

This interpretation has been discussed in detail in another article[75] where it is demonstrated that the sequence theory in its strictest form is unacceptable in the special case of Al-Cu.

(1) There are many differences between the properties of a hardened alloy with and without visible precipitations (see Section 26). The differences cannot be explained in terms of simple variation of the size of precipitates, keeping the same structure.

(2) A detailed analysis of the x-ray scattering figures shows that they cannot be explained by the presence of a precipitated crystalline phase imbedded in the matrix whatever the structure and shape of such particle might be. The most striking example is given by streaks characteristic of the natural aging of Al-Cu (Section 18a).

If we reject the idea of the continuous change of a growing phase during aging, we are lead to the opposite concept that there are several *different and successive stages* in the evolution of the alloy corresponding to distinct atomic structures and distinct properties.

Thus, Dehlinger[76,77] before the development of our knowledge of the atomic structure of aged alloys, had distinguished two stages: the *cold-hardening* and the *warm-hardening* stages which take place at low and moderate temperatures respectively (25°C and 250°C in Al-Cu, for instance). It is now certain that the warm-hardening corresponds to a true precipitation, for the precipitates are easily made visible, even at the very beginning, with the aid of the electron microscope.

The cold-hardening ought then to be of another nature; it is the *pre-precipitation*.[78] We shall show later on how the atomic structure of this stage has been revealed by the x-ray scattering phenomena.

The idea which is now generally accepted is that the variations in the properties of the solid solutions in the pre-precipitation stage are caused by heterogenities in the distribution of the dissolved atoms; however,

[75] A. Guinier, *Acta Cryst.* **5,** 121 (1952).

[76] U. Dehlinger, *Z. Metallk.* **29,** 401 (1937).

[77] U. Dehlinger, "Chemische Physik der Metallen und Legierungen." Akad. Verlagsges., Leipzig, 1939.

[78] A. Guinier, *J. Metals* **8,** 673 (1956).

these occur without the presence of two distinct phases. The excess dissolved atoms have a tendency to form clusters or *"zones"* which are included in the matrix, which always have submicroscopic dimensions and whose structure is that of the matrix, more or less deformed. In a general manner, we shall define a zone as a small region of crystalline irregularity in the matrix resulting from a variation in composition and possibly from displacement of the atoms outside of the nodes of the average matrix.

The essential characteristics of the zone which justify making a distinction between it and a very small precipitate are that it does not have a well-defined lattice, and it does not have precise boundaries. There is perfect *coherence with the lattice of the matrix.* To be more precise, let us say that, as one passes from the matrix to the interior of the zone, the immediate surroundings of each atom vary in a continuous manner if we consider both the nature and position of these neighbors.

Thus, the zones are large faults in periodicity, and can produce anomalous x-ray scattering which is easy to detect if the zones are sufficiently numerous.

The major part of this chapter will be devoted to the description of the arrangement of the atoms in the zones as determined with x-rays. With the aid of this structure, it will be possible to explain certain of their properties and the laws governing their growth and transformation.

The zones found in different alloys are of various types. It is convenient to classify them into two groups.

16. Zones with or without Distortion

Suppose the dissolved (B) atoms diffuse to coalesce into a zone in the solid solution. The simplest possible zone in which lattice distortion is negligible (Fig. 16a) is obtained if the B atoms can remain at the nodes of the matrix in spite of their very strong concentration. The zone, then, is a region having a geometrically perfect lattice where the concentration of B atoms is anomalously high. This case ought to be found in alloys where the radii of the two atoms A and B are very similar. The most typical example is that of Al and Ag (both have face-centered cubic lattices with parameters 4.04 and 4.08 A, respectively). This is the example of *zones without distortion,* which we will treat in detail in the next section.

This is an exceptional case, however; in general, the accumulation of B atoms produces local distortions in the lattice (Fig. 16b). The problem is then much more complex since both kinds of disorder, substitution and displacement, are present in the zone; it is difficult to determine simultaneously and completely the distribution of the B atoms and the

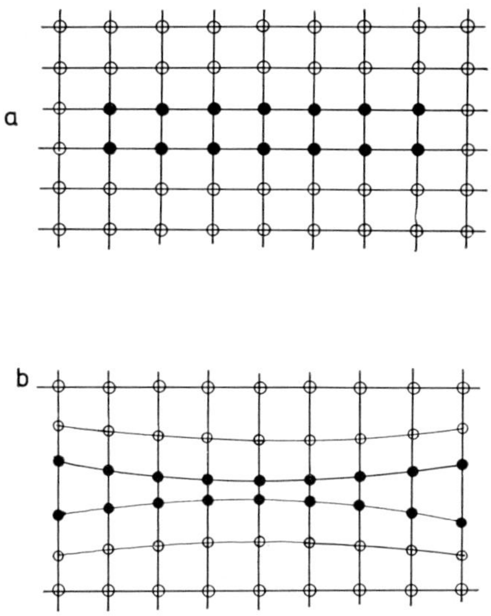

FIG. 16. Schematic of a zone. a. Without distortions. b. With distortions.

deformation of the lattice. To illustrate this case, the Al-Cu alloy will be chosen, for it is the age-hardening system which has been studied the most.

17. Zones in an Al-Ag Alloy

Numerous experimental investigations are now available concerning Al-Ag alloys having Ag contents varying up to the maximum of the stability of a single phase namely 38 weight per cent of Ag (13 atomic per cent). The structure has been studied as a function of thermal treatments in states associated with pre-precipitation and precipitation; diverse physical and mechanical properties have been examined as well.[79–103]

[79] C. S. Barrett, A. H. Geisler, and R. F. Mehl, *Trans. AIME* **152,** 182 (1943).
[80] A. H. Geisler, C. S. Barrett, and R. F. Mehl, *Trans. AIME* **143,** 134 (1941).
[81] A. Guinier, *J. phys. radium* **3,** 124 (1942).
[82] C. B. Walker and A. Guinier, *Acta Met.* **1,** 568 (1953).
[83] W. Köster and F. Braumann, *Z. Metallk.* **43,** 193 (1956).
[84] W. Köster, H. Steinert, and J. Scherb, *Z. Metallk.* **43,** 202 (1952).
[85] R. Glocker, W. Köster, J. Scherb, and G. Ziegler, *Z. Metallk.* **43,** 208 (1952).
[86] G. Ziegler, *Z. Metallk.* **43,** 213 (1952).
[87] A. Guinier, *Z. Metallk.* **43,** 217 (1952).
[88] W. Köster and H. Dietrich, *Z. Metallk.* **43,** 449 (1952).
[89] W. Köster and H. A. Schell, *Z. Metallk.* **43,** 454 (1952).

Generally, the alloy is homogenized first by heating at a temperature in the neighborhood of 550°C, and is then quenched to room temperature. Subsequently the metal is held for varying times at varying temperatures. Thus it is possible to control the rate of evolution of the alloy; furthermore, it can be studied at room temperature, after treatment at other temperatures, without appreciable variations in its state.

The characteristic scattering arising from the irregularities of the lattice in the pre-precipitation stage can be described in the reciprocal lattice in the following way. The most intense regions of scattering are located in the immediate neighborhood of each node of the matrix in the

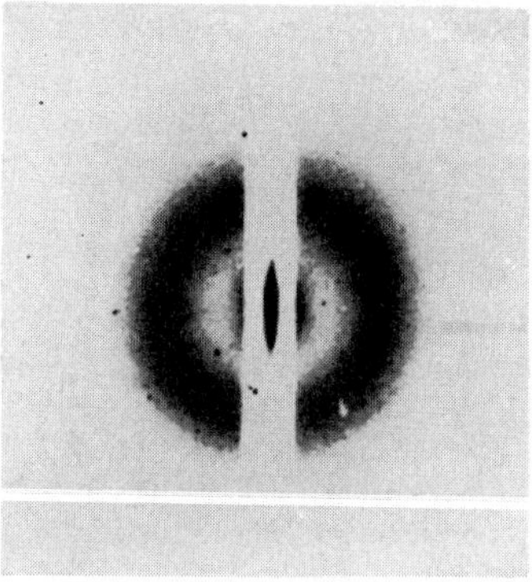

FIG. 17. Small-angle pattern of an Al-Ag alloy (20% weight), aged at 20°C [C. B. Walker and A. Guinier, *Acta Met.* **1,** 568 (1953)].

reciprocal lattice. The *same domains* of scattering are found around *each of the nodes,* including the 000 node, i.e., the center of the reciprocal space. They have the shape of blurred spherical shells centered on the node.

The easiest region to examine is that around the direct beam (the node 000). The small-angle scattering pattern reproduced in Fig. 17 corresponds to a section of the spherical scattering domain by a diametral plane. The curve of Fig. 19 gives the quantitative variation of the

[90] K. Hirano, *J. Phys. Soc. Japan* **8,** 603 (1953).

[91] G. Borelius, *Rept. Bristol Conf. on Defects in Crystalline Solids, 1954* p. 169 (1955).

[92] W. Köster and A. Knödler, *Z. Metallk.* **46,** 632 (1955).

[93] B. Belbeoch and A. Guinier, *Acta Met.* **3,** 370 (1955).

[94] D. Turnbull and H. N. Treaftis, *Acta Met.* **5,** 534 (1957).

[95] A. M. Elistratov, *Doklady Akad. Nauk S.S.S.R.* **69,** 337 (1949); **101,** 473 (1955).

[96] P. N. Buinov and R. M. Lerinman, *Izvest. Akad. Nauk S.S.S.R., Ser. Fiz.* **15,** 358 (1951).

[97] V. Gerold, *Z. Metallk.* **46,** 623 (1955).

[98] W. Köster and A. Knödler, *Z. Metallk.* **46,** 632 (1955).

[99] K. Hirano, *J. Phys. Soc. Japan* **8,** 603 (1953).

[100] K. Hirano and J. Sakai, *J. Phys. Soc. Japan* **10,** 23 (1955)

[101] K. Tanaka, H. Abe, and K. Hirano, *J. Phys. Soc. Japan* **10,** 454 (1955).

[102] R. H. Beton and E. C. Rollason, *J. Inst. Metals* **86,** 85 (1957).

[103] G. Borelius and L. E. Larsson, *Arkiv Fysik* **11,** 137 (1956).

scattered intensity as a function of the distance from the center, measured with a G-M counter.

There is no indication of anisotropy in the scattering. Therefore the small-angle pattern is the same for a single crystal, regardless of its orientation, and for a powder of small unoriented crystals.

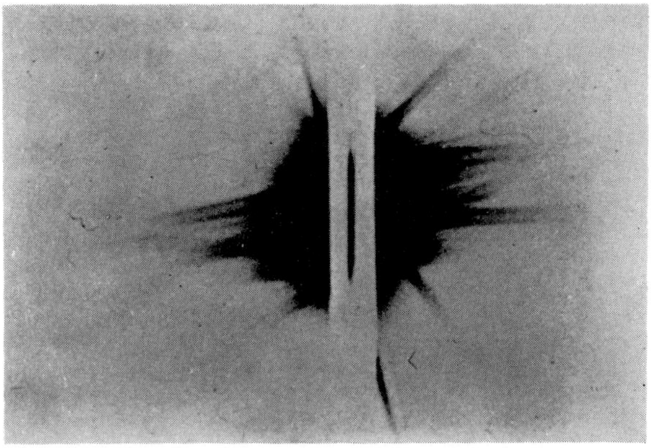

FIG. 18. Same alloy pattern taken in the same conditions after aging at 170°C.

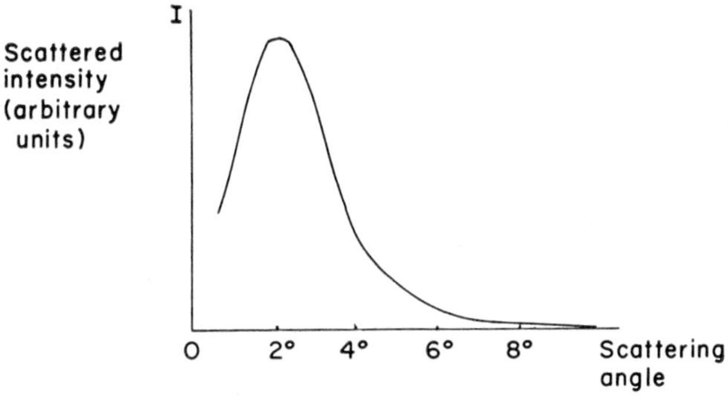

FIG. 19. Scattered intensity corresponding to the diagram of Fig. 17 measured with a G-M counter [C. B. Walker and A. Guinier, *Acta Met.* **1,** 568 (1953)].

Let us consider a small crystal having an undistorted lattice identical to that of the matrix. In reciprocal space, its diffraction pattern is composed of identical domains centered on each node of the reciprocal lattice of the matrix; the shape of the domain is determined by the form factor or dimensions of the crystal.[104] Suppose now that this small crystal, composed of silver atoms, is extended in such a way that the external nodes

[104] P. P. Ewald, *Proc. Phys. Soc.* **52,** 167 (1946).

of its lattice are occupied by aluminum atoms. The shape of the diffraction pattern is not changed: its intensity is changed but remains considerable, because of the strong contrast between the scattering factors of the two atoms. More generally, a sharp transition from pure Al to pure Ag can be replaced by a progressive variation in the lattice of the concentration of Ag, without changing the aspect of the diffraction figure.

It is evident that the figure has the chief characteristic of that observed for the Al-Ag alloy in the stage of pre-precipitation. Thus we may interpret the pre-precipitation pattern as arising from undeformed zones in which the Ag atoms are concentrated. It is possible to deduce the shape of the zone, or more precisely the distribution of silver in the zone,

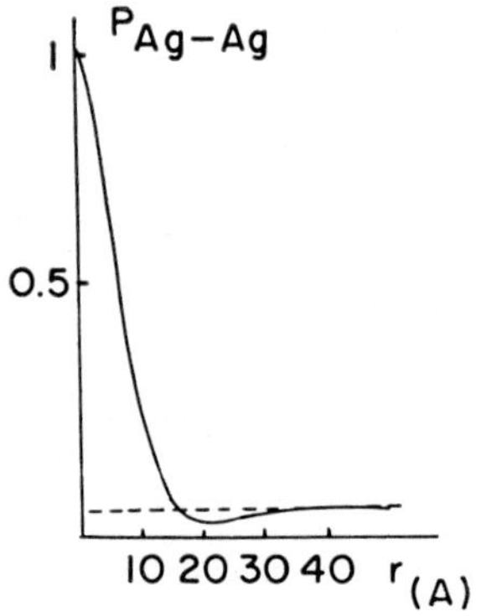

FIG. 20. Repartition of silver atom deduced from the scattered intensity (Fig. 17).

from the distribution of the intensity around each node. For experimental reasons, the small-angle scattering region is the least troublesome to investigate for quantitative results.[93,97]

Two methods of calculation are possible. As in the study of order in alloys, it is possible, without an *a priori* hypothesis, to calculate a SRO parameter, that is, the probability that an Ag atom will be at a given lattice site relative to an Ag atom; or it is possible to calculate the electronic density, provided the zone is assumed to be spherically symmetrical.[105] For example, the concentration of silver can be obtained as a function of the distance from the center of the zones. In the first case, it is necessary to take the Fourier transform of $I(s)$;[78] in the second, one requires the transform of the scattered amplitude $\sqrt{I(s)}$. The curve of Fig. 20 provides an example of the results. A concentration curve with a maximum at the origin surrounded by a *negative region* corresponds to the ring of scattering characteristic of these diagrams (Fig. 19). This curve contains *two* indeterminate factors. The absolute value of the electronic density is not known, for absolute measurements were not made. Further-

[105] R. Smoluchowski and Y. Y. Li, *Phys. Rev.* **94**, 866 (1954).

more, it is possible to add a constant quantity to the atomic density without changing the observed diagram.[106] Thus we may interpretate the curve of Fig. 20, in the following way: the limiting density is that of the average supersaturated solid solution. The zone is formed by a central core, in which the silver atoms are gathered. This is surrounded by a halo-like region from which the silver atoms have diffused and which is left nearly empty of silver. It is possible to account satisfactorily for the experimental curve in relative magnitude by a simple schematic model composed of an inner sphere of radius R_1, having a concentration c_1, which is surrounded by a sphere of radius R_2. It is supposed that all the silver atoms contained in the sphere of radius R_2 are assembled in the inner sphere R_1. Thus the silver concentration is zero between the two spheres. The bigger sphere is surrounded by the solid solution which has the average initial concentration, c_0.

It is possible to adjust the parameters R_1 and R_2 of this model so that its diffraction pattern has the same form as that observed. However, even if absolute measurements of the scattered intensity are made, it is necessary to choose c_1 arbitrarily in order to find the number of zones per unit volume. Taking the reasonable, but somewhat arbitrary, value 0.5 for c_1, Belbeoch and Guinier[93] find that the proportion of segregated silver atoms after long aging at room temperature is 40%.

Table III gives the values of R_1 and R_2 found for an alloy with 38% Ag as a function of thermal treatment.

TABLE III. DIMENSIONS OF THE ZONES IN AGED AL-AG ALLOY

State of the alloy	R_1(A)	R_2(A)
Quenched	26	70
Annealed at 130°C	30	65
Annealed at 175°C	44	92

Without adopting models or making an *a priori* hypothesis, Gerold[97] has determined what he terms a "degree of heterogeneity" from the experimental measurements. This is the proportion of silver atoms in the alloy that have been displaced by diffusion and concentrated in the cores of the zones. This parameter is deduced from the intensity curve by the integral

$$x = \frac{4\pi}{n} \int_0^\infty s^2 I(s)\, ds. \tag{17.1}$$

Here n is the number of atoms per unit volume and $s = (2 \sin \theta)/\lambda$.

[106] The Fourier transform of a constant is a peak function at the origin, which is not observable.

Gerold finds the value 0.2 for the alloy containing 20% of Ag and fully hardened at room temperature. This figure is of the same order of magnitude as the result given by Belbeoch and Guinier. It confirms the important conclusion that the proportion of dissolved atoms involved in the zones, even in the pre-precipitation stage, is quite large and is comparable to the proportion of segregated atoms in the precipitation (second stage). The evolution of the alloy must not be regarded as a progressive elimination of the atoms in excess from the matrix but as the result of changes in the mode of distribution of the atoms eliminated from the matrix.

To give an adequate idea of results concerning the development of this type of zone, we can make use of the experiments carried out on the alloy Al-Zn[107–114] as well as on Al-Ag, for the two systems behave in a similar way in many respects. One observes the shape and an intensity distribution for the small-angle scattering or, more simply, the diameter of the ring of scattering.

(1) The zone is always detected even when the investigation is made as soon as possible after the quench. At least thirty minutes are required for a photographic exposure, but measurements with a counter can be made in only a few minutes.

(2) The dimensions of these initial zones vary with the conditions of the quench, the temperature of homogenization and the rate of cooling. For example, the "diameter" of the zone in an alloy Al-10% Zn (calculated roughly from the diameter of the ring) is 64 A when quenched from 250°C and 94 A when quenched from 400°C.[112]

(3) During the annealing of the quenched samples, the diameter of the ring of diffuse scattering diminishes and its intensity increases. Figure 21 shows that the rate of decrease is greater, the higher the annealing temperature.[82] The ring obtained for long annealing at a temperature higher than 150°C becomes so small that only a blurred spot is observable around the trace of the direct beam. The diameter of the ring cannot be measured. Thus it is found that the zones grow during annealing and reach a diameter which increases with the temperature of annealing.

(4) After a certain period of annealing and when the central spot is small, short streaks which definitely depend on the orientation of the

[107] A. H. Geisler, C. S. Barrett, and R. F. Mehl, *Trans. AIME* **152,** 209 (1943).
[108] A. Guinier, *Métaux, corrosion, usure* **18,** 209 (1943).
[109] P. Lacombe, *Rev. mét.* **41,** 180, 217, 2259 (1944).
[110] J. Hérenguel and G. Chaudron, *Rev. mét.* **41,** 33 (1944).
[111] E. C. W. Perryman and J. C. Blade, *J. Inst. Metals* **77,** 263 (1950).
[112] R. Graf, *Compt. rend.* **246,** 1544 (1958).
[113] K. Hirano, *J. Phys. Soc. Japan* **10,** 995 (1955).
[114] G. Borelius and L. E. Larsson, *Arkiv Mat. Astron. Fysik* **35A,** No. 13, 1 (1948).

crystal (Fig. 18) appear. They are not visible with fine-grained specimens. Belbeoch and Guinier,[93] using an Al-38% Ag alloy, have determined the annealing times necessary for the appearance of these streaks as a function of the temperature of annealing. The streaks appear after 72 hr at 130°C and after $\frac{1}{2}$ hr at 185°C. These times agree very well with the times for the beginning of the second stage of age-hardening, that is warm-hardening, which Köster and colleagues[83–85] determined from the curves giving the variations of various physical properties such as hardness, conductivity . . . . Thus, the streaks are not related to the zones but to the precipitate which is the source of warm-hardening. This constitutes a very important experimental observation for the study of the age-hardening of these alloys; however, it lies outside the range of our subject. Therefore, we will give only brief indications of its interpretation, which is

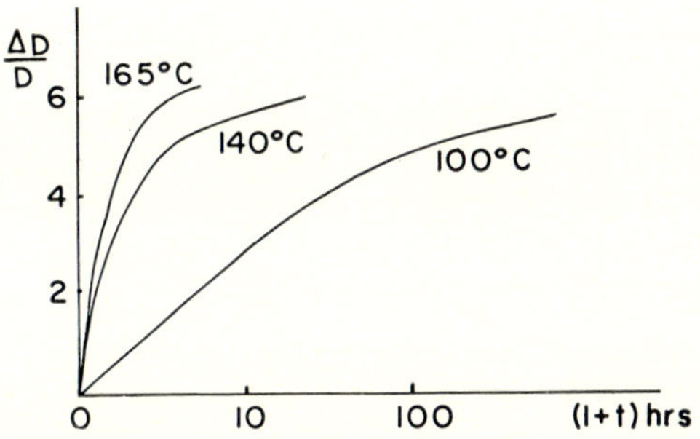

FIG. 21. Relative decrease of diameter of the ring of diffuse scattering for several annealing times and temperatures [C. B. Walker and A. Guinier, *Acta Met.* **1**, 568 (1953)].

now quite clear. The streaks which are parallel to <111> axes are caused by thin platelets of the γ' precipitate (Al_2-Ag) parallel to the {111} planes of the matrix; the thickness of the platelets is of the order of 30–100 A. Additional streaks, also parallel to <111> axes but not around the center are observed (Fig. 22). They pass through the matrix nodes but may have a length considerably larger than the length of the central streaks. These streaks are caused by imperfections in the lattice of the γ' particles which contain stacking faults.[115] It is important not to confuse the streaks in the Al-Ag diagrams, which are related to the precipitation stage, with streaks, characteristic of a pre-precipitation structure which we shall describe in Al-Cu. It is not correct to call them "Guinier-Preston streaks"

[115] Ziegler[86] postulates the existence of stacking faults but does not localize them in the platelets of precipitate. His theory was introduced in part to explain the absence of {111} streaks around the 000 node. In actual fact, such streaks have been found and even are reported in the paper[87] following Ziegler's in the same issue of the *Zeitschrift für Metallkunde.*

for, as Glocker *et al.* point out, they have nothing to do with the cold-hardening. The strongest argument in support of the view that they are not associated with cold-hardening is that the streaks observed in Al-Ag do not vanish as a result of reversion, but, on the contrary, are reinforced (see Section 23).

Another anomalous scattering additional to the small rings of scattering around the nodes of the matrix is observed in the stage of pre-precipitation. The scattering pattern has been named "double cross" by Geisler

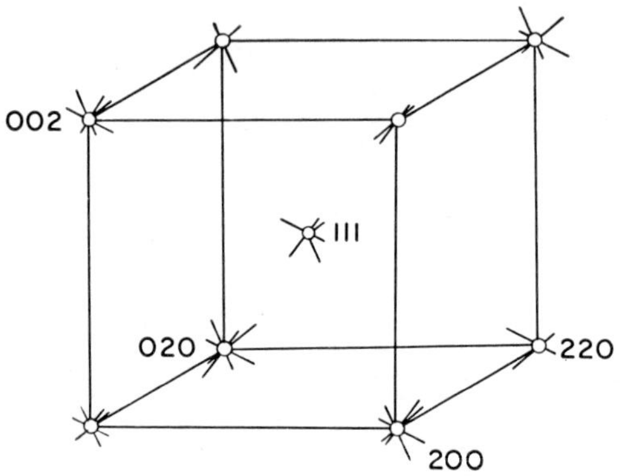

FIG. 22. Schemes of the repartition of scattering in the reciprocal space of the Al-Ag crystal in the second stage of hardening [G. Ziegler, *Z. Metallk.* **43,** 213 (1952)].

and Hill. This phenomenon is observed in Al-Ag,[81] but not in Al-Zn.[108] A similar but more intense effect is found in some ternary alloys. We shall postpone the description and the interpretation of this phenomenon until after the discussion of the Al-Mg-Zn alloys (see Section 20).

18. Zones in an Al-Cu Alloy

Another type of zone whose structure is now known in considerable detail is found in the Al-Cu alloy containing about 4 weight per cent of Cu (2 atomic per cent).[116–140,75] It has been studied by numerous authors.

[116] P. P. Merica, R. G. Waltenberg, and J. R. Freeman, *Trans. AIME* **64,** 3 (1920).
[117] J. B. Friauf, *J. Am. Chem. Soc.* **49,** 3107 (1927).
[118] G. D. Preston, *Proc. Roy. Soc.* **A167,** 526 (1938).
[119] M. L. V. Gayler and G. D. Preston, *J. Inst. Metals* **41,** 191 (1929); **48,** 197 (1932).
[120] W. L. Fink and D. W. Smith, *Trans. AIME* **128,** 223 (1938); **137,** 95, 107 (1940).
[121] D. A. Petrov, *J. Inst. Metals* **62,** 63 (1938).
[122] G. Wassermann and J. Weerts, *Metallwirtschaft* **14,** 605 (1935).
[123] J. Calvet, P. Jacquet, and A. Guinier, *J. Inst. Metals* **65,** 121 (1939).
[124] A. Guinier, *Ann. phys.* **12,** 161 (1939).
[125] C. H. Samans, *Trans. AIME* **137,** 85 (1940).

In this alloy the existence of the zones was first established by Preston[118] and Guinier.[124] This type of zone is now generally referred to as a "Guinier-Preston zone" or a "G.P. zone."

The mechanism of the decomposition of the quenched supersaturated solid solution of Al-Cu shows four different stages whose relative roles depend strongly on the annealing temperature.

Of these four stages, two correspond to true precipitates of composition Al_2Cu, called the θ and θ' phases. The stage we are chiefly interested in is that of pre-precipitation characterized by the G.P. zones. Between this first stage and the two kinds of true precipitate, is an intermediate stage in which a structure, which has been called either the θ'' phase or G.P. II zone, appears. (The first type of zone is the G.P. I zone.) In reality, this structure is very similar to that of a precipitate; we shall regard only the first stage (G.P. I zone) as a genuine pre-precipitation stage here.

The G.P. I zones appear alone both when the annealing temperature is not above 100°C and at the very beginning of treatments between 100° and 180°C.

a. X-Ray Scattering Characteristics of the G.P. Zones

Figure 23 gives an example of the characteristic x-ray pattern of an Al-4% Cu alloy aged at low temperature (10 minutes at 100°C). The pattern shows several regions of scattering in the form of streaks, all of which are directed along <100> axes. A succession of such patterns, obtained with monochromatic radiation and a single crystal that is either immobile or oscillating, determines the position of the regions of scattering in the reciprocal space and also permits an approximate

126 A. Guinier, *J. phys. radium* **3,** 124 (1942).
127 M. L. V. Gayler, *J. Inst. Metals* **72,** 243, 543 (1946).
128 I. A. Bagaryatskii, *Zhur. Tekh. Fiz.* **18,** 827 (1948).
129 F. Rohner, *J. Inst. Metals* **73,** 285 (1947).
130 H. Jagodzinski and F. Laves, *Z. Metallk.* **40,** 296 (1949).
131 I. A. Bagaryatskii, *Doklady Akad. Nauk S.S. S.R.* **77,** 261 (1951).
132 J. M. Silcock, T. J. Heal, and H. K. Hardy, *J. Inst. Metals* **82,** 239 (1953).
133 V. Gerold, *Z. Metallk.* **45,** 593, 599 (1954).
134 K. Toman, *Czechoslov. J. Phys.* **5,** 556 (1955).
135 R. Graf, *Publs. sci. et tech. ministère air* No. 315 (1956).
136 K. Toman, *Acta Cryst.* **10,** 187 (1957).
137 V. Gerold, *Acta Cryst.* **11,** 236 (1958).
138 A. Guinier, *Physica* **15,** 148 (1949).
139 W. L. Fink and D. W. Smith, *Trans. AIME* **124,** 162 (1937).
140 F. Sébilleau, "Achromatisme en rayonnement X. Application à l'étude des raies de diffraction de la solution solide Aluminium-Cuivre à 4% au cours du durcissement structural." Publications ONERA, No. 87, Paris, 1957.

evaluation of the relative intensities. Detailed measurements of intensity have been made photographically or with a G-M counter especially along the <100> axis. The synthesis of these experiments is given by the intensity distribution in the reciprocal space shown in Fig. 24.

It is possible to use the observed distribution of scattered intensity to deduce *a priori* certain facts about the structure of the alloy that can be accepted with confidence and can form the basis for developing structural models.

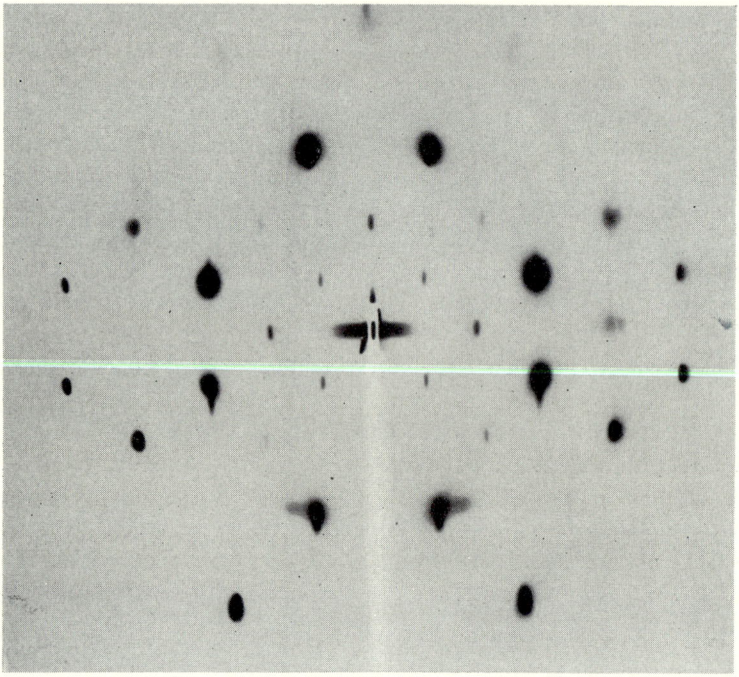

FIG. 23. Scattering of Al-Cu alloy aged 18 hr at 100°C. Single crystal oscillating around [001] axis. Monochromated Mo $K\alpha$ radiation [R. Graf, *Publ. sci. et tech. ministère air* No. 315 (1956)].

(1) The streaks pass precisely through the nodes of the matrix and the scattering is concentrated along the <100> rows of the reciprocal lattice (<100> relrods). This proves that the defects in periodicity are planar defects, which are parallel to the {100} planes (Section 5). The individual {100} planes in the zones conserve, without change, the structure of a {100} plane of the matrix.

It is natural, using considerations of symmetry to imagine that there are three systems of independent zones, parallel to each of the {100} planes respectively. This interpretation is suggested by photomicrographs of the alloy after precipitation has started. A mixture of three systems of platelets parallel to the cube faces is observed (Fig. 25). The diffraction spots of the precipitates, in the x-ray pattern of such a precipitated alloy, ought to be separated into three systems, each given by the particles in one orientation among the three existing ones. In a similar way, the observed regions of scattering must be divided into three systems, corresponding to the three orientations of the zones.

(2) Since the scattering intensity is a maximum in the neighborhood of the center of the diagram, it is *certain* that a *substitutional* disorder

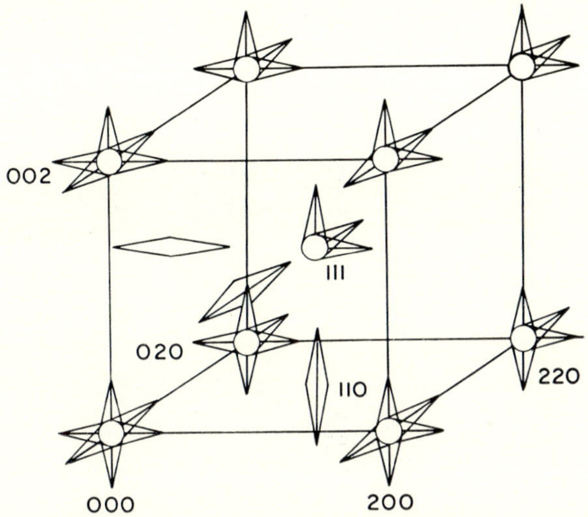

FIG. 24. Scheme of the repartition of the scattering in the reciprocal space of Al-Cu.

exists. Inasmuch as the copper atoms are known to have a tendency to segregate, it is reasonable to assume that they gather into the zones responsible for the observed scattering. In accordance with the last paragraph these zones must be parallel to the {100} planes. Let us suppose, first of all, that the Al atoms are replaced by Cu atoms on a large area of a single (100) plane. This irregularity in the lattice would give rise to scattering streaks along [100] relrods (Section 5). The intensity would be relatively uniform, however, or, more exactly, decrease slowly with the scattering angle as the square of the difference between the Cu and Al scattering factors. This is *not* what is observed.

If now the zone is supposed to be formed of a few (100) planes occupied by Cu atoms, the distribution of the scattered intensity is modified. In any case, however, the scattering would repeat itself identically from cells in the reciprocal lattice and the streaks around a Bragg spot would be centrosymmetrical around the corresponding node. This also is *not* observed. On the contrary, as is shown in Fig. 23, the streaks are visible only on the high-angle side.

To explain such an observation, it is necessary to suppose that there also are atomic displacements in the zone. From the general theory, one

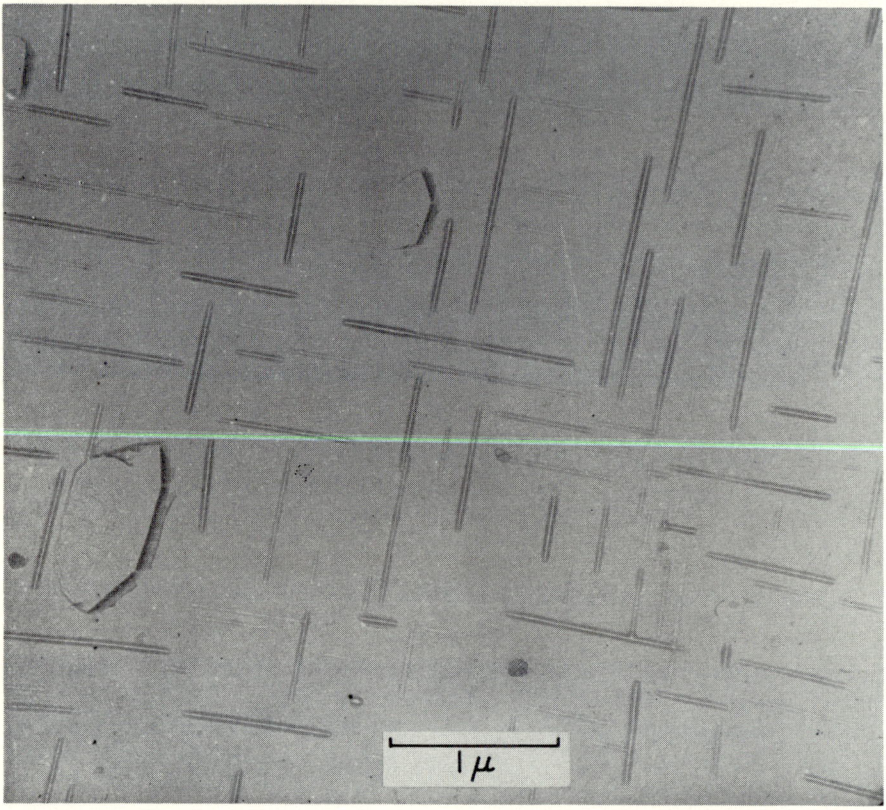

FIG. 25. Electron micrograph of an Al-Cu alloy containing θ' precipitates aged 27 hr at 250°C.

can deduce, in addition, that there ought to be a correlation between the substitutional and displacement disorders. These observations lead to the concept that a zone is a gathering of copper atoms which strongly perturbs the lattice of the matrix. The zones in Al-Cu thus are *zones with distortions*. This hypothesis is supported by the fact that the copper atom is substantially smaller than the aluminum atom.

Preston[118] was the first to show that the observed asymmetry in the streaks could be explained in terms of a variation of the distance between successive (100) planes. The copper-rich planes have an increased scattering power and also are closer to the neighboring planes than the normal aluminum planes, for the Cu atoms have simultaneously a greater scattering factor and a smaller radius than the Al atoms [2.56 A compared to 2.86 A and $(f_{Cu}/f_{Al}) = (27/13)$].

(3) It is easy to find the order of magnitude of the dimensions of the zones. The dimensions of any one of the streaks emerging from the center (Fig. 23) is determined by the external shape of the average copper-rich cluster. For an alloy fully aged at room temperature, the streak has a length of the order of magnitude of the parameter of the cubic cell. Now, the length of the streak, taking the dimensions of the cell as unity, is of the order of $1/N$, N being the number of (100) planes of the zone. Thus, the zones, parallel to the (100) planes, ought to be very thin, having a thickness of one or two atomic planes. Otherwise, the scattering would be concentrated in the neighborhood of the node. It must be pointed out, however, that there probably is no sharply defined thickness for a zone; it is necessary to describe its structure more fully; this will be discussed later on.

It is possible, in a more meaningful way, to deduce the diameter of the zone from the width of the streaks when they are produced by a primary beam with a narrow cross section. This is not the case in Fig. 23, for the beam has a linear trace of finite height. Thus the width of the horizontal streak is simply equal to the height of the beam. The width of the vertical streak is significant, however: it is narrower, the larger the diameter of the zone. It is found that the diameter of the zones increases from 30 A at 20°C to 80°A at 100°C[135] when the annealing temperature increases.

(4) By making an appropriate hypothesis concerning the composition of the zones and giving them dimensions compatible with the shape of the streaks, it is possible to determine the number per unit volume from the absolute value of the scattered intensity measured at small angles. This is a very approximate calculation in view of the necessary but arbitrary hypotheses; however, it leads to the interesting result that an important fraction of the copper atoms are assembled in the zones. For instance, the average distance between neighboring zones is found to be less than 100 A at room temperature in a fully aged alloy.

This conclusion seems to contradict the well-ascertained fact that the lattice parameter of the matrix does not vary appreciably during the cold aging.[139] Yet the depletion of the matrix would produce a measurable change of parameter if a considerable fraction of the copper atoms have left the matrix. In addition, Sébilleau[140] has found that the Debye-Scherrer lines are very sharp in the Al-Cu alloy containing G.P.I zones.

For instance, he obtained a measured width of 1.7′ for the (200) line, which is equal to the width of the line given by well-crystallized pure Al crystal. The size of the coherent diffracting domain calculated by the Scherrer formula[141] is found to be 3200 A. A large volume of this order of magnitude contains a great number of zones, so that zones are to be regarded as local disturbances of the lattice completely coherent with the matrix. Thus the parameter, which is measured by x-rays, corresponds to an average over the matrix *and* the zones. The situation is similar to that of a solid solution; however, the centers of distortion are the zones instead of the isolated atoms. The methods of calculating the average parameter (Huang,[50] Eshelby[142]) may be used in the case of coherent zones. The constant value of the parameter and the constant sharpness of the diffraction lines of the matrix are characteristic of the zones and are a consequence of their coherence with the matrix. On the other hand, the lines originating from the matrix, are displaced and become broad when a precipitate is formed.

b. Structure of a G.P. Zone

The first model, proposed independently by Preston[118] and Guinier,[124] consisted of a single (100) plane, rich in copper and surrounded by Al planes at a distance a little smaller than the normal parameter of the matrix. The agreement with experiment was only qualitative. An alternate model proposed by Jagodzinski and Laves,[130] can now be eliminated. They attributed the scattering chiefly to the lattice distortions and supposed that the concentration of copper in the zones was not large. Bagaryatskii[131] calculated the scattering to be expected from this model and it does not agree at all with the observed intensity distribution. The principal discrepancy is the absence of small-angle scattering in the calculated pattern.

Other structural models that have been proposed more recently are more complicated.[133,136,137] For example, it has been proposed that in a series of normal (100) planes of scattering factors f and separation a, additional planes are inserted having a scattering factor f_n and at a distance $na + da_n$ from the origin, where da_n is a function of the number n of the plane. The zone is supposed, for simplicity, to be symmetrical.

Gerold[133,137] has proposed a model consisting of a single plane of copper between two planes of aluminum which are nearer the origin than in the regular crystal. These displacements attenuate in such a manner that at the extremities of the zone, the 14th plane has retaken

[141] A. Guinier, "Théorie et Technique de la Radiocristallographie," p. 462. Dunod, Paris, 1951.

[142] J. D. Eshelby, *J. Appl. Phys.* **25**, 255 (1954).

its regular position (Fig. 26). Table IV gives the structure of the model which provides the best agreement with Gerold's experimental measurements of the scattered intensity.

Toman[136] has used a more general method of calculation for obtaining f_n and da_n. He supposes that several planes contain more copper than the

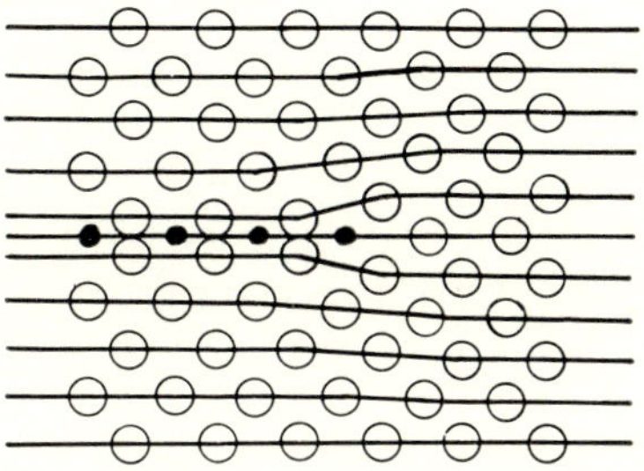

FIG. 26. Model of the G.P. zone according to V. Gerold [*Z. Metallk.* **45,** 593, 599 (1954)].

TABLE IV. STRUCTURE OF A G.P. ZONE IN AL-4% CU ACCORDING TO GEROLD

$N°$ of the plane n	Composition of the nth plane	Relative displacement of the nth plane da_n/a
0	Cu	−0.10
1	Al	. . .
2	Al	−0.09
3	Al	−0.08
4	Al	−0.07
5	Al	−0.06
6	Al	−0.05
7	Al	−0.04
8	Al	−0.03
9	Al	−0.02
10	Al	−0.02
11	Al	−0.01
12	Al	−0.01
13	Al	−0.01
14	Al	−0.00

average and assumes that the displacements are very small. This hypothesis permits him to calculate the composition and the location of the planes in the zone.

There is a certain similarity between the two models; the principal difference between them lies in the fact that many planes are displaced in Gerold's model and only one is filled with copper, whereas very few planes are displaced in Toman's but a large number are filled with copper.

TABLE V. STRUCTURE OF A G.P. ZONE IN AL-4% CU ACCORDING TO TOMAN

N° of the plane n	Composition of the nth % plane % Cu	Relative displacement of the nth plane da_n/a
0	98	0
1	50	−0.039
2	39	−0.015
3	30	−0.003
4	23	0
5	17.5	0
6	13	0
7	9	0
8	6	0
9	4	0
10	2.5	0

It must be stated that the agreement of the two models with experiment is of the same order; however, the results of the measurements are not very precise. In particular, the value of the scattered intensity in the immediate neighborhood of a node of the solid solution (a region which is critical for the calculations) is difficult to determine because this scattering is superimposed on the scattering arising from the thermal agitation of the atoms, which becomes strong around the node.

It may be added that Toman has found a strong increase of the scattered intensity at small angles, which has not been observed by others and which does not seem to be real. Gerold[137] points out that the calculations of Toman are correct and would lead to a result similar to his own if the starting points were the same.

c. *Disappearance of the G.P. Zones*

The streaks of the G.P.I zones gradually disappear after a given time, which decreases with increasing temperature of annealing at a temperature above 100°C. A new pattern appears: it is characteristic of the structure θ'' (or G.P.II zones) at temperatures below 200°C, and it is the diagram of the precipitate at temperatures above 200°C. It is not obvious whether θ'' must be considered as a true phase or as a zone.

The diagram of θ'' is composed of a series of diffuse spots; it can be described as the diffraction pattern of a crystal with a well-defined lattice, the broadening of the spots being a consequence of the small size of the θ'' particles. In this case it is not necessary to assume distorted structures to explain all the features of this pattern. Therefore, it is perhaps more convenient to consider the structure simply as a phase. The difference between it and an ordinary precipitate is that the particles generally do not grow (they are platelets less than 30 A in thickness).[135] They become unstable and dissolve before they can reach a size sufficient to give sharp diffraction spots, except after very special heat treatment.[135] Sébilleau[140] has shown that during the development of the θ'' phase, the parameter of the matrix changes, much as during the precipitation of θ'. On the other hand, the θ'' phase has some of the characteristics of the zones; its lattice is very similar to the lattice of the matrix. To a first approximation, it is merely a superlattice of aluminum (tetragonal with parameters a, $c = 2a$, a being the parameter of aluminum). The θ'' phase also produces very strong hardening and undergoes reversion (Section 23), as do the G.P.I zones.

Silcock *et al.*[132] have determined the conditions in which the three metastable phases and the equilibrium phase appear as a function of time and temperature for different concentrations of Cu in Al, in the range from 2% to 4.5%. They have found that the θ' precipitate is the first structure detected above 100°C for the low concentrations of Cu whereas zones are formed in the more concentrated alloys.

d. G.P. Zones in Other Systems

There are other systems in which zones similar to those in Al-Cu are observed. This is true, for example, in the alloy Cu-Be which also is one of the most important hardenable alloys.[143–147] During the pre-precipitation stage, the zones of Cu-Be are parallel to {100} planes. The distortions are considerable because of the large difference in the atomic diameters of Cu and Be. The distortions of the lattice are quite different from the distortions in Al-Cu zones, however; the distribution of the scattered intensity along the [100] axis is not the same. At the beginning of the pre-precipitation, the intensity is nearly uniform along the [100] axis. Then, a broad maximum corresponding to a Bragg spacing of 2.7 A appears. This is the (001) spacing of the crystalline phase Cu-Be (phase γ); however, the width of the maximum is larger than the width of the small angle scattering. From the latter, it follows that the particles must have a thickness of about 30 A. But the small size of the particle cannot explain the width of the maximum observed at large angle. There are irregularities in the spacing of the successive (100) planes inside the platelets. It can be said that the (100) planes tend to have the spacing of the intermetallic compound but they do not reach a perfectly regular structure. The work

[143] A. Guinier and P. Jacquet, *Rev. mét.* **41,** 1 (1944).
[144] A. G. Guy, C. S. Barrett and R. F. Mehl, *Trans. AIME*
[145] A. M. Elistratov, *Doklady Akad. Nauk S.S.S.R.* **69,** 337 (1949).
[146] W. Gruhl and G. Wassermann, *Metall.* **5,** 141 (1951).
[147] A. H. Geisler, J. H. Mallery, and F. E. Steigert, *J. Metals* **4,** 307 (1952).

of Geisler and his colleagues[147] shows the complexity of this system, which is still far from being completely understood.

Plate-like zones, named G.P. zones in analogy to the Al-Cu system, are observed in nonmetallic systems. For example, clusters of interstitial Li atoms appear in neutron-irradiated LiF.[148] Moreover, clusters appear in solid solutions of $CdCl_2$ in NaCl before true precipitation.[149,150] It is curious to observe that the G.P. platelets are generally parallel to {100} planes in face-centered cubic lattices.

Mott and Nabarro[151] have shown that a disk-like particle produces less strain energy in a continuous surrounding medium than a spherical or needle-like one. Smoluchowski[152] tried to calculate the shape corresponding to a minimum strain energy in a precipitate imbedded in a face-centered cubic lattice; however, the calculation does not explain in a simple way why the (100) habit plane is observed so frequently.

19. Zones Having Linear Disorder

A distinctive type of zone has been found in the pre-precipitation stage of the ternary alloys Al-Mg-Si.[153–156] The equilibrium precipitate is the compound Mg_2Si. The alloys which have been studied are "quasibinary," such as Al-Mg_2Si, the proportion of Mg_2Si being 0.8%[155] or 1.5%.[153] This is less than the maximum solubility, i.e., 1.85% at 595°C.

A very similar alloy, Al-Mg-Ge, has been studied recently by Lutts and Lambot.[156] Since the scattering is more intense than for Al-Mg-Si, the study was easier experimentally. The behavior of the Si and Ge alloys is very similar and most of the observations we shall describe in the following paragraphs have been made on both alloys.

After quenching and subsequent aging at room temperature, no abnormal scattering appears in the pattern of the alloy, even though the mechanical properties of the metal change. The first interesting scattering appears during annealing between 130 and 150°C. It is concentrated in {100} planes of the reciprocal lattice (the {100} relplanes). Very thin lines of extremely weak intensity appear in the x-ray pattern (Fig. 27) corresponding to the intersection of {100} relplanes with the Ewald sphere.

[148] M. Lambert and A. Guinier, *Compt. rend.* **244,** 2791 (1957).
[149] S. Miyake and K. Suzuki, *Acta Cryst.* **7,** 514 (1954).
[150] S. Miyake and K. Suzuki, *J. Phys. Soc. Japan* **9,** 702 (1951).
[151] N. F. Mott and R. F. N. Nabarro, *Proc. Phys. Soc.* **52,** 86 (1940).
[152] R. Smoluchowski, *Intern. Conf. on Phys. of Metals, The Hague* p. 179 (1949).
[153] A. H. Geisler and J. K. Hill, *Acta Cryst.* **1,** 238 (1948).
[154] A. Guinier and H. Lambot, *Compt. rend.* **227,** 74 (1948).
[155] H. Lambot, *Rev. mét.* **47,** 709 (1950).
[156] A. Lutts and H. Lambot, *Rev. mét.* **55,** 775 (1958).

At the beginning of annealing (3 hours at 150°C), only the planes passing through the origin exhibit scattering. After 6 hours at the same temperature, the initial streak becomes longer and sharper. In addition, identical streaks which correspond to the two adjacent planes of the family become visible (relplanes passing through 100 and $\bar{1}00$ relpoints). The streaks in the second planes (200 and $\bar{2}00$) appear after 14 hours at 150°C. If the annealing is continued up to 1000 hours, the intensity of the streaks increases slightly but no substantial changes in form take place.

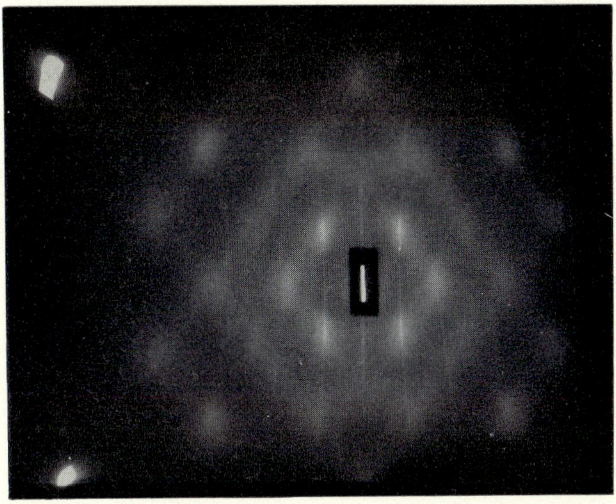

FIG. 27. Scattering by Al-Mg-Ge alloy aged 100 hr at 150°C. Scattering concentrated in (100) planes [H. Lambot and A. Lutts, *Rev. mét.* **55,** 775 (1958)].

At the beginning of the anneals, the scattered intensity seems to be distributed essentially uniformly in the (100) relplanes. The intensity is reinforced in certain regions of the plane and diminishes in others, only after treatments at 200°C. The map of the distribution of scattered intensity is very complex ("checker board" aspect).[155]

Geisler and Hill[153] and Lambot[155] have found diffuse spots at a higher annealing temperature, which can be interpreted as the diffraction spots of very small crystals. This scattering could be produced by an intermediate phase; however, its lattice has not been determined. Finally, the equilibrium precipitate, Mg_2Si or Mg_2Ge, is detected at 300°C. It produces sharp diffraction spots, which seem to be quite independent of the preceding scattering regions.

The various scattering phenomena are very weak and can only be studied qualitatively as yet. The very thin lines of scattering are hardly visible on the photographs; the total scattered intensity per unit length of streak is so small that the streaks could not be detected above the general background by a counting technique.

The comparison between the Si and Ge alloys is very interesting. In the silicon alloys, the three atoms cannot be distinguished by x-rays, their scattering power being too nearly alike; the scattering arises only from displacement disorder and gives no information concerning the distribution of the three atoms. On the contrary, one can expect that clustering of Ge atoms could be detected.[157]

In fact, as we already mentioned, the evolution of the diagrams of the two alloys during aging is very similar. Apart from slight changes between the times or temperatures corresponding to the various stages, the interesting points are the following. In the case of the Si alloy, the intensities of the streaks passing through the nodes 100, $\bar{1}00$, and 200, $\bar{2}00$ are not uniform when they first appear. There are marked reinforcements in certain regions of the relplanes. At the same time, the streak in the vicinity of the center vanishes. It is observable only at distances larger than a half unit cell.

Since scattering is found in the case of the Al-Mg-Si, one must conclude that there is a displacement disorder in the lattice; furthermore, the relatively minor differences in the diffraction of Si and Ge alloys prove that the influence of the lattice disorder on the observed scattering is preponderant.

Of course, we can suppose that the primary cause of this disorder is the accumulation of atoms of the unstable supersaturated lattice into zones. But we do not yet have a clear idea of the structure of these zones. One point is certain, however; they are characterized by a *linear disorder*. Since the scattering is concentrated in $\{100\}$ relplanes, we know that the periodicity of the matrix is maintained only in one direction along the $\langle 100 \rangle$ axes. Thus the idea that the zones could be needle-shaped and parallel to the $\langle 100 \rangle$ axes of the matrix is suggested. This idea is supported by electron microscopic observations[158] of the needles of precipitate which appear in the subsequent stages of the evolution. The case of Al-Cu[159] shows that there can be a strong resemblance between the shape and orientation of the zones and precipitates.

When the streaks begin to be visible, their width indicates that the zones should have a length of the order of 100 A. This dimension ceases to be measurable after subsequent anneals, because the breadth of the

[157] A. Guinier, *Acta Cryst.* **5,** 121 (1952).

[158] R. Castaing and A. Guinier, *Compt. rend.* **229,** 1146 (1949).

[159] R. Castaing and A. Guinier, *Compt. rend.* **228,** 2033 (1949).

streaks is too small, much smaller than the experimental broadening. It is only possible to say that the disordered regions are more than a few hundred angstroms long. Their diameter is still more uncertain. *A priori* the patterns of the alloys Al-Mg-Si cannot give any information concerning the external shape of the zone, for the difference between the electronic densities of the zone and the matrix is too small. The situation is not the same for Al-Mg-Ge. Unfortunately the patterns give no clear indication of the diameter of the zone. The very extended small-angle scattering in the diagram of the Al-Mg-Ge alloy could be caused by quite thin zones formed by isolated rows of atoms along $<100>$ axes or thin bundles of a few rows. The very weak scattered intensity favors this hypothesis. On the other hand, it is not impossible that the zones are much thicker, and that the small-angle scattering has not been observed for some unknown experimental reason. In fact, the published patterns have a rather poor resolution toward the center; in any case, they have very weak intensities.

Only tentative hypotheses have been made concerning the structure of the zone up to the present time. Lambot[155] thinks that the dissolved atoms assemble in columns along the $<100>$ axes with the aid of vacancies at the beginning, so that there is no periodicity along these axes, but only a change in density. The columns of atoms order progressively, the chain taking the period of the unit cell of the matrix, namely 4.04 A. In this way, the successive appearance of scattering regions, first in the plane at the origin and later in parallel planes, is explained. Laterally, the columns of atoms tend progressively to take an arrangement different from that of the matrix, thus causing modulations of the intensity inside a plane of scattering. To know the mutual arrangement of these rows, it would be necessary to know the scattered intensities with sufficient precision to make calculations involving Fourier transformations; however, this is not possible with the existing data.

To guide speculation, it is important to take account of the fact that the final equilibrium precipitate in these ternary alloys is a compound of the dissolved atoms of a well-fixed composition (Mg_2Si or Mg_2Ge). The formation of such a precipitate requires the simultaneous presence of Mg and Si (or Ge) in definite proportions for the compounds are stoichiometric without metallic character. The electronic state of Mg and Ge or Si in these compounds is markedly different from that of these atoms dissolved in the metallic matrix of aluminum, because the interatomic bonds have a different character.

Thus it is necessary that there be a kind of reaction between the dissolved atoms when they encounter one another. It often is said that the atoms precipitate in the form of molecules of Mg_2Si in this alloy. In the continuous series of ternary alloys of Al-Mg-Si the curves describing many properties pass through a somewhat singular point for a composition where there are just 2 Mg atoms for each Si. There is an excess of Mg or Si on one side or the other, in addition to the precipitate Mg_2Si.

Actually, isolated molecules of Mg_2Si should not be assumed to exist; the bonds characteristic of Mg_2Si have to form in the framework of the lattice of the aluminum matrix. By comparing the structures of the two crystals (Al and Mg_2Si) it is found that the chain of atoms in Mg_2Si extending along the <110> axes is similar to an atomic chain along the [100] axis of Al, as is shown in Fig. 28. It is possible to imagine that 2 Mg and 2 Si atoms in the supersaturated matrix occupy the sites A_1, B_1 and C_1, A_2, respectively. The sites B_2,C_2 would be positions of low energy for Mg atoms, so that they would be drawn there in consequence, whereas

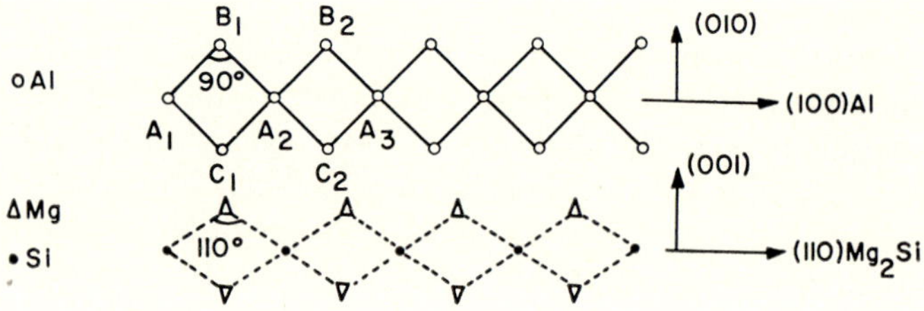

FIG. 28. Chains of atoms [110] in Mg_2Si and along [100] in Al.

the silicon atoms would go to A_3, etc. This would provide an origin for the formation columns directed along the <100> axes of Al. The surrounding matrix would exert a force imposing the linear periodicity associated with the distance of 4.04 A as long as the zone is sufficiently thin.

The atoms of Mg and Si would be drawn laterally as well, because they would find themselves in positions of lower energy. The organization of neighboring chains in Mg_2Si and in Al is very different, however. This explains the lateral disorder of the chains of Mg_2Si in the zone.

Perhaps, the Mg and Si atoms tend to organize in a metastable structure that is strongly coherent with the matrix. This could be the intermediate phase of still unknown structure. This phase may not attain a well-defined lattice before sufficiently large nuclei of Mg_2Si develop and precipitation begins. In support of the schematic initial disposition proposed above, we should mention that the orientation of the Mg_2Si precipitation in the matrix is such that the [110] Mg_2Si axis is parallel to the [100] Al axis.

20. Zones in the Ternary Alloys Al-Zn-Mg

A type of pattern different from those we have described up to now is found in certain ternary alloys such as Al-Cu-Mg[128] and Al-Zn-Mg[160–168] where the relative proportion of the three elements may vary between large limits.

The new type of pattern is more clearly observed in the alloys of Zn and we will emphasize them particularly using the results obtained recently by Graf[162] and Schmalzried and Gerold[163] with two alloys of aluminum containing 7% Zn, 3% Mg, and 9% Zn, 1% Mg (weight per cent).

Pre-precipitation at room temperature in alloys of this type is characterized by the appearance of somewhat round diffuse spots in the diffraction pattern which lie between the Bragg spots of the matrix. A complete, quantitative study of the reciprocal space has not yet been made. It is only possible to give a brief qualitative description. The scattering regions for the alloy Al-Zn-Mg have a nearly spherical shape. The intensity is a maximum at the center of these spheres, and their diameter extends over about $\frac{1}{10}$ of the cell of the reciprocal lattice (Fig. 29). The position of the center of these regions seems to have a simple relation to the lattice of the matrix. Thus the most intense spots are situated exactly at the 100 points in the case of Al-Zn-Mg. Lambot[155] has found scattering spots at the 100 points of the matrix for Al-Cu-Mg and has also observed spots on the [110] axis at $\frac{1}{3}$ and $\frac{2}{3}$ of the distance between the center of the reciprocal lattice and the 220 nodes.

An analogous effect is also observed in the alloy Al-Ag, in addition to the rings around each node described in Section 17.[126] The scattering regions have a flattened ellipsoidal form with a diameter of the order of $\frac{1}{6}$ of the unit cell of the reciprocal lattice. They have [100] or [110] directions as axes; the spots on each of these axes divide the distances between two neighboring nodes into three equal parts. The spots between 000, and 200 or 220 are the more intense.

In contrast to the Al-Ag alloy, this type of scattering does not exist for the Al-Zn alloy which has a very similar behavior in many respects.

160 J. Hérenguel and G. Chaudron, *Métaux, corrosions, usure* **18,** 30, 37 (1943).

161 A. Saulnier and G. Cabane, *Rev. mét.* **46,** 13 (1949).

162 R. Graf, *Compt. rend.* **242,** 1311, 2834 (1956); **244,** 337 (1957).

163 H. Schmalzried and V. Gerold, *Z. Metallk.* **49,** 291 (1958).

164 L. F. Mondolfo, N. A. Gjostein, and D. W. Levinson, *J. Metals* **8,** 378 (1956).

165 K. Hirano and Y. Takagi, *J. Phys. Soc. Japan* **10,** 187 (1955).

166 P. C. Varley, M. K. B. Day, and A. Sendorek, *J. Inst. Metals* **86,** 337 (1958).

167 H. Nishimura and Y. Murakami, *Mem. Fac. Eng. Kyoto Univ.* **12,** 47 (1950).

168 I. J. Polmear, *J. Inst. Metals* **86,** 113 (1957).

We may suppose that this scattering is caused by the zones responsible for the aging at room temperature. The following remarks are important in the quest for a possible interpretation. The scattering phenomena in question seem to be observed only when the zones contain more than one species of atoms. Of course, we have no direct information concerning the composition of a zone, but it is reasonable to suppose that it is not different from the composition of the subsequent precipitate. We know that

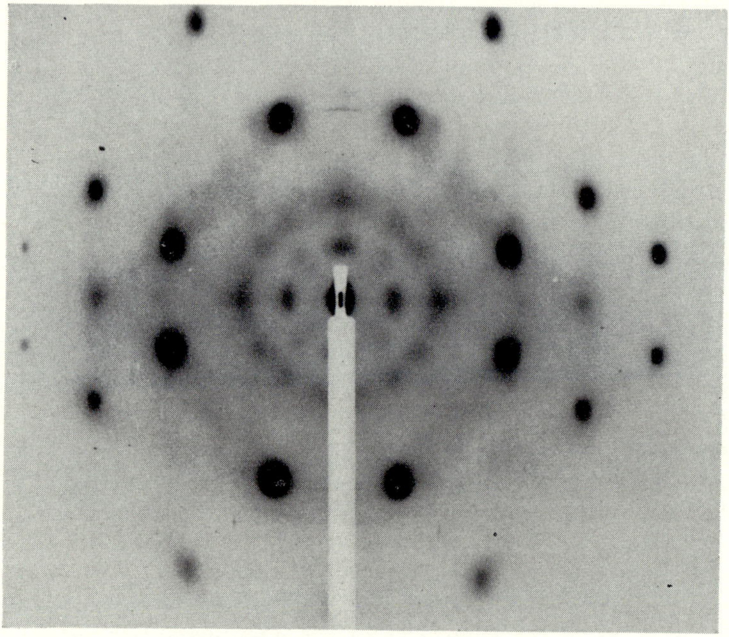

FIG. 29. Scattering by a single crystal Al 7% Mg 3% Zn (aged 240 hr at 20°C) [R. Graf, *Compt. rend.* **242,** 1311, 2834 (1956); **244,** 337 (1957)].

in the ternary alloys and in Al-Ag, the precipitate is a complex compound ($MgZn_2$, $Al_2Mg_3Zn_2$, Cu_2Mg_2Al, Al_2Ag etc. . . .) and, in contrast, pure zinc precipitates in Al-Zn.

A second hint is offered by the analogy between the diffuse spots and the maxima observed in a partially ordered alloy, for example, $AuCu_3$.[26,37] The shape of the spots is similar in both cases. The more intense spots are in the first cells of the reciprocal lattice. Their positions are simply related to the lattice of the matrix.

Thus it is possible to suppose that the cause of the scattering is an ordering of the different atoms *inside* the zone. Schmalzried and Gerold[163] have proposed a structure with alternate (100) planes of Zn and Mg.

These atoms stay approximately at the nodes of the matrix, and the scattering regions may be expected to be centered on the diffraction nodes of a possible superlattice of the matrix. The extension of the scattering regions may be explained by imperfection of the order. The "superlattice spots" then can be much more diffuse than the normal diffuse spots; the width of the latter depends on the size of the zone, whereas the width of the former depends in addition on the average distance between faults of order within the zone.

To develop this idea, it would be necessary to calculate coefficients of order from the experimental data. However, the calculation would have significance only if the distribution of the scattering intensity is known with precision; such measurements have not been published as yet. In any case, it would be necessary to assume the exact composition of the zone in order to carry out this calculation for the ternary alloys.

If this interpretation is accepted, the small-angle scattering ought to give the exterior shape of the zone, regardless of the atomic arrangement at short distances. There is no special difficulty in the case of Al-Ag; the size and the shape of the zones have been well determined from the small-angle scattering (Section 17), whereas scattering at large angles gives supplementary information regarding the internal structure of the zone without modifying the conclusions of Section 17 concerning their external shape. Guinier has described a simple way of achieving the composition Al_2Ag in the zone. Each (111) plane is considered to be a hexagonal lattice, with an Ag atom placed at the centers of contiguous hexagons of Al atoms.[81]

One finds that the [110] rows are occupied by a series of atoms of the type AlAlAgAlAlAg and that the period along these rows is three times that of the normal lattice. Such threefold periods also may occur along [100] axes with the hypothesis of a suitable arrangement of successive ordered (111) planes.

The situation for the ternary alloy is not yet clear because there is doubt about the shape of the zone. Lambot failed to detect any small-angle scattering produced by the zone Al-Cu-Mg.[155] This absence is not easy to understand. Copper, if present in the zone, would most likely produce an appreciable contrast between the electronic densities of matrix and zone. One may wonder if the contribution of the copper is nearly balanced by a decrease arising from the lack of close packing which originates in turn from the largeness of the Mg atom. Or are the zones sufficiently small to reduce the scattered intensity around the center below the observable level?

Small-angle scattering has been observed by Graf[162] in Al-Mg-Zn. This is analogous to the ring-shaped scattering given by Al-Zn alloys.

He has found that the rings are narrower and more intense for identical thermal treatments when the proportion of Zn increased, the proportion (Mg + Zn)/Al remaining constant. Thus the zones responsible for this effect become more voluminous when the proportion of Mg diminishes. The zones which cause the large angle scattering are presumed, however, to be formed by a more or less ordered mixture of Zn and Mg atoms. Thus these zones ought to become smaller or more rare when the Mg content decreases. This is in disagreement with experiment. One could imagine that the composition of the zone is progressively altered and that it contains more and more Zn as the relative amount of Zn in the alloy increases. The distribution of the large angle scattering is the same, however, in the first stage in the case of the 7% Zn and 9% Zn alloy. This means that the internal structure of the zone remains the same.

Another suggestion can be made. Two types of zone may exist at the same time in this alloy, one containing Mg and Zn and characterized by an ordering of Zn and Mg atoms, the other containing pure or nearly pure zinc, analogous to the zones which form in Al-Zn.

The observed ring of small-angle scattering is attributed to the Zn zones, the ternary zones giving rise only to the large-angle scattering. The number of zinc atoms available for forming the zones of pure zinc then decreases as the content of Mg increases. Thus Zn zones would become smaller. The alloy richest in Mg studied by Graf had the atomic composition 3% Zn, 3.5% Mg, which may also be written as 2% Mg, 1.5% $MgZn_2$. Thus there is an excess of Mg if the segregated zones have the composition $MgZn_2$. Nevertheless Graf observed a ring in this alloy. Thus the zinc atoms may be divided between the two possible kinds of zone.

The comparison with Al-Cu-Mg favors the hypothesis that two types of zone appear simultaneously. Lambot[155] observed G.P. zones similar to the Al-Cu zones, with the characteristic central streaks, whereas the ternary zones were detected *only* with the use of the higher angle scattering.

It must be stressed that the absence of small-angle scattering corresponding to ternary zones in this type of alloy (as well as in the Al-Mg-Ge type) is still mysterious at the present time. A satisfactory explanation of this fact could cause considerable change in the interpretations given in the foregoing.

The pre-precipitation stage described in Al-Mg-Zn disappears at 50°C in the case of Al 7% Zn 3% Mg and after a few days at room temperature in Al 9% Zn 1% Mg. Other scattering regions having essentially the same appearance but lying at other places in the reciprocal lattice are observed. Perhaps the internal structure of the zones changes; however, this phenomenon has not yet been analyzed.

Graf[162] has found an intermediate precipitate in Al 9% Zn 1% Mg at high temperatures, before the final $MgZn_2$ precipitate forms. Very diffuse diffraction spots which can be attributed to small $MgZn_2$ crystals appear in Al 7% Zn 3% Mg. These crystals disappear in subsequent annealing. The equilibrium precipitate is the phase $Al_2Mg_3Zn_2$.[161,162]

It is interesting to note that the precipitate in this alloy is perceptible as very small quasi-spherical grains when it first forms. In contrast, the θ' precipitate is in the form of thin platelets in the alloy Al-Cu, whereas the Mg_2Si phase is in the form of needles in the alloy Al-Mg-Si. Further the zones are platelets in Al-Cu, needles in Al-Mg-Si, and one can suppose

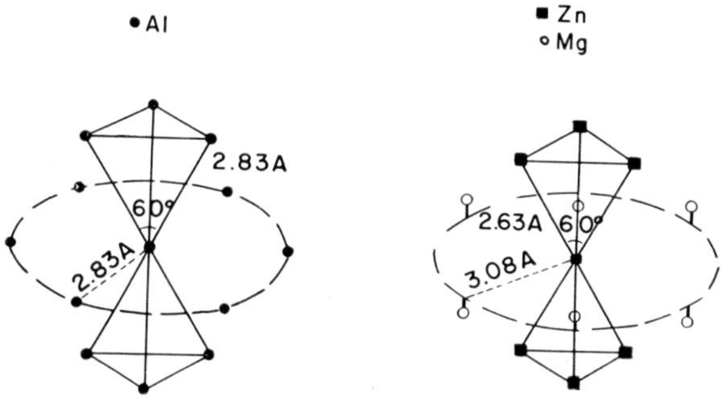

FIG. 30. Group of atoms in $MgZn_2$ crystal compared to a group of Al atoms in Al crystal.

that they are somewhat round in Al-Mg-Zn. This variation can be explained by attempting to compare the structures of the precipitate and the matrix. The θ' crystal and Al have a similar plane of atoms, whereas Mg_2Si and Al have a chain of atoms of similar structures (Fig. 28). Thus the respective nuclei are Cu platelets and Si needles. In the case of $MgZn_2$, the group of zinc atoms and its neighbors have a configuration resembling a group of atoms in the face-centered cubic lattice (Fig. 30). Thus a nucleus of $MgZn_2$ can be accommodated in the matrix without too much distortion. But its development in the matrix in any one direction is not easy. The nucleation is easy but the growth is not, for it implies complete reorganization of the atoms at the boundary of the nucleus. This nucleus grows slowly and nearly isotropically. Consequently, there is a precipitation stage where the crystals are still very small but sufficiently numerous to be detected by their diffraction diagrams.

21. Alloys Displaying the Phenomenon of "Side Bands"

"Side bands" were first observed on Debye-Scherrer patterns of a Cu-Ni-Fe alloy by Bradley[169,170] and Daniel and Lipson.[171] This particular form of scattering is characteristic of a process of pre-precipitation which has now been found in many systems: Cu-Ni-Fe,[169–173] Cu-Ni-Co,[174] Ni-Si,[175] Ni-Cr with Si, Al, or Ti, Ni-Al,[176–180] Au-Pt.[181]

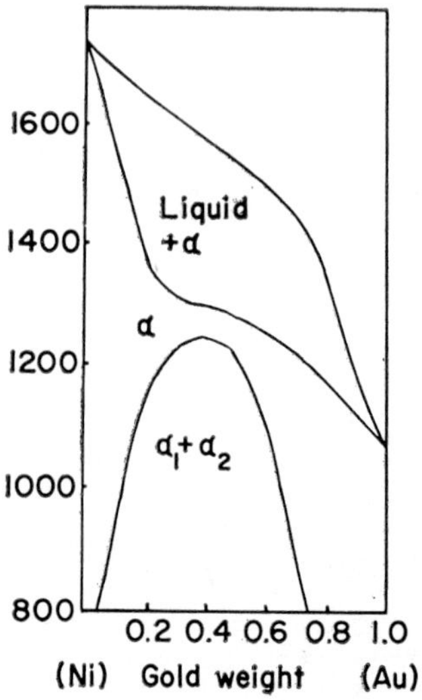

Fig. 31. Equilibrium diagram of Au-Pt with a miscibility gap: single phase at high temperature and two phases at low temperatures.

The common feature of these alloys is the existence at high temperature of both a single phase with a face-centered cubic structure and a miscibility gap at lower temperature (Fig. 31). Thus, the equilibrium state of the alloy at room temperature is a mixture of face-centered cubic lattices, corresponding to different compositions, the parameters of which are not very different.

[169] A. J. Bradley, *Proc. Phys. Soc.* **52,** 80 (1940).
[170] A. J. Bradley, N. L. Bragg, and C. Sykes, *J. Iron Steel Inst.* **141,** 63 (1940).
[171] V. Daniel and H. Lipson, *Proc. Roy. Soc.* **181,** 368 (1943); **182,** 378 (1944); **192,** 57 (1948).
[172] M. E. Hargreaves, *Acta Cryst.* **2,** 259 (1949); **4,** 30 (1951).
[173] A. H. Geisler and J. B. Newkirk, *Trans. AIME* **180,** 101 (1949).
[174] E. Biedermann and E. Kneller, *Z. Metallk.* **47,** 289, 760 (1956).
[175] J. Manenc, *Compt. rend.* **243,** 1119 (1956).
[176] J. Manenc, *Acta Met.* **7,** 124 (1959).
[177] J. Manenc, *Compt. rend.* **240,** 2413 (1955).
[178] J. Manenc, *Compt. rend.* **242,** 2344 (1956).
[179] J. Manenc, *Rev. mét.* **54,** 867 (1957).
[180] J. Manenc, *Acta Cryst.* **10,** 259 (1957).
[181] I. J. Tiedema, J. Bouman, and W. G. Burgers, *Acta Met.* **5,** 310 (1957).

For example, in Cu-Ni-Co alloy,[173] the parameters for the Cu-rich and the Cu-poor phases are 3.53 A and 3.59 A, respectively (3.56 A is the parameter of the homogeneous phase). The single phase, which is stable at high temperature, can be maintained at room temperature by a rapid quench, but it is not stable. On annealing at temperatures below the top of the miscibility gap, the alloy returns to the equilibrium state, passing through two intermediate stages, namely a pre-precipitation stage and an intermediate precipitation. There are two tetragonal phases involved in this precipitation. The parameter a is equal to the parameter of the cubic matrix (whereas c/a is respectively slightly larger and smaller than 1, 0.98 and 1.02 for Cu-Ni-Co).

Let us note immediately that these intermediate stages are observed only if the difference of the parameters of the final precipitates is small. For example, the Au-Ni system also has an equilibrium diagram of the type described above (see Fig. 10); however, the difference between the parameters of the equilibrium phases is large being 3.58 and 3.95 A. The quenched single phase decomposes directly into the mixture of the two cubic phases, without transition stages, by a discontinuous process originating chiefly at grain boundaries.

The parameters of the equilibrium phases in the ternary systems depend on their composition. There exist concentration domains where the conditions for continuous precipitation are fulfilled. This is the reason why the "side-band" phenomenon is observed principally in ternary alloys.

As in the previous sections, we shall discuss only the first stage of the evolution of the quenched alloy, prior to visible precipitation. The pre-precipitation is characterized by the appearance of side bands symmetrically flanking each diffraction line in the Debye-Scherrer patterns originating in the matrix (Fig. 32a). These bands are rather broad and they are much less intense than the normal matrix lines, which remain as sharp as in the quenched stage, where no anomalous scattering is found. Single crystal patterns show that the normal diffraction spots are surrounded by satellites (Fig. 32b).

It appears from the published data[172–180] that there are two classes of alloys which give rise to two different kinds of satellites.

(1) In the most common case (Cu-Ni-Fe, Ni-Al), each node is surrounded by a pair of satellites located symmetrically on the three [100] relrods passing through the node and at a distance from the node which is *the same for each node* (Fig. 33). However, the satellites on the axis normal to this plane are missing when the node is situated in a plane passing through the origin of the reciprocal lattice. Thus, there is no satellite around the origin; there is only one pair for the nodes (h 00)

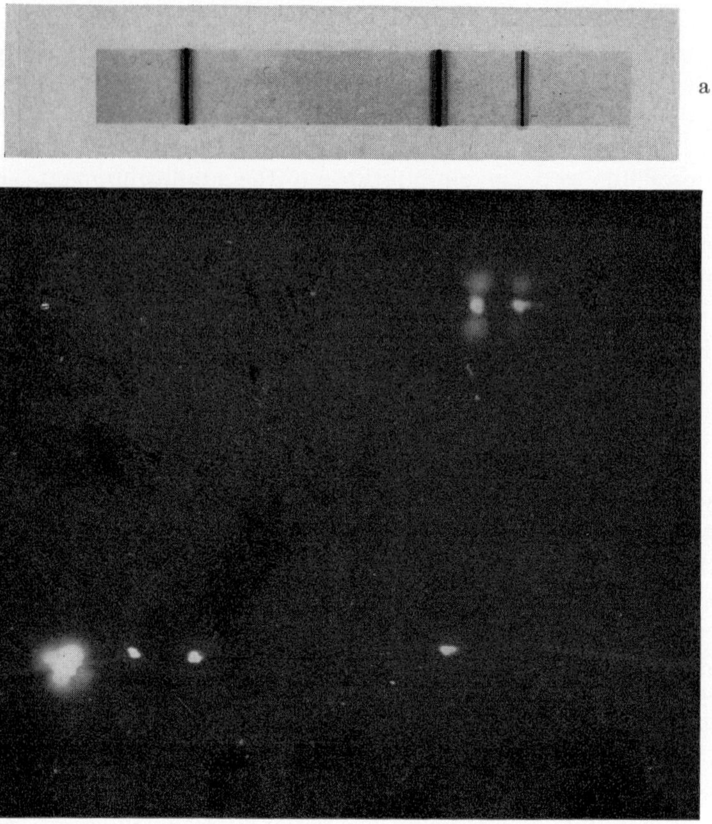

FIG. 32. a. Debye-Scherrer diagram with side bands. Ni-Si aged at 675°C [J. Manenc, *Rev. mét.* **54,** 867 (1957)]. b. Satellites around 111 and 200 spots Ni-Al aged at 700°C [J. Manenc, *Compt. rend.* **242,** 2344 (1956)].

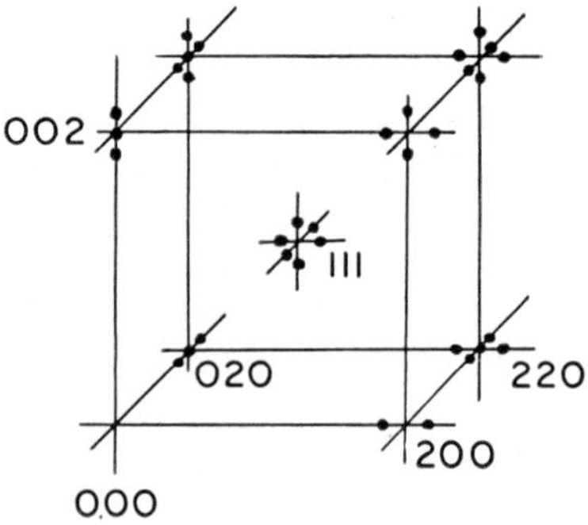

FIG. 33. Scheme of the position of satellites in the reciprocal space.

and two pairs for $(h\ k\ 0)$ and three pairs for $(h\ k\ l)$. The distance between the node and the satellite is of the order of $a^*/15$ to $a^*/30$, a^* being the parameter of the reciprocal lattice. The satellites are always diffuse, but are clearly separated from the normal node. In many cases, the two satellites of a pair on the same relrod do not have the same intensity.

(2) The second type has been observed by Manenc both in Ni-Cr alloys containing Ti and Al, and in the simple binary alloy Ni 7% Si. The normal lines of a powder pattern are also flanked by side bands; however, the pattern of single crystals shows that the disposition of the satellites is not the same as in case (1). Around each node, there are only two diffuse spots, the centers of which are on the line joining the node to the origin of the reciprocal lattice. The scattered intensity vanishes on the plane passing through the node normal to this vector. The intensity of the two satellites also may be different in this case.

It is natural to assume that these scattering phenomena are caused by the formation in the matrix of domains whose composition tends toward that of the two equilibrium phases. During the precipitation the domains remain completely coherent with the matrix lattice. The difference between these cases and the alloys studied in previous sections is that the proportion of the phases may be of the same order in the final stage. Thus the first stage does not originate in small zones of distortion imbedded in an almost unchanged matrix. Here the distorted parts of the lattice are much more extended. This is the reason why the distortions can be detected by x-rays even though they are relatively small.

Let us consider first the example of Cu-Ni-Fe. It behaves as a quasi-binary alloy in which Cu is dissolved in a Ni-Fe matrix. The decomposition is caused by the segregation of the copper. The following conclusions can be deduced from the observed scattering with the aid of the general theory of x-ray diffraction (Part II):

(1) Since the scattering factor for the three atoms is very nearly the same, the scattering originates in displacement disorder. This is confirmed by the absence of scattering in the small-angle region. The disorder must be caused by a contraction of the Cu-rich regions and a dilatation of the Cu-poor regions (the parameter of the corresponding two phases are 3.53 and 3.59 A).

(2) Since scattering is concentrated on the [100] relrods, the disorder is planar; the (100) planes are unchanged but the interplanar spacings are altered. Daniel and Lipson[171] and Hargreaves[172] have shown that a periodic variation of this spacing produces satellites similar to those observed. The wavelength of this variation is equal to the inverse of the distance from the satellite to the node in the reciprocal lattice. It varies from $65a$ to $126a$ according to the heat treatment (a is the parameter of

the cubic cell). The wavelength increases with the annealing temperature. The model developed by Hargreaves consists of lamellas alternatively Cu-rich and Cu-poor in which the interplanar spacing is, respectively, $a(1 - \epsilon)$ and $a(1 + \epsilon)$. The intensity of the satellites is proportional to ϵ. If the lamellas have the same number of planes N, the total thickness, which is the wavelength of this "modulated" structure, is rigorously equal to the thickness of 2 N planes in the normal lattice. For $N = 33$ and $\epsilon = 0.1$, the maximum displacement is only 0.33. Thus coherency can be maintained between the two lamellas and the lattice.

Tiedema[181] and Biedermann and Kneller[174] have shown that the two symmetric satellites have different intensities if the number of the planes in the dilated and contracted lamellas are not equal. They found that the calculated distribution of the planes between the two regions is in good agreement with the proportion of the two phases in the final equilibrium state.

An important confirmation of Hargreaves model has been provided by observation of the Cu-Ni-Fe alloy with the electron microscope. Biedermann and Kneller[174] have resolved in the pre-precipitation stage a periodic structure of parallel lamellas, the period being in good agreement with that calculated from the position of the side-bands in x-ray patterns of the same sample. The number of parallel lamellas generally is found to be less than 4 or 5. The spacing between lamellas in a given sample is fairly uniform (only 15% fluctuation).

There are nevertheless some objections to the Hargreaves model. The theory predicts sharp satellites; actually they are always broad. A periodic structure should give second-order, third-order satellites, etc. . . . , they are very rarely observable. The width of the satellite may be explained by the small number of periods in coherent regions and also by some imperfections in this periodicity, which are not unlikely. It is very difficult, however, to understand how the wavelength of the modulated structure can change during annealing. It seems necessary to postulate destruction of the structure followed by complete reorganization.

Guinier[182] pointed out that it is possible to imagine that a disturbed *zone*, which does not possess periodicity, produces the "side bands." The model is somewhat similar to the zone in the Al-Ag alloy. Consider Cu-Ni-Fe as an example. The Cu atoms are supposed to segregate on a series of (100) planes and produce a lamella with decreased spacing. This central lamella is flanked by two outer lamellas which are left copper-poor after the migration of the copper towards the center. The parameter is increased in the external regions and the zone is marked by a perturbation

[182] A. Guinier, *Acta Met.* **3,** 510 (1955).

of the spacing as shown in Fig. 34. The calculation predicts the formation of two diffuse but clearly distinct satellites on the [100] relrod, except for the nodes lying in a plane passing through the origin. No second order is predicted. The distance from the satellite to the node is inversely proportional to the thickness of the zone.

A growth of the zone is indicated when the satellites come closer to the node. The mechanism of this growth does not raise the same difficulty as a variation of the wavelength of a periodic structure.

In reality, there is no fundamental contradiction between the periodic and the zone model. The latter is merely formed by one period of the Hargreaves model. As is suggested by observations with the electron microscope by Biedermann and Kneller[174] a model intermediate between the zone and the modulated structure is perhaps the more reasonable. A

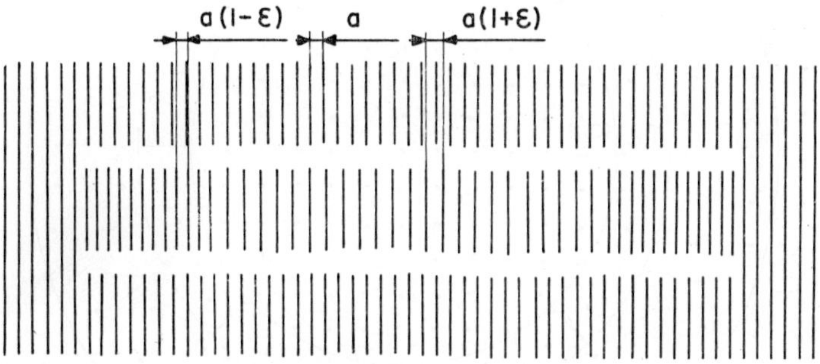

FIG. 34. Model of a zone with dilated planes surrounded by contracted planes [A. Guinier, *Acta Met.* **3**, 510 (1955)].

coherent region would be formed by a very small number of alternate lamellas of approximately uniform thickness.

Let us now consider the second kind of satellite (Ni-Si alloy). Here the periodic model is not valid. On the contrary, the zone model gives a satisfactory explanation of the observed scattering. However, in this case, the disorder is no longer planar; let us imagine that it is isotropic. Also suppose that there is clustering of Si atoms into quasi-spherical regions in which the average parameter is smaller than the normal value. This region is surrounded by a Si-poor shell, having a larger parameter. Thus the shrinkage of the core is exactly compensated by the dilatation of the outer shell (Fig. 35). The deformation is analogous to the distortion of a pure lattice in which one atom has been replaced by another with a different diameter. The calculation of the resultant scattering has been made by Huang[50] and recently by Cochran.[183] The scattering is similar in our case but is concentrated around the nodes, because the distorted

[183] W. Cochran and G. Kartha, *Acta Cryst.* **9**, 944 (1956).

regions are large (from 10 to 100 atomic diameters). This calculation shows that the proposed model of the zone accounts for the characteristic features of the observed scattering. Two maxima of scattering are found on the line joining the node to the center of the reciprocal lattice, and the

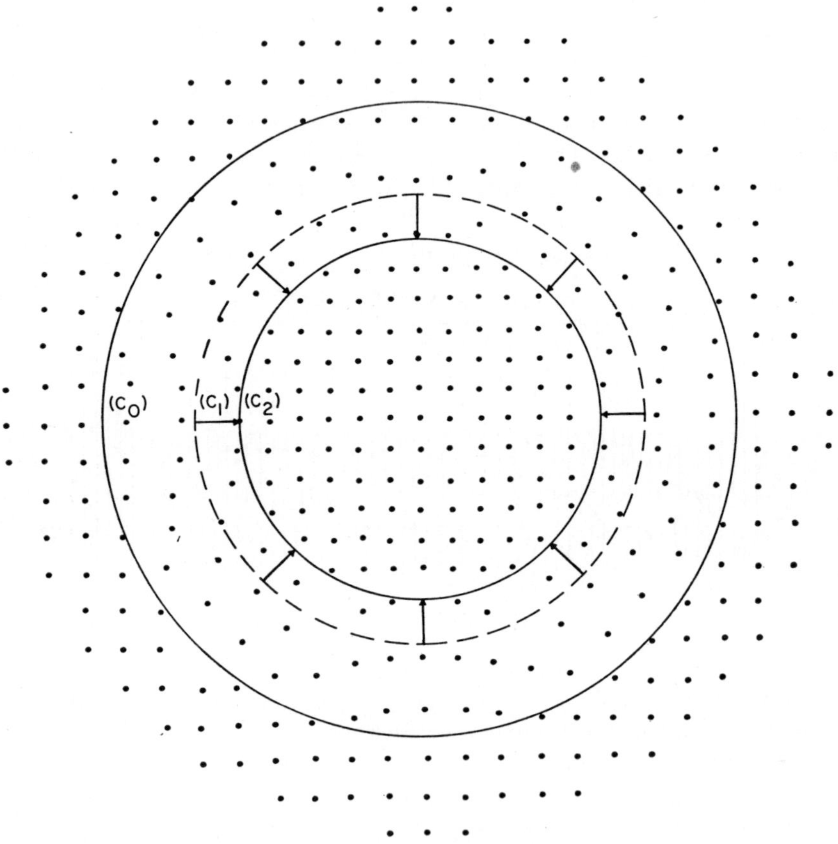

FIG. 35. Model of a spherical zone with an internal contracted zone and outer dilated shell.

intensity vanishes on the plane normal to this line, passing through the node.

Our conclusion is that, in these alloys, the pre-precipitation arises from numerous small round clusters which provoke complex distortions in localized zones, the strains being small outside of these zones.

Such clusters have not yet been seen by electron microscopy, but, as in many other cases, a confirmation of the structure of the zone can be deduced from the shape of the subsequent precipitate. For Ni-Si these precipitates are small isolated cubes dispersed in the matrix,[179] which proves that the precipitation is isotropic. On the other hand, in Cu-Ni-Fe, the precipitates are thin lamellas, parallel to (100) planes.

Manenc[179] points out that the factor which determines the type of pre-precipitation seems to be the difference between the parameters of the equilibrium phases. Isotropic zones are formed only if this difference is very small. For example these parameters are 3.50 and 3.51 A for Ni-Si. When the difference of parameter is larger (0.06 for Cu-Ni-Fe), the first type of pre-precipitation is observed. One might say that the strains in an isotropic zone would become too important. Thin lamellas are formed because the strain energy is diminished by plate-like clusters in regular alternation.

22. Formation and Stability of Zones

A supersaturated solid solution is unstable and evolves toward a system having lower free energy. The one with the minimum energy is made up of a mixture of an impoverished matrix and the stable precipitate, conforming to the equilibrium diagram. However, the theory of precipitation shows that this evolution through the process of nucleation and growth, necessitates an energy of activation W, which is associated with the energy of the interface between the nucleus and the matrix.[184] This energy is larger the more the crystalline structure of the nucleus differs from that of the matrix. The direct passage of the solid solution to the equilibrium structure may be impossible at low temperatures, because the factor $\exp(-W/kT)$ is too small. However, if perturbations of the lattice, such that a group of atoms have an excess of energy dW over the normal, exist locally, nuclei can be created, because the probability of formation of a nucleus, which is of the form $\exp[(-W + dW)/kT]$, is increased.

The influence of deformation of the lattice upon the nucleation is shown by a number of observations.[135,185] Even at the beginning of the annealing, before any generalized precipitation, some isolated grains are visible along the grain boundaries or along sub-boundaries. In this case, the precipitates mark a row of individual dislocation lines.[186] When two forms of precipitate are present, that which forms at the lowest temperature is the one for which the nuclei are created more easily. Let us take the Al-Cu system as an example. The θ' precipitate requires a lower energy of activation than the equilibrium precipitate θ, because the

[184] D. Turnbull, *Solid State Phys.* **3,** 225 (1956).

[185] P. C. Varley, M. K. B. Day, and A. Sendorek, *J. Inst. Metals* **86,** 337 (1958).

[186] H. Wilsdorf and D. Kuhlmann-Wilsdorf, *Rept. Bristol Conf. on Defects in Crystalline Solids,* 1954 p. 175 (1955).

former is coherent with the matrix on the habit plane (100), whereas the latter is completely incoherent. Thus the temperature of appearance is 200°C for θ' and 350°C for θ in the normal undeformed lattice. After slight cold-working (10% reduction in thickness by rolling), the θ' precipitate is visible at 150°C after 20 hours. In a powder obtained by filing, however, the θ precipitate is formed even at room temperature.[135]

The interfacial energy of zones is very weak, since the zones are coherent with the matrix in three dimensions. They can form with a very small activation energy and hence appear at low temperatures, when other processes of decomposition are impossible. Evidently it is necessary that atoms have a smaller free energy in the zones than when in the solid solution.

It can also be said that the dimensions of the stable embryo of a zone are very small. They would be zero if there were no interfacial energy. However, it is sufficiently large for a true precipitate that true precipitation does not have a chance to occur at room temperature. This is the fundamental reason that the pre-precipitation stage exists.

There are two interesting points to discuss. First, it is necessary to prove that the free energy of the whole crystal possessing zones is lower than that of the homogeneous crystal. Moreover, it is necessary to account for the observed kinetics of zone formations.

Our starting point will be the curve giving the free energy of a solid solution as a function of the concentration of solute. It is not known how to calculate this curve in general. Thus the following considerations must be regarded as qualitative.

The thermodynamic theory of alloys will be recalled briefly.[187,188] The shape of the free-energy curve of an alloy whose solubility limit at the temperature considered is C_1, is given by Fig. 36a. The homogeneous solution is unstable when its composition is between C_1 and C_2. The alloy of concentration C may become heterogeneous by progressive diffusion if a decrease in the free energy of the system occurs during its decomposition into two solid solutions having very nearly the same compositions $C - dC$ and $C + dC$. The necessary condition is that the curve $F = f(C)$ be concave downward; that is to say, d^2F/dC^2 is negative.[189] This is the condition for uphill diffusion. The spinoidal is the curve giving the concentrations for which $d^2F/dC^2 = 0$ as a function of temperature. It lies below the solubility curve marking the boundary above which the solid solution is homogeneous (Fig. 36b). The spinodal

[187] H. K. Hardy, *J. Inst. Metals* **77**, 457 (1950).

[188] R. Smoluchowski, "Phase Transformation in Solids," p. 149. Wiley, New York, 1951.

[189] U. Dehlinger and H. Franz, *Z. Metallk.* **48**, 176 (1957).

can be calculated (Borelius[190]) from the solubility curves. The free energy is represented by a semiempirical formula whose coefficients are determined to fit the solubility data.

In states below the spinodal, the solid solution can decompose into two phases having the same lattice but different compositions. According to Dehlinger and Franz,[189] this is the situation for the ternary alloys Cu-Ni-Fe and Cu-Ni-Co (Section 21); however the calculation for Al-Cu

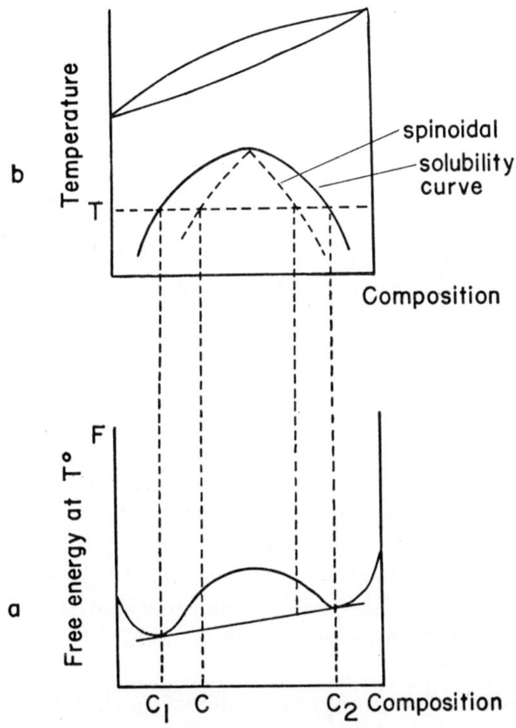

FIG. 36. a. Free energy *versus* composition for a solid solution. b. Limit of solubility of the solid solution and spinodal.

indicates that the domain of temperature and concentration where precipitation is observed is situated above the spinodal.

Figure 36a shows that the mixture of two phases would be less stable than the homogeneous alloy *at the beginning of decomposition* of the alloy under these conditions.

Dehlinger and Knapp[191] point out, however, that the free energy

[190] G. Borelius, *Ann. phys.* **33,** 517 (1938); *Arkiv Mat. Astron. Fysik* **32A,** No. 1 (1945); *J. Metals* **3,** 477 (1951).

[191] U. Dehlinger and H. Knapp, *Z. Metallk.* **43,** 223 (1952).

given by the theoretical curve corresponds to a combination of large volumes of each phase. If the result of segregation is the production of a number of very small "complexes" where two small domains of either composition are in juxtaposition, there is a positive entropy term arising from irregularity of the dispersion of these complexes on the matrix. This term decreases the value of $F = U - TS$. The calculation of Dehlinger and Knapp indicates that the diminution is sufficient to render F smaller for the alloy possessing complexes than for a homogeneous solid solution.

Below the spinoidal, the complexes, or zones, can grow indefinitely and a total separation can be reached. Above it the size of the stable zones decreases and becomes zero at the solubility temperature, where the homogeneous alloy is stable.

Let us note that the zones we have described in Al-Ag alloys are just the complexes suggested by Dehlinger.

A very important point must be stressed. This theory considers the various possible configurations of solid solutions inside the same perfect lattice. It does not take account of possible distortions of the lattice associated with clustering of the dissolved atoms. This restricts the validity of the theory seriously. It is thus best when applied to the cases of Al-Zn or Al-Ag. In fact, as we have indicated above, experiment has verified the existence in those alloys of zones which are similar to the complexes of Dehlinger. But the validity of this type of calculation is a little doubtful, when the zones are strongly deformed portions of the lattice (Al-Cu, for example). The strain energy could be much larger than the variation of energy arising from the changes of the interaction between neighbors. Nevertheless, Meijering[192] has justified the existence of G.P. zones in Al-Cu from thermodynamic data deduced from the equilibrium diagram.

One result of the thermodynamic theory is that the zones are stable only at low temperatures. This is verified by experiment, as will be shown in the next section (reversion phenomena). In contrast, it is found that the zones in Al-Ag, like those in Al-Zn or Al-Cu, are more voluminous when the temperature at which they are formed within their domain of stability (for example below 150°C for Al-Cu), is increased. This result is contrary to Dehlinger's conclusions. However, the calculations give only the limiting stable dimensions. The observed dimensions can be smaller. In effect, it is possible that the zones reach a dimension inferior to the stable limit at low temperature as a consequence of the kinetics of formation.

To sum up, it is necessary to remember that the hypothesis of an

[192] J. L. Meijering, *Rev. mét.* **49**, 906 (1952).

undeformed lattice made thus far in the typical thermodynamic calculation is restrictive. Such calculations do not take account of the energy of deformation. The general case of a lattice deformed by the clustering of atoms has not yet been treated theoretically.[193] We should add that the influence of the strains has been studied by Franz and Kröner[194] in the case of Al-Cu in order to explain the stability of G.P. zones of type II relative to G.P. zones of type I.

23. Reversion of the Zones

A phenomenon which gives very important information concerning the stability of the zones is *reversion*. It was first found (Gayler[195]) in the

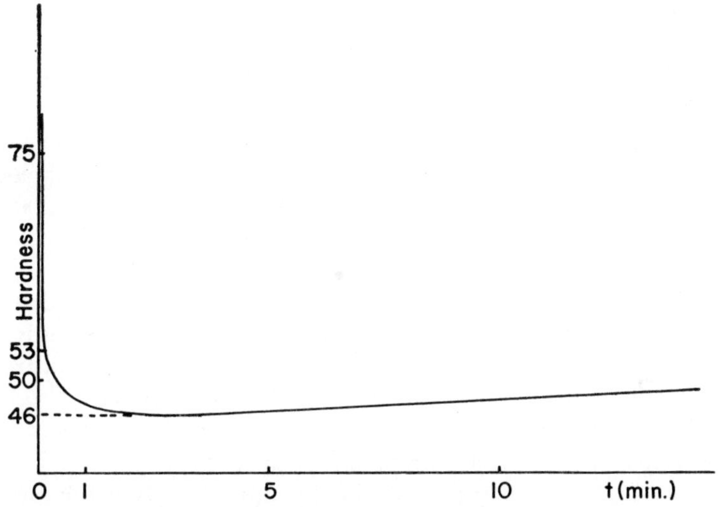

FIG. 37. Hardness of an Al-Cu alloy aged at room temperature annealed at 200°C. Hardness after quenching before first aging: 53 kg/mm².

study of the hardness of age-hardening alloys submitted to complex heat treatments.

Consider an Al-Cu alloy which is quenched and then hardened at room-temperature (25°C): after a time of the order of 24 hours, it reaches a nearly constant hardness. This hardened alloy is then treated at 200°C and its hardness is measured at room temperature immediately after treatment as a function of the time at 200°C. It is found (Fig. 37) that the hardness decreases very rapidly. After 4 minutes, the alloy is

[193] E. Kröner, *Acta Met.* **2,** 302 (1954).
[194] H. Franz and E. Kröner, *Z. Metallk.* **46,** 639 (1955).
[195] M. L. V. Gayler, *J. Inst. Metals* **28,** 213 (1922).

softer than immediately after quenching. Then the hardness increases anew. This phenomenon, called reversion, is a very general one among the cold-hardening alloys.[195–202] It can be defined as the softening of a metal, which has been completely hardened by aging at low temperatures (25° – 100°C) after a very short heating at a temperature of the order of 200°C (these temperatures are for aluminum alloys). The reversion treatment completely cancels the effect of the cold-hardening, not only for the hardness but also for other physical properties.

The first explanations of reversion were based on the idea that cold-hardening was caused by a very fine precipitate (Konobeevski[203]). It was supposed that the structure of these small particles and of the visible precipitates were the same. The particle is stable at a given temperature when it is larger than the critical nucleus, according to the ordinary theory of nucleation and growth of a precipitate. The diameter of the critical nucleus is assumed to increase with temperature. Thus, most of the very small particles formed during aging at room temperature are unstable at 200°C, because they are below the critical size at this temperature. They dissolve except for a few nuclei having sufficient size to develop further. After the first few minutes of treatment at 200°C, the small precipitates are dissolved and the stable nuclei have not yet grown. This accounts for the softening of the metals.

The theory must be modified, for we know that the distinction between the segregation at low and high temperatures is not only one of size but of structure as well. The structure of the low-temperature zones is such that they are unstable at high temperature. Dehlinger[191] predicted this result from his model of the complex. Qualitatively, it can be said that the lattice of the matrix can be maintained inside clusters having an anomalously high concentration of dissolved atoms only if the amplitude of the thermal vibrations of the atoms (and consequently the rate of diffusion) is sufficiently low.

During the reversion treatment, the alloy is heated at a temperature in the region of warm-hardening, that is to say, where the solid solution decomposes normally by true precipitation. If, for example the Al-4% Cu alloy is quenched *directly* from the homogenization temperature to 200°C, it is found that the precipitate θ' develops; however, this precipitation is

[196] H. Auer, *Z. Naturforsch.* **40,** 533 (1949).
[197] A. Saulnier, *Rev. aluminium* **26,** 157, 235 (1949).
[198] G. Wassermann, *Z. Metallk.* **41,** 50 (1950).
[199] M. L. V. Gayler and G. D. Preston, *J. Inst. Metals* **48,** 197 (1932).
[200] H. Borchers, *Z. Metallwirtschaft* **20,** 1161 (1941).
[201] R. Graf, *Recherche aéronaut.* No. 60, 47 (1957).
[202] R. H. Beton and E. C. Rollason, *J. Inst. Metals* **86,** 77 (1957).
[203] S. T. Konobeevski, *J. Inst. Metals* **69,** 397 (1943).

accompanied by appreciable hardening only after 30 minutes. The amount of precipitation is only 10% after a treatment of 10 hours. Thus the time for reversion (of the order of four minutes at 200°C) is quite insufficient to promote precipitation, but it is sufficient for the zones to dissolve. It has been verified with x-rays[199,201] that the streaks characteristic of the zones become progressively less intense after 1, 2, 3 minutes and then disappear after 4 minutes. Diagrams made especially to detect the precipitate Al_2Cu θ' indicate that the amount of precipitation is less than 10^{-3} after the reversion treatment.

By measuring the temperature above which the reversion is complete for Al-Cu alloys of various compositions, Beton and Rollason[202] have

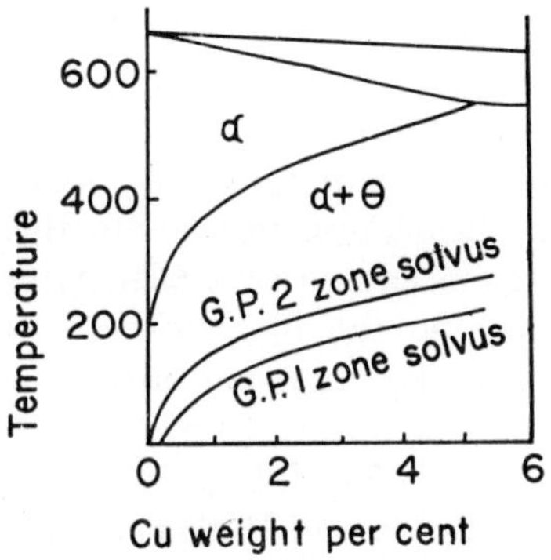

FIG. 38. Domain of stability of G.P. I and G.P. II zones [R. H. Beton and E. C. Rollason, *J. Inst. Metals* **86**, 77 (1957)].

established the curve of solubility of G.P. I zones (Fig. 38) and found that it is represented by the equation

$$x = A - \frac{B}{T}. \tag{23.1}$$

The heat of solution of the zone is given by $\Delta Q = RB$; ΔQ is equal to 5.35 Kcal/mole of Cu. However, the temperature of solubility of a zone is not a fixed temperature for a given composition of the alloy, as for the solubility limit of a precipitate. The stability of a zone strongly depends on its size. For example, the very small zones produced in Al-Cu during cold work at room temperature (Section 25a) are reverted at 100°C, whereas the normal zones are stable and even grow at this temperature.[135]

The reverted state corresponds to the best degree of homogeneity known for the solid solution and thus to a minimum hardness. When the reverted alloy is left to age at room temperature, it hardens anew, but as we shall see in Section 24, at a much smaller rate. That is why the solid solution is conserved for some time in a homogeneous state, whereas the zones form so quickly after the quench that it is practically impossible to make a measurement sufficiently rapid to retain the true quenched state.

The condition necessary for the reversion is the existence at higher temperature of a *distinct* process of decomposition of the solid solution, which is *independent of the low-temperature process* and which is initiated by *a prior* redissolution of the first structure. Thus reversion allows one to make a clear distinction between a zone and a precipitate of very small size and gives a strong argument against the sequence theory (p. 327). The zones dissolve and the precipitates are formed from nuclei, which perhaps are distinct from all zones. It is certain that the number of nuclei is much smaller than the number of zones; the transformation of a zone into a precipitate is, in any case, exceptional.

Graf[162] found some diffuse spots in the x-ray scattering patterns of the Al-7% Zn-3% Mg alloy after aging at 50°C, which could be ascribed to very small particles of $MgZn_2$ precipitate. On the other hand, he found additional scattering regions caused by zones. After the reversion process, only the latter disappear so the reversion is not complete (minimum hardness after reversion 65 kg/mm^2 and after quench 59 kg/mm^2, on a Brinell scale). The small precipitate particles of $MgZn_2$ do not dissolve and are responsible for the irreversible part.

In Al-Cu, Graf[201] and Beton and Rollason[202] have shown that an alloy containing the second structure, G.P. II zones or θ'', may be reverted completely by heating at 250°C for 5 min. Graf verified by x-ray methods that the G.P. II zones vanish, while only a small proportion of θ' precipitation appears.

On the other hand, the same AlCu alloy containing the precipitate θ' cannot be reverted, in spite of the fact that above 300°C, the θ' phase no longer is stable and is replaced by the phase θ. Thus it seems that there is a sequence connecting the two precipitates. It is likely that some θ grains are the results of the allotropic transformation of θ' without redissolution and new nucleation. Gayler[204] has obtained microphotographs of partly transformed θ' grains.

The alloy Al-Ag behaves in an unusual manner which is still difficult to understand. Its physical and mechanical properties show clearly that it undergoes reversion;[83,84] however, the central scattering of x-rays characteristic of the zones do not disappear. A ring of scattering, which has a diameter that is smaller after the reversion treatment[93] rather than before, is always observed. Thus there are more voluminous zones still present. It is certain, therefore, that all the zones are not unstable at

[204] M. L. V. Gayler, *Proc. Roy. Soc.* **A173,** 83 (1939).

180°C. The hardness of the alloy falls to its initial value after quenching without the solid solution becoming homogeneous. Gerold[97] has deduced from a quantitative study of the scattered intensity that reversion lowers what he has defined as the degree of heterogeneity. He suggests that the silver atoms leave the central core of the zone and diffuse toward the surrounding shell. This diminishes the contrast in concentration, but the zone does not disappear. However, it is difficult to explain why a weak variation of concentration produces such a large variation in the mechanical properties. Belbeoch and Guinier[93] remark that certain zones can disappear, without this fact being revealed by x-rays, for smaller zones are difficult to detect in the presence of large ones. The small zones would be more effective in hardening. Beton and Rollason[202] distinguish two stages in the reversion process and conclude that two structures are responsible for the hardening in the cold aged Al-Ag alloys. Using the work of Ziegler,[86] they identify these two phases tentatively with clustering of silver and stacking faults in the matrix. As it has been said, in opposition to the opinion of Ziegler, however, it seems that stacking faults may be present only inside the Al_2Ag γ' precipitate. Thus, stacking faults would not be connected with a cold-aging process.

24. Kinetics of the Formation of Zones

The simplest case is represented by that of the zones of the alloy Al-Zn. Here there is little deformation of the lattice and there is no indication of a special short-range order inside the central zinc-rich core of the zone. The growth of such a zone occurs by the combination of two mechanisms, namely *up-hill diffusion* of atoms of zinc which enlarges the central core, and a *normal diffusion* which allows the zinc atoms to migrate toward the surrounding shell. It is necessary to imagine an initial nucleus of the zone composed of several zinc atoms. To say that there is up-hill diffusion means that a zinc atom which jumps on a site situated at the edge of an enriched core has a large probability of remaining there, because it finds itself in a position of lower energy. On the contrary, the probability of a given jump and its inverse are equal in normal diffusion.

If it is assumed that the concentration of the core is fixed and uniform, and that all atoms arriving on the nucleus collect there, the principal problem associated with the growth of the zone is simply that of the growth of a spherical particle inside an infinite matrix. It has been solved with the use of the diffusion equations by many authors.[205,206] The distribution of the concentration assumed in the calculations (Fig. 39) is closely related to the experimental curve of Fig. 20.

[205] C. Zener, *J. Appl. Phys.* **20,** 950 (1949).
[206] F. C. Frank, *Proc. Roy. Soc.* **A201,** 586 (1950).

The radius of the core is given by the formula:

$$R \simeq \delta \sqrt{Dt} \tag{24.1}$$

where t is the time of aging and D is the diffusion coefficient at the temperature of aging. Here δ is a coefficient which depends on the concentration in the matrix and the shape of the particles, but which is not very different from unity.

The model is very schematic, but it gives a fundamental result concerning the kinetics of the zone formation, even without refinement. The diffusion coefficient that is deduced from the dimensions of the zones observed experimentally is considerably larger than the coefficient deduced by extrapolation of the high-temperature measurements.

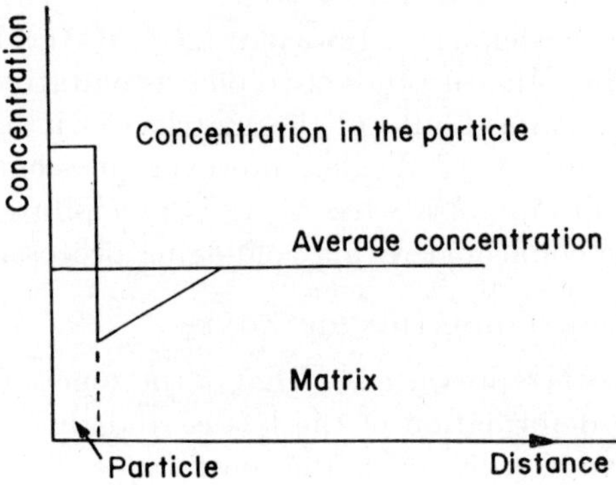

FIG. 39. Variation of the concentration of dissolved atoms around a zone adopted by H. K. Hardy [*J. Inst. Metals* **79,** 363 (1951)] for the calculation of the diffusion coefficient.

Consider, for example, Hardy's calculations for the G.P. zones in Al-Cu.[207] He concludes on the basis of available x-ray data reported in Section 18 that, at 25°C after one day of aging, half of the copper is assembled in zones which are 45 A in diameter and two planes thick. Thus, the zones are at an average distance of 80 A from each other. The copper is assumed to converge toward each side of the zone through a surface a little larger than the surface of the zone. The diffusion coefficient is given by Fick's equation

$$P = D \frac{dc}{dx} \tag{24.2}$$

where P is the flux of copper atoms per unit area per unit of time. The number of copper atoms in the zone is 390. The surface of diffusion, somewhat arbitrarily, is taken equal to 3600 A^2 (2.2 × area of the zone).

[207] H. K. Hardy, *J. Inst. Metals* **79,** 321 (1951).

Thus, the average flux during the aging is

$$P = \frac{390}{2} \frac{10^{16}}{3600} \text{ atoms/cm}^2\text{/day}. \tag{24.3}$$

To evaluate the concentration gradient, Hardy admits that a decrease of concentration equal to half of the initial matrix concentration

$$\Delta C = 5.6 \times 10^{14} \text{ atom/cm}^3$$

occurs in a distance equal to half the distance between zones, 40×10^{-8} cm. Thus the value of D is $\mathit{2.8 \times 10^{-13}\ cm^2/day}$. According to the measurement at high temperature,[208]

$$D = 2.0 \times 10^5 \left(- \frac{34{,}900}{RT} \right) \text{cm}^2\text{/day} \tag{24.4}$$

the extrapolated value at 25°C is $\mathit{4.6 \times 10^{-20}}$, which is $\mathit{10^{-7}}$ smaller than the coefficient of diffusion necessary to account for the observed zones.

This enormous discrepancy, which is far beyond the uncertainties of the measurements or of the calculations, has been confirmed in many examples. de Sorbo and colleagues[209] found, by measurements of resistivity, that significant clustering occurs in AlCu in a period of one minute at a temperature of -45°C. Assuming that the copper atoms moved 3 or 4 atom diameters, the estimated diffusion coefficient is found to be 10^{17} times larger than the one calculated by Eq. (24.4). Turnbull and Treaftis[94] have found a discrepancy of the same order of magnitude in Al-Ag alloys aged at low temperatures.

It follows that the normal mechanism of diffusion is not sufficient to account for the zone formation. Jagodzinski and Laves[130] were the first to emphasize this considerable divergence in the case of the zones in Al-Cu. They concluded that the zones were not the result of an important segregation but originated in a local lattice deformation. However, it has been shown (Section 18) that the model they propose involving very limited diffusion is in disagreement with the x-ray data.

It is thus necessary to find the reason for the considerable acceleration of diffusion which is observed.

a. Action of Dislocations (Turnbull[210]*)*

It is known that foreign atoms can move along dislocation lines easily. Therefore, if gliding dislocation lines sweep a plane, atoms can drain to

[208] R. F. Mehl, F. W. Rhines, and K. A. van den Steinen, *Metals & Alloys* **13** (1), 41 (1941).

[209] W. de Sorbo, H. N. Treaftis, and D. Turnbull, *Acta Met.* **6,** 2 (1958).

[210] D. Turnbull, *Rept. Bristol Conf. on Defects in Crystalline Solids, 1954* p. 203 (1955).

zones along these moving channels. Dislocations may act as short circuits which enhance the diffusion considerably. A detailed mechanism of the movements of dislocations during zone growth has not been developed. The free energy of precipitation would give the driving force for the dislocation glide (Fisher and Hollomon[211]).

This theory which is still qualitative seems to be applicable only in the case of zones accompanied by distortion because the dislocation lines may be very numerous in the stress fields around the zone. The suggested mechanism seems most appropriate for plane zones such as the G.P. zones. Unfortunately, the high rate of the diffusion also must be explained in the case of Al-Ag where the clusters are spherical and where the distortion of the lattice is very weak. Another cause favoring diffusion has to be found.

b. Action of Vacancies

Seitz[212] first suggested that the increase of the rate of diffusion could be the result of the presence of an anomalous density of vacancies in the lattice. Age-hardening immediately follows a quench to low temperature. The number of vacancies is proportional to a Boltzmann factor $\exp(-Q_v/RT)$ in a metal in equilibrium at a given temperature. It reaches a value of the order of magnitude of 10^{-4} at the melting point. During a rapid quench to low temperature, the new equilibrium is not attained immediately; the metal maintains quenched in vacancies in a state of supersaturation. The effect of vacancies on the electrical resistance and on the mechanical properties of pure metals have been observed.[213] Thus there now are good theoretical and experimental reasons to believe that the vacancies play an important role in zone formation (Federighi,[214] de Sorbo,[209] Turnbull and Treaftis,[94] Guinier[215]).

(1) It is possible to explain the order of magnitude of the diffusion coefficient observed during cold-hardening if it is admitted that diffusion takes place by motion of vacancies. In effect the ratio of the coefficient in the metal in equilibrium at T_0 and in the metal quenched from T is equal to the ratio of the number of vacancies in equilibrium at T and T_0, if all vacancies are retained by the quench. In aluminum, the energy of formation of a vacancy is assumed to be equal to 0.76 ev.[216] Taking

[211] J. C. Fisher and J. H. Hollomon, *Acta Met.* **3,** 608 (1955).

[212] F. Seitz, *in* "L'Etat Solide," p. 401. R. Stoops, Bruxelles, 1952.

[213] A. H. Cottrell, *Inst. Metals Monograph and Rept. Ser.* No. 23, 1 (1957).

[214] T. Federighi, *Acta Met.* **6,** 379 (1958).

[215] A. Guinier, *in* "L'Etat Solide," p. 408. R. Stoops, Bruxelles, 1952.

[216] F. J. Bradshaw and S. Pearson, *Phil. Mag.* [8] **2,** 2570 (1957).

$T = 530°C$ and $T_0 = 20°C$, this ratio is $r = 2 \times 10^8$, which is of the order of magnitude desired.

(2) The rate of the zone formation in Al-Cu and Al-Ag (measured by the decrease of conductivity) has been compared at different temperatures of aging from $-50°C$ to $40°C$.[209,94] The energy of activation found is 11 kcal and 10 kcal, which is very near to the energy of activation for annealing of the point defects in aluminum (12 kcal). Actually this calculation is not very precise, because the energy of activation may be changed considerably if there is association of vacancies.

The vacancies disappear during the aging. Thus the rate of diffusion must decrease and tend towards its normal value at the aging temperature. This may explain why the dimensions of the zones at a given temperature seem to reach a limiting value.

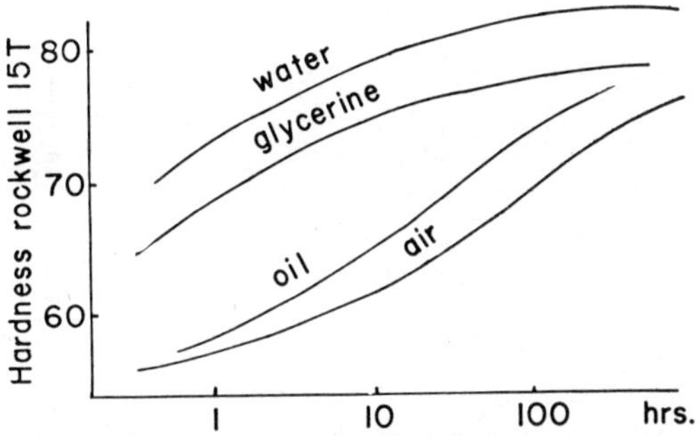

FIG. 40. Variation of rate of hardening of an Al-Cu alloy with the mode of quenching.

(3) When a given alloy is quenched from the same temperature in different media, it is observed that the rate of hardening at room temperature, that is to say, the rate of formation of zones, is higher the more severe the quench (Fig. 40).[217] This is easily explained, for the vacancies in equilibrium at high temperatures are not preserved in the lattice, when the quench is slow (in oil or air); the number of vacancies which remain decreases as the rate of quench decreases.

(4) Graf[217] measured the rate of hardening of an Al-Zn alloy containing 10% Zn and quenched in a given manner from different temperatures above the homogenization temperature. The Al-Zn alloy is particularly attractive for these experiments, for it is possible to vary the temperature before the quench from 180°C to 616°C. Figure 41 shows that the rate of hardening increases as the temperature prior to quench increases from

[217] R. Graf, *Compt. rend.* **246,** 1544 (1958).

200°C to 460°C.[218] At the same time the diameter of zones, measured by the small-angle scattering of x-rays, increases. There is a small decrease when the temperature is above 460°C. However, it is possible that this is the result of a decrease in the speed of cooling of the metal in the quench liquid. The metal, being very hot, produces bubbles of vapor which partially insulate it. Another explanation is possible.[220] It has been observed that the vacancies disappear more rapidly in pure aluminum when the metal is quenched from a high temperature, because they can be trapped easily by the defects created by the quench. Thus the vacancies would be less efficient when the alloy is quenched from a very high temperature.

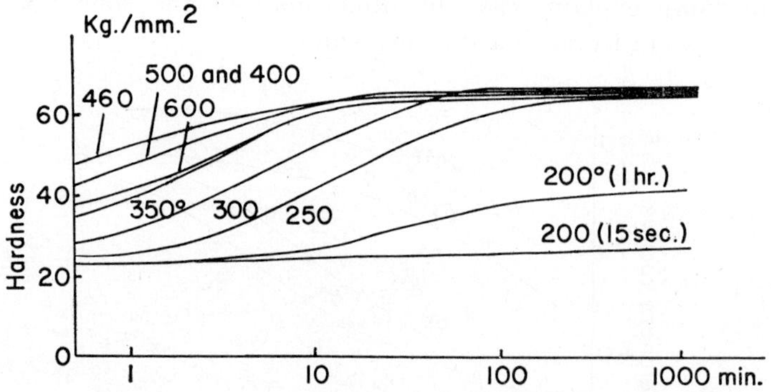

FIG. 41. Hardening of Al-Zn quenched from different temperatures [R. Graf, *Compt. rend.* **246,** 1544 (1958)]. The alloy was heated 1 hr at each temperature and at 200°C for 1 hr and 15 sec.

A value for the heat of formation of the zones can be deduced from a plot representing the logarithm of the time required to attain a given hardness as a function of the reciprocal of the temperature before the quench. The curve is a straight line in the range from 250–350°C. The value found, namely 0.64 ev, is not very far from the heat of formation of vacancies (0.76 ev).

(5) One might believe that after the resolution of the zones by reversion, a homogeneous solid solution whose constitution is identical with that of the solid solution immediately after quenching would be obtained. Nevertheless their properties are different. When compared by aging at the same temperature, the reverted alloy hardens considerably

[218] According to Gane and Parkins,[219] this is one of the facts, which shows that the mechanism of the cold-hardening of Al-Zn is not related to pre-precipitation or precipitation; but the authors do not offer another explanation.

[219] N. Gane and R. N. Parkins, *Nature* **181,** 1198 (1958).

[220] This suggestion is due to M. Winterberger.

more slowly than the quenched alloy (Fig. 42). Moreover, its initial hardness is lower because the quenched alloy has already had time to undergo some hardening.

The facts are particularly clear for Al-Zn. The hardening is practically suppressed through reversion at 200°C. The hardness after 30 days is still equal to the hardness after quench. X-rays reveal no zones, that is, the solid solution remains homogeneous. If there are nuclei of zones, they are not able to grow: the apparent diffusion coefficient is smaller than in the alloy after quenching. The vacancy mechanism offers an easy explanation. After reversion, the zones have disappeared; moreover, the number of vacancies in equilibrium is negligible at the temperature of reversion of the light alloys (200°C). The absence of vacancies constitutes the difference between the reverted alloy and the alloy quenched from a high temperature. Thus the diffusion is so slow that the alloy does not age-harden.

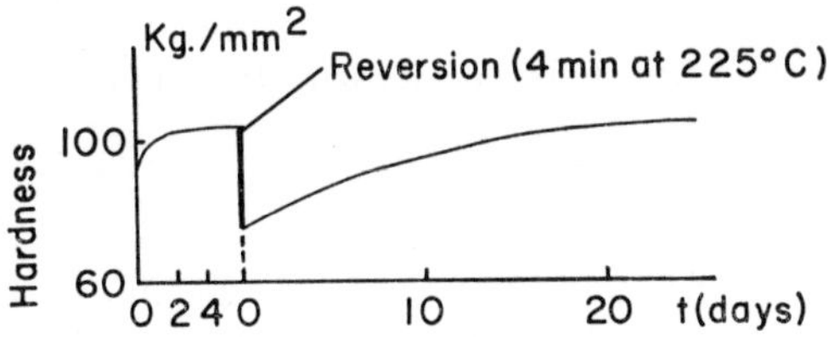

FIG. 42. Comparison of the age-hardening after quench and after reversion.

We see then that the intervention of vacancies in the lattice provides at least a quantitative explanation of a large number of facts. Nevertheless there are still certain difficulties with this theory. It is known that the excess vacancies maintained by the quench disappear very rapidly, in a pure metal, as is shown by the decrease in the resistivity observed immediately after quenching. For aluminum at ordinary temperatures, this recovery time is the order of 1 hour.[221–223]

However the formation of the zones occurs over a much longer interval, of the order of several hours. It is necessary therefore to suppose that vacancies have a longer lifetime in the aluminum alloy. This fact has been verified by resistivity measurements of aluminum alloys containing magnesium and enriched in vacancies by quenching.[222,224] There appears to be an interaction or association between the vacancies and dissolved atoms which prevents the vacancies from disappearing rapidly at grain boundaries or on dislocations.

[221] M. Winterberger, *Compt. rend.* **242,** 128 (1956); **244,** 2800 (1957).
[222] G. Panseri, F. Gatto, and T. Federighi, *Acta Met.* **6,** 198 (1958).
[223] M. Winterberger, *Compt. rend.* **238,** 2175 (1954).
[224] E. C. W. Perryman, *Acta Met.* **3,** 412 (1955).

Quenching is not the only possible way of introducing vacancies into the lattice.[213] Plastic deformation, which accompanies the formation of certain zones, is also able to generate vacancies and thus to maintain a high rate of diffusion during the growth of the zone.

An attempt has been made to link the rate of hardening to the quenching stresses. One of the objections to this is that it is possible to reduce these stresses considerably by using very thin samples without reducing the rate of evolution of the alloy. Moreover, the quenching stresses ought to produce an effect analogous to cold work after the quench. This is not the case, as it will be shown later.

25. Influence of Lattice Faults on the Kinetics of Zone Development

Since it has been shown that the vacancies present after a quench play a role in the kinetics of hardening, it is interesting to examine the influence of other lattice defects created in the matrix after the quench.

The simplest technique is to deform the metal immediately after quenching.

a. *Influence of Deformation*

A study of the hardening curves of specimens of Al-Cu alloys, deformed after quenching, leads to apparently complex results. Gayler[225] concluded that there is an acceleration of hardening, whereas Berghezan[226] found a retardation in certain cases. A recent study of the development of the G.P. zones with x-rays (Graf[135]) now provides a coherent explanation of these phenomena.

The deformation of the matrix favors the formation of nuclei of the precipitate, as we have already indicated. This explains why very large deformation suppresses pre-precipitation in favor of true precipitation (p. 364).

Thus if pre-precipitation after cold-work is to be studied, it is necessary to limit the deformation. Graf[135,227] used a single-crystal of Al-Cu which was stretched less than 30%. He found that G.P. zones were formed during the deformation, for patterns, obtained immediately after the deformation, contained the streaks characteristic of G.P. zones. Zones are also apparent in an undeformed alloy after quench, however, for they develop at room temperature during the time of the x-ray exposure. To render the experiment decisive, Graf[135] compared two crystals quenched under the same conditions. One had been cold-worked

[225] M. L. V. Gayler, *J. Inst. Metals* **72,** 543 (1946).
[226] A. Berghezan, Thesis, University of Paris (1952).
[227] R. Graf and A. Guinier, *Compt. rend.* **238,** 2175 (1954).

after the quench. The two crystals were kept at −25°C during the x-ray exposure. The cold-worked sample definitely gives stronger streaks. Thus it is demonstrated that zones develop during and immediately after cold work.

Furthermore, the streaks are much less intense and more blurred than those given by zones which develop at room temperature in an undeformed single crystal. Thus the zones in the deformed alloy have the same nature but are smaller. Their diameter was estimated to be about 15 A. These zones produce a considerable increase in hardness immediately after deformation. After quenching, the hardness $H = 56.7$ kg/mm^2, whereas after 10% cold work, H is 82.7 kg/mm^2, an increase of 26 kg/mm^2. This hardening is not a simple consequence of lattice deformation for the same cold-work produces an increment of hardness of only 7 kg/mm^2 in an alloy of Al-Cu which has the same composition but is in a structural state which is stable at room temperature (after complete precipitation).

The hardness of the cold-worked sample increases very slightly during aging at room temperature, and reaches its maximum value after 2 hours (Fig. 43).

These facts can be interpreted in the following manner. The development of zones is accelerated because the rate of diffusion is increased as a result of the faults and vacancies created by the deformation. Moreover, the number of zone nuclei increases. It is well known from observations of the platelets of Al-Cu precipitate that the individual dislocations of a subgrain are nucleation sites. Thus it is natural to believe that they can also initiate the formation of the zones. Since dislocations are very numerous in a deformed metal, the result of these two effects is such that a large number of zones form rapidly; as a consequence, they should remain quite small, as is observed. They promote the rapid initial hardening observed during and just after deformation. A major fraction of the excess copper atoms is involved in these small zones and only a few zones can form in aging at room temperature. Therefore, there is hardly any subsequent hardening.

If the alloy is quenched, deformed, and is then annealed at 100°C or 150°C, the hardness decreases slightly, then increases. Compared to an undeformed alloy, however, the hardening is retarded at the beginning (Fig. 44). X-rays show that the small zones formed during cold-hardening disappear and that zones of the size normal for the aging temperature then form, but occur more slowly than for the quenched alloy. There is thus a kind of reversion of the deformed alloy which does not exist for the quenched alloy. The small zones produced by the deformation are not stable; at 100, or 150°C, they dissolve. Normal zones, of a larger dimension, develop but do so more slowly, as one expects after a reversion, because the hardening takes place in a metal in which many of the excess vacancies have disappeared.

These examples at least show that notable progress has been made in our qualitative understanding of the phenomena of hardening in complex conditions. This advance originates from precise information concerning the existence of zones and from the fertile theory of the role played by vacancies in the evolution of the metal during hardening.

b. Influence of Irradiation

Since vacancies can be produced in metals by irradiation, for example with neutrons, it is important to study the influence of irradiation on the

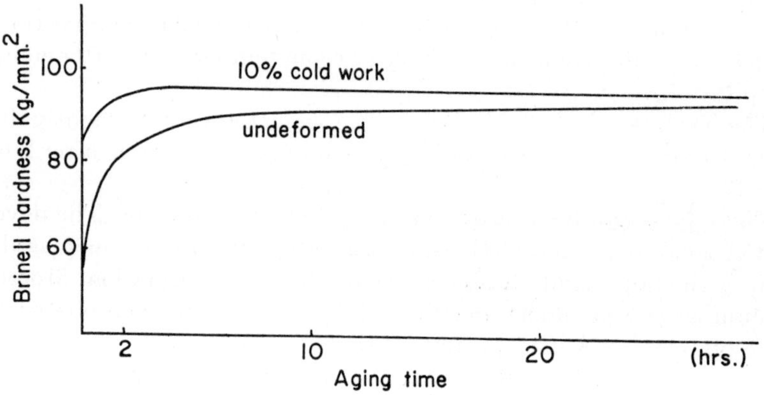

FIG. 43. Age-hardening at 20°C of Al-Cu after cold work.

development of zones. Experiments are difficult, however, for the times of irradiation required are long, and it is necessary in order to isolate the effect of irradiation to cool the alloy to very low temperatures. Various experiments have been reported, but research is only at a preliminary stage; at the moment the results do not seem to be definitive.

The first age-hardening alloy to be studied after irradiation was a Cu-Be alloy. Murray and Taylor[228] irradiated a 2.2% Be alloy with an integrated neutron flux between 10^{17} and 5×10^{19} n/cm^2, the temperature being maintained between 0 and 40°C. They found changes in hardness and resistivity similar to the changes produced by aging at low temperatures (100°C). These changes can be removed by carrying out a reversion treatment at 200–300°C for the resistivity and at 350°C for hardness. Subsequent aging at the reversion temperature is accelerated.

[228] G. T. Murray and W. E. Taylor, *Acta Met.* **2**, 52 (1954).

In addition, shifts of the powder diffraction lines and changes of density have been observed following the irradiation but have not been explained.

The authors conclude that the irradiation induces "precipitation nuclei." This hypothesis is supported by the results of irradiation at low temperature (120°K).[229] The change of resistivity then is one-fourth as large; however, the properties assume values which would result from irradiation at room temperature when the temperature is subsequently raised to room temperature. Thus nucleation proceeds rapidly, at the higher temperature, at all the damage centers introduced by the irradiation regardless of the temperature of irradiation.

The structure of the alloy has not yet been determined, but it is probable that the changes observed are the result of a pre-precipitation which is indicated by the effect of reversion treatment. The experimental

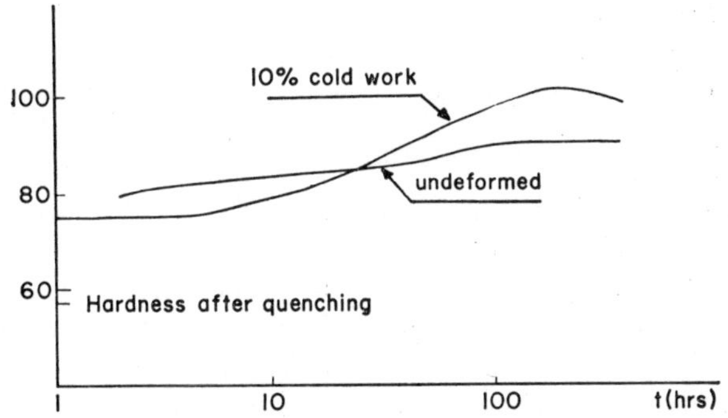

FIG. 44. Age-hardening at 100°C of Al-Cu after cold work.

results may be explained by an increase in the rate of diffusion produced by radiation damage. The pre-precipitation that occurs at 100°C in normal alloys can occur at room temperature in irradiated alloys.

Silcock[230] irradiated single crystals of Al 4% Cu for 25 days. The crystals were first examined with x-rays only 7 to 8 weeks after the end of the irradiation. Therefore, it was only possible to study the effect of irradiation upon those structures that are stable after aging for a few weeks at 80°C. It was found that large G.P. I zones (formed at 130°C) and large G.P. II zones (formed at 165°C) are destroyed by irradiation. In contrast, θ' precipitates (formed at 250°C) are not destroyed. The effect of irradiation on subsequent aging behavior is negligible. Irradiation appears to be very nearly equivalent to a brief heat treatment at 250°C (dissolution of G.P. I and G.P. II zones without formation of θ').

[229] J. W. Cleland, D. S. Billington, and J. A. Crawford, *Phys. Rev.* **91,** 238 (1953).
[230] J. M. Silcock, *Research Rept., Fulmer Research Inst.* R 10/56 (1957).

The opposite effects observed with Cu-Be and Al-Cu can be explained if this equivalence is admitted, for the zones are respectively formed and dissolved in these two alloys at 250°C. It is difficult, however, to justify the idea that an "average temperature" is equivalent to the "thermal spikes" produced by neutron bombardment.

In Al-Ag, Jamison[231] followed the structure of the zones with small-angle x-ray scattering after irradiation. He found that irradiation caused a retardation of zone growth during a heat treatment following irradiation. It was also observed in some cases that the ring of scattering in the small-angle pattern characteristic of the zones in Al-Ag disappears after irradiation. This would indicate a change in the distribution of the Ag atoms in the clusters.

Many experimental questions remain to be answered before a plausible and general mechanism is achieved. Nevertheless, it is beyond doubt that studies of irradiation of age-hardening alloys will produce very interesting results for the understanding of both the mechanism of hardening and radiation damage.

c. *Influence of Impurities in the Matrix*

Hardy[232] has studied the aging of an Al-Cu alloy containing very small additions of lead, cadmium, or indium (of the order 0.05% by weight, i.e., 0.012 atomic per cent of lead). The kinetics of the hardening is altered profoundly by this apparently minute modification. The rate of natural hardening at 25°C and its final value are reduced. By contrast, the rate of hardening and the value attained at 165°C are increased in an appreciable way (Fig. 45). These results indicate that the formation of G.P. zones is both retarded and diminished, whereas the formation of θ'' phase (or G.P. II zones), which is responsible for the hardening at 165°C, is accelerated and increased.[233]

Hardy has shown that, to be effective, the additional element ought to be soluble in aluminum. However, the number of dissolved Cd or In atoms is too small to provide a direct effect by precipitation of a ternary compound of Cd or In with Al and Cu. It must be concluded that the additional elements are capable of influencing the pre-precipitation and normal precipitation of copper in aluminum.

The explanation proposed by Hardy is that the dissolved atoms (Cd, In, Sn) in the matrix, being larger than Al, tend to surround the copper atoms, in order to reduce the lattice distortion. These small groups should be very stable and should diminish the quantity of copper

[231] R. E. Jamison, *Solid State Progr. Rept.* p. 18 (Aug. 1957).

[232] H. K. Hardy, *J. Inst. Metals* **78,** 169 (1950).

[233] A. H. Sully, H. K. Hardy, and T. J. Heal, *J. Inst. Metals* **76,** 269 (1949).

available for the formation of zones. However, even if each atom of indium were surrounded by a shell of 12 neighbors of copper, the number of copper atoms in these groups would only be 5% of the total number. This is insufficient to explain the notable decrease in the hardening. On the other hand, it is difficult to imagine that the number of copper atoms in one group is larger, for the defect in the volume of the copper atoms would be much larger than the excess volume of one impurity atom; thus the group of Cu and impurity atoms would no longer have a reason for being stable.

It is possible to offer another explanation, making use of vacancies if we accept the premise that the diameter of the impurity atom must be

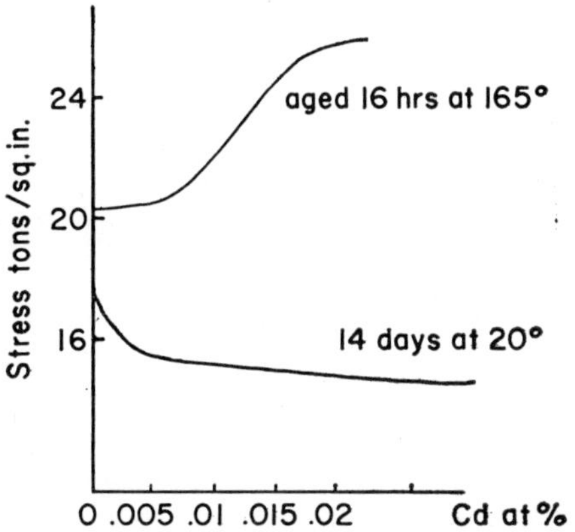

FIG. 45. Influence of Cd on the hardening of Al-4% Cu (and 0.15% Ti) at 20°C and 165°C. Maximum tensile test [H. K. Hardy, *J. Inst. Metals* **78,** 169 (1950)].

larger than the average. We suppose that these impurities find themselves in a position of minimum energy when they are associated with a vacancy. Thus a part of the vacancies are annihilated and are not able to activate the diffusion of copper atoms. The number of impurity atoms (about 10^{-4}) is of the order of magnitude of the number of vacancies in the lattice retained by quenching. In this way, it is possible to explain why a very small amount of impurity is able to have relatively large effect on the rate of diffusion, and hence on the rate of growth of zones.

Hardy[232] indicated that magnesium counter-balances the effect of indium. If the atoms of indium have a larger affinity for magnesium atoms than for vacancies, they will form molecules of an In-Mg complex and release the vacancies, which can then play their normal role in zone formation.

By contrast, the formation of the θ'' phase at 160–180°C would be accelerated because: (1) the effect of vacancies on the rate of diffusion is smaller at this temperature; (2) the nucleation of the θ'' phase is easier since a smaller number of copper atoms is involved in the G.P. I zones, which should dissolve before the θ'' precipitate form.

26. Properties of the Zones

In this section, certain properties imparted to the alloy by the zones will be reviewed and compared with those of an alloy containing true precipitates.

a. Changes in Internal Energy

The formation of zones, like the formation of precipitates, is accompanied by a release of heat. Using calorimetric measurements, it is possible to determine the heat of evolution of the alloy during the various stages of decomposition. Such measurements are delicate because the quantities of heat involved are small, and it is difficult to evaluate the corrections arising from the energy absorbed in the formation of strains.

Experimentally, one measures the instantaneous specific heat during heating of the specimen at a uniform rate[234] or the evolution of heat *versus* time, the alloy being maintained at a constant temperature.[235–239] The first method shows clearly the succession of different processes. However, the interpretation of the specific heat curve is quantitatively difficult. One difficulty originates in the fact that the structures of the alloy during the experiment are uncertain because the rate of heating generally is rapid (1°C per minute).

As an example, consider the curves for the specific heat *versus* temperature obtained by Hirano[239] with Al-38% Ag (Fig. 46). The alloy (curve a) is heated immediately after quench. Up to 150°C there is an evolution of heat (A) which is followed by an absorption (B) (150 − 190°). Then there is a new evolution (C) (210–320°C). The interpretation is clear according to the structures which have been described (Section 17): the first stage corresponds to the formation of zones, the second to their redissolution, the third to the formation of the precipitate. This interpretation is confirmed by the following experiments.

(1) The alloy is aged fully at room temperature before heating

[234] N. Swindells and C. Sykes, *Proc. Roy. Soc.* **A168**, 237 (1938).

[235] G. Borelius, J. Andersson, and K. Gullberg, *Ing. Vetenskaps Akad., Handl.* **169** (1943).

[236] G. Borelius and L. Strom, *Arkiv Mat. Astron. Fysik* **32A**, No. 21 (1945).

[237] J. Suzuki, *Sci. Repts. Research Insts. Tôhoku Univ.* **A1,** 183 (1949).

[238] W. Köster and H. A. Schell, *Z. Metallk.* **43,** 454 (1952).

[239] K. Hirano, *J. Phys. Soc. Japan* **8,** 603 (1953).

(curve b). Here (A) is less important because more zones have been formed previously (B) and (C) are nearly the same.

(2) The alloy is aged at 110°C (curve c). (A) disappears because there, no more zones are formed during heating.

(3) The alloy is annealed at 330°C one hour (curve d). It contains no zones at all and does possess an appreciable amount of precipitate: (A) and (B) are no longer visible and (C) is notably reduced.

Similar but more intricate results have been found with Al-Cu, because there are four successive processes.

Another method consists of measuring the heat evolved by the alloy

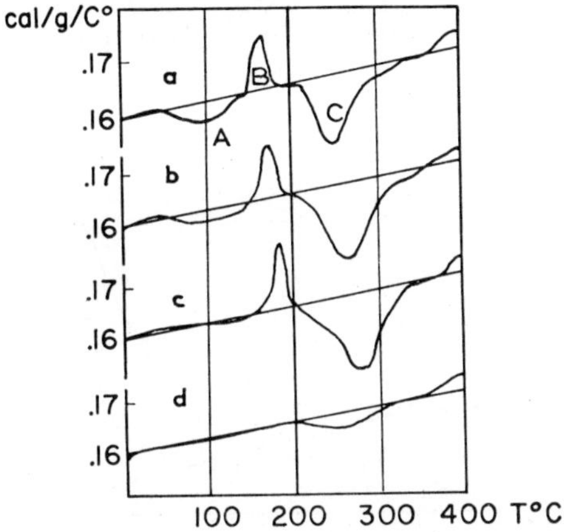

FIG. 46. Specific heat *versus* temperature, for an Al-Ag (38 weight per cent alloy). The base line is calculated using the Neumann-Kopp law [K. Hirano, *J. Phys. Soc. Japan* **8**, 603 (1953)]. a. Alloy after quench. b. Fully aged at room temperature. c. Fully aged at 110°C. d. Annealed 1 hour at 330°C.

during an isothermal transformation. Thus it is possible to measure the heat of formation of the different stages as well as the heat of dissolution of the zones by reversion. According to the results obtained by Suzuki[237] with Al-4.3% Cu, the heat of formation is 1.3 cal/g for the G.P. I zone and 2 cal/g for G.P. II. The heat absorbed during reversion in Al-38% Ag is 1.54 cal/g or 58 (cal/g-atom) whereas the heat of precipitation is 4.0 cal/g or 151 (cal/g-atom).

b. Dimensional Variations of the Crystal

The zone is only a local deformation of the matrix. Thus, theoretically, the zones leave the macroscopic crystal unchanged and, as has been shown (p. 341) do not cause a variation of the parameter of the solid solution measured with x-rays. In fact, very slight dimensional changes are observed. Wassermann and Lanke[240] found a contraction correlated with the development of the zones in an Al-Cu alloy containing 4% Cu. The contraction attained a maximum value of 5×10^{-5} (Fig. 47). This result indicates that the contraction of the planes within the zones is not compensated exactly in the surrounding matrix. Thus there should be a change in lattice parameter. Since the precision of the parameter measurements is only about 10^{-4}, however, a negative result,[139,140] does not contradict the dilatometric result.

The possibility of hardening a metal without appreciable deformation is one of the major advantages of the age-hardening of light alloys. The piece can be worked in the soft state and then hardened; in the ideal case all the dimensions remain constant to very good accuracy. Nevertheless,

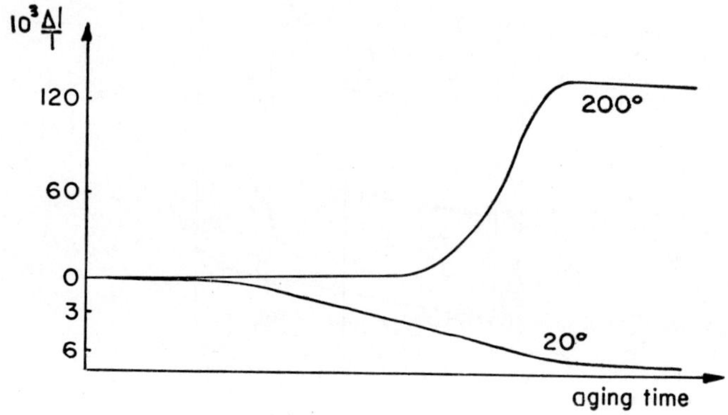

FIG. 47. Variation of length of an Al 4% Cu alloy during aging at 20°C (zones) and 200°C (precipitate).

very light superficial deformations have been observed in the case of Al-Cu, after natural aging under appropriate conditions of polishing.[241]

The situation is different for Cu-Be alloys. At the stage corresponding to the maximum hardness, the surface of the crystal is covered with several systems of parallel ripples which are easily visible with the naked eye.[242,243]

On the other hand, the formation of a true precipitate changes the volume of the metal appreciably to a value which can be considered simply as equal to the sum of the volume of the remaining matrix and of the precipitate. Thus, it is possible to calculate the dilatation from the

[240] J. C. Lankes and G. Wassermann, *Z. Metallk.* **41,** 381 (1950).
[241] J. Calvet, A. Guinier, and P. Jacquet, *J. Inst. Metals* **45,** 121 (1939).
[242] J. S. Bowles and W. J. Tegart, *Acta Met.* **3,** 590 (1955).
[243] J. S. Bowles and J. K. MacKenzie, *Acta Met.* **2,** 129 (1954).

known structures of the two phases. For example, in the Al-Cu alloy aged at 200°C, the θ' precipitate is formed and causes an increase in volume (Fig. 47). When precipitation is complete, the measured dilatation is 0.136%, in good accord with the calculated dilatation, namely 0.117%.[240] The proportion of precipitate can thus be calculated from the measured dilatation and is found to be in good agreement with the proportion determined from the relative intensity of the x-ray diffraction spots.[244]

In other words, zones have an effect which is quite distinct from that of precipitates. The difference can be detected at temperatures where the two processes succeed each other. The two stages are then clearly apparent in the dilatometric curve.

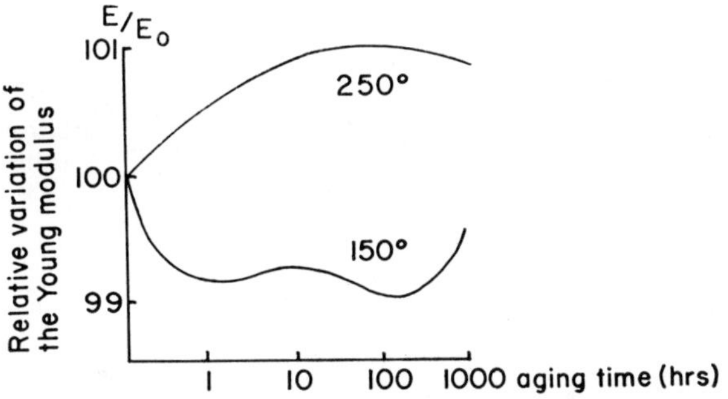

FIG. 48. Variation of the Young modulus during aging of Al 20% Ag alloy at 150°C and 250°C [K. Tanaka, H. Abe, and K. Hirano, *J. Phys. Soc. Japan* **10,** 454 (1955)].

c. Mechanical Properties

The change of the mechanical properties of the alloy due to zone formation gives rise to the age-hardening phenomenon which is of practical importance.

(*1*) *Elasticity.* Hirano[245] has studied the variations of the Young modulus of an Al-20% Ag alloy during aging. It appears very clearly that the pre-precipitation decreases the modulus whereas the precipitation increases it. The aging curves obtained at 250°C (pure precipitation) and at 150°C (pre-precipitation followed by precipitation) are shown in Fig. 48.

(*2*) *Plasticity.* In most cases, there is an evident correlation between the presence of zones as revealed by small angle scattering of x-rays and the hardness of the metal. Thus, the shape of the (100) streaks is practically unchanged during aging of Al-Cu at room temperature, but their

[244] A. Guinier, *Z. Elektrochem.* **56,** 468 (1952).
[245] K. Tanaka, H. Abe, and K. Hirano, *J. Phys. Soc. Japan* **10,** 454 (1955).

absolute intensity increases. The shape and the average dimensions of the G.P. I zones are constant but their number increases in proportion to the intensity of the scattering. Figure 49 shows that the curve giving hardness as a function of time is parallel to the curve showing the number of zones. On the other hand, as mentioned previously the zones disappear during reversion of Al-Cu or Al-Zn at the same time that the metal becomes soft.

The hardness *versus* time curves exhibit different aspects during aging at various temperatures. Gayler and Parkhouse[246] have demonstrated

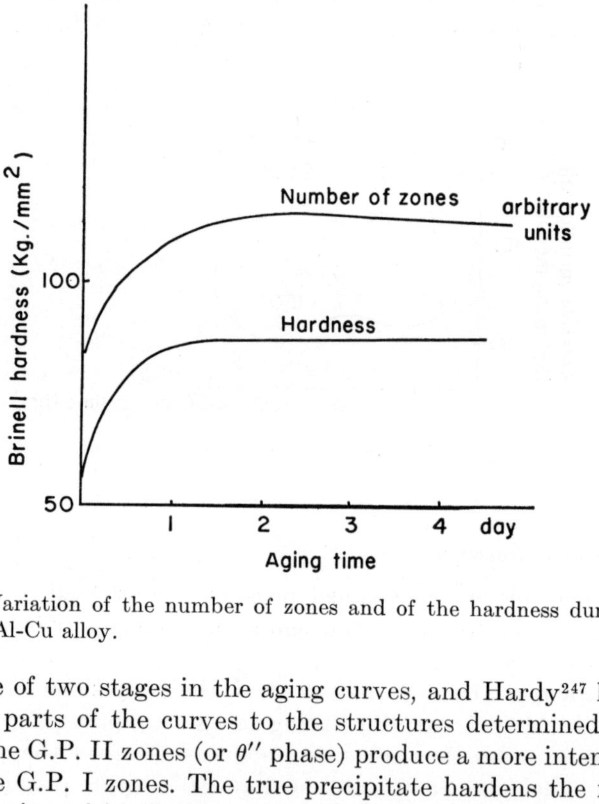

FIG. 49. Variation of the number of zones and of the hardness during the age-hardening of Al-Cu alloy.

the presence of two stages in the aging curves, and Hardy[247] has related the various parts of the curves to the structures determined by x-rays (Fig. 50). The G.P. II zones (or θ'' phase) produce a more intense hardening than the G.P. I zones. The true precipitate hardens the metal only at the beginning of its development when the precipitate particles are very numerous and very small. The coalescence of the θ' precipitate into large grains corresponds to a softening of the metal. The θ precipitate, which is not coherent with the matrix, also produces a softening.

The isochronal curves for hardness (Fig. 51) in Al-Ag alloy (Köster

[246] M. L. V. Gayler and R. Parkhouse, *J. Inst. Metals* **66,** 67 (1940).

[247] H. K. Hardy, *J. Inst. Metals* **79,** 321 (1951); **82,** 236 (1953).

and Braumann[83]) also show clearly the two stages corresponding to zones and precipitates. These also correspond to cold- and warm-hardening. Zones have a much smaller influence than small precipitates in this alloy probably because the deformation of the lattice in the zones is particularly weak.

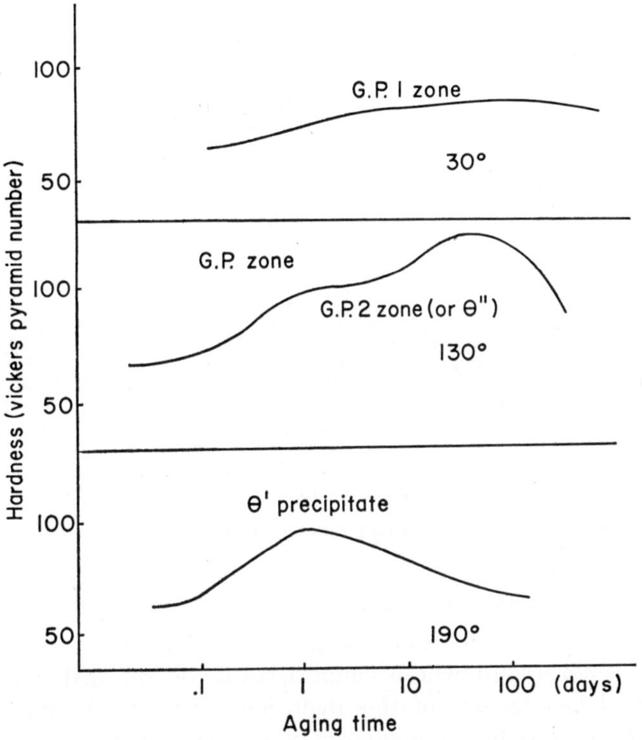

FIG. 50. Hardening curves of Al-Cu alloy in correlation with the structures [H. K. Hardy, *J. Inst. Metals* **79,** 321 (1951); **82,** 236 (1953)]. Al-4% Cu alloy aged at 30°C, 130°C, and 190°C.

Hardness is the mechanical property observed most widely in age-hardening studies, for it is very simple to measure. However, information of more direct value for quantitative work is given by the stress-strain curve during deformation, for example, during tension.

Figure 52 gives the tensile curves of a single crystal Al-3.5% Cu crystal in three states, namely, just after quenching, after complete aging (zones), and after complete precipitation (Carlsen and Honeycombe[248]).

In the quenched state, isolated atoms of copper are dispersed in the

[248] K. M. Carlsen and H. W. K. Honeycombe, *J. Inst. Metals* **83,** 449 (1955).

solid solution. The copper atoms are gathered in very numerous and very small zones in the cold-hardened state, whereas they are assembled in large grains incoherent with the matrix and separated from each other by several tens of microns in the precipitated state. Since the stress-strain curves are different in these three states, the resistance to movements of dislocations depends on the mode of segregation of the copper atoms for the same average concentration of copper. This has been explained by Mott and Nabarro[249–251,151] and later by Friedel.[252]

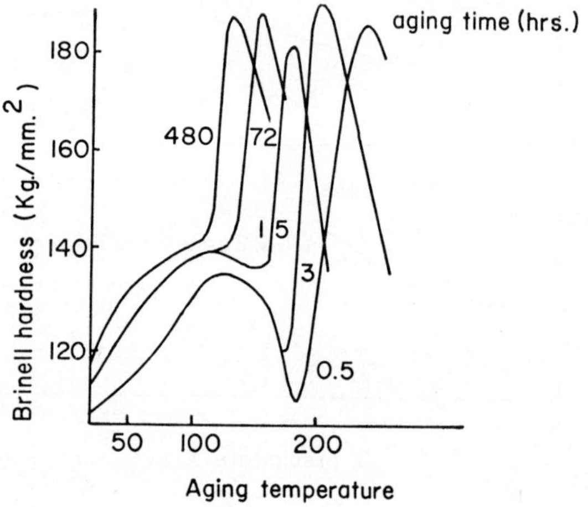

FIG. 51. Isochromal curves of hardness of Al-38% Ag alloy [W. Köster and F. Braumann, *Z. Metallk.* **43,** 193 (1952)].

In the region around each inclusion, whether an isolated atom or G.P. zone, a constraint whose value depends on the distance from the inclusion and on the size of this inclusion exists. If the inclusions are assumed to be distributed regularly in the crystal, the constraints are periodic with a period equal to the average distance between two neighboring inclusions. The maximum constraining stress is of the order of

$$\sigma_0 = \mu \eta f \tag{26.1}$$

where μ is the shear modulus, η the relative difference of the diameters of Cu and Al atoms, and f the fraction of Cu atoms segregated in the inclusions. Thus σ_0 is independent of the mode of segregation for a given value of f. However, the propagation of dislocation lines depends on the

[249] N. F. Mott and F. R. N. Nabarro, *Rept. Bristol Conf. on Strength of Solids, 1947* p. 11 (1948).
[250] N. F. Mott, *Phil. Mag.* [7] **43,** 1151 (1952).
[251] F. R. N. Nabarro, *Proc. Phys. Soc.* **58,** 669 (1946).
[252] J. Friedel, "Les Dislocations," p. 242. Gauthiers Villars, Paris, 1956.

distance between inclusions. This distance is so short in the homogeneous solid solution that the dislocation line remains linear during its motion and encounters an average constraint which is nearly zero.

When G.P. zones are separated by a distance of the order of 25–05 atomic distances, the dislocation line is deformed by the zones and bulges between them. To pass through a zone the line must jump from one side to the other (Fig. 53). The shear necessary would be of the order of σ_0, i.e., $\sim\frac{1}{2}\ 10^{-2}$ for Al-4% Cu. Experiment gives a critical shear of this order of magnitude.

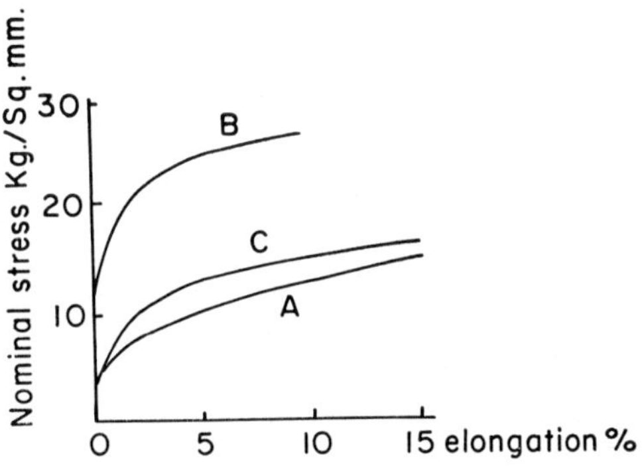

FIG. 52. Tensile curves of a single Al-Cu crystal as quenched (A), containing zones (B) and precipitates (C) [K. M. Carlsen and H. W. K. Honeycombe, *J. Inst. Metals* **83,** 449 (1955)].

In the case of incoherent precipitates, constraints no longer exist in the matrix, for the disorder in the lattice is concentrated at the interface between the matrix and precipitate. The dislocation line moves more easily under the relatively weak constraints in the matrix; it is simply anchored by the precipitates which cut across its glide plane. Friedel[252] has pointed out that these precipitates are likely to be separated by a distance much larger than the average distance of nearest neighboring grains.

According to the preceding theory, the hardening is more important the greater the difference in diameter between solute and solvent atoms. Thus the hardening in Al-Ag is much less than in Al-Cu or Cu-Be. It does exist, however (see Fig. 51), even though the relative difference of the diameters of Al and Ag is only 0.005. Thus the size factor is not the only one defining the stresses which oppose the movement of dislocations. Let us note, however, that the correlation between hardening and zones is much less clear for Al-Ag than for Al-Cu or Al-Zn. Thus the hardness can diminish during reversion, or analogous thermal treatments, whereas the small-angle scattering shows an increase in the dimensions of the zones; it is possible in Al-Ag that all zones do not affect the mechanical properties in the same way. As suggested by Belbeoch and Guinier,[93] the small zones could be more effective than the large ones, or, according to Gerold,[97] the more effective zones might be the ones in which the contrast in silver concentration between the core and surrounding shell is strongest.

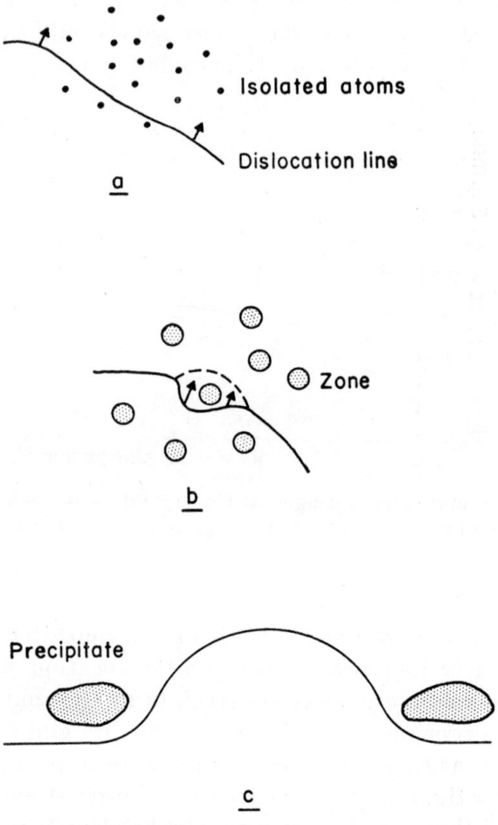

FIG. 53. Influence of obstacles on dislocations lines. a. Isolated atoms. b. G.P. zones. c. Precipitates [J. Friedel, "Les Dislocations," p. 242. Gauthier-Villars, Paris, 1956].

An interesting experiment carried out by Jan[253] reveals information concerning the mechanism of deformation of an aged solid solution containing zones. He submitted an Al-30% Ag (and also 30% Zn) alloy, fully aged at room temperature, to cold-rolling (50 to 95%) and compared the small angle x-ray scattering before and after cold-work in order to study the shapes of the zones in the two states. He found that the scatter-

[253] J. P. Jan, *J. Appl. Phys.* **26**, 1921 (1955).

ing in the reciprocal space is no longer spherically symmetrical after deformation. Instead of being bounded by a sphere, the region of scattering has the form of an oblate ellipsoid, the small axis of which is directed along the rolling direction. Furthermore, the ring in the diffuse spot is no longer visible after rolling; the scattered intensity does not seem to vanish at zero angle. Thus the shape of the zones as well as the distributions of the Ag (or Zn) atoms inside the zone has been altered by plastic deformation. Jan found that the zone is elongated in the rolling direction, flattened in the direction normal to the rolling plane, and is not changed in the transverse direction. He made a simple and rough calculation of the three principal "radii of gyration" of the ellipsoid for the average zone after cold-working.[254] For Al-30% Ag and 50% of deformation by cold rolling, Jan found $R_{R.D.} = 14.1$ A, $R_{N.D.} = 9.0$ A, $R_{T.D.} = 11.6$ A. Before deformation, the radius of gyration of the spherical zone was 11.6 A.

These results can be explained by the action of glide planes on zones. Let us suppose first that slip is concentrated on discrete planes whose separation corresponds to the distance often observed between slip lines in slip bands (Heidenreich and Shockley[255]), that is about 200 – 400 A. Some zones are cut in two distinct clusters, but most will be untouched. Jan has shown that this hypothesis does not account for his experimental results.

On the other hand, the hypothesis of homogeneous slip[256] in layers sufficiently thick to contain a few entire zones is acceptable because the zone is elongated in the direction of slip (Fig. 54).

This interpretation is confirmed by observations on coherent cubic precipitates in the alloy Ni-Al which are sufficiently large to be visible by electron microscopy.[257] Some cubes are cut by slip planes and some are deformed by homogeneous shear.

Such an effect has not been found in the cases of the platelike zone in Al-Cu. When a single crystal containing G.P. zones is deformed by tension, the central streaks characteristic of the zones are not modified.

This result has been confirmed by the electron microscopic observations of Thomas and Nutting.[258] The visible slip lines are separated widely in the G.P. I stage and the slip lines are not visible in the G.P.

[254] A. Guinier, "Théorie et Technique de la Radiocristallographie," p. 645. Dunod, Paris, 1956.

[255] R. D. Heidenreich and W. Shockley, *Rept. Bristol Conf. on Strength of Solids, 1947* p. 57 (1948).

[256] T. Suzuki, *Sci. Repts. Research Insts. Tôhoku Univ.* **A6,** 309 (1954).

[257] Y. Baillie, *Rev. universelle mines* **12,** 507 (1956).

[258] G. Thomas and J. Nutting, *J. Inst. Metals* **86,** 7 (1957).

II stage. Some of the precipitate particles are deformed by a slip line, when the θ' phase is present.

d. Electrical Properties

Measurements of resistivity, which are simple and rapid, have often been used to follow the kinetics of the evolution of age-hardening alloys. It is known that the resistivity of a homogeneous solid solution increases with the proportion of solute atoms because the electronic waves are scattered by the atoms substituted in the lattice. The latter constitute defects in the periodicity. When there is a segregation of atoms, their influence depends on the mode of segregation, but this influence is

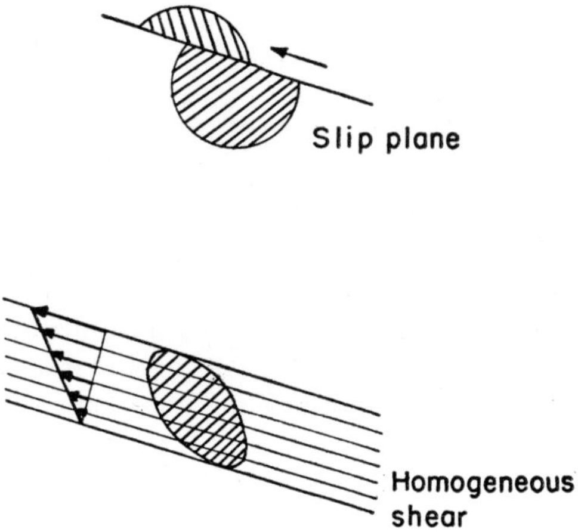

FIG. 54. Spherical zone deformed by slip [J. P. Jan, *J. Appl. Phys.* **26,** 1921 (1955)] or homogeneous shear.

complex. The variation of resistivity per substituted atom is variable. It depends not only on the dimensions of the cluster, but also on the lattice distortions produced by the cluster. Thus it is possible to understand why the zones possessing distortions (Al-Cu type) have different electrical properties from those present in Al-Ag which do not.

In Al-Cu, it has been found that the zones increase the resistivity slightly.[259] On the other hand, the resistivity decreases when the precipitates form (Fig. 55). The same effect is observed with Cu-Be[260,228] in which the zones are also accompanied by large distortion.

The resistivity in the Al-Ag alloys is always smaller than the resistivity of the homogeneous solid solutions. In the isochronal curves determined by Köster and colleagues[83] (Fig. 56), however, the two stages of zone

[259] W. L. Fink and D. W. Smith, *Trans. AIME* **137,** 95 (1940).
[260] A. G. Guy, C. S. Barrett, and R. F. Mehl, *Trans. AIME* **175,** 216 (1948).

formation and precipitation appear clearly and are similar to the isochronal curves for hardness (Fig. 51). Turnbull and Treaftis[94] have succeeded in quenching certain Al-Ag alloys sufficiently rapidly to obtain the resistivity corresponding to the homogeneous solution at low temperatures. The alloy evolves very quickly, however. Its resistivity

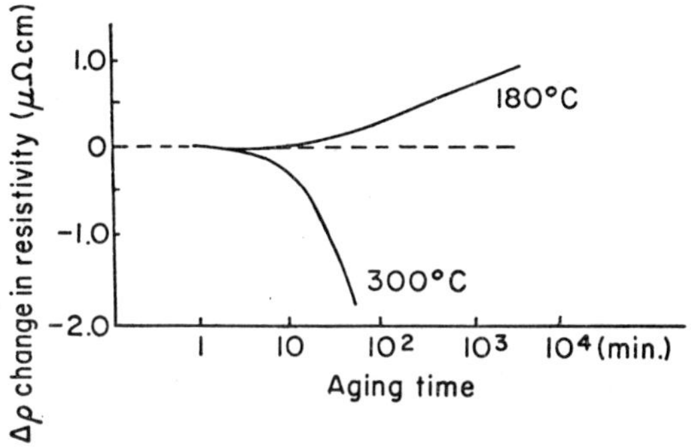

FIG. 55. Resistivity changes during aging of Cu-Be at 125° and 300°C [G. T. Murray and W. E. Taylor, *Acta Met.* **2,** 52 (1954)].

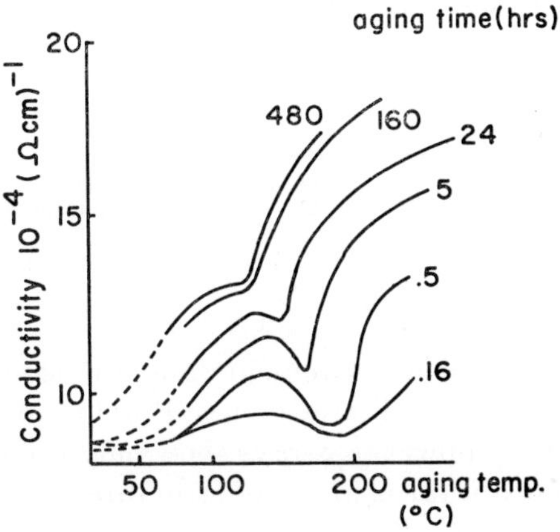

FIG. 56. Isochronal curves for conductivity of Al-Ag alloy [W. Köster and F. Braumann, *Z. Metallk.* **43,** 193 (1952)].

decreases in several minutes even at temperatures between −60°C and 0°C (Fig. 57). These results show it is possible that zones can form at low temperatures. The zones probably are very small and their structure has not yet been confirmed by x-rays.

Borelius[261] measured the resistivity of an Al-Ag alloy as a function of temperature in the region where the homogeneous solid solution is at equilibrium. When the alloy was quenched to 20°C, he found a resistivity

[261] G. Borelius, *Rept. Bristol Conf. on Defects in Crystalline Solids, 1954* p. 169 (1955).

clearly smaller than the value for the homogeneous solution obtained by extrapolating to 20°C. The quenching did not prevent the formation of some zones. The resistivity of the quenched alloy remains constant during aging at room temperature and varies reversibly, as does that of a metal with a stable structure if it is heated to moderate temperatures (~100°C).

Borelius also noted that the resistance decreases, tending toward a certain limit if the alloy is held at 100°C. Finally, the resistivity decreases after a treatment at 250°C. These results show that the zones of Al-Ag diminish the resistivity more when they are more voluminous. Furthermore, precipitates have a still larger influence. Although there is a general

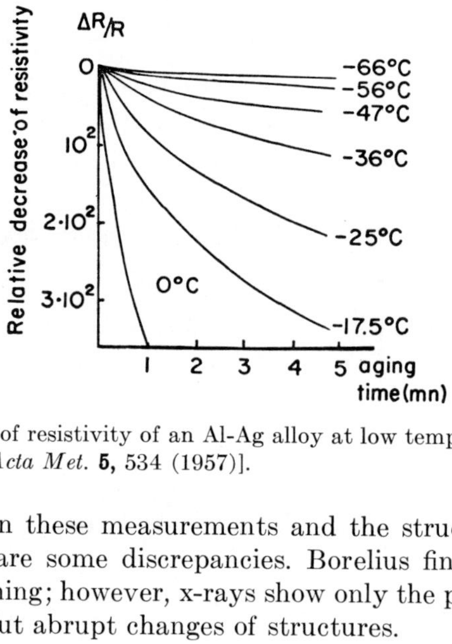

FIG. 57. Decrease of resistivity of an Al-Ag alloy at low temperature [D. Turnbull and H. N. Treaftis, *Acta Met.* **5**, 534 (1957)].

agreement between these measurements and the structures determined by x-rays, there are some discrepancies. Borelius finds distinct stages during cold-hardening; however, x-rays show only the progressive growth of the zones without abrupt changes of structures.

e. Magnetic Properties

They are interesting in the case in which the decomposition products of solid solution are ferromagnetic. This occurs in ternary alloys such as Cu-Ni-Fe, Fe-Ni-Al, Cu-Ni-Co, the decomposition of which is characterized by the occurrence of the structure which produces "side bands." The description of the successive stages (precipitation, metastable and stable precipitates) has been given in Section 21.

In a recent study Biedermann and Kneller[262] have drawn some important conclusions concerning the mechanism of evolution of the

[262] E. Biedermann and E. Kneller, *Z. Metallk.* **47**, 760 (1956).

alloy Cu-Ni-Fe (or Cu-Ni-Co) from the variations of their magnetic properties.

(1) One of the two equilibrium phases is ferromagnetic (Ni-Fe) and the other, namely copper-rich Cu-Ni-Fe, is not. The magnetic moment of the mixture of the two phases at saturation is simply $I = c_1 I_1$, c_1 being the proportion of the magnetic phase whose saturation moment is I_1. The magnetic moment I does not depend on the size of the ferromagnetic grains. Experiment shows that, when the alloy is quenched from the single-phase domain (Fig. 58) and annealed at 600°C, the saturation moment is practically constant. In fact, just after quenching,

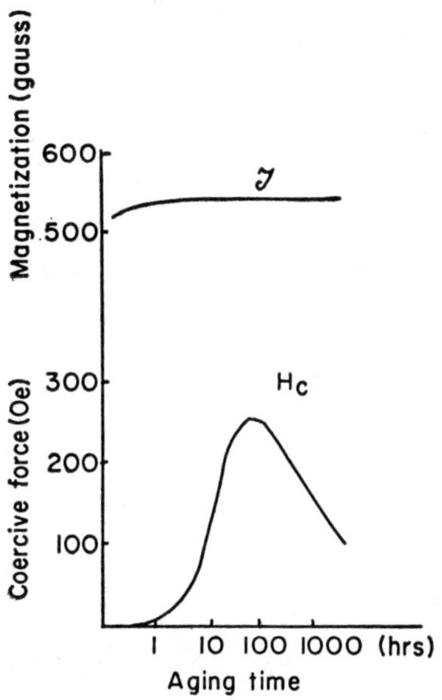

FIG. 58. Variation of the magnetization and of the coercive force [E. Biedermann and E. Kneller, *Z. Metallk.* **47,** 760 (1956)].

the measured value is in agreement with the one calculated with use of the hypothesis of total decomposition.

(2) The Curie temperature of the alloy changes notably from the value 200°C just after quench to a value greater than 500°C after 24 hours of annealing at 600°C. The Curie temperature is a function of the size of the ferromagnetic grain when it is very small. Such variations have been observed for layers of pure nickel, the thickness of which varies from 20 to 200 A.

The conclusion of Biedermann and Kneller is that the segregation of copper is effective just after the quench. They have shown that the diffusion is sufficiently rapid to account for the formation of clusters which are 50 to 100 A in diameter during the quench. These initial clusters grow during annealing and transform into the alternate lamellas parallel to (100) which give rise to "side bands" in the x-ray patterns. The total volume of the ferromagnetic phase does not vary; however, the thickness of the ferromagnetic lamellas increases with the annealing time.

(3) The coercive force which is very weak (0.5 gauss) after the quench (Fig. 58) increases during annealing to a value as high as 280 gauss. It then decreases again. The apparent remanent magnetization also passes through a maximum. Thus, after suitable heat treatment, the alloy may be used for permanent magnets. The maximum of the coercive force does not correspond to a well-defined structure. It occurs during the pre-precipitation in some alloys. In other compositions it occurs when the two coherent tetragonal phases are present. The high value of the coercive force is associated with the small size of the ferromagnetic particles, which constitute a single magnetic domain with the magnetic anisotropy of the thin platelets. The increase in coercive force is associated with the growth of these platelets from the initial clusters which are nearly isotropic. The decrease originates in the thickening of the platelets, which cease to be single domains. The coercive force calculated from the dimensions of the platelets deduced from side-bands patterns (Section 21) and electron micrographs are in good agreement with experiment.

The important new feature revealed by magnetic measurements is the existence of clusters immediately after quenching.

Unfortunately, it is not possible to verify this conclusion with x-rays, because the contrast between the scattering factors of the different atoms is far too small to produce small-angle scattering of observable intensity. Furthermore, the situation is about the same in all the known alloys of the same type. One could expect to observe scattering arising from the deformation of the lattice inside the clusters, but nothing of the kind actually is observed. It is likely that the displacement of the atoms in the initial clusters is too irregular, and thus the scattering regions are very extended and of very weak intensity. Thus nothing is observable before the formation of the regular lamellas parallel to the {100} planes, which have successively enlarged and diminished interplanar spacings. Such a regular alternation reduces the lattice strains. This decrease in energy may be the source of the structure.

V. Acknowledgments

The author wishes to express his sincere thanks to M. R. Graf for helpful discussions on age-hardening alloys and Dr. J. B. Cohen who very kindly translated the French manuscript. He is deeply grateful to Professor F. Seitz for his important contribution in bringing the text into its final form.

Electronic Spectra of Molecules and Ions in Crystals

Part II. Spectra of Ions in Crystals

DONALD S. MCCLURE
RCA Laboratories, Princeton, New Jersey

I. Theory of Spectra of Ions in Crystals

There are several types of ions whose optical spectra in crystals are best described in the tight binding approximation. One such class is that of transition metal ions. These have a partially filled inner *d* shell in their ground state, and their excited states of interest in solids arise principally from a rearrangement of the coupling between the inner electrons. The crystal field produces splittings of these states on the order of 10,000 cm^{-1}. The lanthanide and actinide elements are similar to the transition elements except that they have inner *f* orbitals which are even less influenced by the crystal field than are *d* orbitals. The crystal field splittings are on the order of 100 to 1000 cm^{-1}. Spectra arising from inner electrons are called crystal field spectra. A second class are complex ions consisting of a group of atoms. They are perhaps more properly considered as molecules, yet there is a continuous transition from so-called single ions to molecular ions. It is interesting to explore this transition region. The principal type of spectrum encountered in ions of this type is called a charge transfer spectrum, since it corresponds roughly to the transfer of charge from a "central ion" to an outer one, or from an outer one inward. Optical spectra also arise from the ions at the ends of the transition groups, which have filled inner shells, but have somewhat localized outer *s* and *p* orbitals.

The coupling between identical ions in the crystal will not be considered explicitly in this article. For methods that may be generalized to include cases of weak coupling between ions, see Vol. 8 of this series.* In a pure compound, if the coupling is strong, the methods of band theory, which are outside the scope of this article, must be used.

The following theoretical sections contain sufficient information for the interpretation of spectra and will provide a guide to the literature where fuller accounts and details may be found.

1. Theory of Crystal Field Spectra

Crystal field spectra are recognized by their low intensity and their occurrence in the low-energy spectral region (below 50,000 cm^{-1}). Their low intensity is a consequence of their origin in $d \rightarrow d$ or $f \rightarrow f$ transitions and the resulting violation of the selection rule, $\Delta l = \pm 1$, which applies to electric dipole transitions in the free atoms. The spectra occur in the region of the low-energy transitions of the free ion to which they are related. The more intense and energetic charge transfer spectra obscure high-energy crystal field transitions.

* D. S. McClure, *Solid State Phys.* **8,** 1 (1958).

The crystal field theory was first treated in detail by Bethe[1] and his paper remains an important reference. Schlapp and Penney[2] and Van Vleck[3] used it at an early date to explain magnetic properties. Van Vleck[4] demonstrated later the relationships between Pauling's valence bond theory and the crystal field theory. The first application to electronic absorption spectra in the visible was also made by Van Vleck,[5] and by Finkelstein and Van Vleck.[6] They explained the Zeeman effect in chrome alum and the position of the sharp weak absorption bands showing the Zeeman effect. Hartmann and his group became interested in the spectral work in the late 1940's and carried out several calculations for the purpose of interpreting optical absorption spectra in $3d$ transition metal ions.[7,8] The rising interest in paramagnetic resonance spectra stimulated more work on crystal field theory in this period. Applications of crystal field theory to this area are reviewed by Bleaney and Stevens.[9] Further theoretical work directed toward interpretation of spectra was carried out by Orgel,[10–12] Tanabe and Sugano,[13] Jorgensen,[14,15] Ballhausen[16,17] and Owen.[18] More recently there have been many others to which reference will be made later. Most of these workers concentrated on the $3d$ transition ions. Work on the rare earths seems to have been pursued almost independently by others, although basically, everything follows from Bethe's paper.[1] Hellwege[19] carried out an extensive theoretical analysis of the crystal field effect in rare earths. He extended Bethe's work and established a system of notation for symmetry groups which he and some others have used in discussing experimental results. A most helpful extension

[1] H. A. Bethe, *Ann. Physik* [5] **3,** 133 (1929).
[2] R. Schlapp and W. G. Penney, *Phys. Rev.* **41,** 194 (1932); **42,** 666 (1932).
[3] J. H. Van Vleck, *Phys. Rev.* **41,** 208 (1932).
[4] J. H. Van Vleck, *J. Chem. Phys.* **3,** 803, 807 (1935).
[5] J. H. Van Vleck, *J. Chem. Phys.* **8,** 787 (1940).
[6] R. Finkelstein and J. H. Van Vleck, *J. Chem. Phys.* **8,** 790, (1940).
[7] F. E. Ilse and H. Hartmann, *Z. physik. Chem. (Leipzig)* **197,** 239 (1951).
[8] F. E. Ilse and H. Hartmann, *Z. Naturforsch.* **6a,** 751 (1951).
[9] B. Bleaney and K. W. H. Stevens, *Repts. Progr. in Phys.* **16,** 108 (1953).
[10] L. E. Orgel, *J. Chem. Phys.* **23,** 1004 (1955).
[11] L. E. Orgel, *J. Chem. Phys.* **23,** 1819 (1955).
[12] L. E. Orgel, *J. Chem. Phys.* **23,** 1824 (1955).
[13] Y. Tanabe and S. Sugano, *J. Phys. Soc. (Japan)* **9,** 753, 766 (1954).
[14] C. K. Jorgensen, *Kgl. Danske Videnskab. Selskab Mat.-fys. Medd.* **29,** No. 7 (1955).
[15] C. K. Jorgensen, *Kgl. Danske Videnskab. Selskab Mat.-fys. Medd.* **30,** No. 22 (1956).
[16] C. J. Ballhausen, *Kgl. Danske Videnskab. Selskab Mat.-fys. Medd.* **29,** No. 4 (1954).
[17] C. J. Ballhausen, *Kgl. Danske Videnskab. Selskab Mat.-fys. Medd.* **29,** No. 8 (1955).
[18] J. Owen, *Proc. Roy. Soc. (London)* **A227,** 183 (1955).
[19] K. H. Hellwege, *Ann. Physik* [6] **4,** 95, 127, 136, 143 (1948); **4,** 357 (1949).

of crystal field theory in this area is the work of Stevens,[20] Elliott and Stevens,[21–23] and Judd.[24]

The extensive literature on crystal field theory and its applications has been reviewed in the recent article by Moffitt and Ballhausen.[25]

a. A Simple Example

The nature of the crystal field approximation is illustrated by a simple but important example. Let us consider a single electron in a 3*d* orbital as in Ti^{+++}. In the free atom the *d* electron may occupy any of the five *d* orbitals with equal energy. These orbitals may be described as d_{xy}, d_{xz}, d_{yz}, $d_{x^2-y^2}$, d_{z^2} and are pictured in Fig. 1. They are appropriate in this form for cubic symmetry rather than for axial symmetry which is used in atomic spectroscopy. If negative electric charges are now placed at $\pm x$, $\pm y$, and $\pm z$ at an equal distance from the Ti^{+++} ion, an electron in one of the *d* orbitals is repelled by an amount depending upon the distribution of the orbital in space. Since the first three, the *t* orbitals, point away from the coordinate axes, electrons in them are repelled less than if they were in the two *e* orbitals. As a result, the five *d* orbitals are no longer degenerate in the field of the six point charges, but are split into a lower group of three, called *t* orbitals and an upper group of two, called *e* orbitals.

TABLE I. CHARACTER TABLE FOR THE GROUPS T_d AND O

T_d / O	E / E	$8C_3$ / $8C_3$	$3C_2$ / $3C_2$	$6S_4$ / $6C_4$	$6\sigma_d$ / $6C_2'$	Representations of some tensor components
A_1	1	1	1	1	1	$xx + yy + zz$
A_2	1	1	1	-1	-1	
E	2	-1	2	0	0	$xx + yy - 2zz,\ xx - yy$
T_1	3	0	-1	1	-1	x, y, z in 0; $R_xR_yR_z$
T_2	3	0	-1	-1	1	x, y, z in T_d; xy, xz, yz

Spinor representations

	E	R	$8C_3$	$8RC_3$	$3C_2 + 3RC_2$	$6C_4$	$6RC_4$	$6C_2' + 6RC_2'$
$E_{\frac{1}{2}}$	2	-2	1	-1	0	$\sqrt{2}$	$-\sqrt{2}$	0
$E_{\frac{5}{2}}$	2	-2	1	-1	0	$-\sqrt{2}$	$\sqrt{2}$	0
G	4	-4	-1	1	0	0	0	0

Notes to table: $O_h = O \times i$; E = identity; C_3 = threefold axis; C_4 = fourfold axis; $C_2 = C_4^2$; C_2' = twofold (110) axes; R = rotation by 2π.

[20] K. W. H. Stevens, *Proc. Phys. Soc.* (*London*) **A65**, 209 (1952).
[21] R. J. Elliott and K. W. H. Stevens, *Proc. Roy. Soc.* (*London*) **A215**, 437 (1952).
[22] R. J. Elliott and K. W. H. Stevens, *Proc. Roy. Soc.* (*London*) **A218**, 553 (1953).
[23] R. J. Elliott and K. W. H. Stevens, *Proc. Roy. Soc.* (*London*) **A219**, 387 (1953).
[24] B. R. Judd, *Proc. Roy. Soc.* (*London*) **A227**, 552 (1955).
[25] W. Moffitt and C. J. Ballhausen, *Ann. Rev. Phys. Chem.* **7**, 107 (1956).

The labels t and e indicate the representation to which the orbital belongs in the group of the octahedron, $O_h (= O \times i)$. Character tables are given for the groups O and T_d in Table I. The t orbitals belong to T_2 and the e orbitals to E in group O, or to T_{2g} and E_g in O_h. The t orbitals are

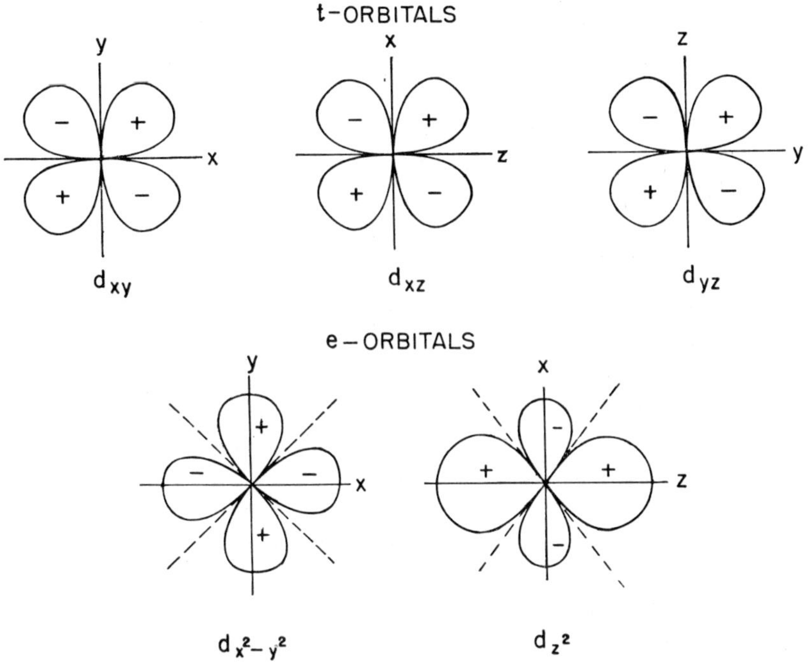

FIG. 1. Contours of constant amplitude of the d functions. Above are the three t orbitals, which differ from each other only in orientation, and below are the two e orbitals. These contours do not reach the origin, as they appear to do in these drawings, but for the rather small amplitude found in the bonding region, the contour approaches the nucleus closely. The actual d functions are:

$$\begin{aligned} d_{xy} &= 1/\sqrt{2}\,(\phi_{22} - \phi_{2-2}) = R_{3d}(r) \cdot \sqrt{15}/2\,\sqrt{\pi}\; xy \\ d_{xz} &= 1/\sqrt{2}\,(\phi_{21} + \phi_{2-1}) = R_{3d}(r) \cdot \sqrt{15}/2\,\sqrt{\pi}\; xz \\ d_{yz} &= 1/\sqrt{2}\,(\phi_{21} - \phi_{2-1}) = R_{3d}(r) \cdot \sqrt{15}/2\,\sqrt{\pi}\; yz \\ d_{x^2-y^2} &= 1/\sqrt{2}\,(\phi_{22} + \phi_{2-2}) = R_{3d}(r) \cdot \sqrt{15}/4\,\sqrt{\pi}\; (x^2 - y^2) \\ d_{z^2} &= \phi_{20} = R_{3d}(r) \cdot \sqrt{5}/4\,\sqrt{\pi}\; (2z^2 - x^2 - y^2) \end{aligned}$$

where x, y, and z are to be interpreted as the coordinates of a point at unit distance from the origin; $R_{3d}(r)$ is the radial wave function for a $3d$ electron; and the ϕ_{22}, etc., are the $\phi(lm_l)$.

triply degenerate, and may contain six electrons; the doubly degenerate e orbitals may contain four. Table I includes the spinor representations, and is actually the cubic double group. These additional representations are needed to classify the states having odd J values.[1,26] The use of these

[26] W. Opechowski, *Physica* **7**, 552 (1940).

representations in crystal field theory is discussed in Section 1e. The notation is Mulliken's extended in an obvious way to include the spinor representations.

The importance of this example arises from the fact that the effect of the octahedral coordination sphere around a transition metal atom is qualitatively similar to the effect of the six point charges. In particular Ti^{+++} surrounded by six H_2O molecules has a single absorption band at 20, 300 cm^{-1} which may be interpreted as the transition of an electron from the lower to the upper orbitals.[7]

This example may be generalized in several ways. First, suppose the field remains cubic, but that the anions are at the corners of a cube instead of at the corners of an octahedron. The *e* orbitals point toward the cube faces and they do not encounter the fields of the anions, contrary to the case in the former example. The *t* orbitals point toward cube edges, and consequently feel the effect of the anions more than do the *e* orbitals. The result is that the energy of the *e* orbital is below that of the *t* orbital. Figure 2 illustrates this difference between the octahedral and "cubal" field.

If the anions are removed from alternate cube corners, the central transition metal cation becomes tetrahedrally coordinated. Figure 2 also illustrates this case. The order of *e* and *t* orbitals is evidently unchanged from the case of cubal coordination. However, since there are now half as many anions to supply the crystal field, the splitting between *e* and *t* energy levels is expected to be half as great. This is approximately confirmed by experiment.

The inversion of the *d*-shell splitting under these changes in geometry has important physical consequences which will be pointed out in later sections.

A second generalization of our simple example is to remove the restriction that the electron be in a *d* orbital. An *s* orbital is of no great interest since it is nondegenerate. A *p* orbital does not split in a cubic field, since its three partners are all contained in the T_{1u} representation of the cubic group. In a field of lower symmetry, splitting must occur. However, the approach represented by crystal field theory is not applicable to *p* electrons because they participate so strongly in bonding processes. The excited states of Tl^+, Pb^{++}, etc., in which *p* electrons are involved require a different approach which is discussed in Section 6. The *f* orbitals are therefore the only others to which crystal field theory may be applied with profit.

The representations subtended by an *f* orbital in a crystal field may be determined from Table IX. In a cubic field these are A_{1u}, T_{1u}, T_{2u}. These states do not determine the optical spectra of an ion having a single *f*

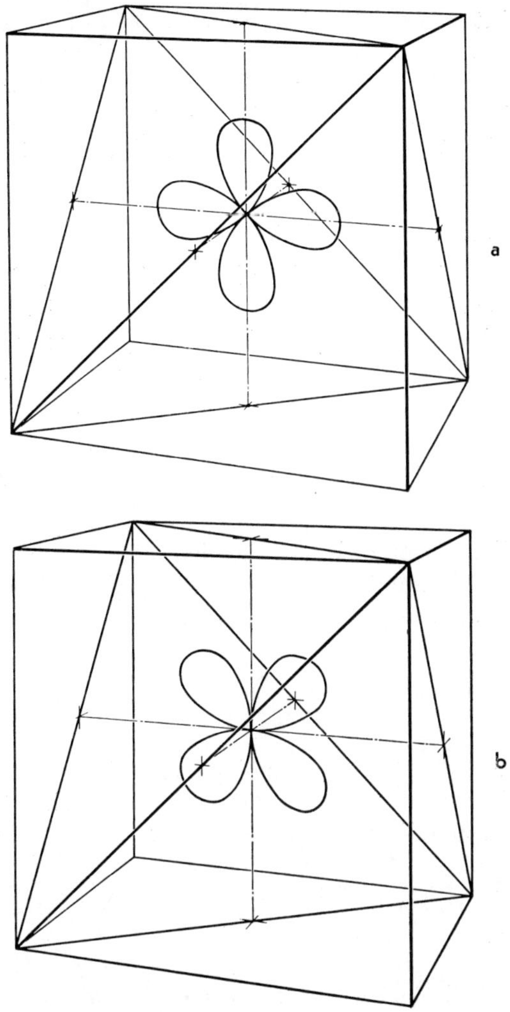

FIG. 2. a. e orbital inscribed within a cube, tetrahedron and octahedron. b. Inscribed t orbital.

The cube and tetrahedron edges are shown. The octahedron corners are the centers of cube faces. Note that the e orbitals point toward octahedron corners and that the t orbitals avoid them. The e orbitals point toward tetrahedron edges or cube faces, and thus avoid the corners more effectively than do the t orbitals.

electron, since the magnitude of the crystal field splitting may be only on the order of 100 cm^{-1} for a lanthanide element, or 1000 cm^{-1} for an actinide. Instead the crystal field splitting determines the structure of transitions between states characteristic of the free ion.

This last point suggests a third generalization of our simple example: namely, to take account of the presence of more than one electron, and to work out the crystal field splitting of the many-electron states of an atom. This problem will be taken up in Section 1d. It will be advantageous before reaching this point to examine the quantitative aspects of crystal field theory more closely.

b. Electrostatic Charge Distributions

The field of the six point charges in the vicinity of the central ion in the example of the last section may be represented by an expansion in spherical harmonics about the central ion as origin. It must be assumed in making this expansion that the potential obeys Laplace's equation, and therefore that the d orbitals of the ion do not overlap the charges. The potential is

$$V = \frac{2e_i}{R}\left[3 + \frac{35}{8R^4}(x^4 + y^4 + z^4 - \tfrac{3}{5}r^4) + \cdots\right] \qquad (1.1)$$

where R is the distance from ion to the charges; x,y,z,r are coordinates of the electron; and e_i is the charge at each external point. An expression of this type for the potential of a more general distribution of charges is[21]

$$V(r,\theta,\phi) = \sum_{n=0}^{\infty} \sum_{m=-n}^{n} A_n^m r^n Y_n^m(\theta,\phi) \qquad (1.2)$$

where the A_n^m are constants determined by all the parameters of the charge distribution and the Y_n^m are normalized spherical harmonics. The Hamiltonian corresponding to this potential is

$$H = e\sum_j V(r_j,\theta_j,\phi_j) = e\sum_{n=0}^{\infty}\sum_{m=-n}^{n} A_n^m \sum_j r_j^n Y_n^m(\theta_j,\phi_j) \qquad (1.3)$$

where j runs over all electrons of the central ion and e is the electronic charge. This Hamiltonian is added to the Hamiltonian of the free ion as a perturbation term.

In the interpretation of spectra by means of crystal field theory, the A_n^m are treated as constants to be determined by experiment. The real charge distributions in crystals are too complicated to permit a reliable calculation of the field constants and furthermore the d orbitals actually overlap the ligating atoms appreciably.[27] The largest term of the potential is the constant term to which the "core electrons" contribute. This factor is assumed to shift all spectroscopic levels equally. It gives the major

[27] Y. Tanabe and S. Sugano, *J. Phys. Soc.* (*Japan*) **11,** 864 (1956).

share of the thermodynamic heat of formation of a compound so it is not "negligible" except in certain applications.

Despite the fact that distributions of point charges and dipoles cannot be used to calculate a crystal field accurately, they are useful for gaining a qualitative understanding of the behavior of the various terms in a complicated potential function. Such a function may be constructed easily by using the equation

$$V_i(r,\theta,\phi) = \frac{1}{R_i} \sum_{n=0}^{\infty} \frac{2}{2n+1} \left(\frac{r}{R_i}\right)^n e_i \left[\pi_n{}^0(\theta_i)\pi_n{}^0(\theta) + 2 \sum_{m=1}^{n} \pi_n{}^m(\theta_i)\pi_n{}^m(\theta) \cos m(\phi - \phi_i) \right] \quad (1.4)$$

where $\pi_n{}^m$ is the normalized associated Legendre polynomial of degree n. This equation gives the potential of a charge at point $i = (\theta_i, \phi_i, R_i)$ at the field points r, θ, ϕ. The potential of an arbitrary distribution of charges is $\sum_i V_i$. To convert this potential to that of a *radially directed* point dipole, e_i is replaced by $(n + 1)P_i/R_i$, where P_i is the dipole at point i.

c. Energy Level Calculations

We shall apply the foregoing results to the calculation of the matrix elements of the suitability modified potential of the octahedral field. Thus we write

$$V_4 = D(x^4 + y^4 + z^4 - \tfrac{3}{5}r^4) + \text{higher terms} \quad (1.5)$$

where D is considered to be an empirical parameter replacing the coefficient $35e_i/4R^5$ of Eq. (1.1). The potential, in this case, has the symmetry of the cubic group O_h, and the wave functions d_{xy}, etc., have been chosen because they belong to irreducible representations of this group. The two different d orbitals thus have the energies

$$\begin{aligned} E_t &= e\int (d_{xy})^2 V_4 r^2 \sin\theta \, dr \, d\theta \, d\phi \\ E_e &= e\int (d_{x^2-y^2})^2 V_4 r^2 \sin\theta \, dr \, d\theta \, d\phi. \end{aligned} \quad (1.6)$$

The integrals may be carried out by quadrature giving

$$\begin{aligned} E_t &= -4Dq \\ E_e &= +6Dq \end{aligned} \quad (1.7)$$

where $q = 2e/105\int R_{3d}{}^2(r) r^4 \cdot r^2 \, dr = 2\overline{r^4}e/105$ and $R_{3d}(r)$ is the normalized radial $3d$ function of the free ion. The radial integral q is separable from the integrals over the angles. The latter give the coefficients -4 and 6 of Eq. (1.7).

A general result illustrated by this example is that the radial integral q, must be treated as another empirical constant since the radial wave functions cannot be known accurately. However q always appears multiplied by the potential constant, and only the product Dq is determined by experiment. In the general case, the product $A_n{}^m\overline{r^n}$ is determined.

The higher terms in the potential do not contribute in this case because of considerations of symmetry in the full rotation group. The d functions transform in the same way as second-order spherical harmonics, and their squares, which appear in the matrix elements, contain no terms higher than those of the fourth-order. Because of the orthogonality relations, V_4, but not V_5 or V_6 etc., may contribute to the integral. In fields of lower symmetries V_2 appears, but V_1 and V_3 do not contribute because the d-orbital product contains only even orders. This is true of course for any product of equivalent electrons. In general, the potential terms contributing to a diagonal matrix element $(SLJJ_z|V_n|SLJJ_z)$ must have $n \leqslant 2J$, $n \leqslant 2L$, $n \leqslant 2l$, and n even.

The relations among the Dq values for the three cubic figures are obtainable from Eq. (1.4) and for a point charge model they are: $-\frac{9}{4}Dq$ (tetrahedron) $= Dq$ (octahedron) $= -\frac{9}{8}Dq$ (cube). These relations are approximately confirmed by experiment. The most significant result is the small value of Dq for tetrahedral fields. This result may be rationalized easily by an examination of Fig. 2.

We are now ready to consider the energy levels when more than one electron is present.

d. Weak, Strong, and Intermediate Crystal Fields

There are obviously two limiting cases which must be considered. An atom having several equivalent d or f electrons: (1) may have energy levels which are determined predominantly by the crystal fields; or (2) the energy levels may be hardly affected by the crystal field and be determined by the forces which are already present in the free atoms. In order to achieve the first condition, the crystal field must be strong enough to overcome the coupling force which operates in the free atom. This force is the electrostatic interaction between equivalent electrons. Between the limiting cases is an intermediate region in which the two forces are comparable.

In the limiting case of a weak crystal field, the problem which must be solved is the first-order effect of the Hamiltonian given in Eq. (1.3) upon the states of the free ion. The symmetry defined by the perturbation term of the Hamiltonian determines the number and degeneracy of each substate, and in any given case this may be found from Table IX (discussed in Section 2, f). For example, the ^3F state of V^{+++} splits into $^3A_2$, $^3T_2$, and $^3T_1$.

In the limiting case of a strong crystal field a many-electron state belongs to a crystal field configuration. In the example of the cubic field, the configurations are obtained by assigning electrons to e or t orbitals. Thus the possible 2-electron configurations in any cubic field are t^2, et, and e^2. Equation (1.7) shows that in an octahedral field the energy of an electron in a t orbital is $10Dq$ lower than when it is in an e orbital. If the zero of energy is chosen as the et configuration, then the t^2 configuration has the energy $-10Dq$ and the e^2 configuration $+10Dq$. In general the energy of a configuration $t^n e^{N-n}$, where N is the total number of d electrons in the ion, is

$$E(n, N - n) = 10nDq + 6NDq \tag{1.8}$$

where a choice of the zero of energy is implied.

The number of states arising from a configuration, and the symmetry of each, is determined by reducing the direct product of the representations of each occupied orbital, taking proper account of the Pauli principle. For example the configuration te leads to the reducible representation $T_{2g} \times E_g$ in O_h. This may be reduced to $T_{1g} + T_{2g}$, but since the electrons are in different orbitals, the permissible states are

$$^1T_{1g} + {}^1T_{2g} + {}^3T_{1g} + {}^3T_{2g}.$$

The configuration t^2 requires the exclusion of certain states because here the electrons occupy the same orbital. This may sometimes be done by inspection. Bethe[1] reduces the symmetry of the system in order to separate the partners of the degenerate orbitals, and accomplishes the reduction in this way.

The number and types of states must be the same in the weak field limit as in the strong field limit. In fact it must be possible to draw a correlation diagram relating the states in the limiting cases to each other by passing through the region of intermediate crystal fields. Such a diagram is shown in Fig. 3 for V^{+++}, an ion having two d electrons.

Actual energy level calculations may be divided into three types: a perturbation calculation in which states of the free ion are split by the crystal field, but are not caused to interact with each other; a perturbation calculation in which the electrostatic interaction splits the strong-field configurations, but does not cause mixing of different configurations; and a complete calculation in which the two perturbation terms are present from the start.

These calculations will be illustrated in brief by using the example of d^2, whose free-atom states and strong field configurations are listed in Fig.

3. The free-atom wave functions are easily set up by the methods given by Condon and Shortley; for example, by the method of Gray and Wills (p. 226 in ref. 28). For a given term, say ^3F, the procedure gives the partners classified by their M_l value in terms of products of one-electron functions, $\phi(l,m_l)$. These functions may be combined in such a way that they transform under the representations of the cubic group. Once these combinations are found, the diagonal matrix element of the crystal field potential using each of these functions gives the desired energies. The

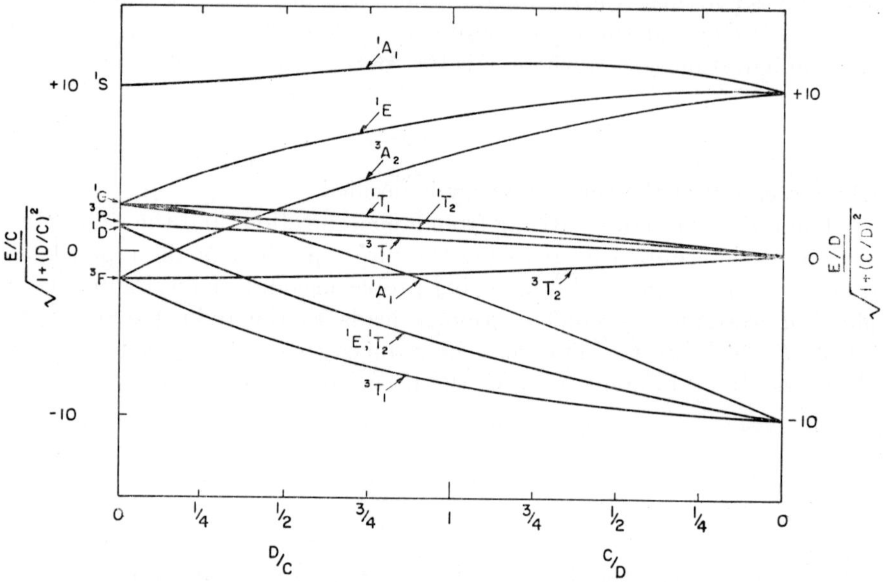

FIG. 3. The correlation between the states of weak and strong crystal fields for the case of two d electrons in an octahedral field. The energy levels to the left are the calculated levels for $B/C = 4.5$. The cubic field parameter is denoted by D. The left side of the diagram shows the crystal field splitting of free ion levels, while the right side shows the exchange splitting of the strong-field levels. The lowest level on the right is the t^2 configuration; above it is et and at the top is e^2.

details of crystal field splitting calculations are well explained in the review article by Bleaney and Stevens,[9] and further description of them therefore will be omitted.

Near the strong-field limit, the wave functions for the problem of two d electrons may be constructed from the properly antisymmetrized products of the d_{xy}, d_{z^2}, etc. Since these one-electron functions already belong to cubic representations, it is a simple matter to combine their products into

[28] E. U. Condon and G. H. Shortley, "The Theory of Atomic Spectra." Cambridge University Press, London and New York, 1951.

functions belonging to cubic representations. In the case of the e^2 configuration the wave functions are:

$$\psi(^1A_1) = \tfrac{1}{2}[E_a(1)E_a(2) + E_b(1)E_b(2)](\alpha_1\beta_2 - \alpha_2\beta_1)$$

$$\psi(^3A_2) = \frac{1}{\sqrt{2}}[E_a(1)E_b(2) - E_a(2)E_b(1)]\begin{cases}\alpha_1\alpha_2 \\ \frac{1}{\sqrt{2}}(\alpha_1\beta_2 + \alpha_2\beta_1) \\ \beta_1\beta_2\end{cases} \quad (1.9)$$

$$\psi(^1E) = \tfrac{1}{2}\begin{Bmatrix}[E_a(1)E_a(2) - E_b(1)E_b(2)] \\ [E_a(1)E_b(2) + E_a(2)E_b(1)]\end{Bmatrix}(\alpha_1\beta_2 - \alpha_2\beta_1)$$

where $E_a = \dfrac{\phi(l = 2,\, m = 2) + \phi(l = 2,\, m = -2)}{\sqrt{2}} = d_{(x^2-y^2)}$

$$E_b = \phi(l = 2,\, m = 0) = d_{z^2}.$$

The first-order splitting of the e^2 configuration arises from the electrostatic interaction. Since there are no two functions having the same symmetry, the problem is simply to calculate the diagonal matrix elements of the electrostatic interaction for each state. By converting the E_a, E_b to $\phi(l,m_l)$ the integrals needed may be obtained with the help of the tables given by Condon and Shortley (pp. 179, 180 in ref. 28). The resulting electrostatic interaction integrals for the e^2 configuration are:

$$\int \psi(^1A_1)\frac{e^2}{r_{12}}\psi(^1A_1)\,d\tau = F_0 + 8F_2 + 51F_4 = A + 8B + 4C$$

$$\int \psi(^3A_2)\frac{e^2}{r_{12}}\psi(^3A_2)\,d\tau = F_0 - 8F_2 - 9F_4 = A - 8B \quad (1.10)$$

$$\int \psi(^1E)\frac{e^2}{r_{12}}\psi(^1E)\,d\tau = F_0 + 21F_4 = A + 2C.$$

The F_0, F_2, and F_4 are the radial integrals over the interelectronic distance, as defined by Condon and Shortley (p. 177 in ref. 28). The A, B, and C are the corresponding parameters defined by Racah.[29] The relations between them are

$$A = F_0 - 49F_4$$
$$B = F_2 - 5F_4$$
$$C = 35F_4.$$

Since Racah's notation has been used extensively, it is introduced here.

In the two limiting cases, there is only one perturbation term. In the intermediate case both the crystal field and the electrostatic interaction are considered simultaneously. In order to carry out the intermediate field calculation, one must start with a set of wave functions diagonal either in the electrostatic interaction or in the crystal field. It seems somewhat

[29] G. Racah, *Phys. Rev.* **62**, 438 (1942).

simpler to begin with the latter set, as has been done in extensive calculations by Tanabe and Sugano.[13] One begins with a set of functions, such as those for e^2, Eq. (1.9), for each crystal field configuration. States of the same symmetry character from different configurations interact when the electrostatic interaction is added. Such interactions provide the off-diagonal elements of the secular equation for the states of the given symmetry. For example, the $^1A_1$ state of the e^2 configuration perturbs the $^1A_1$ state of the t^2 configuration, giving rise to a matrix element $\sqrt{6}\,(2B + C)$. This matrix element may be found in the same way as described for the diagonal elements in the last paragraph.

The complete solution to the crystal field problem is provided by the solutions to the set of secular equations, one for each representation involved. The secular equations for d^2 are shown in Table II. Their solutions make it possible to draw a diagram correlating the weak field and strong field limits, and showing the course of the energy levels throughout the intermediate region. The diagram for d^2 is given in Fig. 3.

TABLE II. ENERGY MATRICES FOR d^2

$^1A_1$	t^2	$10B + 5C - 10Dq$	$\sqrt{6}\,(2B + C)$
	e^2	$\sqrt{6}\,(2B + C)$	$8B + 4C + 10Dq$
$^3A_2$	t^2	$-8B + 10Dq$	
^1E	t^2	$B + 2C - 10Dq$	$-2\sqrt{3}\,B$
	e^2	$-2\sqrt{3}\,B$	$2C + 10Dq$
$^1T_1$	et	$4B + 2C$	
$^3T_1$	t^2	$-5B - 10Dq$	$6B$
	et	$6B$	$4B$
$^1T_2$	t^2	$B + 2C - 10Dq$	$2\sqrt{3}\,B$
	et	$2\sqrt{3}\,B$	$2C$
$^3T_2$	et	$-8B$	

The methods we have used for obtaining the matrix elements of the crystal field and the electrostatic interaction are tedious when applied to systems having more than two electrons. A great deal of time can be saved in the calculation of the matrix elements of the crystal field by using the "operator equivalents" of crystal field potential functions, developed by Stevens[20] and extended by Stevens and others.[9,22–24] Through an extension of Racah's methods to nonspherically symmetric systems, Tanabe and Sugano[13] developed a convenient method for obtaining the matrix elements of electrostatic interaction in a cubic crystal field. The latter authors have made extensive calculations for d electrons in cubic fields.

Since ions having incomplete d- shells are common, and since the crystal field determines the most important aspects of their spectra, these are the cases for which crystal field theory has its widest applications. For this reason, we have included the full secular equations and energy diagrams given by Tanabe and Sugano for all the d-electron configurations. Most of the important information concerning the theory of spectra of the transition metal ions is contained in the secular equations, Tables III, IV, and V.

The equations for d^n apply to d^{10-n} when the sign of Dq is changed. The reason is that electrons are equivalent to "holes" except for charge, and Dq depends upon the first power of the electronic charge. The matrix elements for electrostatic interaction depend upon the second power of the electronic charge, and therefore do not change sign.

The equations also apply to cubal and tetrahedral as well as to octahedral fields, when the sign of Dq is changed appropriately. This sign change was discussed in Section 1a and illustrated by Fig. 2.

In the tetrahedral case, however, the array of charges does not have a center of symmetry, and an odd term appears in the potential expansion. It is a term containing third order spherical harmonics, and may be written $Txyz$ where T is a constant. The value of T for point charges is $20Ze/\sqrt{3}\,R^4$. This term cannot have matrix elements between two pure d-electron states, but may connect d and p states, d_{xy} and p_z for example. The configurations $d^{n-1}p$ lie on the order of 10^5 cm^{-1} above the d^n configurations, so that this mixing is a second order effect. However, Low has presented theoretical evidence that the effect is nevertheless quite large.[30] This problem awaits experimental study.

Some of the solutions of the equations given in Tables II, III, IV, and V are shown graphically in Figs. 4 through 10. The ordinate of these diagrams is energy in units of B. The abscissa is Dq/B. The notation for the states and configurations is that of Tanabe and Sugano; the conversion to ours is given in the caption to Fig. 4. The ground state is always the abscissa line. A change in ground state is indicated by a vertical line at the value of Dq/B where the change occurs. These diagrams do not show the complete transition from weak to strong fields (compare Figs. 3 and 4) but they are carried far enough ($Dq/B = 4$) so that the states occur in the same order as in a strong field.

Both B and C determine the energy levels of an ion, but they are in a nearly constant ratio, namely $C/B = 4.5$ for all ions of the transition

[30] W. Low and M. Weger, to be published. The writer wishes to thank Dr. Low for showing him the manuscript before publication.

Table III. Energy Matrices for the Configuration d^3 in a Cubic Field (Tanabe and Sugano)

$^2T_2(a^2D, b^2D, {}^2F, {}^2G, {}^2H)$

t^3	$-12Dq + 5C$	$-3\sqrt{3}\,B$	$-5\sqrt{3}\,B$	$4B + 2C$	$2B$
$t^2(^3T_1)e$		$-2Dq - 6B + 3C$	$3B$	$-3\sqrt{3}\,B$	$-3\sqrt{3}\,B$
$t^2(^1T_2)e$			$-2Dq + 4B + 3C$	$-\sqrt{3}\,B$	$\sqrt{3}\,B$
$te^2(^1A_1)$				$8Dq + 6B + 5C$	$10B$
$te^2(^1E)$					$8Dq - 2B + 3C$

$^2T_1(^2P, {}^2F, {}^2G, {}^2H)$

t^3	$-12Dq - 6B + 3C$	$-3B$	$3B$	0	$-2\sqrt{3}\,B$
$t^2(^3T_1)e$		$-2Dq + 3C$	$-3B$	$3B$	$3\sqrt{3}\,B$
$t^2(^1T_2)e$			$-2Dq - 6B + 3C$	$-3B$	$-\sqrt{3}\,B$
$te^2(^3A_2)$				$8Dq - 6B + 3C$	$2\sqrt{3}\,B$
$te^2(^1E)$					$8Dq - 2B + 3C$

$^2E(a^2D, b^2D, {}^2G, {}^2H)$

t^3	$-12Dq - 6B + 3C$	$-6\sqrt{2}\,B$	$-3\sqrt{2}\,B$	0
$t^2(^1A_1)e$		$-2Dq + 8B + 6C$	$10B$	$\sqrt{3}\,(2B + C)$
$t^2(^1E)e$			$-2Dq - B + 3C$	$2\sqrt{3}\,B$
e^3				$18Dq - 8B + 4C$

$^4T_1(^4P, {}^4F)$

$t^2(^3T_1)e$	$-2Dq - 3B$	$6B$
$te^2(^3A_2)$		$8Dq - 12B$

$^4A_2(^4F)t^3$	$-12Dq - 15B$
$^4T_2(^4F)t^2(^3T_1)e$	$-2Dq - 15B$
$^2A_1(^2G)t^2(^1E)e$	$-2Dq - 11B + 3C$
$^2A_2(^2F)t^2(^1E)e$	$-2Dq + 9B + 3C$

TABLE IV. ENERGY MATRICES FOR THE CONFIGURATION d^4 IN A CUBIC FIELD (TANABE AND SUGANO)

$^1E(a^1D, b^1D, a^1G, b^1G, {}^1I)$

t^4	$-16Dq - 9B + 7C$	$6B$	$\sqrt{2}\,(2B + C)$	$-2B$	$-4B$
$t^3(^2E)e$		$-6Dq - 6B + 6C$	$-3\sqrt{2}\,B$	$-12B$	0
$t^2(^1E)e^2(^1A_1)$			$4Dq + 5B + 8C$	$10\sqrt{2}\,B$	$-10\sqrt{2}\,B$
$t^2(^1A)e^2(^1E)$				$4Dq + 6B + 9C$	0
$t^2(^1E)e^2(^1E)$					$4Dq - 3B + 6C$

$^3T_2(^3D, a^3F, b^3F, {}^3G, {}^3H)$

$t^3(^2T_1)e$	$-6Dq - 9B + 4C$	$-5\sqrt{3}\,B$	$\sqrt{6}\,B$	$\sqrt{3}\,B$	$-\sqrt{6}\,B$
$t^3(^2T_2)e$		$-6Dq - 5B + 6C$	$-3\sqrt{2}\,B$	$3B$	$\sqrt{2}\,(3B + C)$
$t^2(^3T_1)e^2(^3A_2)$			$4Dq - 13B + 4C$	$-2\sqrt{2}\,B$	$-6B$
$t^2(^3T_1)e^2(^1E)$				$4Dq - 9B + 4C$	$3\sqrt{2}\,B$
te^3					$14Dq - 8B + 5C$

$^1T_1(^1F, a^1G, b^1G, {}^1I)$

$t^3(^2T_1)e$	$-6Dq - 3B + 6C$	$5\sqrt{3}\,B$	$3B$	$\sqrt{6}\,B$
$t^3(^2T_2)e$		$-6Dq - 3B + 8C$	$-5\sqrt{3}\,B$	$\sqrt{2}\,(B + C)$
$t^2(^1T_2)e^2(^1E)$			$4Dq - 3B + 6C$	$-\sqrt{6}\,B$
te^3				$14Dq - 16B + 7C$

$^3E(^3D, {}^3G, {}^3H)$

$t^3(^4A_2)e$	$-6Dq - 13B + 4C$	$-4B$	0
$t^3(^2E)e$		$-6Dq - 10B + 4C$	$-3\sqrt{2}\,B$
$t^2(^1E)e^2(^3A_2)$			$4Dq - 11B + 4C$

$^3A_2(a^3F, b^3F)$

$t^3(^2E)e$	$-6Dq - 8B + 4C$	$-12B$
$t^2(^1A_1)e^2(^3A_2)$		$4Dq - 2B + 7C$

TABLE IV. ENERGY MATRICES FOR THE CONFIGURATION d^4 IN A CUBIC FIELD (TANABE AND SUGANO) (*Continued*)

$^1A_2(^1F,^1I)$

$t^3(^2E)e$	$-6Dq - 12B + 6C$	$6B$
$t^2(^1E)e^2(^1E)$		$4Dq - 3B + 6C$

$^5E(^5D)t^3(^4A_2)e$	$-6Dq - 21B$
$^5T_2(^5D)t^2(^3T_1)e^2(^3A_2)$	$4Dq - 21B$
$^3A_1(^3G)t^3(^2E)e$	$-6Dq - 12B + 4C$

$^3T_1(a^3P,b^3P,a^3F,b^3F,^3G,^3H)$

t^4	$-16Dq - 15B + 5C$	$-\sqrt{6}\,B$	$-3\sqrt{2}\,B$	$\sqrt{2}\,(2B + C)$	$-2\sqrt{2}\,B$	0	0
$t^3(^2T_1)e$		$-6Dq - 11B + 4C$	$5\sqrt{3}\,B$	$\sqrt{3}\,B$	$-\sqrt{3}\,B$	$3B$	$\sqrt{6}\,B$
$t^3(^2T_2)e$			$-6Dq - 3B + 6C$	$-3B$	$-3B$	$5\sqrt{3}\,B$	$\sqrt{2}\,(B + C)$
$t^2(^3T_1)e^2(^1A_1)$				$4Dq - B + 6C$	$-10B$	0	$3\sqrt{2}\,B$
$t^2(^3T_1)e^2(^1E)$					$4Dq - 9B + 4C$	$-2\sqrt{3}\,B$	$-3\sqrt{2}\,B$
$t^2(^1T_2)e^2(^3A_2)$						$4Dq - 11B + 4C$	$\sqrt{6}\,B$
te^3							$14Dq - 16B + 5C$

$^1T_2(a^1D,b^1D,a^1G,b^1G,^1F,^1I)$

t^4	$-16Dq - 9B + 7C$	$3\sqrt{2}\,B$	$-5\sqrt{6}\,B$	0	$-2\sqrt{2}\,B$	$\sqrt{2}\,(2B + C)$	0
$t^3(^2T_1)e$		$-6Dq - 9B + 6C$	$-5\sqrt{3}\,B$	$3B$	$-3B$	$-3B$	$-\sqrt{6}\,B$
$t^3(^2T_2)e$			$-6Dq + 3B + 8C$	$-3\sqrt{3}\,B$	$5\sqrt{3}\,B$	$-5\sqrt{3}\,B$	$\sqrt{2}\,(3B + C)$
$t^2(^3T_1)e^2(^3A_2)$				$4Dq - 9B + 6C$	$-6B$	0	$-3\sqrt{6}\,B$
$t^2(^1T_2)e^2(^1E)$					$4Dq - 3B + 6C$	$-10B$	$\sqrt{6}\,B$
$t^2(^1T_2)e^2(^1A_1)$						$4Dq + 5B + 8C$	$\sqrt{6}\,B$
te^3							$14Dq + 7C$

$^1A_1(a^1S,b^1S,a^1G,b^1G,^1I)$

t^4	$-16Dq + 10C$	$-12\sqrt{2}\,B$	$\sqrt{2}\,(4B + 2C)$	$2\sqrt{2}\,B$	0
$t^3(^2E)e$		$-6Dq + 6C$	$-12B$	$-6B$	0
$t^2(^1A_1)e^2(^1A_1)$			$4Dq + 14B + 11C$	$20B$	$\sqrt{6}\,(2B + C)$
$t^2(^1E)e^2(^1E)$				$4Dq - 3B + 6C$	$2\sqrt{6}\,B$
e^4					$24Dq - 16B + 8C$

TABLE V. ENERGY MATRICES FOR THE CONFIGURATION d^5 IN A CUBIC FIELD (TANABE AND SUGANO)

$^2E(a^2D, b^2D, c^2D, a^2G, b^2G, {}^2H, {}^2I)$

$t^4(^1A_1)e$	$-10Dq - 4B + 12C$	$10B$	$6B$	$6\sqrt{3}\,B$	$6\sqrt{2}\,B$	$-2B$	$4B + 2C$
$t^4(^1E)e$		$-10Dq - 13B + 9C$	$-3B$	$3\sqrt{3}\,B$	0	$2B + C$	$2B$
$t^3(^2E)e^2(^1A_1)$			$-4B + 10C$	0	0	$-3B$	$-6B$
$t^3(^2E)e^2(^3A_2)$				$-16B + 8C$	$2\sqrt{6}\,B$	$-3\sqrt{3}\,B$	$6\sqrt{3}\,B$
$t^3(^2E)e^2(^1E)$					$-12B + 8C$	0	$6\sqrt{2}\,B$
$t^2(^1E)e^3$						$10Dq - 13B + 9C$	$-10B$
$t^2(^1A_1)e^3$							$10Dq - 4B + 12C$

$^2A_1(^2S, a^2G, b^2G, {}^2I)$

$t^4(^1E)e$	$-10Dq - 3B + 9C$	$-3\sqrt{2}\,B$	0	$6B + C$
$t^3(^2E)e^2(^1E)$		$-12B + 8C$	$-4\sqrt{3}\,B$	$3\sqrt{2}\,B$
$t^3(^4A_2)e^2(^3A_2)$			$-19B + 8C$	0
$t^2(^1E)e^3$				$10Dq - 3B + 9C$

$^2A_2(a^2F, b^2F, {}^2I)$

$t^4(^1E)e$	$-10Dq - 23B + 9C$	$3\sqrt{2}\,B$	$-2B + C$
$t^3(^2E)e^2(^1E)$		$-12B + 8C$	$-3\sqrt{2}\,B$
$t^2(^1E)e^3$			$10Dq - 23B + 9C$

$^4T_1(^4P, {}^4F, {}^4G)$

$t^4(^3T_1)e$	$-10Dq - 25B + 6C$	$-3\sqrt{2}\,B$	C
$t^3(^2T_2)e^2(^3A_2)$		$-16B + 7C$	$-3\sqrt{2}\,B$
$t^2(^3T_1)e^3$			$10Dq - 25B + 6C$

$^4T_2(^4E, {}^4G, {}^4D)$

$t^4(^3T_1)e$	$-10Dq - 17B + 6C$	$\sqrt{6}\,B$	$4B + C$
$t^3(^2T_1)e^2(^3A_2)$		$-22B + 5C$	$-\sqrt{6}\,B$
$t^2(^3T_1)e^3$			$10Dq - 17B + 6C$

$^4E(^4D, {}^4G)$

$t^3(^2E)e^2(^3A_2)$	$-22B + 5C$	$-2\sqrt{3}\,B$
$t^3(^4A_2)e^2(^1E)$		$-21B + 5C$

$^6A_1$	(^6S)	$t^3(^4A_2)e^2(^3A_2)$	$-35B$
$^4A_1$	(^4G)	$t^3(^4A_2)e^2(^3A_2)$	$-25B + 5C$
$^4A_2$	(^4F)	$t^3(^4A_2)e^2(^1A_1)$	$-13B + 7C$

TABLE V. ENERGY MATRICES FOR THE CONFIGURATION d^5 IN A CUBIC FIELD (TANABE AND SUGANO) (*Continued*)

$^2T_2(a^2F, b^2F, a^2G, b^2G, {}^2H, {}^2I, a^2D, b^2D, c^2D)$

t^5	$-20Dq-20B+10C$	$3\sqrt{6}\,B$	$\sqrt{6}\,B$	0	$-2\sqrt{3}\,B$	$4B+2C$	$2B$	0	0	0
$t^4(^3T_1)e$		$-10Dq-8B+9C$	$3B$	$\sqrt{6}/2B$	$-3\sqrt{2}/2B$	$3\sqrt{6}/2B$	$3\sqrt{6}/2B$	0	$4B+C$	0
$t^4(^1T_2)e$			$-10Dq-18B+9C$	$3\sqrt{6}/2B$	$-3\sqrt{2}/2B$	$5\sqrt{6}/2B$	$-5\sqrt{6}/2B$	C	0	0
$t^3(^2T_1)e^2(^3A_2)$				$-16B+8C$	$2\sqrt{3}\,B$	0	0	$-3\sqrt{6}/2B$	$-\sqrt{6}/2B$	0
$t^3(^2T_1)e^2(^1E)$					$-12B+8C$	$-10\sqrt{3}\,B$	0	$3\sqrt{2}/2B$	$3\sqrt{2}/2B$	$-2\sqrt{3}\,B$
$t^3(^2T_2)e^2(^1A_1)$						$2B+12C$	0	$-5\sqrt{6}/2B$	$-3\sqrt{6}/2B$	$4B+2C$
$t^3(^2T_2)e^2(^1E)$							$-6B+10C$	$-5\sqrt{6}/2B$	$3\sqrt{6}/2B$	$-2B$
$t^2(^1T_2)e^3(^2E)$								$10Dq-18B+9C$	$3B$	$-\sqrt{6}\,B$
$t^2(^3T_1)e^3(^2E)$									$10Dq-8B+9C$	$-3\sqrt{6}\,B$
te^4										$20Dq-20B+10C$

$^2T_1(^2P, a^2F, b^2F, a^2G, b^2G, {}^2H, {}^2I)$

$t^4(^3T_1)e$	$-10Dq-22B+9C$	$-3B$	$-3\sqrt{2}/2B$	$3\sqrt{2}/2B$	$-3\sqrt{2}/2B$	$-3\sqrt{6}/2B$	0	C
$t^4(^1T_2)e$		$-10Dq-8B+9C$	$3\sqrt{2}/2B$	$3\sqrt{2}/2B$	$15\sqrt{2}/2B$	$5\sqrt{6}/2B$	$4B+C$	0
$t^3(^2T_1)e^2(^1A_1)$			$-4B+10C$	0	0	$10\sqrt{3}\,B$	$3\sqrt{2}/2B$	$-3\sqrt{2}/2B$
$t^3(^2T_1)e^2(^1E)$				$-12B+8C$	0	0	$-3\sqrt{2}/2B$	$-3\sqrt{2}/2B$
$t^3(^2T_2)e^2(^3A_2)$					$-10B+10C$	$2\sqrt{3}\,B$	$15\sqrt{2}/2B$	$-3\sqrt{2}/2B$
$t^3(^2T_2)e^2(^1E)$						$-6B+10C$	$5\sqrt{6}/2B$	$-3\sqrt{6}/2B$
$t^2(^1T_2)e^3$							$10Dq-8B+9C$	$-3B$
$t^2(^3T_1)e^3$								$10Dq-22B+9C$

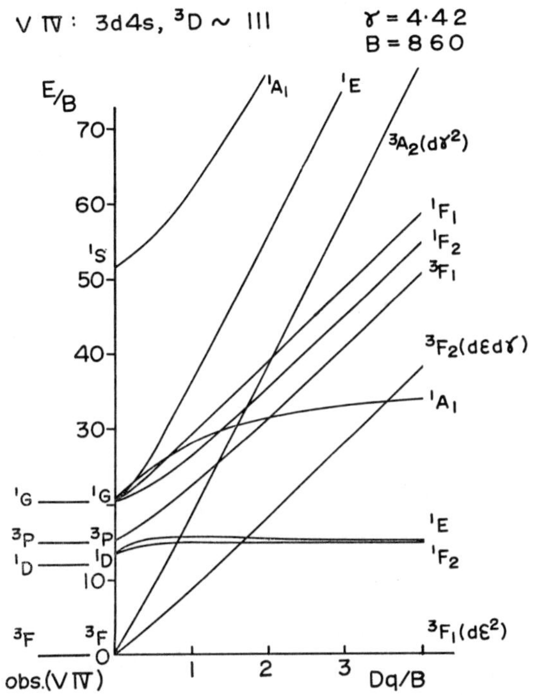

FIG. 4. Splitting of states of the d^2 configuration by an octahedral field (Tanabe and Sugano[13]). The observed levels of V^{+++} are shown to the left for comparison with the calculated positions of ^1D, ^3P, and ^1G as shown in the diagram. The value of $\gamma = C/B$ used in constructing the diagram is 4.42, the best value for V^{+++}. The importance of interference from the next higher configuration of the same parity as $3d^2$ *viz.* $3d4s$, may be judged from the fact that its lowest level ^3D is at $111B$ on the energy scale of the diagram. At the right of the diagram one notes that there are three sets of lines each characterized by a certain slope. These correspond to the three strong-field states of Fig. 3. The notation used in these figures is that of Tanabe and Sugano; its relation to the notation used here is as follows:

$$d\epsilon \rightarrow t$$
$$d\gamma \rightarrow e$$
$$^3F_1,\ ^1F_2, \text{ etc.} \rightarrow\ ^3T_1,\ ^3T_2, \text{ etc.}$$
$$\text{V(IV), etc.} \rightarrow \text{V}^{+++}, \text{ etc.}$$

series, so that either one may be used alone in the diagrams. Values of B and C are given in Table VI. The B values are all close to 1000 cm^{-1}, in the first transition series, and are smaller for the second and third. Values of Dq determined experimentally for divalent ions of the first transition series are about 1000, and are about 1700 cm^{-1} for the trivalent ions. They become larger in the second and third series. Values of Dq for ions in various environments are given in Table VII.

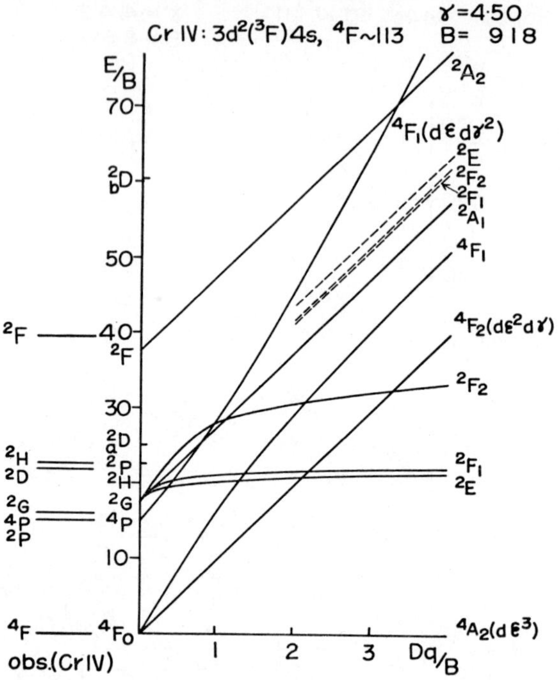

FIG. 5. Splitting of states of the d^3 configuration by an octahedral field (Tanabe and Sugano[13]).

TABLE VI. VALUES OF B AND C

	B	C	C/B		B	C	C/B
Ti^{++}	694	2910	4.19	V^{+++}	861	3814	4.43
V	755	3257	4.31	Cr	918	4133	4.50
Cr	810	3565	4.40	Mn	965	4450	4.61
Mn	860	3850	4.48	Fe	1015	4800	4.73
Fe	917	4040	4.41	Co	1065	5120	4.81
Co	971	4497	4.63	Ni	1115	5450	4.89
Ni	1030	4850	4.71				

Notes to table: The values quoted are the ones given by Y. Tanabe and S. Sugano [*J. Phys. Soc.* (*Japan*) **9,** 753, 766 (1954)]. Similar values were obtained by M. A. Catalan, F. Rohrlich, and A. G. Shenstone [*Proc. Roy. Soc.* (*London*) **A221,** 421 (1954)] and by L. E. Orgel [*J. Chem. Phys.* **23,** 1819 (1955)]. The dependence of the numbers on ionicity and atomic number are more significant than any individual value.

An important feature of the diagrams is the change from the ground state given by Hund's rule to a different ground state in strong fields. This change occurs, for example, in the configurations d^4, d^5, d^6, and d^7 when in octahedral fields. The Dq/B value is marked by a vertical line where the crossing over occurs. A physical interpretation of this crossover has been given by Orgel[11] and by Griffith,[31] and is important for understanding the chemical stability and magnetic properties of these ions. The

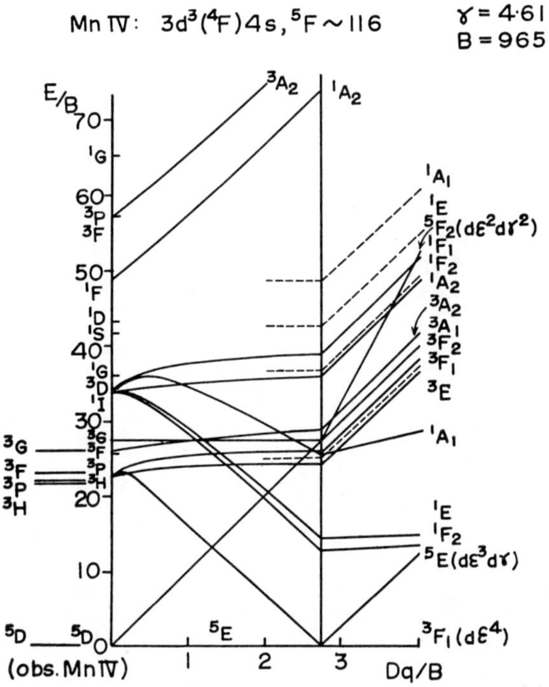

FIG. 6. Splitting of states of the d^4 configuration by an octahedral field (Tanabe and Sugano[13]).

essence of their interpretation is as follows. In the Hund rule ground states (weak fields), the energy required for pairing electrons in the lower or *t* orbitals is greater than the energy required to promote an electron from the *t* to the *e* orbital where it may remain unpaired. In strong fields, the *t*-*e* separation is so great that it requires less energy to pair the electrons in the lower *t* orbital than to promote one into the *e*-orbital. For example in d^4, the t^3e configuration lies deeper in weak fields than does t^4 since it contains the ^5D (Hund rule) ground state. The configuration t^4 can at

[31] J. S. Griffith, *J. Inorg. & Nuclear Chem.* **2,** 1 (1956).

most contain a triplet, since it is equivalent to two holes in a filled shell. In strong fields it becomes more stable than t^3e and a $^3T_1$ ground state results. Recently the same authors have calculated the pairing energies for electrons in the actinide ions.[32]

When competition between pairing energy and crystal field energy occurs, the ion may be rather unstable toward chemical oxidation or

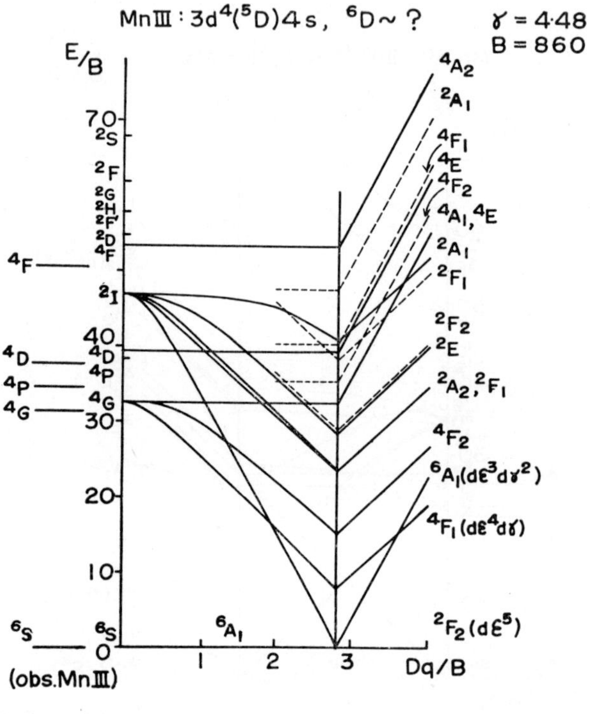

FIG. 7. Splitting of states of the d^5 configuration by an octahedral field (Tanabe and Sugano[13]).

reduction to a neighboring ion. For example Mn^{+++} having the d^4 configuration disproportionates into Mn^{++} and Mn^{++++}, partly for the reason that Mn^{++}, d^5, has a low Dq and is stable in the Hund rule ground state, and Mn^{++++} has a d^3 configuration enabling it to accommodate all its electrons unpaired in the t orbitals. Thus the crystal field theory partially accounts for the stability of the oxidation states of transition metal ions.[32a]

[32] J. S. Griffith and L. E. Orgel, *J. Chem. Phys.* **26,** 988 (1957).

[32a] For a discussion of the relation between crystal field theory and thermodynamic properties of compounds, see P. E. George and D. S. McClure, *Progr. in Inorg. Chem.* **1,** to be published. Interscience, New York.

By finding the ground state of an ion from the diagrams of Figs. 4 through 10 one can determine which ions will suffer the Jahn-Teller effect.[33] Van Vleck[34] showed that in an octahedral field, ions in E_g states are unstable in the symmetrical configuration, and should gain on the order of 1000 cm^{-1} of energy through certain distortions of the octahedron

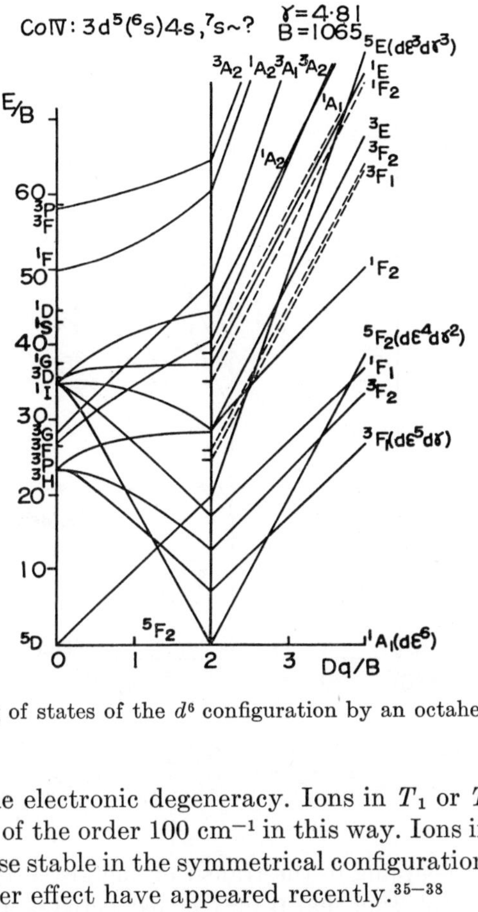

FIG. 8. Splitting of states of the d^6 configuration by an octahedral field (Tanabe and Sugano[13]).

which remove the electronic degeneracy. Ions in T_1 or T_2 ground states should gain only of the order 100 cm^{-1} in this way. Ions in nondegenerate states are of course stable in the symmetrical configuration. Several papers on the Jahn-Teller effect have appeared recently.[35–38]

[33] H. A. Jahn and E. Teller, *Proc. Roy. Soc.* (*London*) **A161,** 220 (1957).
[34] J. H. Van Vleck, *J. Chem. Phys.* **7,** 61, 72 (1939).
[35] U. Opik and M. H. L. Pryce, *Proc. Roy. Soc.* (*London*) **A238,** 425 (1957).
[36] M. H. L. Pryce, U. Opik, H. C. Longuet-Higgins, and R. A. Sack, *Proc. Roy. Soc.* (*London*) **A244,** 1 (1958).
[37] W. Moffitt and A. D. Liehr, *Phys. Rev.* **106,** 1195 (1957).
[37a] W. Moffitt and W. Thorson, *Phys. Rev.* **108,** 1251 (1957).
[38] A. D. Liehr and C. J. Ballhausen, *Ann. Phys.* (*N.Y.*) **3,** 304 (1958).

Although in this section we have emphasized crystal fields having cubic symmetry, other symmetries are often important, and will be discussed in connection with the experimental data to which they apply in Sections 3 and 4. Tanabe and Kamimura[39] have given a general method for evaluating the perturbations resulting from a distortion from cubic symmetry.

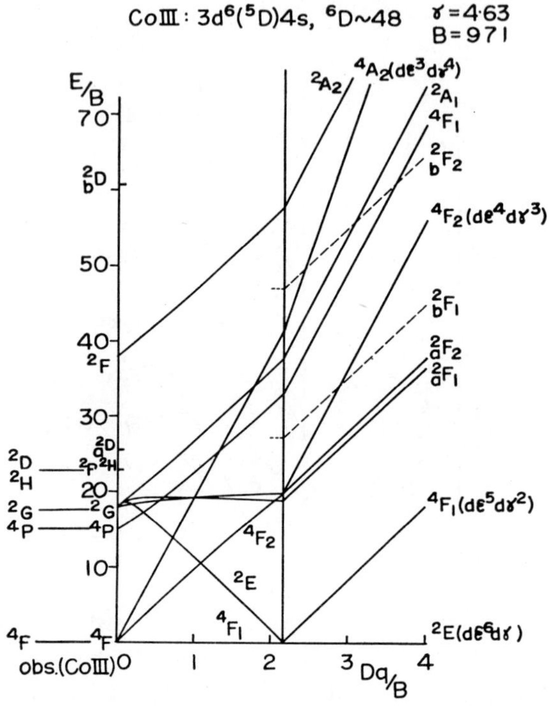

FIG. 9. Splitting of states of the d^7 configuration by an octahedral field (Tanabe and Sugano[13]).

e. Spin-Orbit Coupling

The spin-orbit coupling in the transition metals is so small compared to the crystal field splitting that it may be neglected in a first approximation. On the other hand in the rare earth ions it is so large compared to the crystal field splitting that the crystal field operator may be applied to the individual multiplet components as a first order perturbation. In the first case, one determines the splitting of the L value in the crystal field, and in the second case the J value. There are intermediate cases in which neither L nor J are good quantum numbers. One such example is that of

[39] Y. Tanabe and H. Kamimura, *J. Phys. Soc. Japan* **13**, 394 (1958).

the d-levels of Ce^{+++}. The $f \rightarrow d$ transition occurs in the near ultraviolet. The spectrum is determined by a spin orbit interaction $\zeta_{5d}(L \cdot S)$, in which $\zeta_{5d} = 996\ cm^{-1}$, and by the crystal field splitting of the d-level, which is approximately 10,000 cm^{-1} (Section 3).

Spin-orbit splitting has been observed in the spectra of transition metal ions only recently. In only a few cases has a reliable value of the spin-orbit interaction parameter been obtained from the spectrum. In

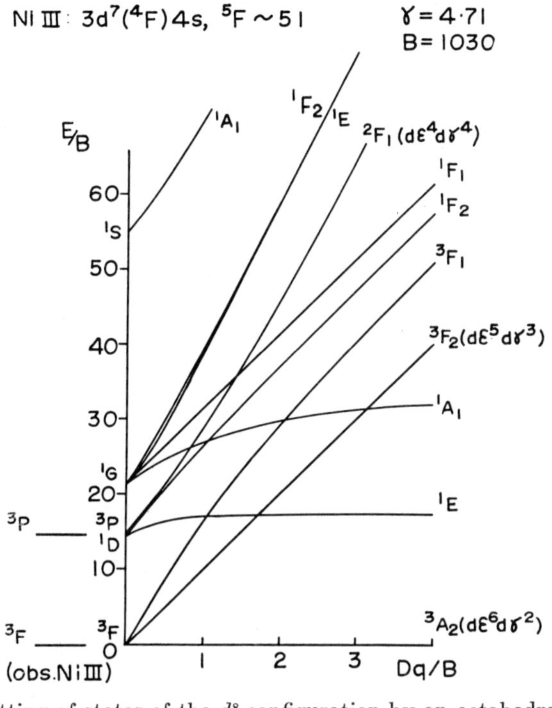

FIG. 10. Splitting of states of the d^8 configuration by an octahedral field (Tanabe and Sugano[13]).

many cases, the spectra are broad, and the multiplet structure is poorly resolved. It seems probable however, that with more careful work, the spin-orbit splitting parameters of excited states will be derived from the spectra.

The states which arise from spin-orbit splitting of a degenerate state in a crystal field are given by the reducible product representation of the spin and orbital parts of the wave function. For example, the $^3T_2$ state in the 2-electron octahedral case splits into $T_1 \times T_2 = A_2 + E + T_1 + T_2$, four spin-orbit product states. When the number of electrons is odd, the spinor representations of the group must be used to find the resultant spin-orbit states. In the $^4T_1$ state of octahedral d^5 for example, the spin part belongs to the G representation of the cubic double group, Table I, and the product states are $G \times T_1 = E_{\frac{1}{2}} + E_{\frac{5}{2}} + 2G$. The double groups used here are shown in Tables I, XIII and XIX. A complete collection of them is to be found in the article by Koster in volume V of this series.[39a]

TABLE VII. VALUES OF Dq

The values given here are for complex ions in solution, but they do not differ markedly from the values in the solid state (see Section 4). The ion is surrounded by a regular or nearly regular octahedron of ligands in each case except for d^4 and d^9 configurations (see text). The values have been taken from the compilations of C. K. Jorgensen [Rept. to 10th Solvay Council, Brussels, May 1956] and O. Holmes and D. S. McClure [*J. Chem. Phys.* **26,** 1686 (1957)]. (Units are cm^{-1}.)

		$6Br^-$	$6Cl^-$	$6H_2O$	$6NH_3$
$3d^1$	Ti^{+++}			2030	
$3d^2$	V^{+++}			1800	
$3d^3$	V^{++}			1180	
	Cr^{+++}		1330	1760	2160
$4d^3$	Mo^{+++}		1920		
$3d^4$	Cr^{++}			1400	
	Mn^{+++}			2100	
$3d^5$	Mn^{++}			750	
	Fe^{+++}			1400	
$3d^6$	Fe^{++}			1000	
	Co^{+++}			1910	
$4d^6$	Rh^{+++}	1930	2080	2770	3460
$5d^6$	Ir^{+++}	2340	2530		
	Pt^{++++}	2500	3000		
$3d^7$	Co^{++}			1000	1050
$3d^8$	Ni^{++}	600		860	1080
$3d^9$	Cu^{++}		650	1260	1510

The notation used in these tables is an extension of Mulliken's. The double groups differ from the ordinary groups in having additional representations which we will call spinor representations (more properly, representations of spinors of half integral order). The other representations are identical to those of the ordinary group.

A first approximation to the splitting pattern produced by spin-orbit coupling in a cubic crystalline field may be obtained through the use of the Landé formula using a fictitious L value, $L' = 1$ for T_1 or T_2 states. A and E states are not split. The splitting pattern in first order is then given by:

$$E_{J'} = a[J'(J' + 1) - L'(L' + 1) - S(S + 1)] \qquad (1.11)$$

[39a] G. F. Koster, *Solid State Phys.* **5,** 173 (1957).

where a is a constant, and J' is $L' + S, L' + S - 1, \cdots |L' - S|$. The constant a must be calculated in each particular case. We will take the lowest $^4T_1$ state of Mn^{++} as an example, since Clogston has recently made detailed calculations of the properties of this state.[39b] The components of $^4T_1$ are $G \times T_1 = E_{\frac{1}{2}} + E_{\frac{5}{2}} + 2G$. Letting $S = \frac{3}{2}$, $L' = 1$, we find from (1.11):

$$E = 3a \text{ (6-fold)}, \ -2a \text{ (4-fold)}, \ -5a \text{ (2-fold)}$$

The 6-fold level must be $G + E_{\frac{5}{2}}$, the 4-fold level G and the 2-fold level $E_{\frac{1}{2}}$. The fact that the 6-fold level is accidentally degenerate means that higher order perturbations may remove the degeneracy and cause a further splitting. In his detailed calculations, Clogston has found the value of a and the additional splitting of the 6-fold level.

Spin-orbit splitting calculations have been made by Opechowski[40] for d^2, d^8, Liehr and Ballhausen[41] for d^2, d^8; Low[30,42] for d^3, d^7, d^4, d^6; and Clogston[39b] for d^5. A general method for finding the matrix elements of spin-orbit interaction has been developed by Tanabe and Kamimura.[39] They have also developed the spin-hamiltonian formalism for degenerate excited states.

Spin orbit interaction is also responsible for the appearance of intercombination bands. The probability of an intercombination is determined by the mixing of states having different multiplicity. The mixing of a state having spin a with a state having spin b may be found from second order perturbation theory and is given by:

$$\psi' = \psi^0(s = a) + \sum_j \frac{\int \psi^0(s = a) \left[\sum_i \zeta_{nl} l_i s_i\right] \psi_j^0(s = b)\, d\tau}{E_a - E_j} \psi_j^0(s = b). \tag{1.12}$$

The mixing coefficient may be abbreviated to $\frac{k\zeta_{nl}}{\Delta E}$ where k is on the order of unity. The square of this coefficient determines the transition probability. In terms of the f-number we have $f_{ab} = f^0 \left(\frac{k\zeta_{nl}}{\Delta E}\right)^2$ where f^0 is the oscillator strength of an allowed transition between states of the same multiplicity. Jorgensen[42a] has made detailed studies of intercombination strengths in several transition metal ions.

[39b] A. M. Clogston, *J. Phys. Chem. Solids* **7**, 201 (1958).
[40] J. Becquerel and W. Opechowski, *Physica* **6**, 1039 (1939).
[41] A. D. Liehr and C. J. Ballhausen, *Ann. Phys.* **6**, 134 (1959).
[42] W. Low, *Phys. Rev.* **109**, 256 (1958).
[42a] C. K. Jorgensen, *Acta Chem. Scand.* **8**, 1502 (1954).

TABLE VIII. VALUES OF THE ONE-ELECTRON SPIN-ORBIT COUPLING PARAMETER ζ_{nl}[a]
Neutral atom is I, etc.

3d	I	II	III	IV	Ref.
Ti	$3d^2\,4s^2\,{}^3F$ 110		$3d^2\,{}^3F$ 121	$3d\,{}^2D$ 154	*b*
V	$3d^3\,4s^2\,{}^4F$ 159		$3d^3\,{}^4F$ 168	$3d^2\,{}^3F$ 209	*b*
Cr	$3d^4\,4s^2\,{}^5D$ 224		$3d^4\,{}^5D$ 236	$3d^3\,{}^4F$ 276	*b*
Mn		$3d^6\,{}^5D$ 224	$3d^4\,4s\,{}^6D,\,{}^4D$ 335	$3d^4\,{}^5D$ 360	*b*
Fe	$3d^6\,4s^2\,{}^5D$ 380		$3d^6\,{}^5D$ 404		*b*
Co	$3d^7\,4s^2\,{}^4F$ 510		$3d^7\,{}^4F$ 528		*b*
Ni	$3d^8\,4s^2\,{}^3F$ 628		$3d^8\,{}^3F$ 644		*b*
Cu	$3d^9\,4s^2\,{}^2D$ 818		$3d^9\,{}^2D$ 829		*b*
4d	**I**	**II**	**III**	**IV**	**Ref.**
Y	$4d\,5s^2\,{}^2D$ 225		$4d\,{}^2D$ 290		*b*
Zr	$4d^2\,5s^2\,{}^3F$ 355		$4d^2\,{}^3F$ 424	$4d^2\,D$ 500	*b*
Nb	$4d^3\,5s^2\,{}^4F$ 470		$4d^3\,{}^4F$ 554	$4d^2\,{}^3F$ 670	*b*
Mo	$4d^4\,5s^2\,{}^5D$ 540			$4d^3\,{}^4F$ 840	*b*
Tc		$4d^6\,{}^5D$ 620			
Ru		$4d^7\,{}^4F$ 860	$4d^6\,{}^5D$ 932		
Rh	$4d^9\,{}^2D$ 940		$4d^7\,{}^4F$ 1190		*b*
Pd	$4d^9\,5s$ 1410	$4d^9\,{}^2D$ 1420			*b*
Ag			$4d^9\,{}^2D$ 1813		*b, c*
5d	**I**	**II**	**III**	**IV**	**Ref.**
Hf		$5d\,6s^2\,{}^2D$ 1220			*b*
Ta					
W					
Re					
Os					
Ir					
Pt	$5d^9\,6s$ 4060				*b*
Au		$5d^8\,6s^2\,{}^3F$ 5000			*b*
4f	**I**	**II**	**III**	**IV**	**Ref.**
La		$4f^2\,{}^3H$ 315			*d*
Ce	$4f^2\,6s^2\,{}^3H$ 623	$4f^2\,6s\,{}^4H$ 622		$4f\,{}^2F$ 643	*e*
Pr	$4f^3\,6s^2$			$4f^2$ 800	*f*
Nd	$4f^4\,6s^2$ 777			$4f^3$ 900	*g, h*
Pm	$4f^5\,6s^2$				
Sm	$4f^6\,6s^2$ 1150		$4f^6\,{}^7F$ 1090	$4f^5$ 1200	*g, i, j*
Eu				$4f^6$ 1415	*k*
Gd					
Tb				$4f^8$ 1620	*l*
Dy				$4f^9$ 1820	*m*
Ho				$4f^{10}$ 2080	*j*
Er				$4f^{11}$ 2360	*j*
Tm	$4f^{13}\,6s^2$ 2500			$4f^{12}$ 2800	*g, n*
Yb				$4f^{13}$ 2940	*j*

TABLE VIII. VALUES OF THE ONE-ELECTRON SPIN-ORBIT COUPLING PARAMETER ζ_{nl} (*Continued*)

$5f$	III		IV		V		VI		Reference
Ac									
Th	$5f^2$	1035	$5f$	1236					*o, p*
Pa									
U			$5f^3$	1700	$5f^2$	1600			*q, r*
Np					$5f^3$	2100	$5f^2$	1900	*q, s*
Pu	$5f^2$(VII)	2500	$5f^5$	2300	$5f^4$	2300	$5f^3$	2700	*s, q*
Am			$5f^6$	3500			$5f^4$	3000	*q*
Cm									

[a] The ζ_{nl} values computed from the spectra are based on one or two multiplets of the configuration shown. Most of the $5d$ spectra require a full theoretical analysis, and ζ values are therefore not given.

[b] C. E. Moore, Atomic Energy Levels, *Natl. Bur. Standards (U.S.) Circ.* **467,** Vol. I, 1949, Vol. II, 1952, Vol. III, 1958. Washington, D.C.

[c] C. K. Jorgensen, "Energy Levels of Complexes and Gaseous Ions." Jul. Gjellerups Forlag, Copenhagen, 1957.

[d] R. Lang, *Can. J. Research* **A13,** 1 (1935); **A14,** 127 (1936).

[e] W. F. Meggers, *Revs. Modern Phys.* **14,** 96 (1942).

[f] F. H. Spedding, *Phys. Rev.* **58,** 255 (1940).

[g] P. F. A. Klinkenberg, *Physica* **13,** 1 (1947).

[h] R. A. Satten, *J. Chem. Phys.* **21,** 637 (1953).

[i] F. D. S. Butement, *Trans. Faraday Soc.* **44,** 617 (1948).

[j] H. Gobrecht *Ann. Physik* [5] **31,** 755 (1938).

[k] E. V. Sayre and S. Freed, *J. Chem. Phys.* **24,** 1211, 1213 (1956).

[l] H. F. Geisler and K. H. Hellwege, *Z. Physik* **136,** 293 (1953).

[m] G. H. Dieke and S. Singh, *J. Opt. Soc. Am.* **46,** 495 (1956).

[n] H. A. Bethe and F. H. Spedding, *Phys. Rev.* **52,** 454 (1937).

[o] P. F. A. Klinkenberg and R. Lang, *Physica* **15,** 774 (1949).

[p] G. Racah, *Physica* **16,** 651 (1950).

[q] C. K. Jorgensen, *Kgl. Danske Videnskab. Selskab Mat.-fys. Medd.* **29,** No. 11 (1955).

[r] C. K. Jorgensen, *Kgl. Danske Videnskab. Selskab Mat.-fys. Medd.* **29,** No. 7 (1955).

[s] D. Gruen, *J. Chem. Phys.* **20,** 1818 (1952).

The atomic constants which determine the magnitude the foregoing effects are the ζ_{nl}, defined by Condon and Shortley (p. 195 in ref. 28). These are tabulated for the free ions in Table VIII. It will be shown in Section 2 that the free-ion values are somewhat too large when applied to the transition ions in the solid state. They appear to be satisfactory for the rare earths in the few cases where comparisons have been made. The majority of the values for the tripositive rare earth ions have been determined in the crystalline state. The trends of the ζ values shown in Table VIII are: (1) ζ increases within each transition group by about one order of magnitude; (2) $\zeta_{5d} = 2\zeta_{4d} = 5\zeta_{3d}$; (3) $2\zeta_{4f} = \zeta_{5f}$.

f. Selection Rules

The states which arise when the free ion is subjected to a crystal field may be found by means of group theory, as shown by Bethe.[1] The $2J + 1$- or $2L + 1$-dimensional representation of the rotation group to which the atomic state belongs is considered to be a reducible representation of the group of the crystal field, and is therefore expressible in terms of the irreducible representations of the crystal field group. If the spin-orbit coupling is large compared to the crystal field splitting, as in the case of the rare earths, the J value of the state is used, but if it is smaller, as in transition metal ions, the L value is used. Table IX gives the numbers and representations of the states arising from an atomic state characterized by J or L and subjected to a crystal field of given symmetry. Mulliken's notation for the representations is preferred over Bethe's[42b] and is used in Table IX.

The selection rules for transitions of various multipole orders between the component states may be determined in the usual way. If the transition is allowed, the product of the representations in the appropriate symmetry group of the two combining states must contain the representation of the transition moment operator. The types of transitions which have been considered in the past[43] are electric dipole, magnetic dipole, and electric quadrupole.

The components of the dipole moment operator transform like translations and those of the magnetic dipole operator like rotations. The six components of the quadrupole moment operator transform like x^2, y^2, z^2, xy, xz, and yz. These are the same as the transformation properties of the polarizability components used in deriving selection rules for the Raman effect. The selection rules for quadrupole radiation are therefore the same as for the Raman effect. The transformation properties of the transition moment operators may be found with the character tables for the point groups in several textbooks.[44] They are also shown for a few important groups in Tables I and II of "Electronic Spectra of Molecules,"[45] and in Tables I, XIII and XIX of this article.

In the case of a very weak field, the selection rules for crystal field spectra will resemble those of the free atom. Only electric quadrupole and magnetic dipole transitions are allowed in the free atom, since all states of the same configuration have the same parity, and only an

[42b] This is explained in Section 1b of "Electronic Spectra of Molecules," D. S. McClure, *Solid State Phys.* **8,** 11 (1958).

[43] J. H. Van Vleck, *J. Phys. Chem.* **41,** 67 (1937).

[44] G. Herzberg, "Infrared and Raman Spectra." Van Nostrand, New York, 1945; E. B. Wilson, Jr., J. C. Decius, and P. C. Gross, "Molecular Vibrations." McGraw-Hill, New York, 1955.

[45] D. S. McClure, *Solid State Phys.* **8,** 1 (1958).

TABLE IX. REDUCTION OF REPRESENTATIONS IN SPHERE GROUP INTO REPRESENTATIONS OF POINT GROUPS*

Part A. Cubic groups, integral J values

Even states (g)

Symmetry group	$J = 0$	1	2	3	4	5	6
T	A	T	$E + T$	$A + 2T$	$A + E + 2T$	$E + 3T$	$2A + E + 3T$
T_d	A_1	T_1	$E + T_2$	$A_2 + T_1 + T_2$	$A_1 + E + T_1 + T_2$	$E + 2T_1 + T_2$	$A_1 + A_2 + E + T_1 + 2T_2$
O	A_1	T_1	$E + T_2$	$A_2 + T_1 + T_2$	$A_1 + E + T_1 + T_2$	$E + 2T_1 + T_2$	$A_1 + A_2 + E + T_1 + 2T_2$
O_h	A_{1g}	T_{1g}	$E + T_{2g}$	$A_{2g} + T_{1g} + T_{2g}$	$A_{1g} + E_g + T_{1g} + T_{2g}$	$E_g + 2T_{1g} + T_{2g}$	$A_{1g} + A_{2g} + E_g + T_{1g} + 2T_{2g}$

Odd states (u)

T no change
T_d replace subscript 1 by 2, and 2 by 1
O no change
O_h change g to u

Part B. Cubic groups, half-integral J values

Even states (g)

Symmetry group	$J = \frac{1}{2}$	$\frac{3}{2}$	$\frac{5}{2}$	$\frac{7}{2}$	$\frac{9}{2}$	$\frac{11}{2}$	$\frac{13}{2}$
T	$E_{\frac{1}{2}}$	G	$E_{\frac{1}{2}} + G$	$2E_{\frac{1}{2}} + G$	$E_{\frac{1}{2}} + 2G$	$2E_{\frac{1}{2}} + 2G$	$3E_{\frac{1}{2}} + 2G$
T_d	$E_{\frac{1}{2}}$	G	$E_{\frac{5}{2}} + G$	$E_{\frac{1}{2}} + E_{\frac{5}{2}} + G$	$E_{\frac{1}{2}} + 2G$	$E_{\frac{1}{2}} + E_{\frac{5}{2}} + 2G$	$E_{\frac{1}{2}} + 2E_{\frac{5}{2}} + 2G$
O	$E_{\frac{1}{2}}$	G	$E_{\frac{5}{2}} + G$	$E_{\frac{1}{2}} + E_{\frac{5}{2}} + G$	$E_{\frac{1}{2}} + 2G$	$E_{\frac{1}{2}} + E_{\frac{5}{2}} + 2G$	$E_{\frac{1}{2}} + 2E_{\frac{5}{2}} + 2G$
O_h	$E_{\frac{1}{2}g}$	G_g	$E_{\frac{5}{2}g} + G_g$	$E_{\frac{1}{2}g} + E_{\frac{5}{2}g} + G_g$	$E_{\frac{1}{2}g} + 2G_g$	$E_{\frac{1}{2}g} + E_{\frac{5}{2}g} + 2G_g$	$E_{\frac{1}{2}g} + 2E_{\frac{5}{2}g} + 2G_g$

Odd states (u)

T no change
T_d replace subscript $\frac{5}{2}$ by $\frac{1}{2}$, and $\frac{1}{2}$ by $\frac{5}{2}$
O no change
O_h replace g by u

* See Table I for explanation of notation.

TABLE IX. REDUCTION OF REPRESENTATIONS IN SPHERE GROUP INTO REPRESENTATIONS OF POINT GROUPS (*Continued*)

Part C. Noncubic point groups

Integral J values. The states corresponding to a given J are the sum of all the states given in the columns up to and including the Jth. In the $J = 0$ column, one uses either one state or the other depending upon whether J is an *even number* or an *odd number*. In some of the point groups, the states resulting from sphere group states depend upon the parity of the state. These are distinguished by the labels g or u on the rows.

		$J = 0$							
		Even J	Odd J	± 1	± 2	± 3	± 4	± 5	± 6
C_i	g	A	A	$2A_g$	$2A_g$				
	u	A_u	A_u	$2A_u$	$2A_u$				
C_{2h}	g	A_g	A_g	$2B_g$	$2A_g$	$2B_g$	$2A_g$		
	u	A_u	A_u	$2B_u$	$2A_u$	$2B_u$	$2A_u$		
D_2		A	B_1	$B_2 + B_3$	$A + B_1$	$B_2 + B_3$	$A + B_1$		
D_4		A_1	A_2	E	$B_1 + B_2$	E	$A_1 + A_2$	E	$B_1 + B_2$
D_3		A_1	A_2	E	E	$A_1 + A_2$	E	E	$A_1 + A_2$
C_{3h}	g	A'	A'	E''	E'	$2A''$	E'	E''	$2A'$
	u	A''	A''	E'	E''	$2A'$	E''	E'	$2A''$
D_{3h}	g	A_1'	A_2'	E''	E'	$A_1'' + A_2''$	E'	E''	$A_1' + A_2'$
	u	A_1''	A_2''	E'	E''	$A_1' + A_2'$	E''	E'	$A_1'' + A_2''$
D_6		A_1	A_2	E_1	E_2	$B_1 + B_2$	E_2	E_1	$A_1 + A_2$

Part D. Noncubic point groups, half-integral J values

		$\frac{1}{2}$	$\frac{3}{2}$	$\frac{5}{2}$	$\frac{7}{2}$	$\frac{9}{2}$	$\frac{11}{2}$	$\frac{13}{2}$
C_i	g	$E_{\frac{1}{2}g}$	$E_{\frac{1}{2}g}$					
	u	$E_{\ u}$	$E_{\frac{1}{2}u}$					
C_{2h}	g	$E_{\frac{1}{2}g}$	$E_{\frac{1}{2}g}$	$E_{\frac{1}{2}g}$				
	u	$E_{\frac{1}{2}u}$	$E_{\frac{1}{2}u}$	$E_{\frac{1}{2}u}$				
D_2		$E_{\frac{1}{2}}$	$E_{\frac{1}{2}}$	$E_{\frac{1}{2}}$				
D_4(a)		$E_{\frac{1}{2}}$	$E_{\frac{3}{2}}$	$E_{\frac{3}{2}}$	$E_{\frac{1}{2}}$	$E_{\frac{1}{2}}$	$E_{\frac{3}{2}}$	$E_{\frac{3}{2}}$
D_3(b)		$E_{\frac{1}{2}}$	$E_{\frac{3}{2}}$	$E_{\frac{1}{2}}$	$E_{\frac{1}{2}}$	$E_{\frac{3}{2}}$	$E_{\frac{1}{2}}$	$E_{\frac{1}{2}}$
D_{3h}(c)	g	$E_{\frac{1}{2}}$	$E_{\frac{3}{2}}$	$E_{\frac{5}{2}}$	$E_{\frac{5}{2}}$	$E_{\frac{3}{2}}$	$E_{\frac{1}{2}}$	$E_{\frac{1}{2}}$
	u	$E_{\frac{5}{2}}$	$E_{\frac{3}{2}}$	$E_{\frac{1}{2}}$	$E_{\frac{1}{2}}$	$E_{\frac{3}{2}}$	$E_{\frac{5}{2}}$	$E_{\frac{5}{2}}$
D_6(d)		$E_{\frac{1}{2}}$	$E_{\frac{3}{2}}$	$E_{\frac{5}{2}}$	$E_{\frac{5}{2}}$	$E_{\frac{3}{2}}$	$E_{\frac{1}{2}}$	$E_{\frac{1}{2}}$

(a) $E_{\frac{1}{2}} = \Gamma_6$, $E_{\frac{3}{2}} = \Gamma_7$ of tetragonal double group [H. A. Bethe, *Ann. Physik* [5] **3,** 133 (1929)].

(b) $E_{\frac{1}{2}} = {}^2\Gamma_6$, $E_{\frac{3}{2}} = {}^1\Gamma_4 + {}^1\Gamma_5$ of rhombohedral double group [W. Opechowski, *Physica* **7,** 552 (1940)].

(c) See Table XIII.

(d) $E_{\frac{1}{2}} = \Gamma_7$; $E_{\frac{3}{2}} = \Gamma_9$; $E_{\frac{5}{2}} = \Gamma_8$ of hexagonal double group [H. A. Bethe, *Ann. Physik* [5] **3,** 133 (1929)].

operator of even parity may cause a transition. An unsymmetrical electric crystal field may of course destroy the inversion symmetry enough to permit electric dipole transitions.

These facts show: (1) the symmetry group which determines the selection rules for electric dipole transitions is that of the local crystal field at the site of the ion; (2) magnetic dipole transitions and electric quadrupole transitions are determined both by the site group and by the selection rules for the free atom. The appropriate free atom selection rules are:

Magnetic dipole: $\Delta J = 0, \pm 1 \qquad J = 0 \rightarrow J = 0$ forbidden.
Electric quadrupole: $\Delta J = 0, \pm 2.$

The selection rules for the free atom have the most force in the lanthanide and actinide series where $f \rightarrow f$ transitions occur. Many magnetic dipole transitions are observed in spectra of these ions (Section 3). In the transition groups, practically all of the $d \rightarrow d$ transitions appear to be electric dipole, and therefore occur through environmental perturbations.

g. Intensities

The intensities of spectral transitions provide valuable information about the nature of the combining states. Electric dipole transitions are involved in the majority of crystal field spectra. The combining states ψ_1 and ψ_2 must therefore differ in their symmetry properties by at least one component of a translation vector. In the free ion, as already mentioned, direct inner shell transitions, $d \rightarrow d$ and $f \rightarrow f$ are therefore forbidden. When the ion is in a crystal, this prohibition is removed by the mixing of orbitals having opposite parity. Such mixing may be induced by (a) absence of a center of symmetry of the crystal field; (b) destruction of the center of symmetry by vibrations. Van Vleck[43] was the first to have pointed out these facts.

The first of these conditions appears in most rare earth salts and in many transition metal salts, such as those containing tetrahedrally coordinated ions. Ballhausen and Liehr[45a] have calculated intensities in tetrahedral complexes resulting from the mixing of $3d$- and $4p$-orbitals and from the mixing of the ligand orbitals with the $3d$-orbitals.

The $4p$-$3d$ mixing in a tetrahedral field is brought about by potential terms of the form $V_3 = Txyz$, and it can be seen that d-orbitals transforming like xy are mixed with p-orbitals transforming like z, etc. Therefore the t-orbitals are perturbed by this mixing whereas the e-orbitals are not. The actual magnitude of the mixing may be estimated by the use of a table of ligand field integrals published by Ballhausen and Ancmon.[45b]

[45a] C. J. Ballhausen and A. D. Liehr, *J. Molec. Spectrosc.* **2,** 342 (1958).

[45b] C. J. Ballhausen and E. M. Ancmon, *Kgl. Danske Videnskab. Selskab Mat.-fys. Medd.* **31,** No. 9, 1–38 (1958).

Hydrogen-like radial wave functions are used in these tables, and the integrals are tabulated for various values of the effective nuclear charges. The intensities of the absorption spectra of $CuCl_4^=$ and $CoCl_4^=$ ions were estimated by Ballhausen and Liehr using the perturbed wave functions:

$$\psi'(3d) = \psi^0(3d) + \sum_{4p} \frac{(\psi^0(3d)|V_3|\psi^0(4p))}{E_p{}^0 - E_d{}^0} \psi^0(4p). \tag{1.13}$$

The intensities from this source were found to be too small by a factor of ten or more. A similar intensity calculation could be made for rare-earth ions, for which the $4f$-$5d$ mixing would be used. The results should be better for this case. For transition metal ions, however, Ballhausen and Liehr found that the $3d$-ligand orbital mixing was needed in order to explain the observed intensities.

The new parameter introduced by considering the effect of the ligands is the intensity of a one-electron transition from a ligand orbital into a d-orbital, i.e. a charge transfer transition. These transitions are observed directly, and are discussed in Section 2. In the present case however, they affect the intensity of crystal field bands by a second order effect. Since the intensities calculated in this way are of the correct order of magnitude, it appears that the mixing of crystal field transitions with charge transfer transitions is the main source of intensity in tetrahedral complexes, and presumably also in other non-centrosymmetric complexes.

Mixing of orbitals with opposite parity is also brought about by odd vibrations. This coupling permits mixing between p- and d-orbitals in transition metal ions, or between f- and d-orbitals in lanthanide and actinide ions. A detailed theoretical treatment for transition ions has been given recently by Liehr and Ballhausen,[46] and for rare earth ions by Satten.[47]

Liehr and Ballhausen applied their theory to transition strengths in $Ti^{+++}(H_2O)_6$ and $Cu^{++}(H_2O)_6$, and obtained results in good agreement with experiment. The vibrationally perturbed wave functions are of the form $\psi'(d) = \psi^0(d) + \sum_k \sum_p Q_k \gamma_k{}^d(p) \psi^0(p)$ in which the summations are carried out over all normal coordinates Q_k and all excited p-states having odd parity, $\psi_u{}^0(p)$. The mixing coefficient is given by

$$\gamma_k{}^d(p) = -\frac{\int \psi_u{}^0(p) H(Q_k) \psi_g{}^0\, d\tau}{E_p - E_d}.$$

The perturbing potential for vibrational-electronic interaction in mole-

[46] A. D. Liehr and C. J. Ballhausen, *Phys. Rev.* **106,** 1161 (1957); also unpublished work.
[47] R. A. Satten, *J. Chem. Phys.* **27,** 286 (1957); **29,** 658 (1958).

cules $H(Q_k)$, is discussed in an important paper by Liehr and Moffitt.[48] In the case treated by Liehr and Ballhausen, the $X(H_2O)_6{}^{n+}$ complex ion is simplified to a XY_6 molecule having octahedral symmetry. Of the six normal modes of vibration of such a molecule the most effective one is a T_{1u} mode since this produces the most unsymmetrical potential in the region of the central ion. A similar type of motion occurs as one of the optical mode branches in cubic crystals.

Koide and Pryce[48a] have made even more detailed calculations of the induced intensity. They treated the case of Mn^{++} ion in cubic fields, where it is necessary to consider vibrational-electronic interaction and spin-orbit coupling simultaneously. They calculated the ${}^6A_{1g}({}^6S) \rightarrow {}^4A_1,\ {}^4E({}^4G)$ transition probability. Furthermore they estimated the individual vibronic band intensities. These estimates were of the correct magnitude and some of the features of the 25000 cm^{-1} region of the Mn^{++} spectrum (see Fig. 23) are explained by the calculation.

Satten[47] has recently discussed the effects of lattice vibrations on the spectra of the rare earths. He has considered the altered selection rules and the enhanced intensities which may result from the presence of low frequency lattice vibrations. He has concluded that certain low frequency lattice modes should have at least one-percent of the intensity of the pure electronic (electric-dipole) transitions. The reason that they are so nearly comparable in rare earths is that the transitions allowed by symmetry are permitted because of $4f$-$5d$ mixing, and this is also the source of the intensity of the vibrationally induced transitions. The energy denominators are therefore the same for both perturbations. The direct transitions are caused by the terms of odd power, $V_3{}^3$ and $V_5{}^3$, in a D_{3h} potential field, but the vibrational-electronic interaction may make use of the $V_2{}^0$ term. The contributions to the latter from distant atoms may be considerable, and a relatively large vibrational-electronic interaction could result.

Satten suggests that vibrationally induced lines may be the source of extra structure in rare earth spectra. The "extra lines" are those appearing in excess of the number required by group theory for the splitting of an atomic level with a given J value, as given in Table IX.

If a transition is electronically forbidden, but is permitted to occur through vibrational-electronic interaction, the corresponding absorption strength should be strongly dependent upon temperature, because the permitting vibration of the ground state may be "frozen out" at sufficiently low temperatures. The temperature dependence of the absorption

[48] A. D. Liehr and W. Moffitt, to be published; A. D. Liehr, Thesis, Harvard University (1955).

[48a] S. Koide and M. H. L. Pryce, *Phil. Mag.* [8] **3**, 607 (1958).

strength of the $d \rightarrow d$ transitions in several crystalline hydrates was studied by Holmes and McClure.[49] The integrated band intensity expressed as f-number for two absorption bands of $NiSO_4 \cdot 7H_2O$ is plotted against temperature in Fig. 11. The temperature dependence of the

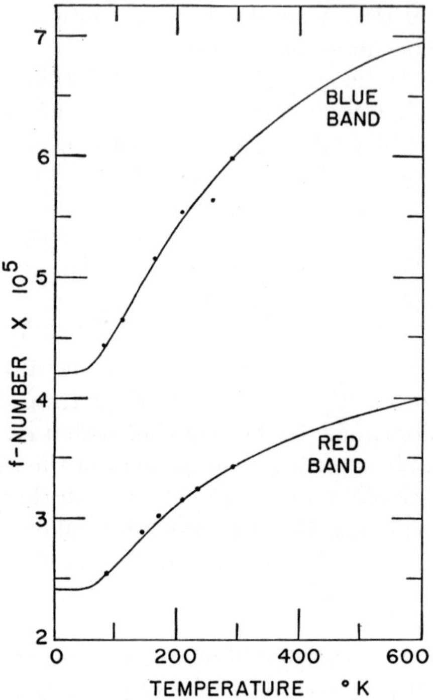

FIG. 11. The f-number of two absorption bands of $NiSO_4 \cdot 7H_2O$. The "red band" is at 14,000 cm^{-1} and the blue band at 25,000 cm^{-1} (Fig. 20). The points are experimental, and the curve is Eq. (1.14). The parameter θ is 250°K. The f-number is proportional to the area under the absorption band.

oscillator strength, f, was fitted to the formula:

$$f = f_0(1 + \exp(-\theta/T)) \qquad (1.14)$$

where θ is the frequency of the non-totally-symmetric vibration in temperature units ($\theta/1.44 = \omega$ cm^{-1}), and f_0 is the value of the oscillator strength at 0°K. The data on each band of $NiSO_4 \cdot 7H_2O$ gave the same θ value, 250°K, or 173 cm^{-1}. This is a reasonable frequency for a T_{1u} vibration of the $Ni(H_2O)_6$ group. The more highly charged Cr^{+++} ion would be expected to polarize the water molecules surrounding it more

[49] O. Holmes and D. S. McClure, *J. Chem. Phys.* **26**, 1686 (1957).

than Ni^{++} does and therefore to raise the vibration frequency. The θ value for $KCr(SO_4)_2 \cdot 12H_2O$ was, in fact, found to be 400°K or 277 cm^{-1}.

The values of f_0 were not obtained, although the f value at 77°K should not be much larger than f_0. In the typical case of $NiSO_4 \cdot 7H_2O$, the f number dropped by 40% between 300°K and 77°K. This shows that a very large fraction of the oscillator strength comes from the vibrational perturbation. If the upper state vibration is as effective as the lower state vibration in causing *p-d* mixing, only a small percentage of the transition strength can arise from any other perturbation, such as that due to unsymmetrical crystal fields.

The weak and sharp intercombination bands observed in some transition metal ions may be due to magnetic dipole transitions, but no experimental work has been carried out to find if this is true. Many rare earth

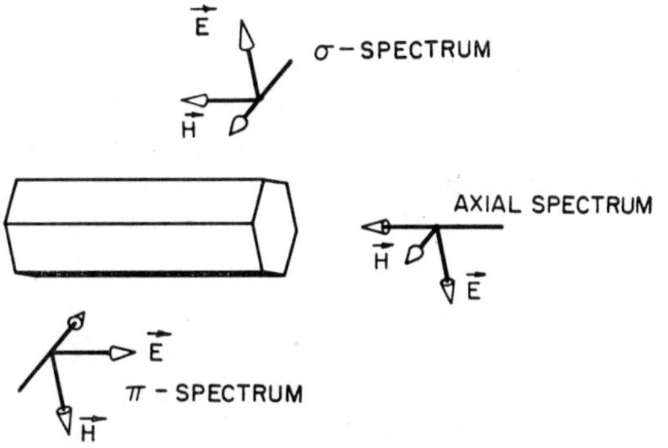

FIG. 12. Three ways in which to observe the spectrum of a uniaxial crystal.

ions have magnetic dipole transitions that are as strong as the electric dipole transitions.

Electric and magnetic transition processes may be distinguished experimentally in a simple way (see, for instance, Sayre *et al.*[50]). Take for example, a uniaxial crystal. Referring to Fig. 12 the absorption spectrum of the crystal could be observed in three ways:

(1) with propagation vector along the optic axis (*a* spectrum)
(2) with propagation vector and E vector perpendicular to optic axis (σ spectrum)
(3) with propagation vector and H vector perpendicular to optic axis (π spectrum).

If the σ spectrum coincides with the *a* spectrum, the electric vector must have been active, and the transition is either electric dipole or electric

[50] E. V. Sayre, K. Sancier, and S. Freed, *J. Chem. Phys.* **23**, 2060 (1955).

quadrupole. If the π spectrum coincides with the a spectrum, the magnetic vector must have been active, and the transition is magnetic dipole. The distinction between electric dipole and electric quadrupole processes could be made on the basis of the polarization behavior in known transitions if the symmetry of the system is high enough, but quadrupole processes are so weak that no one has yet identified one with certainty.

2. Molecular Orbital Theory and Charge Transfer Spectra

a. The Need for Extending Crystal Field Theory

Crystal field theory in its simplest form deals with an atom in a static electric field. This approximation cannot be strictly correct, and there are several experiments which demonstrate the limits of its usefulness. For the series of $3d$ ions, Owen[18] has shown that both the positions of the visible absorption bands and the g factor observed in electron spin resonance are not in exact agreement with the static theory. Stevens[51] had shown earlier that the extra hyperfine structure in the spin-resonance spectrum of $IrCl_6^=$ comes from the Cl^- ions. He interpreted this as evidence that the orbitals involved in the resonance are mixed with the Cl^- orbitals. More recent work on Mn^{++} and Fe^{++} in solid solution in ZnF_2 by Tinkham[52] has given a detailed picture of the orbital mixing between Mn^{++} (or Fe^{++}) and F^-. This work was based mainly upon resolving the F^{19} hyperfine structure in the electron spin-resonance spectrum.

The simplest way in which to include the ligands, or nearest neighbor anions, in the electronically excited system is to use molecular orbital theory.[4] This theory is grafted easily onto crystal field theory in the strong field limit by adding ligand orbitals of the appropriate symmetry to each d orbital. For example, one partner of e_g is

$$\psi_e = bd_{(x^2-y^2)} + \frac{\sqrt{1-b^2}}{2}(\sigma_x - \sigma_{\bar{x}} - \sigma_y + \sigma_{\bar{y}}) \tag{2.1}$$

where b is a mixing coefficient of nearly unity, and σ_x, etc., are ligand orbitals to be described more fully later. The departures from strict crystal field theory depend upon the value of b. Since b must be less than one, all physical properties depending on d orbitals are reduced by this factor, or some power of it. The occurrence of the factor b is one major difference between crystal field theory and molecular orbital theory. It makes the latter theory more flexible because of the presence of an additional arbitrary parameter.

Owen[18] showed that the term separations of the free ion are altered

[51] K. W. H. Stevens, *Proc. Roy. Soc.* (*London*) **A219,** 542 (1953).
[52] M. Tinkham, *Proc. Roy. Soc.* (*London*) **A236,** 535, 549 (1956).

upon forming the solid. In the hydrated ions which he investigated, the ^nF-^nP separation in Cr^{+++}, Ni^{++}, V^{++}, V^{+++}, Co^{++} was smaller in the solid by 10 to 30%. This separation equals $15B$, and therefore gives a direct measure of B in the crystal. The spin-orbit interaction parameter λ was also found to be smaller than in the free ion.[52a] Since $g = 2[1 - (4\lambda/10Dq)]$ for those ions having nondegenerate normal states, the g value may be used to measure λ, whose value depends upon the amount of time which the d electron spends near heavy nuclei. The contributions to λ from oxygen nuclei are negligible compared to that from a Cr or Ni nucleus. In order to agree with experiment, the value used to calculate g must be about 20% smaller than the value measured in the free ion. Thus the d electron is only on the metal ion about 80% of the time.

The reduction of the width of the hyperfine structure of the central ion in complex ions or solids relative to its value in the free ion may also be interpreted in terms of molecular orbital formation.[53] Tanabe and Sugano's work on the direct calculation of Dq[27] also shows that the d orbitals mix with the ligand orbitals. They found that, without including the orbitals of the surrounding ions explicitly, the calculated Dq value has the wrong sign. Unfortunately, their calculations of λ and of the electrostatic interaction parameters do not agree with the empirical estimates.

b. *Formal Treatment of Molecular Orbital Theory*

The mathematical treatment of the molecular orbital approximation is best started from the theory of the strong crystal field (Section 1d).[4] The d orbitals in an octahedral field belong to the E_g and T_{2g} representations of O_h. The orbitals of the nearest surrounding ions may also be classified in this group, provided that the effect of the ions in the next neighboring shell are ignored and it is assumed that the coupling between the metal atom and its nearest neighbor entirely determines the symmetry orbitals. Each nearest neighbor is assumed to have a p orbital directed along the bond called a σ orbital and a pair perpendicular to the bond, called π-orbitals.

One of the combinations of ligand σ orbitals has the symmetry E_g and may therefore mix with the E_g d orbitals to form a bonding and an antibonding combination. One of the combinations of π-orbitals has the symmetry T_{2g}, and may mix with the T_{2g} d-orbital. Since the π-orbitals do not point toward the central atom, they form much weaker bonding and antibonding orbitals than do the σ-orbitals. They may sometimes be con-

[52a] The relation between λ and the one-electron parameters ζ_{nl} for the Hund rule ground states (maximum S and L) is $\lambda = \pm(1/2S)\zeta_{nl}$ where the + sign applies to the first half of the transition series and the − sign to the second half.

[53] J. S. van Wieringen, *Discussions Faraday Soc.* **19,** 118 (1955).

sidered as nonbonding. In both the crystal field theory and the molecular orbital theory, the difference in energy between the E_g and T_{2g} orbitals is a result of the same geometrical factors, but it is called an electrostatic effect in the former, and an antibonding effect of σ-ligand orbitals in the

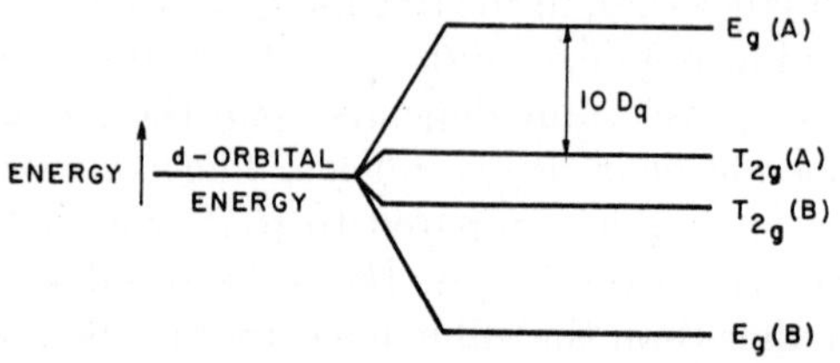

FIG. 13. The interpretation of Dq on the basis of molecular orbital theory. The upper orbitals are antibonding (A), the lower ones bonding (B); 10 Dq is the difference in energy between the strongly antibonding e_g orbital and the weakly antibonding t_{2g} orbital.

latter. The interpretation of Dq on the basis of molecular orbital theory is illustrated by Fig. 13.

The other ligand orbitals do not perturb the d but may act upon the s- or p-orbitals of the central ion.

The symmetry orbitals are given in Table X. A p-orbital is called positive if its positive lobe points along the positive axis. A diagram show-

TABLE X. ORBITALS FOR OCTAHEDRAL ARRAY OF ATOMS AROUND A CENTRAL ATOM

The outer atom orbitals are explained in Fig. 14. In the A_{1g}, E_g, and T_{1u} classes there is more than one orbital in the symmetry class, and the correct "combined orbitals" are mixtures. The description σa_{1g}, sa_{1g} are no longer appropriate after mixing, and they are called $1A_{1g}$, $2A_{1g}$, etc., in the text and in Fig. 16. The coefficients of the central atom and ligand orbitals in the combined orbital are determined in principle by the variation method.

Symmetry	Central atom	Outer atoms (ligands)	Combined orbital
A_{1g}	s	$\frac{1}{\sqrt{6}}(\sigma_x - \sigma_{\bar{x}} + \sigma_y - \sigma_{\bar{y}} + \sigma_z - \sigma_{\bar{z}})$	σa_{1g}, sa_{1g}
E_g	$de = (2x^2 - y^2 - z^2)$,	$\frac{1}{2\sqrt{3}}[2(\sigma_x - \sigma_{\bar{x}}) - (\sigma_y - \sigma_{\bar{y}} + \sigma_z - \sigma_{\bar{z}})]$	de_g, σe_g
	$(y^2 - z^2)$	$\frac{1}{2}[\sigma_y - \sigma_{\bar{y}} - \sigma_z + \sigma_{\bar{z}}]$	
T_{1u}	$p = (x,y,z)$	$\frac{1}{2}(\sigma_x + \sigma_{\bar{x}})$, etc.	pt_{1u}, πt_{1u}, σt_{1u}
		$\frac{1}{2}[(x_z + x_{\bar{z}}) + (x_y + x_{\bar{y}})]$, etc.	
T_{1g}		$\frac{1}{2}[(x_z - x_{\bar{z}}) - (z_x - z_{\bar{x}})]$, etc.	πt_{1g}
T_{2u}		$\frac{1}{2}[(x_z + x_{\bar{z}}) - (x_y + x_{\bar{y}})]$, etc.	πt_{2u}
T_{2g}	$dt = (xz$, etc.)	$\frac{1}{2}[(x_z - x_{\bar{z}}) + (z_x - z_{\bar{x}})]$, etc.	πt_{2g}, dt_{2g}

ing the coordinates and explaining the notation and the phases of the orbitals is given in Fig. 14.

The only attempt to use the molecular orbital theory quantitatively to explain the deficiencies of crystal field theory has been made by Tanabe and Sugano.[27] They have calculated Dq, λ, and the ${}^nF$-${}^nP$ separation in $Cr(H_2O)_6^{+++}$ using the self-consistent field d function of Cr^{++} for the Cr^{+++} ion, and Slater orbitals for the $O^=$ ion. The central ion and oxygen orbitals were made orthogonal in the first part of this calculation, rather than being mixed as in the MO method. This is therefore a purely ionic model. Although the sign and order of magnitude of Dq were found correctly, the B integral of Racah (determining the ${}^nF$-${}^nP$ separation) was

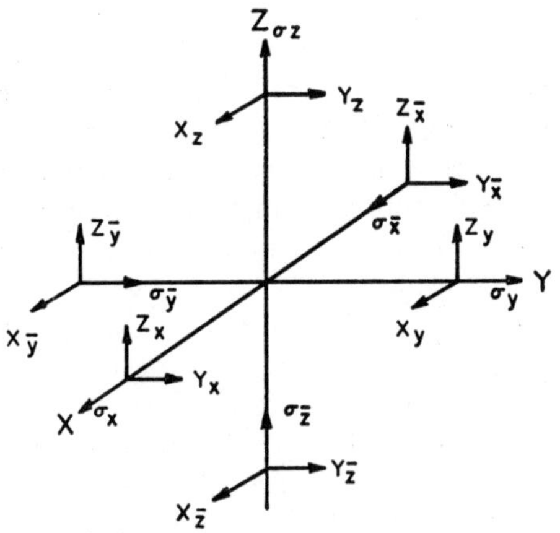

FIG. 14. To illustrate the notation and the phases for the ligand orbitals of an octahedral array of atoms. X_z, etc., are π orbitals, σ_z, etc., are σ orbitals.

not significantly smaller than in the free ion. The second step was to determine a molecular orbital by the variation principle but even this did not produce the reduction in B (or λ) which is needed.

The authors believe that the Cr^{++} d-wave function they use is too compressed, and quote the results of Kuroda and Itoh[54] in support of this diagnosis. The latter workers used Pauling's electroneutrality principle which is based on the assumption that the Cr atom is neutral and the charge of the complex is spread over the six ligands. When the d orbital of neutral Cr is used, the d orbitals spread out and overlap more with the ligand orbitals. The use of O^- orbitals instead of the more diffuse $O^=$ orbitals, reduces the overlap with the central ion. In Tanabe and Sugano's calculation, these orbitals repelled the d orbitals from the ligand direction

[54] Y. Kuroda and K. Itoh, *J. Chem. Soc. Japan* **76,** 766 (1955).

too much, and resulted in too large a value for Dq. Kuroda and Itoh obtained a reasonable Dq value, but did not report any calculations of B and λ.

These calculations illustrate some of the important constituents of a crystal field and suggest directions for further progress. In spite of the complications which arise when a detailed interpretation of the measured quantities is attempted, the molecular orbital theory gives a consistent semiquantitative account of the observations and is valuable for the interpretation of spectra.

c. *Charge Transfer Processes*

A second major difference between crystal field theory and molecular orbital theory is that the latter predicts the occurrence of more states. When the ligand orbitals are included explicitly, there arises the possibility of charge transfer from ligand to ion or the reverse. In the language of molecular orbitals, transitions may occur from orbitals having predominantly d character to those having predominantly ligand character. There is considerable experimental evidence to show that these transitions are a very large class (Section 5).

The term charge transfer should not be accepted literally. In the absence of exact knowledge of the wave functions of complicated electronic systems we cannot know the true charge distribution and how it changes upon electronic excitation. The principal reasons for placing an observed spectrum in this category are: (1) the intensity is hundreds or thousands of times greater than crystal field spectra; (2) a loosely bound electron and a low-energy hole are available for a transition. We will consider only the low-energy charge transfer transitions at first, since their identification in observed spectra appears to be quite unambiguous.

The molecular orbital theory also permits transitions to occur which involve a redistribution of electrons on the ligands. Again, it is not correct to regard the central ion as unaffected; a component of the charge transfer process must also be involved. These transitions may occur without benefit of the available holes on a central transition metal ion, whereas the low-energy charge transfer transitions could not occur without such states.

A third type of transition process, not explicitly considered in crystal field theory, is the $d \rightarrow s$ or $d \rightarrow p$ transition within the metal ion. The upper p-orbitals of the metal ion must be strongly involved with the ligand orbitals, so that a distinction between these and charge transfer processes may not be easy. Furthermore, the excited states of these two kinds of transitions have the same symmetry properties, and therefore mix through configuration interaction. The inner $d \rightarrow s$ or p transitions will therefore be assumed to be not essentially different from charge transfer processes of the type metal ion $\rightarrow$ ligand. It is probable that a finer classification of charge transfer processes will be made in the future when more experimental data are available.

(*1*) *Octahedral coordination.* In order to understand the details of the charge transfer process in a semiquantitative way we must examine the orbitals of Table X in more detail. The nodal patterns of the ligand part of these orbitals are shown in Fig. 15. The electrons in the ligands will interact in such a way as to make the different ligand orbitals move apart. Those orbitals having the greatest number of nodes will be highest in energy. The order of the orbitals having the same number of nodes is determined qualitatively by the overlap in the nonbonding region. The splitting of the ligand orbitals due to the interaction of electrons on different ligands is shown on the left of Fig. 16. These ligand orbital energies are modified by their interaction with the metal ion whose unperturbed orbitals are shown at the right of Fig. 16. A final orbital energy diagram is shown in the center of the figure. The scale of the figure is arbitrary.

In crystals containing transition metal ions or complex ions, the orbitals up to $1t_{1g}$ are filled, and $2e_g$, $2t_{2g}$ may be partially filled depending upon the number of d electrons present. Crystal field transitions take place between states differing only in the occupation of $2e_g$ and $2t_{2g}$. The initial and final states of a charge transfer transition differ by the transfer of an electron from one of the orbitals $1t_{1g}$ and below, to an orbital above.

There are two nonbonding ligand orbitals. Since they are the least tightly bound, electrons in them may make low-energy transitions into the antibonding d orbitals. Judging from the nodal patterns, the $1t_{1g}$ orbital should be the higher one. However, since the d orbitals are g, this charge transfer process is forbidden, and the first strong transition should appear at a higher energy, corresponding to $1t_{2u} \rightarrow 2e_g$ or $2t_{2g}$.

Many states arise from a given electron configuration because the orbitals are degenerate. The ground state may be degenerate as it is in a weak crystal field, but if the ion-ligand interaction is strong, most of degeneracy is removed as in the strong-field limit of crystal field theory. In case there are no d electrons, the ground configuration and ground state are:

$$\cdots\ 1e_g{}^4\ 1t_{2u}{}^6\ 1t_{1g}{}^6 \qquad {}^1A_{1g}$$

and the first excited configurations and states are:

$$\cdots\ 1e_g{}^4\ 1t_{2u}{}^6\ 1t_{1g}{}^5\ 2e_g \qquad {}^{1,3}(T_{1g} + T_{2g})$$
$$\cdots\ 1e_g{}^4\ 1t_{2u}{}^5\ 1t_{1g}{}^6\ 2e_g \qquad {}^{1,3}(T_{1u} + T_{2u}).$$

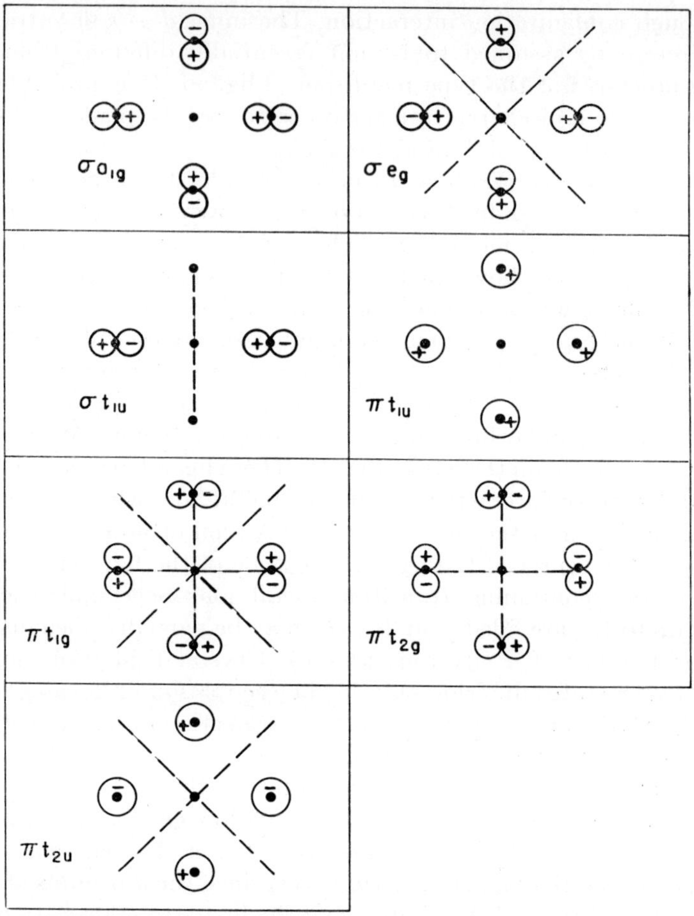

FIG. 15. Nodal patterns of the ligand orbitals of an octahedral array of atoms. The ligand atoms are assumed to have filled p orbitals which are oriented in each of the three coordinate directions. The phases are indicated by + and −, and the nodal planes by dotted lines. In πt_{1u} and πt_{2u}, the plane of the paper is a nodal plane. Only one partner of a degenerate orbital is shown. The s orbitals have a set of nodal diagrams like the $p\sigma$ orbitals.

A detailed comparison of the molecular orbital theory of octahedral complex ions with experimental data has been made only for the case of $Co(NH_3)_6^{+++}$ and closely related complexes. It will be discussed in detail in Section 5. Less detailed comparisons have been made for $IrCl_6^{=}$, $RhCl_6^{≡}$, $IrCl_6^{-}$, $PtCl_6^{-}$, and many others. The complex cyanides such as $Cr(CN)_6^{≡}$ should also be amenable to this kind of treatment. The weak transitions characteristic of crystal field spectra appear at low energies in many of these complexes, and the strong charge transfer bands at higher energies.

(*2*) *Tetrahedral coordination.* The molecular orbital theory of tetrahedral complexes was worked out several years ago by Wolfsberg and Helmholz,[55] and applied to the interpretation of the charge transfer spectra of MnO_4^- and $CrO_4^=$. The molecular orbitals were constructed as described for octahedral complexes, and a fairly detailed calculation of

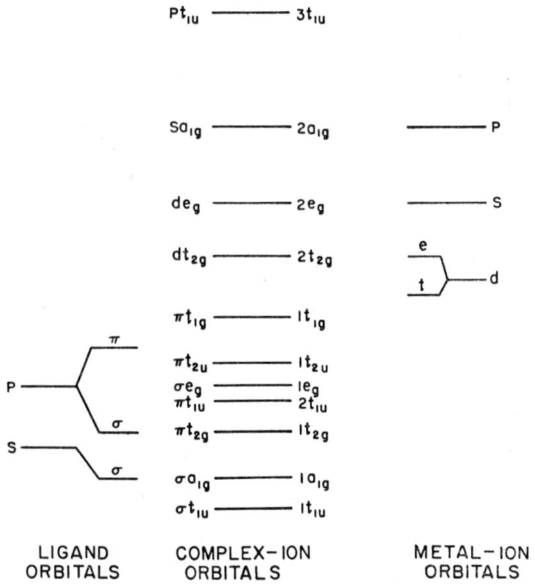

FIG. 16. Illustrating the energy levels of an octahedral complex ion. The order of the ligand orbitals in the complex ion is determined from Fig. 15. The metal ion orbitals in the complex ion are assumed to be antibonding. There are usually several orbitals in each symmetry class, and the orbitals resulting from the solution of a secular equation are labeled on the right of the central column in order of their energy; e.g., $1t_{1u}$, $2t_{1u}$, $3t_{1u}$.

the orbital energies was made. The molecular orbitals of a tetrahedral system are shown in Table XI.

The energy level diagram for the molecular orbitals of a tetrahedral ion is shown in Fig. 17. The order of the energy levels given here is appropriate for the molecular ion MnO_4^- according to the work of Wolfsberg and Helmholz.

There are some important differences between tetrahedral and octahedral systems. One is that the orbitals of the central ion d_{xy}, d_{xz}, d_{yz}

[55] M. Wolfsberg and L. Helmholz, *J. Chem. Phys.* **20**, 837 (1952).

TABLE XI. SYMMETRY ORBITALS FOR A TETRAHEDRAL ARRAY OF ATOMS ABOUT A CENTRAL ATOM
[M. Wolfsberg and L. Helmholz, *J. Chem. Phys.* **20,** 837 (1952)]

The σ orbitals of the ligand atoms are directed from the ligand toward the central atom in certain of the 111 directions: $\sigma_1(\overline{111})$, $\sigma_2(\overline{11}1)$, $\sigma_3(\bar{1}11)$, $\sigma_4(11\bar{1})$. One set of the π-ligand orbitals is perpendicular to the plane formed by the y axis and the threefold axis. They are denoted π_y. The other set is perpendicular to the threefold axis and the first set and is denoted π_x. Each set transforms into a positive member of the same set under the $2x$ axes 100, 010, 001. These orbitals may be pictured with the help of Fig. 2.

Symmetry	Central atom	Outer atoms
A_1	s	$(\frac{1}{2})[\sigma_1 + \sigma_2 + \sigma_3 + \sigma_4]$
E	d_{z^2}, $d_{x^2-y^2}$	$(\frac{1}{4})[\pi_{x1} + \pi_{x2} + \pi_{x3} + \pi_{x4} - 3^{\frac{1}{2}}(\pi_{y1} + \pi_{y2} + \pi_{y3} + \pi_{y4})]$
		$(\frac{1}{4})[\pi_{y1} + \pi_{y2} + \pi_{y3} + \pi_{y4} + 3^{\frac{1}{2}}(\pi_{x1} + \pi_{x2} + \pi_{x3} + \pi_{x4})]$
T_2	p_x, d_{yz}	$(\frac{1}{2})[\sigma_1 + \sigma_3 - \sigma_2 - \sigma_4]$, $(\frac{1}{4})[\pi_{x4} + \pi_{x2} - \pi_{x1} - \pi_{x3} + 3^{\frac{1}{2}}(\pi_{y4} + \pi_{y2} - \pi_{y1} - \pi_{y3})]$
	p_y, d_{xz}	$(\frac{1}{2})[\sigma_1 + \sigma_2 - \sigma_3 - \sigma_4]$, $(\frac{1}{2})[\pi_{x1} + \pi_{x2} - \pi_{x3} - \pi_{x4}]$
	p_z, d_{xy}	$(\frac{1}{2})[\sigma_1 + \sigma_4 - \sigma_2 - \sigma_3]$, $(\frac{1}{4})[\pi_{x3} + \pi_{x2} - \pi_{x1} - \pi_{x4} + 3^{\frac{1}{2}}(\pi_{y4} + \pi_{y1} - \pi_{y2} - \pi_{y3})]$
T_1		$(\frac{1}{4})[\pi_{y2} + \pi_{y4} - \pi_{y3} - \pi_{y1} + 3^{\frac{1}{2}}(\pi_{x1} + \pi_{x3} - \pi_{x2} - \pi_{x4})]$
		$\frac{1}{2}[\pi_{y1} + \pi_{y2} - \pi_{y3} - \pi_{y4}]$
		$(\frac{1}{4})[\pi_{y2} + \pi_{y3} - \pi_{y1} - \pi_{y4} + 3^{\frac{1}{2}}(\pi_{x2} + \pi_{x3} - \pi_{x1} - \pi_{x4})]$

belong to the same representation as do the orbitals p_x, p_y, p_z, namely T_2. In the octahedral system, the presence of a center of symmetry prevents these orbitals from mixing. One result of *p-d* mixing is an increase in the intensity of the crystal field transitions, $3t_2 \rightarrow 2e$ in Fig. 17. The reason is that the Laporte rule has reduced significance since a $d \rightarrow p$ component

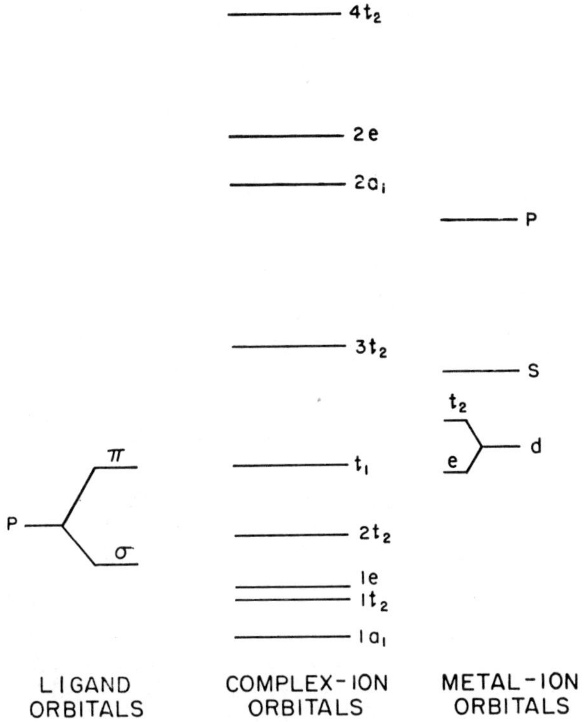

FIG. 17. Illustrating the energy levels of a tetrahedral array of atoms about a central atom. The diagram is appropriate for the MnO_4^- ion, according to the work of Wolfsberg and Helmholz.[55]

occurs in the transition moment integral. The observed intensities are about 100 times greater than in octahedral systems.

Another difference, which has already been mentioned (Section 1d) is the smaller value of Dq in tetrahedral systems for which

$$Dq = E(3t_2) - E(2e),$$

using the notation of Fig. 17. The reduction of Dq has two origins, according to molecular orbital theory: (1) the mixing of p and d orbitals reducer the energy of the $3t_2$ orbital; (2) the π- and σ-ligand orbitals are no longes in separate symmetry classes as in octahedral systems, and the energy differences between them are not as great. The result is a reduction in the energy difference between e and t_2 orbitals. Wolfsberg and Helmholz found that the sum of these effects in MnO_4^- actually produced a reversal in the order of the $2e$ and $3t_2$ orbitals.

In a recent paper, Liehr and Ballhausen[56] have disputed this reversal found by Wolfsberg and Helmholz.[55] In their view, the lowest energy transition should be $t_1 \rightarrow 2e$, since with this choice their intensity calculations agree better with experiment than do those of Wolfsberg and Helmholz. These two viewpoints lead to different interpretations of the observed spectra. In the first case the excited states arise from the configuration $(t_1)^5(3t_2)$ and are $^{1,3}(A_2, E, T_1, T_2)$. In the second case $(t_1)^5e$ leads to $^{1,3}(T_1, T_2)$. In either case one allowed $(^1A_1 \rightarrow {}^1T_2)$ and several forbidden transitions are predicted. Only careful experimental investigation of the forbidden transitions can distinguish these two possibilities.

(3) *The classical point of view.* Charge transfer spectra have often been considered from a classical point of view.[57] A recent example of the application of this method is the work of Schlaefer.[58] He considers metal-water complexes such as $Ce(H_2O)_6^{++++}$ as electrostatic complexes and finds the energy required to transfer an electron from the water molecules into the metal atom. This energy is $U = -\Delta I + I_L + \Delta A$ where ΔI is the difference between the n and n-1 ionization potentials of the central ion, I_L is the ligand ionization energy, and ΔA is the repulsion between the ligands resulting from the charge left on them plus the change in the ligand-ion attraction. The results were not in good quantitative agreement with experiment, but usually showed the correct trends from one ion to another.

d. Intensity of Charge Transfer Processes

Transitions involving charge transfer usually have greater intensities than crystal field transitions. Only one of the combining states is predominantly a d-electron state, so that the forbiddeness of a $d \rightarrow d$ transition process is not involved as it is in a crystal field transition.

The selection rules for dipole transitions follow in the usual way from the character table for the point group involved.

The dipole transition moment integral may be written in a straightforward way, and reduces in the usual approximation to a one-electron integral over the excited and unexcited orbitals. The important facts about the intensity integral are illustrated by considering the general case

[56] A. D. Liehr and C. J. Ballhausen, *J. Molec. Spectrosc.* **2,** 342 (1958).
[57] E. Rabinowitch, *Revs. Modern Phys.* **14,** 112 (1942).
[58] H. L. Schlaefer, *Z. physik. Chem.* (*Frankfurt*) [N.S.] **3,** 222 (1955).

of a transition between two orbitals of mixed central ion and ligand character, namely, $\psi_1 = a\psi_d + b\psi_L$, $\psi_2 = f\psi_d' + g\psi_L'$ where ψ_d and ψ_d' are either the same or orthogonal d orbitals, and ψ_L and ψ_L' are orthogonal ligand orbitals. The transition moment integral is

$$M = af\int\psi_d' m\psi_d \, d\tau + gb\int\psi_L m\psi_L' \, d\tau + ag\int\psi_d m\psi_L' \, d\tau + fb\int\psi_d' m\psi_L d. \quad (2.2)$$

The first term, an integration over the d orbitals contributes nothing. The second is a transition within the ligand atoms, while the last two are the charge transfer integrals. The contribution due to the ligands alone may be comparable to the charge transfer contribution in highly covalent complexes.

e. The Change from Localized to Nonlocalized Orbitals

It has been assumed in the foregoing discussion that the ligand electrons are acted upon only by the central ion and the ligand atoms. This must be a reasonable assumption in crystals such as $KMnO_4$ or $K_3Fe(CN)_6$, which are molecular in the sense that electrons of different complex anions in the crystal do not interact strongly with one another. Evidence for their molecular nature is the fact that spectra of aqueous or alcoholic solutions of these compounds resemble the spectra of the crystal strongly.

A part or all of the nonbonding electrons of the nearest neighbor anions may be made less available for charge transfer processes by the formation of bonds to hydrogen. Thus in the compound $Co(NH_3)_6Cl_3$, the $Co(NH_3)_6^{+++}$ ion is a molecular unit similar to $Fe(CN)_6^{\equiv}$ or $CoCl_6^{\equiv}$, but the electrons on the nitrogen atom are occupied in binding the hydrogen atoms. Charge transfer processes involving these electrons are more energetic than those involving the nonbonding electrons of Cl^- ions in the otherwise similar $CoCl_6^{\equiv}$ complexes. The correctness of this view is demonstrated by recent work which shows that the energy of the charge transfer spectrum is reduced by replacing an NH_3 molecule by a Cl^- ion in the $Co(NH_3)_6^{+++}$ molecular ion (Section 5). The lowest energy charge transfer process therefore involves the Cl^- electrons.

The availability of ligand electrons for charge transfer processes is further reduced by bonding with the other cations of the crystal. In an extreme case the ligand or anion electrons are shared equally by the cations surrounding them, and the crystal is no longer one made up of molecular ions. An illustration of the gradual transition from a crystal of *molecular ions* to an *ionic* crystal in an "isoelectronic series" is afforded by the tetrahedral ions XO_4^{n-} where X is Mn, Cr, V, or Ti. As the nuclear charge decreases in this series, the oxygen atoms become less tightly bound to the X ion, and the originally nonbonding electrons of the oxygen atoms become more tightly bound to the other cations in the lattice. Another effect is that the *d* orbitals become less tightly bound to the central ion as the nuclear charge decreases. The energy of the charge transfer transition rises for both of these reasons as the nuclear charge decreases. This trend is shown for the $1t_1 \rightarrow 3t_2$ transition of XO_4^{n-} in Fig. 26.

At the same time that the nonbonding electrons are becoming more tightly bound by neighboring cations, the originally σ-bonding electrons are becoming less tightly bound, until in the ionic crystal limit the π and σ classes merge into one.

We have imagined the transition from a lattice of molecular ions to an ionic lattice to have occurred in a cubic crystal. If the crystal is not cubic, the electrons on the ions in their ground state may fall into different symmetry classes and be of different energies. This is not important for the present discussion. It leads to a more complicated band structure but no new principles are involved.

In a molecular ionic crystal, the neighboring cations such as the K^+ ions in $KMnO_4$ perturb the MnO_4^- ion states only slightly. Such perturbations have been observed by Teltow,[59] and have been analyzed in part by Wolfsberg and Helmholz.[55] The perturbations may be classified as site group and factor group perturbations (see McClure[45]). Only the former have been studied, but the occurrence of the latter is made evident by the band broadening in the MnO_4^- spectrum when the ion is in pure $KMnO_4$ compared to the spectrum in dilute solid solution with, for example, $KClO_4$.[60] Factor group splitting may possibly be resolvable in the pure crystals at low temperatures.

The transition from a lattice of molecular ions to an ionic lattice may be visualized as an increase in the site group and factor group perturbations, accompanied by a decrease in the binding forces to what is originally called the central ion. One advantage of this viewpoint is that the process of charge transfer may easily be recognized as occurring in the molecular members of the series and an extrapolation from them may permit such an identification in the more ionic members. Charge transfer in the latter contributes to the formation of an exciton or conduction band. The band width becomes an important feature of the spectrum as the ionic crystal limit is approached and band theory must be applied to calculate it.

The applicability of the term charge transfer spectrum must decrease as the initial and final orbitals become less localized. Thus, when the

[59] J. Teltow, *Z. physik. Chem.* (*Leipzig*) **B43,** (B3) 198, (B5) 375 (1939).
[60] J. Teltow, *Z. physik. Chem.* (*Leipzig*) **B40,** 397 (1938).

transition involves a *d* orbital, the term is somewhat justified, but as shown by the isoelectronic series of XO_4^{n-} ions, the localization of the *d* orbitals varies considerably from one ion to another. The localization of unfilled or nonbonding *d* orbitals must increase with an increase in the charge of the ion core. However, there is an opposing tendency to reduce the volume available to a *d* electron since the surrounding atoms are more tightly bound at the same time. In our example the first of these factors was found to predominate. Further study of experimental data should provide more examples of the factors which influence the localization of the *d* orbitals.

Configuration interaction has not been discussed in this section because no theoretical work has been carried out for the systems of interest here, and because our present understanding of the data is not sufficiently advanced that we could recognize its occurrence. It is always present, however, and must reduce the applicability of one-configuration theories.

f. Conclusions

We have adopted a semiempirical and descriptive method for the interpretation of the spectra of ions in crystals. This is justified not only because the Schroedinger equation is too complicated to solve in detail for the systems of interest, but also because it seems to be the best way to make progress at present. There is so much information at hand that our first task is to classify it on the basis of the simplest reasonable theories. The beginnings of such a classification have been made in the sections on crystal field theory and molecular orbital theory. The following sections will illustrate how the experimental data may be accommodated in this framework. It will be seen that crystal field spectra and charge transfer spectra may be identified and understood in a great many cases.

The next step in making the semiempirical method more useful is to derive constants from charge transfer spectra similar to the field constants of crystal field theory. These numbers should be characteristic of an ion and its environment, and should be transferable from one system to another. The electronegativity is a somewhat primitive example of one of the kinds of parameters needed.

Although we have concentrated on ions having *d* shells available for a transition, the ions after the completion of a transition group in the periodic table have *s* or *p* orbitals which may serve as the final state of a transition. The development of a semiempirical theory of these spectra could follow the same lines as has been developed for the *d*-shell ions.

II. Experimental Studies of the Spectra of Ions in Crystals

The central problems of spectroscopy are to characterize the excited states giving rise to spectral bands and to provide appropriate theoretical explanations of the results. The first characteristic of any electronic state which should be determined is its behavior under symmetry operations. However, in many crystalline or molecular systems the symmetry is not high, and other characteristics such as the transition moment from the ground state are necessary to distinguish between states having the same symmetry behavior. Other characteristics may be revealed by perturbing the system in some way.

The spectra of the rare earth ions are very similar to atomic spectra, and the most important problems in this area at present are to identify the parent atomic states which give rise to the crystal field multiplets (Section 3). The present methods for calculating atomic spectra are not sufficiently accurate to be used alone for the identification of observed lines, and the symmetry properties of excited states must be determined experimentally. The methods used are the Zeeman effect and the crystal field splitting.

The parent atomic states giving rise to crystal field spectra of the transition metal ions (Section 4) are fewer in number than for the rare earth ions, and are known in most cases. The main problems are to identify the crystal field components of these states. This has been done chiefly by comparing the observed with the calculated spectrum. It would be desirable to establish the symmetry properties independently of the theory in at least a few cases, but there does not seem to be much doubt that the present results are correct. Crystal field states are now being further characterized by determining the extent to which the ligand electrons participate in crystal field transitions.

In all but a very few cases, charge transfer bands (Section 5) have been identified by their position and intensity alone. A rigorous determination of the symmetry properties of the excited states, such as the work of Teltow and of Wolfsberg and Helmholz on the spectrum of MnO_4^-, would be highly desirable. The experimental requirements may be met by use of low temperatures, single crystals, polarized light, and a spectrograph of high dispersion. Furthermore the spectra must be suitably perturbed by altering the crystalline environment. The interpretation requires the full use of group theory and at least the theoretical framework outlined in Section 2. These features have often been brought together for the study of molecular spectra, but seldom for the study of charge transfer spectra of molecular ions in crystals.

Finally in Section 6 the experimental work on the spectra of metal ions just beyond the transition groups in the periodic table will be discussed. The energy levels have been identified by reference to the spectra of the gaseous ions in a few cases. Only for one ion, Tl^+, have physical properties of excited levels other than energy and intensity been measured. There is a great need for experimental spectroscopy in this area, especially as it appears possible to learn something about the nature of exciton levels through the interaction of the metal ion and the lattice.

3. Spectra of the Lanthanides

a. *The Chemical and Physical Properties of the Lanthanides*

Two groups of atoms in the periodic table of the elements are characterized by having partially filled *f* shells in their ground states. They are called the lanthanides and the actinides because their electron configurations are related to those of the elements lanthanum and actinium respectively. Table XII lists the elements of the lanthanide series and some of the basic information on them to which we shall subsequently refer.

TABLE XII. THE LANTHANIDE ELEMENTS

The number of *f*-electrons is given for each oxidation state known. The radii are given for M^{+++}, and are taken from T. Moeller, "Inorganic Chemistry." Wiley, New York, 1952.

	Atomic no.	M^{++}	M^{+++}	M^{++++}	$r(M^{+++})$ A
La	57		$4f^0$		1.22
Ce	58		$4f^1$	$4f^0$	1.18
Pr	59		$4f^2$	$4f^1$	1.16
Nd	60		$4f^3$		1.15
Pm	61		$4f^4$		
Sm	62	$4f^6$	$4f^5$		1.13
Eu	63	$4f^7$	$4f^6$		1.13
Gd	64		$4f^7$		1.11
Tb	65		$4f^8$	$4f^7$	1.09
Dy	66		$4f^9$		1.07
Ho	67		$4f^{10}$		1.05
Er	68		$4f^{11}$		1.04
Tm	69		$4f^{12}$		1.04
Yb	70		$4f^{13}$		1.00
Lu	71		$4f^{14}$		0.99

The lanthanides usually occur in chemical compounds and in solution as tripositive ions. Cerium +3 has a single $4f$ electron, gadolinium +3 a half-filled shell (seven), and ytterbium +3 a single hole in the $4f$ shell. The +3 ions are chemically similar to lanthanum. Their (Goldschmidt) ionic radii are close to that of lanthanum 1.22 A, but fall nearly regularly to 0.99 A at lutecium (pp. 141, 142 in ref. 61).

Other oxidation states occur for certain of these elements. Their occurrence is determined by the extra stability gained if the ion can reach the configurations $4f^0$, $4f^7$, or $4f^{14}$. Thus $Ce^{+3}(4f^1)$ may be oxidized to $Ce^{+4}(4f^0)$; $Eu^{+3}(4f^6)$ may be reduced to $Eu^{+2}(4f^7)$; $Tb^{+3}(4f^8)$ to $Tb^{+4}(4f^7)$; and $Yb^{+3}(4f^{13})$ to $Yb^{+2}(4f^{14})$. The chemistry of these elements is reviewed in the book by Yost *et al.*[62]

The ground terms of the free atoms and the free +1 ions are known for somewhat over half the rare earths from the analysis of arc and spark spectra. The data are reviewed by Klinkenberg.[63]

The ground states of the tripositive ions have been determined from magnetic susceptibility measurements.[62] In every case, the state predicted from Hund's rule for a $4f^n$ configuration is found, although the suceptibility must be corrected for the effect of low-lying multiplet components in several cases.

The absorption spectrum of a given tripositive rare earth ion is remarkably constant from one compound to another and from solid state to solution. Each such characteristic spectrum consists of many very weak sharp bands from the near infrared through the near ultraviolet. Much stronger bands are found farther in the ultraviolet. These are no longer so characteristic of the positive ion, but depend upon its environment. The strong bands could be caused by charge-transfer processes and by $f \rightarrow d$ transitions. It is well established that the weak bands arise from transitions between states of the $4f^n$ configurations.

b. *Spectra of the Free Ions*

Meggers[64] and Klinkenberg[63] have reviewed the subject of line spectra of the rare earths. A compilation of data is to be found in ref. 65 and a recent bibliography in ref. 65*a*, vol. III. The most comprehensive recent review is to be found in the book on rare earth spectra by El'yashevich.[66] The data are rather scarce and not a great deal of it has been interpreted because of the great complexity of the line spectra. We will review here some of what is known from line spectra that is helpful in interpreting

[61] T. Moeller, "Inorganic Chemistry." Wiley, New York, 1952.

[62] D. M. Yost, H. Russell, and C. S. Garner, "The Rare Earth Elements and Their Compounds." New York, Wiley, 1947.

[63] P. F. A. Klinkenberg, *Physica* **13,** 1 (1947).

[64] W. F. Meggers, *Revs. Modern Phys.* **14,** 96 (1942).

[65] Landolt-Börnstein Tabellen, [6] Vol. I, Part 1 (1950).

[65a] C. E. Moore, Atomic Energy Levels, *Natl. Bur. Standards (U.S.) Circ.* **467,** Vol. I, 1949; Vol. II, 1952; Vol. III, 1958. Washington, D.C.

[66] M. A. El'yashevich, "Spectra of the Rare Earths" (in Russian). Gosudarst. Izdatel'stua Tekh.-Teoret. Lit., Moscow, 1953.

the spectra of rare earth ions in solids. The most important configurations for this purpose are $4f^n$ and $4f^{n-1}d$.

(1) $n = 1$. The cerium IV spectrum was studied in detail by Lang[67] who gave a fairly complete energy level diagram. The energy levels of $4f^1$ are $^2F_{\frac{5}{2}}$ (ground state) and $^2F_{\frac{7}{2}}$ at 2253 cm^{-1}, from which one may obtain the spin-orbit coupling coefficient for a $4f$ electron in cerium: $\zeta_{4f}(Ce) = 643\ cm^{-1}$. The $5d$ levels $^2D_{\frac{3}{2}}$ at 49737 cm^{-1} and $^2D_{\frac{5}{2}}$ at 52,226 are 2489 cm^{-1} apart, from which $\zeta_{5d}(Ce) = 996\ cm^{-1}$. The ^2D levels are of importance in the interpretation of some crystal spectra. The Ce II spectrum[64] is well analyzed. From the $4f^26s$ configuration, another value of ζ_{4f} may be derived. The ground multiplet, $^4H_{\frac{7}{2}} - {}^4H_{\frac{13}{2}}$, shows large deviations from the interval rule, but a reliable value should be obtained from the total multiplet width, 3113.56. Thus $\zeta_{4f}(Ce) = 622$, in fair agreement with the other estimate.

(2) $n = 13$. The Tm I spectrum[64] has a $4f^{13}6s^2$ ground configuration which gives rise to the inverted ^2F doublet having a splitting of 8771.25 cm^{-1}. Thus $\zeta_{4f}(Tm) = 2500\ cm^{-1}$. The lowest multiplet of Tm II in the $4f^{13}6s$ configuration is close to the j-j coupling limit and yields almost the identical ζ_{4f} value. The Yb^{+++} ion in the solid state ($4f^{13}$) has a slightly larger doublet separation, 10,300 cm^{-1} (Dieke and Crosswhite[68]) and $\zeta_{4f}(Yb) = 2940\ cm^{-1}$.

(3) $n = 2$. The states of the $4f^2$ configuration of La II were located by Meggers.[69] Many other configurations with all the expected states were identified. The term diagram of the $4f^2$ configuration is shown in Fig. 18 and on p. 207 in Condon and Shortley[28] where it is compared to the energies calculated assuming Russell-Saunders coupling. The three electrostatic interaction parameters needed to determine the intervals between the seven allowed terms are $F_2 = 93.3$, $F_4 = 21.6$, $F_6 = 0.26\ cm^{-1}$. They reproduce the spectrum fairly well. The multiplets are distorted, however, showing there are appreciable deviations from strict Russell-Saunders coupling even at the beginning of the rare earth series. From the analysis of the multiplets it is found that $\zeta_{4f} = 340\ cm^{-1}$. This spectrum and the theoretical analysis of it have played important parts in attempts to analyze spectra of other rare earth ions in the solid state, especially those of the isoelectronic Pr^{+++} and Tm^{+++} ions ($4f^{12}$).

The $4f^26s$ configuration of Ce II[70] gives rise to terms whose parents are the same as those of La II (Fig. 18). The term values of $4f^26s$ may be averaged so as to yield the positions of the parent terms ^3H, ^3F, and

[67] R. Lang, *Can. J. Research* **A13,** 1 (1935); **A14,** 127 (1936).
[68] G. H. Dieke and H. M. Crosswhite, *J. Opt. Soc. Am.* **46,** 885 (1956).
[69] W. F. Meggers, *J. Research Natl. Bur. Standards* **9,** 239 (1932); H. Russell and W. F. Meggers, *ibid.* **9,** 625 (1932).
[70] G. R. Harrison, W. E. Albertson, and N. F. Hosford, *J. Opt. Soc. Am.* **31,** 439 (1941).

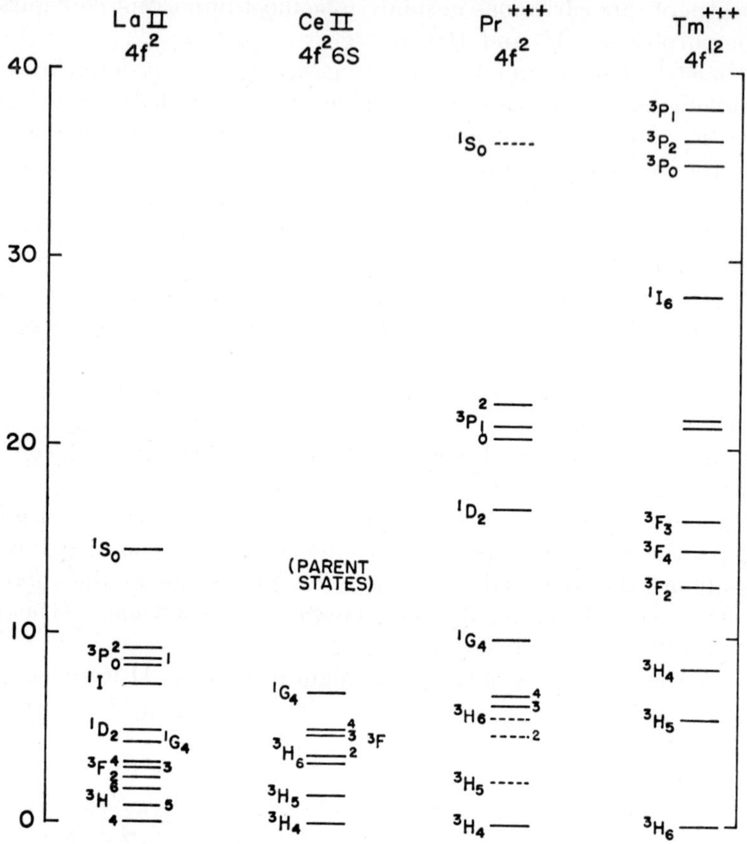

FIG. 18A

FIG. 18. Energy levels of the rare earth ions. The positions of all known multiplet components are shown and all of the assignments which seem fairly certain are indicated. The broad bands of Ce^{+++} and of other ions are also indicated. The vertical scale is in cm^{-1}.

Notes to the energy level diagrams in Figs. 18A,B,C,D (see text for details).

(a) La II and Ce II. These levels are from arc and spark spectra. All others from spectra of solids.

(b) Pr^{+++}. The $^1I_6$ level is between $^3P_2$ and $^3P_1$. The lower dotted levels have been observed recently in fluorescence. The $^3F_2$ and $^3H_6$ labels should be interchanged.

(c) Tm^{+++}. This energy level diagram is based on Spedding's calculations and some inadequate experimental work.

(d) Ce^{+++}. The *d*-bands shown here are for the case of 9-coordination in D_{3d} symmetry. They will change position in other environments.

(e) Yb^{+++}. The *d*-bands shown are not well substantiated.

(f) Yb^{++}. The two levels shown are broad bands arising from the $f \rightarrow d$ transitions.

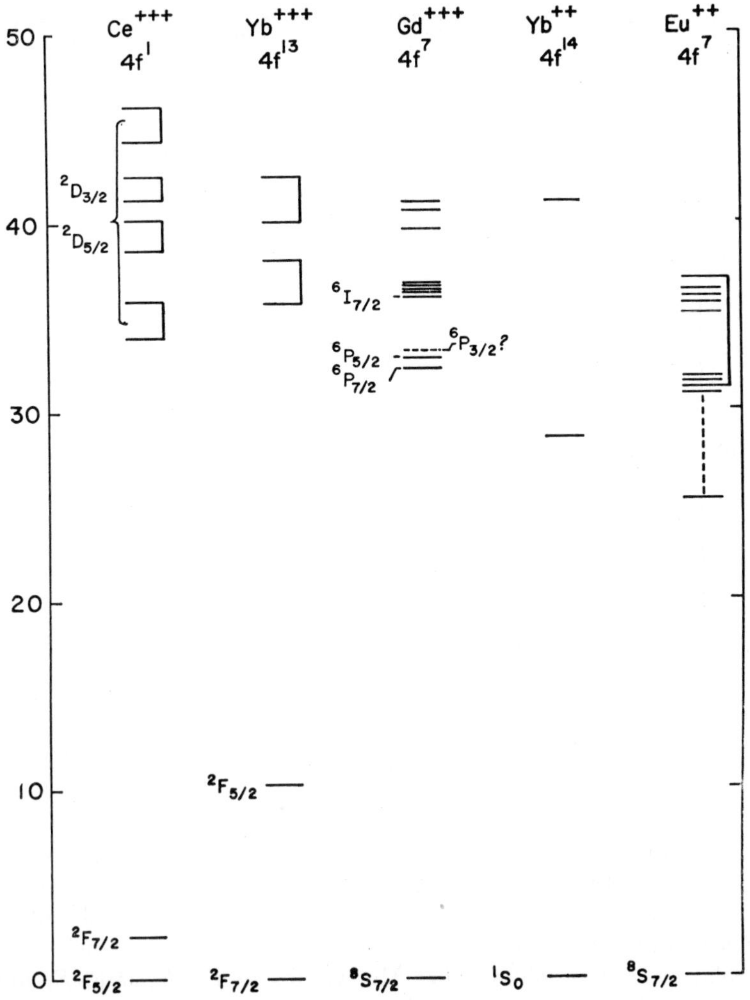

FIG. 18B

(g) Eu^{++}. The lowest line shown is a sharp level from which fluorescence to the ground state may be observed. Above it (dotted line) lie many other sharp levels. The bracketed region has sharp bands plus a continuum and above this at about 40,000 cm^{-1} lies a strong broad band.

(h) Nd^{+++}. $^4F_{\frac{3}{2}}$ level at 7400 cm^{-1} should be shown on the diagram. $^4I_{\frac{11}{2}}$ and $^4I_{\frac{13}{2}}$ lie at 2050 and 4000 cm^{-1}.

(i) Sm^{++}. The band at 17,040 cm^{-1} is diffuse. The others shown are sharp. There are many more diffuse bands at higher energies, not shown here. Fluorescence begins at levels 15,936 and 14,514 cm^{-1}.

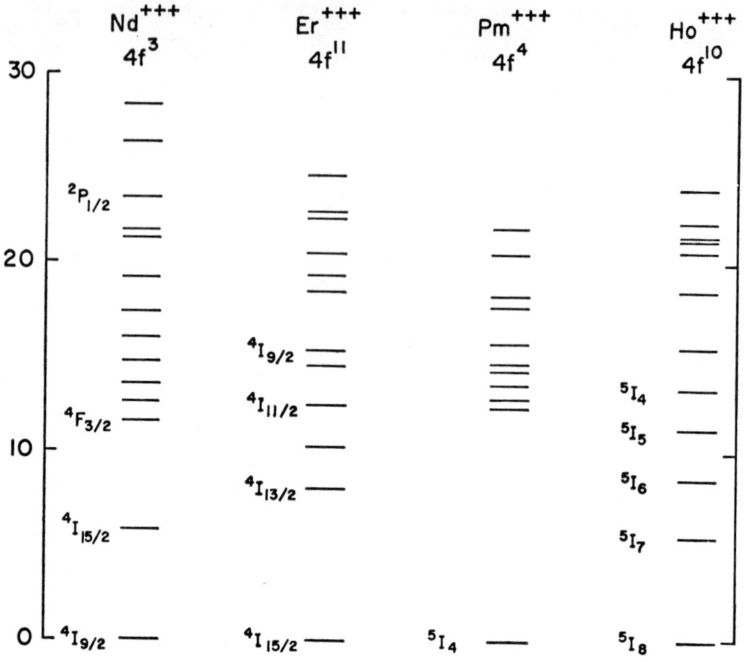

FIG. 18C

${}^1G$ of f^2. In spite of the fact that $4f^26s$ is overlapped and perturbed by $4f^25d$, it is found that the parent terms are in a nearly constant ratio, 1.65 ± 0.09, greater in Ce II than in La II. This difference must be due to the increased nuclear charge. A further increase in the value of the electrostatic interaction integrals from the La II values is observed in the spectra of solid compounds of Pr^{+++}. The ratio is about 2.4[71] but some terms are quite different from the values predicted by such a simple extrapolation.[72] This shows that the values of F_2, F_4, and F_6 are not increased by the same factor, so that certain terms very sensitive to their ratios will be misplaced.[73,74]

The data on the line spectra of the $4f^n6s^2$ configurations of neodymium, samarium, and thulium, to which references may be found in Klinkenberg's article,[63] are well enough understood so that ζ_{4f} values may be obtained from them. Klinkenberg also estimated the total multiplet widths and screening constants for the other rare earths. His results are

[71] F. H. Spedding, *Phys. Rev.* **58,** 255 (1940).
[72] A. M. and K. H. Hellwege, *Z. Physik* **130,** 549 (1951).
[73] E. Trefftz, *Z. Physik* **130,** 561 (1951).
[74] B. R. Judd, *Proc. Roy. Soc.* (*London*) **A228,** 120 (1955).

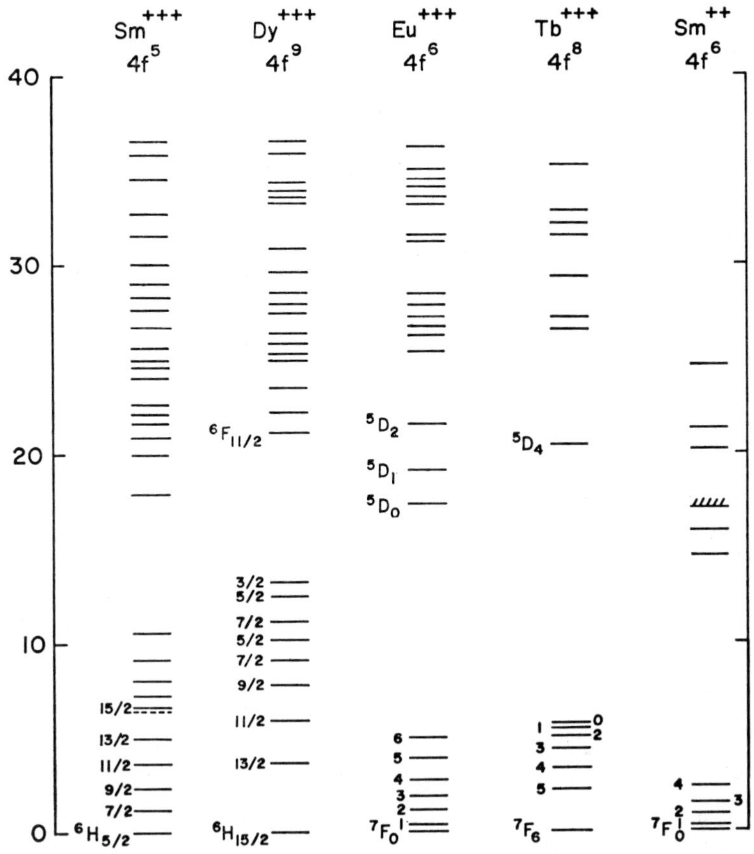

FIG. 18D

presented here as part of Table VIII. The values for ζ_{4f} configurations are obtainable from fluorescence and absorption spectra of the solid compounds of the rare earth ions. Table VIII also includes these data. The least certain values are those for Ho and Er, although they cannot be very far out of line. It can be seen from the larger ζ_{4f} values in the solid that the shielding caused by the overlap of surrounding atoms is less than that caused by the $6s^2$ shell in the un-ionized free atoms.

c. Spectra of the Individual Ions in Crystals

A comprehensive review of the spectra of the rare earths, both in the gaseous state and in crystals, has been made by El'yashevich.[66] This book was very helpful in covering the literature before 1953. We will concentrate in this review on the most recent work.

(*1*) *Ce*$^{+++}$ ($Z = 58$, f^1, $^2F_{\frac{5}{2}}$). Since the doublet splitting of the only term in the $4f^1$ configuration is 2253 cm^{-1}, cerium has no visible absorption bands. Absorption bands are found in Ce^{+++} compounds in the ultraviolet above 35,000 cm^{-1}.[75–77] These are illustrated in Fig. 18. Emission bands are also found,[77] and they apparently correspond to transitions which end on either of the levels $^2F_{\frac{5}{2}}$ and $^2F_{\frac{7}{2}}$. The Ce^{+++} doublet splitting is observed in the solid in this way to be about 1900 cm^{-1}.

Freed[75] was the first to interpret the ultraviolet absorption in Ce^{+++} and similar bands in other rare earths as $4f$-$5d$ transitions. The bands are hundreds of times stronger than $4f$-$4f$ transitions and are much wider, namely 1000–2000 cm^{-1}, compared to less than one cm^{-1} for many $4f$-$4f$ lines. They do not sharpen on lowering the temperature. These facts, and the knowledge that the Ce IV spectrum has its $^2D_{\frac{3}{2}} - {}^2D_{\frac{5}{2}}$ doublet at 49,737 and 52,226 cm^{-1} constitute good proof of Freed's identification of the bands as $f \rightarrow d$ transitions.

If the observed spectrum includes all bands, the total crystal field splitting of the d-states in Ce^{+++} is less than is observed for transition element ions. The splitting for Ti^{+++}, an ion having the same charge as Ce^{+++}, is 20,300 cm^{-1}, whereas it is 10,000 cm^{-1} for Ce^{+++}. There are at least two obvious factors which may account for this difference: the larger radius of Ce^{+++}, leading to a lower ionic potential, and a larger coordination number which leads to a smoother potential and smaller V_4 terms.

(*2*) *Praseodymium, Pr*$^{+++}$ ($Z = 59$, $4f^2$, $^3H_4$), $^1(S\ D\ G\ I)$, $^3(P\ F\ H)$. (*a*) *Experimental work.* The most recent experimental work has been carried out by groups in Hellwege's,[78–80] in Dieke's,[80a] and in Freed's[50,81] laboratories. It has been carried out with such close attention to detail that very little ambiguity in the interpretation of the Pr^{+++} spectrum remains.

The general appearance of the absorption spectrum is the same as the energy level diagram shown in Fig. 18, since all transitions take place from the lowest level. The three bands at about 20,000 cm^{-1} were identified by Ellis[82] as a normal $^3P_{0,1,2}$ multiplet. The recent work which supports this assignment provides a good example to illustrate the experimental identification of energy levels.

[75] S. Freed, *Phys. Rev.* **38,** 2122 (1931).
[76] R. W. Roberts, L. A. Wallace, and I. T. Pierce, *Phil. Mag.* [7] **17,** 934 (1934).
[77] F. A. Kröger and J. Bakker, *Physica* **8,** 628 (1941).
[78] A. M. and K. H. Hellwege, *Z. Physik* **130,** 549 (1951).
[79] A. M. and K. H. Hellwege, *Z. Physik* **133,** 174 (1952).
[80] A. M. and K. H. Hellwege, *Z. Physik* **135,** 92 (1953).
[80a] G. H. Dieke and R. Sarup, *J. Chem. Phys.* **29,** 741 (1958).
[81] E. V. Sayre, K. Sancier, and S. Freed, *J. Chem. Phys.* **23,** 2066 (1955).
[82] G. B. Ellis, *Phys. Rev.* **49,** 875 (1936).

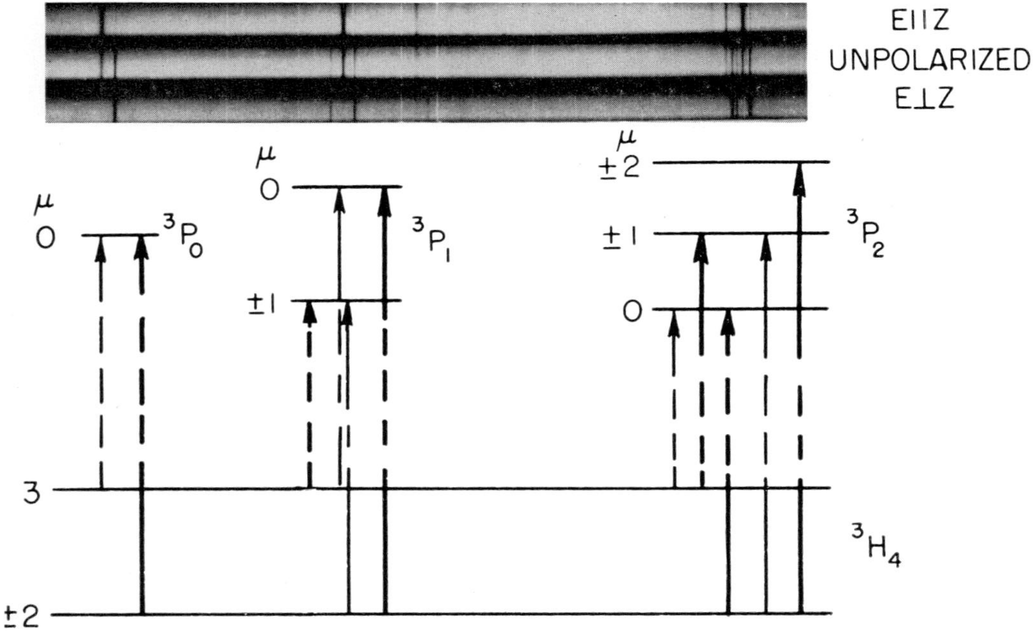

FIG. 19. Energy level diagram and absorption spectrum for transitions from ground states to $^3P_{0,1,2}$ triplet of $PrCl_3$ in $LaCl_3$. Spectra of single crystal at 77°K; light transmitted perpendicular to unique, Z, axis of crystal. The average energies of the components are: $^3P_0$ 20,476, $^3P_1$ 21,077, $^3P_2$ 22,231 cm^{-1}. The crystal field components are labeled by Hellwege's notation for C_{3h}, explained in Table XIII (Sayre *et al.*[50]).

Figure 19 (from the paper of Sayre *et al.*[50]) shows the polarized spectrum in the region of 20,000 cm^{-1}. The crystal used is $LaCl_3$ in which a small amount of $PrCl_3$ has been dissolved. It is known from x-ray work that $LaCl_3$ has the UCl_3 structure in which the metal ion is surrounded by nine Cl^- ions in such a way that the symmetry at the site of the metal ion is C_{3h}, and nearly D_{3h}. The absorption spectrum shown in Fig. 19 was taken with the incident light perpendicular to the optic axis (Z axis) and the electric vector either parallel or perpendicular to it. The axial spectrum coincides with the σ spectrum, showing that all the transitions observed are electric dipole.

The ground state is known to be $^3H_4$ ($J = 4$) which must split in a field of symmetry C_{3h} into the six states A', $2A''$, $2E'$, E'', as shown in Table IX. (The notation used in the figure is Hellwege's. Table XIII

TABLE XIII. CHARACTER TABLES FOR POINT GROUPS C_{3h} AND D_{3h}

	C_{3h}	E	C_3	C_3^2	σ_h	S_3	S_3^2	
(0)	A'	1	1	1	1	1	1	R_z
(3)	A''	1	1	1	-1	-1	-1	z
(± 2)	E	1	ω	ω^2	1	ω	ω^2	x, y
		1	ω^2	ω	1	ω^2	ω	
(± 1)	E'	1	ω	ω^2	-1	$-\omega$	$-\omega^2$	R_x, R_y
		1	ω^2	ω	-1	$-\omega^2$	$-\omega$	
			($\omega = e^{2\pi i/3}$)					

C_{3h}	D_{3h}	E	σ_h	$2C_3(z)$	$2S_3$	$3C_2$	$3\sigma_v$	
A'	A_1'	1	1	1	1	1	1	$x^2 + y^2$, z^2
	A_2'	1	1	1	1	-1	-1	R_z
A''	A_1''	1	-1	1	-1	1	-1	
	A_2''	1	-1	1	-1	-1	1	z
E'	E	2	2	-1	-1	0	0	x, y; xy, $(x^2 - y^2)$
E'	E'	2	-2	-1	1	0	0	R_x, R_y, xz, yz

Spinor representations, D_{3h}

	E	R	σ_h	$2C_3$	$2RC_3$	$2S_3$	$2RS_3$	$3C_2 + 3RC_2$	$3\sigma_v + 3R\sigma_v$
$E_{\frac{1}{2}}$	2	-2	0	1	-1	$\sqrt{3}$	$-\sqrt{3}$	0	0
$E_{\frac{5}{2}}$	2	-2	0	1	-1	$-\sqrt{3}$	$\sqrt{3}$	0	0
$E_{\frac{3}{2}}$	2	-2	0	-2	2	0	0	0	0

shows the correlation to the notation ordinarily used by spectroscopists in the United States.) The P states must split as follows: $^3P_0(A')$, $^3P_1(A' + E'')$, $^3P_2(A' + E' + E'')$, as also shown in Table IX. By using the selection rules for dipole radiation, which may be derived from Table XIII and which are shown in Table XIV, and the experimental data shown in Fig. 19, the symmetry character of the crystal field levels of both lower and upper states may be found. It may be seen, for example, that, since there are no missing lines, the ground state must belong to the E' representation. This state can make transitions to all the possible upper states, A', E', E'', of the ^3P multiplet. The identity of the upper states may be partially established by their polarization, but for a complete identification, transitions from another of the crystal field components of the ground state must be observed. The next higher component of $^3H_4$ must be A'' since its transition to $^3P_0(A')$ is π-polarized (see Table XIV). This level occurs at 33 cm^{-1} above the ground state so that it is well populated at 77°K, the temperature at which the spectrum of Fig. 19 was photographed. Other crystal field levels of $^3H_4$ which have been identified are 97 cm^{-1}, E' and 131 cm^{-1}, E'', leaving an A' and an A'' level unobserved.

TABLE XIV. SELECTION RULES FOR C_{3h} ($\pi = \mathbf{E} \parallel Z$, $\sigma = \mathbf{E} \perp Z$)

	A'	E''	E'	A''
A'			σ	π
E''		σ	π	σ
E'	σ	π	σ	
A''	π	σ		

While Freed's experiments on $PrCl_3$ and PrF_3 were done with crystals of known structure, it is sometimes possible both to determine the local symmetry and to assign the levels when the crystal structure is not known. Such a program was carried out by the Hellweges[80] for crystals of $Pr_2Mg_3(NO_3)_{12} \cdot 24H_2O$.

Dieke and Sarup[80a] have identified the $^1I_6$ state of Pr^{+++}, and have confirmed this and the other identifications in the spectrum by the Zeeman effect. The $^1I_6$ level had not been found until this work. It lies between the $^3P_1$ and $^3P_2$ levels and has apparently been mistaken for vibrational additions to the $^3P_1$ level. For the first time, they have obtained the high resolution fluorescence spectrum of Pr^{+++} in a solid, and by this means have been able to identify the ^3F and ^3H multiplet components, some of which had not been observed in absorption. The emitting levels at 4°K were found to be the lowest crystal field components of $^3P_0$, $^3P_1$ and $^1D_2$. Fluorescence of Pr^{+++} had been observed in some of the early work of Deutschbein, Tomaschek, Gobrecht and others[83,84] in so-called phosphors, solid solutions of Pr^{+++} in various salts such as K_2SO_4 and CaO.

[83] H. Gobrecht, *Ann. Physik* [5] **28,** 673 (1937).
[84] R. Tomaschek and O. Deutschbein, *Z. Physik* **82,** 309 (1933).

(*b*) *Calculation of energy levels.* The Pr^{+++} and Tm^{+++} ions are the simplest of the rare earth ions on which to test atomic energy level calculations, and several such calculations have been reported. Spedding[71] employed the LaII($4f^2$) spectrum as a starting point in his calculation of the levels of Pr^{+++} (see Fig. 18). He simply multiplied the Russell-Saunders term energies of LaII by a factor of 2.4 to take account of the higher nuclear charge and state of ionization, and then applied the spin-orbit interaction, including second order effects, to the resulting terms. The best value of ζ_{4f}, namely 800 cm^{-1}, was also about 2.4 times that in LaII. The agreement with the observed spectrum was quite good, except that the calculation led to the misidentification of $^1D_2$ as $^1I_6$. Hellwege showed experimentally that this identification was incorrect,[72] and Trefftz[73] attempted to explain the discrepancy. She showed that the ^1D and ^1G terms are very sensitive to the ratio of the Slater-Condon parameters F_4/F_2, and attempted to calculate the spectrum using the $F_2; F_4; F_6$ ratios demanded by hydrogen-like wave functions. A numerical error occurred in this work, as pointed out by Judd,[74,85] who, using the same methods was able to obtain agreement with experiment, and to predict the position of the then unidentified $^1I_6$ term. Judd[85] has also presented convincing evidence that the ^3F term is not inverted, contrary to Freed's suggestion.[50]

The position of $^1I_6$ as now observed by Dieke and Sarup agrees within a few per cent with the position calculated by Judd,[85a] using a Fermi-Thomas potential.

(*c*) *Crystal field splitting in Pr^{+++} salts.* The work of Sayer *et al.* in extending the identifications in the Pr^{+++} spectrum offers some interesting possibilities for deriving a crystal field for $PrCl_3$. Because of its high symmetry only four potential constants are necessary to describe the field: V_2^0, V_4^0, V_6^0, V_6^6. Furthermore, not all of these affect every spectroscopic term, since, as was mentioned in Section 1, the potential term V_n^m only contributes if $n \leq 2L$ and $n \leq 2J$. Thus the $^3P_1$ and $^3P_2$ levels are affected only by V_2^0; $^1D_2$ is only affected by V_2^0 and V_4^0, and $^3F_3$ and $^3F_4$ are affected by all. The splittings have been calculated by Judd[85] with results in excellent agreement with experiment.

Judd has recently calculated a crystal field for all of the rare earth double nitrates, and for $Pr_2(Mg,Zn)_3(NO_3)_{12}\cdot 24H_2O$ in particular.[86] In this crystal, the Pr^{+++} is at a site of C_{3v} symmetry. Two terms in addition to those needed for D_{3h} are necessary, namely, V_4^3 and V_6^3. The calculated splitting patterns fit the experimental results quite well for $^3P_1$ and $^3H_4$

[85] B. R. Judd, *Proc. Roy. Soc.* (*London*) **241,** 414 (1957).
[85a] B. R. Judd, private communication, April 1957.
[86] B. R. Judd, *Proc. Roy. Soc.* (*London*) **A232,** 458 (1955).

but not so well for $^1D_2$ and $^3P_2$ when the same set of parameters is used for each. The trouble seems to be that crystal field mixing between $^1I_6$, $^1D_2$, and $^3P_2$ occurs. Since the position of $^1I_6$ was not known, accurate calculations of this effect were not then possible. This is an example of a second-order crystal field perturbation. Other examples will probably be found, especially in salts in which the crystal field splitting is large. The second-order effects may be responsible for part of the small term displacements observed from one salt to another (Section 3d).

(3) Neodymium (Nd^{+++}, $Z = 60$, $4f^3$, $^4I_{\frac{9}{2}}$) $^4(S\ D\ F\ G\ I)$, $^2(P\ 2D\ 2F\ 2G\ 2H\ I\ K\ L)$. Both the experimental and the theoretical interpretation of the Nd^{+++} spectrum are much more complicated than for the Pr^{+++} spectrum. The theoretical work has been of great aid in identifying transitions, and the most successful attacks on this spectrum have combined theory with experiment.

Satten has published the formulas for electrostatic interaction in the f^3 configuration.[87] He has also done extensive experimental work, and using this with theory, has made a number of assignments of observed absorption lines. This is apparently the first attempt to use quantum theoretical calculations of *LSJ* levels to identify lines in the spectra of rare earths having more than two electrons or holes. Jorgensen[88] has given an assignment of the water solution spectrum of Nd^{+++} based entirely upon energy level calculations, using Slater's methods. The experimental comparison is not limited to the Nd^{+++} spectrum, but by adjusting the parameters, comparison to other f^3 spectra were made. These included U^{+3} and Np^{+4}.

Satten showed that there are simple relations among the energies of the quartet terms which reduce the number of independently variable parameters determining their relative spacing. Slater-Condon parameters were estimated with the aid of the Pr^{+++} spectrum and the approximate positions of the doublet terms were calculated. The ^4F-^4I interval could be identified with some assurance as being about 10,000 cm^{-1} by means of these calculations. Experimental work on $Nd(BrO_3)_3 \cdot 9H_2O$ led to the identification of the 8600 A line group as a $^4I_{\frac{9}{2}} \rightarrow {}^4F_{\frac{3}{2}}$ transition (Fig. 18).

Work on the Zeeman effect by Dieke and Heroux[89] has led to confirmation of the $^4F_{\frac{3}{2}}$ assignment of Satten. For two salts, $Nd_2Mg_3(NO_3)_{12} \cdot 24H_2O$ and $Nd(EtSO_4)_3 \cdot 9H_2O$, the 8600 A lines were sharp. The symmetry of the crystal was sufficiently high (trigonal) that first-order Zeeman effects of the crystal field components were the same as for the free atom. The observed and calculated g factors were in very good agreement.

[87] R. A. Satten, *J. Chem. Phys.* **21,** 637 (1953).
[88] C. K. Jorgensen, *Kgl. Danske Videnskab. Selskab Mat.-fys. Medd.* **29,** No. 11 (1955).
[89] G. H. Dieke and L. Heroux, *Phys. Rev.* **103,** 1227 (1956).

The number of Zeeman components for the other levels assigned to the ^4F term were found, but there was no obvious agreement with the number expected. The data in the paper of Dieke and Heroux both for the ordinary absorption spectrum and for the Zeeman effect are of very high quality, and afford material for exact theoretical interpretation. The journal article describing this work is highly condensed, and information vital for anyone attempting an analysis of the data is missing. A complete set of data may be obtained, however, from the Library of Congress.[90]

Other assignments in this spectrum were made by Satten using the two criteria of agreement between calculated and observed lines, and the number of crystal field levels in each group. The number of lines is not always a reliable criterion, however, for more lines sometimes occur than can be accounted for by the J value in some well-identified transitions. In the group of lines at 7400 A, two of the four upper levels were split into pairs about 5 cm^{-1} apart. This seems too small to be crystal field splitting, and was ignored in the count of upper levels. Thus four levels were assumed, making $J = \frac{7}{2}$. The upper state was identified as $^4F_{\frac{7}{2}}$. In a recent paper, Satten has discussed vibrational-electronic interaction as a possible cause of some of the extra lines (Section 1h).[47]

Two other investigations of the Nd^{+++} spectrum were quite extensive and led to a great deal of good data, but not to any assignments.[91,92]

Carlson and Dieke[92a] have recently reported that they have seen the fluorescence of Pr, Nd, Sm, Eu, Gd, Tb, Dy, Er and Tm in $LaCl_3$ at low temperatures. The Nd^{+++} spectrum was discussed in some detail, and the position of all but the $^4I_{\frac{15}{2}}$ components of ^4I was established. They also found most of the crystal field components of the ^4I multiplet. Calculations by Judd[92b] show that these splittings may be fitted with the four parameters $A_2{}^0\overline{r^2}$, $A_4{}^0\overline{r^4}$, $A_6{}^0\overline{r^6}$, $A_6{}^6\overline{r^6}$, to an average deviation of 1.6 cm^{-1}.

It is interesting to compare the ground levels obtained in experiments on the electron spin resonance of $Nd(EtSO_4)_3 \cdot 9H_2O$[23] with those found by optical spectra. The only data on the ethyl sulfate are those of Dieke and Heroux, but not all the ground levels can be derived from this work. In the trigonal bromate crystal, there must be the same number of levels. The actual arrangement of nearest neighbor atoms certainly will not

[90] Document No. 4886, A.D.I. Auxiliary Publications Project, Photoduplication Service, Library of Congress, Washington 25, D.C. (A copy may be secured by remitting $13.75 in advance for photoprints or $4.50 for microfilm. Checks must be payable to: Chief, Photoduplication Service, Library of Congress.)

[91] H. Ewald, *Ann. Physik* [5] **34,** 209 (1939).

[92] Y. K. Chow, *Z. Physik* **124,** 52 (1947).

[92a] E. Carlson and G. H. Dieke, *J. Chem. Phys.* **29,** 229 (1958).

[92b] B. R. Judd, *Proc. Roy. Soc. (London)* **A251,** 134 (1959).

differ much from that in the ethyl sulfate. The crystal field components are actually very similar, as shown in Table XV.

(4) Promethium (Pm^{+++}, $Z = 61$, $4f^4$, $^5I_4$) $^5(S\ D\ F\ G\ I)$, $^3(3P\ 2D\ 4F\ 3G\ 4H\ 2I\ 2K\ L\ M)$, $^1(2S\ 4D\ F\ 4G\ 2H\ 3I\ K\ 2L\ N)$. The electrostatic energy matrices of the configuration f^4 have been published by Reilly.[93]

Jorgensen, using Reilly's results,[93] has made energy level calculations assuming $F_4 = 0.2F_2$ and $F_6 = 0.02F_2$. These results were compared with the existing data on f^4 spectra but without making any specific assignments. The calculations reported by Jorgensen have been largely superceded by those of Elliott, Judd, and Runcimen.[118]

Since Pm^{+++} does not occur naturally, very little experimental work has been done with it. The solution spectrum[94] and the spark spectrum[95] have been studied, but it has not been investigated in the solid state.

(5) Samarium (Sm^{+++}, $Z = 62$, $4f^5$, $^6H_{5/2}$) $^6(PFH)$, $^4(S\ 2P\ 3D\ 4F\ 4G\ 3H\ 3I\ 2K\ L\ M)$, $^2(4P\ 5D\ 7F\ 6G\ 7H\ 5I\ 5K\ 3L\ 2M\ N\ O)$. The salts of the rare earths near the center of the group have the property that

TABLE XV

$Nd(BrO_3)_3 \cdot 9H_2O$ (optical)[a]	$Nd(EtSO_4)_3 \cdot 9H_2O$ (resonance)[b]
0	0
115	130
184	170
363	340
382	350

[a] R. A. Satten, *J. Chem. Phys.* **21,** 637 (1953).
[b] R. J. Elliott and K. W. H. Stevens, *Proc. Roy. Soc. (London)* **A219,** 387 (1953).

they fluoresce efficiently. One contributing reason may be that there are large gaps between certain excited states and all lower states. Thus radiationless processes, which become less efficient the more energy they must carry away, are slowed down to the point where radiative processes may compete.[96] The ions commonly observed to fluoresce are Sm^{+++}, Eu^{+++}, Tb^{+++}, Gd^{+++}, and Dy^{+++}. Fluorescence is an extremely useful property because the components of the ground multiplet, and other low-lying states may be observed as visible fluorescence. In absorption, these multiplet components would appear in the infrared where vibrational spectra could interfere.

[93] E. Reilly, *Phys. Rev.* **91,** 876 (1953).
[94] G. W. Parker and P. M. Lantz, *J. Am. Chem. Soc.* **72,** 2834 (1950).
[95] W. Meggers, B. Scribner, and W. Bozman, *J. Research Natl. Bur. Standards* **46,** 85 (1951).
[96] G. H. Dieke and L. A. Hall, *J. Chem. Phys.* **27,** 465 (1957).

Spedding and his associates[97] studied the absorption spectra of a great many samarium salts. This work was carried out at low temperatures, with high dispersion. Single crystals were used, but the light was not polarized.

The fluorescence of samarium compounds has attracted considerable investigation. The study of fluorescence is simpler than absorption since powdered samples may be investigated easily without hindrance from scattering.

Tomaschek and Deutschbein[85] have reported weak, broad emission in the blue in solid solutions of $CaSO_4$-Sm and a stronger sharp line emission in the red and infrared. The latter is also reported by Gobrecht[84] for $Sm_2(SO_4)_3 \cdot 8H_2O$. He interpreted the sharp emission spectrum as transitions from an unidentified upper level to the levels of the ^6H term and to some of the levels of another unidentified term. Apparently no one has yet measured the polarized spectra nor observed the Zeeman effect in order to verify the assignment of the ^6H components, although the assignment seems reasonable. The theoretical work of Judd[98] supports the identifications made. He calculated the distortion of the ^6H multiplet by its interaction with excited multiplets, and obtained good agreement with Gobrecht's assignments.

(6) Europium (Eu^{+++}, $Z = 63$, $4f^6$, $^7F_0$) ^7F, $^5(S\ P\ 3D\ 2F\ 3G\ 2H\ 2I\ K\ L)$, $^3(6P\ 5D\ 9F\ 7G\ 9H\ 6I\ 6K\ 3L\ 3M\ N\ O)$, $^1(4S\ P\ 6D\ 4F\ 8G\ 4H\ 7I\ 3K\ 4L\ 2M\ 2N\ Q)$. The large number of terms for Eu^{+++} has prevented a detailed theoretical treatment. Nevertheless the spectrum in the visible region is rather simple because the ^7F term is well below the others and only a few of the quintet terms appear in this range. Judd[74,98] has calculated the positions of the most important of these quintets and has given a detailed theory of the spectra of Eu^{+++} salts.

A great deal of experimental work has been carried out on both the absorption and fluorescence spectrum of Eu^{+++}.[98a] Two terms and all their components are definitely identified; the J values have been assigned to many levels; and the crystal field splitting has been found for many salts.

The level diagram is shown in Fig. 18. The low-temperature absorption must originate from the $^7F_0$ ground state, which is not split in the crystal field. The spectrum is thereby made quite simple, but as Freed points out, information is lost as a result, since many transitions are forbidden. Sayre and Freed[99] have studied Eu^{+++} in $LaCl_3$ solution where

[97] F. H. Spedding and R. S. Bear, *Phys. Rev.* **39,** 948 (1932); **42,** 58, 76 (1932); **44,** 287 (1933); **46,** 308 (1934).

[98] B. R. Judd, *Proc. Phys. Soc.* (*London*) **A69,** 157 (1956).

[98a] K. H. Hellwege and W. Schrock-Vietor, *Z. Physik* **138,** 449 (1954).

[99] E. V. Sayre and S. Freed, *J. Chem. Phys.* **24,** 1211, 1213 (1956).

the crystal field definitely is known to have the symmetry C_{3h}. The fact that the lines of the $EuCl_3$ spectrum are the same in number and polarization as that of $Eu(EtSO_4)_3 \cdot 9H_2O$ also confirms that the local field about the Eu^{+++} ion is truly C_{3h} in the latter salt.

The lowest excited states in absorption are the $^5D_{0,1,2}$ multiplet components occurring at 17,300, 19,000, and 21,500 cm^{-1}. In the C_{3h} field, as reference to Table IX shows, there will be 1, 2, and 3 components, respectively. These are observed. An effect of the nondegenerate ground state is that $^7F_0 \rightarrow {}^5D_0$ is not observed in every crystal. The reasons are that $J = 0 \rightarrow J = 0$ transitions are forbidden in magnetic dipole radiation and since both states are totally symmetric, electric dipole radiation also may be forbidden if the crystal symmetry is high enough.

The other components of the ^7F ground term are observed in fluorescence. The emitting states are the crystal field components of $^5D_0$ and of $^5D_1$. It is rather surprising that radiationless processes cannot depopulate the $^5D_1$ state, only 1700 cm^{-1} above $^5D_0$ within the radiative lifetime of $^5D_1$. Fluorescence lines from $^5D_1$ are only slightly less intense than those from $^5D_0$. The lifetimes of emission are on the order of 2×10^{-4} sec[100] and depend upon the particular salt. Sayre and Freed have assigned the fluorescence lines from $Eu(EtSO_4)_3 \cdot 9H_2O$, and have found nearly all the crystal field components of each member of the ^7F term.

It is also worth noting that the selection rules obeyed were usually those for D_{3h} instead of the lower, exact symmetry C_{3h}. The nearest neighbors, that is, the oxygen atoms of the water molecules or the Cl^- ions, are in D_{3h} positions. Only more distant atoms reduce the symmetry to C_{3h}. Some lines forbidden in D_{3h} appeared weakly when allowed in C_{3h}. Such observations are interesting because they indicate the range of the crystal field.

(7) Sm^{++} $Z = 62$, $4f^6$, $^7F_0$. The divalent samarium ion is isoelectronic with the trivalent europium ion, and their spectra are similar in many respects. The lower nuclear charge and state of ionization of Sm^{++} results in a scaling down of the energy levels by a factor of about 0.85. The spectrum of Sm^{++} is shown in Fig. 18.

The spectra of Sm^{++} in solids have been studied by Butement.[101] Sharp lines were found in the absorption spectrum of $SmCl_2$ in solid solution with $BaCl_2$ or $SrCl_2$ in four regions: 14,800, 16,700, 21,000, and 24,000 cm^{-1}. These lines were found between strong diffuse absorption peaks beginning at 17,040 cm^{-1}. There are at least 13 peaks of this kind extending to 44,250 cm^{-1}. Fluorescence occurs from three upper states; 14,514, 15,936, and 22,452 cm^{-1} above the ground state. These do not cor-

[100] B. Rinck, *Z. Naturforsch.* **3a,** 406 (1948).
[101] F. D. S. Butement, *Trans. Faraday Soc.* **44,** 617 (1948).

respond well to observed absorption lines, but the latter may be obscured by the diffuse absorption.

Butement's interpretation is that the sharp lines are transitions within the $4f^6$ configuration. The fluorescence lines correspond very well to what one expects for transitions to the ^7F ground term, and are analogous in appearance to the Eu^{+++} fluorescence. Table XVI shows a comparison between the ^7F multiplets of the two ions. The ratio of corresponding levels is fairly constant and is only slightly higher than the ratio $\zeta_{4f}(Eu^{+++})/\zeta_{4f}(Sm^{+++})$. This is the direction expected if the shielding of the $4f$ electrons is to be greater in Sm^{++} than in Sm^{+++}.

TABLE XVI. GROUND MULTIPLETS IN Eu^{+++} AND Sm^{++}
$\zeta_{4f}(Eu^{+++})/\zeta_{4f}(Sm^{+++}) = 1415/1200 = 1.18.$

	Eu^{+++}	Sm^{++}	Ratio
$^7F_0$	0 cm^{-1}	0	
$^7F_1$	370	269	1.42
$^7F_2$	1058	834	1.25
$^7F_3$	1906	1493	1.28
$^7F_4$	2872	2296	1.25
$^7F_5$	3936	...	
$^7F_6$	4964	...	

The diffuse absorption bands are attributed to transitions between $4f^6(^7F_0, {}^7F_1)$ and states of the $4f^55d$ configuration. This is very reasonable. The crystal field splitting of a $5d$ orbital in Ce^{+++} was shown to be rather small, and one expects an even smaller splitting in Sm^{++} because of the smaller ionic charge. Thus the $5d$ terms should not be spread out over the entire spectrum, in agreement with observation. The $5d$ orbitals in the rare earths must be less perturbed than those of the $3d$ transition series, and this leads to sharper bands. The d bands in Sm^{++} are 500–2000 cm^{-1} broad as contrasted with 2000-6000 in the $3d$ series.

(8) *Gadolinium* (Gd^{+++}, $Z = 64$, $4f^7$, $^8S_{\frac{7}{2}}$) ^8S, $^6(P\ D\ F\ G\ H\ I)$, $^4(2S, 2P, 6D, 5F, 7G, 5H, 5I, 3K, 3L, M, N)$ $^2(2S, 5P, 7D, 10F, 10G, 9H, 9I, 7K, 5L, 4M, 2N, O, Q)$. The ^8S term lies far below the other terms in the f^7 configuration. The sharp line absorption indicating transitions within the $4f$ shell consists of seven absorption regions, beginning at 32,200 cm^{-1}, as shown in Fig. 18. Since the f shell is half-filled in Gd^{+++}, the spin-orbit interaction[28] and the crystal field splitting[102] are effective only in the second order. The multiplets are therefore irregular and narrow, and the crystal field splitting is small. Judd[74] and also Jorgensen[88] have calculated

[102] G. J. Kynch, *Trans. Faraday Soc.* **33**, 1402 (1937).

the positions of some of the multiplets and have found that ^6P should lie approximately 31,000 cm^{-1} above the ^8S state. This is near the position of the first absorption group.

Fluorescence has been observed from the 32,200 cm^{-1} group at 300°K.[96,103] The four strong lines from $Gd_2(SO_4)_3 \cdot 8H_2O$, 32,160, 32,135, 32,710, and 32,083 were found within 1 cm^{-1} of the value given by Nutting and Spedding in absorption.[104] The efficiency of the fluorescence is quite high, as recently shown by Dieke and Hall,[96] and the lifetime is on the order of 7500 μsec. In a recent study of the absorption spectrum of $GdCl_3 \cdot 6H_2O$, Dieke and Leopold[105] have given good evidence for the identification of $^6P_{\frac{7}{2}}$, $^6P_{\frac{5}{2}}$, $^6P_{\frac{3}{2}}$, and $^6I_{\frac{7}{2}}$.

(9) $Eu^{++}(4f^7, {}^8S_{\frac{7}{2}})$. The ground state of Eu^{++} in the vapor has been found to be $4f^7$, $^8S_{\frac{7}{2}}$,[64] the same as for Gd^{+++} with which it is isoelectronic. Magnetic measurements show that its ground state is the same as for Gd^{+++} in crystalline compounds.[62] The spectra of solid solutions of Eu^{++} in $SrCl_2$ and $BaCl_2$ were studied at room temperature by Butement[95] and at temperatures down to 20°K by Katcoff and Freed.[106] The water solution spectrum given by Butement shows the main features of interest: two broad bands at 31,200 cm^{-1} ($f = 0.0062$), and 40,320 ($f = 0.0312$), with sharp lines superposed on the first one.

In the solid solutions at 20°K, Freed and Katcoff found many sharp lines beginning at 25,041 cm^{-1} and extending to at least 34,000 cm^{-1}. Fluorescence occurred from the first absorption band.

Freed has suggested that the broad bands are $f \rightarrow d$ transitions, similar to the ones found in Ce^{+++}. The spectra are very complex, and no analyses of them have been made.

(10) Terbium (Tb^{+++}, $Z = 65$, $4f^8$, $^7F_6$). From terbium to the end of the $4f$ series, the spectrum of each element resembles the one in the first half of the series which has 14-n electrons. In a first approximation, the terms are in the same order of energy and the multiplets are inverted. The analogy between terbium and europium is especially clear. The fluorescence of Tb^{+++}, which is analogous to that of Eu^{+++}, is from $^5D_4$ to the components of ^7F.[107] The fluorescence studies were carried out with the bromate. There is no extensive work on the absorption spectrum in the solid, and the only term assignment is the one made on the basis of the fluorescence.

Judd[98] has calculated the distortion of the ^7F multiplet through spin-

[103] R. Tomaschek and E. Mehnert, *Ann. Physik* [5] **29,** 306 (1937).
[104] See, for references, G. C. Nutting and F. H. Spedding, *J. Chem. Phys.* **5,** 33 (1937).
[105] G. H. Dieke and L. Leopold, *J. Opt. Soc. Am.* **47,** 944 (1957).
[106] S. Katcoff and S. Freed, *Physica* **14,** 17 (1948).
[107] H. F. Geisler and K. H. Hellwege, *Z. Physik* **136,** 293 (1953).

orbit interactions with higher excited quartets. The centers of gravity of these quartets were calculated and could serve as a guide to further assignments in this ion.

(*11*) *Dysprosium* (Dy^{+++}, $Z = 66$, $4f^9$, $^6H_{15/2}$). The energy level diagram of Dy^{+++} shows a qualitative similarity to Sm^{+++} as it should (Fig. 18). The recent work of Dieke and Singh on $DyCl_3 \cdot 6H_2O$ has led to several term assignments and to considerable information about the crystal field.[108] They have used carefully prepared single crystals, high dispersion, and liquid helium temperatures for the absorption experiments. No polarization data are reported. The fluorescence spectrum from a powder showed no surplus lines in regions where comparison with absorption was possible. Fluorescence was observed from an upper state, later identified by its absorption spectrum, to all but the $\frac{5}{2}$ component of the ^6H ground term. The missing level was observed in absorption. The number of crystal field lines should be $J + \frac{1}{2}$ because the number of electrons is odd. This number was found in almost every case in which J was known and was used to make assignments in absorption and fluorescence to unassigned levels. They thus identified the emitting level in fluorescence as $^6F_{11/2}$ and identified all components of the ^6H state in the near infrared.

The ethyl sulfate was studied in absorption by Rosa[109] who reported the polarized spectra from 10,000 to 24,000 cm^{-1}. Both axial and polarized transverse spectra are given, and it is shown what the transition mechanism is in each case. No assignments were attempted but the data appear sufficiently good to warrant it. In particular they should be used to test Dieke's assignments. One first might assign the numerous magnetic dipole transitions, since the possibilities for such transitions are very limited. The spectra of this salt and of the others he studied showed much structure due to lattice vibrations, at the temperature of liquid nitrogen. Some of these were identified by comparing spectra of crystals containing light and heavy water.

(*12*) *Holmium* (Ho^{+++}, $Z = 67$, $4f^{10}$, $^5I_8$). The most recent experimental work on holmium is that of Severin,[110] who recorded the spectrum of $Ho_2(SO_4)_3 \cdot 8H_2O$ and $Ho(C_2H_5SO_4)_3 \cdot 9H_2O$, from 15,000 to 25,000 cm^{-1}. He points out that earlier workers[111] used holmium contaminated with dysprosium and erbium. He therefore sorted the lines of these three elements from each other carefully and showed where the regions of possible

[108] G. H. Dieke and S. Singh, *J. Opt. Soc. Am.* **46,** 495 (1956); S. P. Cook and G. H. Dieke, *J. Chem. Phys.* **27,** 1213 (1957).

[109] A. Rosa, *Ann. Physik* [5] **43,** 162 (1943).

[110] H. Severin, *Z. Physik* **125,** 455 (1947).

[111] E. J. Meehan and G. C. Nutting, *J. Chem. Phys.* **7,** 1002 (1939).

interference are. The work was carried out under high dispersion with single crystals and polarized light. Measurements of both the axial and the polarized transverse spectrum were made, and it can be said that all observed transitions are electric dipole. No assignments were made, but it seems that this should be possible especially in the ethyl sulfate.

Gobrecht[83] reports absorption in the region of 8300–14,000 cm^{-1} and one line at 5150 cm^{-1}. He plausibly identifies many of these lines with transitions from the $^5I_8$ ground state to $^5I_4$ through $^5I_7$.

(13) Erbium (Er^{+++}, $Z = 68$, $4f^{11}$, $^4I_{15/2}$). The only investigation reported so far which employs polarized light and high dispersion over an extensive spectral region is the work of Severin.[110,112] This work was carried out with $Er(C_2H_5SO_4)_3 \cdot 9H_2O$, $Er_2(SO_4)_3 \cdot 8H_2O$, $Er(NO_3)_3 \cdot 6H_2O$, and $Er_2Mg_3(NO_3)_{12} \cdot 24H_2O$. The latter had an unexplainably diffuse spectrum at 77°K, and was not measured. Enough polarization data were taken to determine that the transition goes by electric dipole radiation for a large number of lines in the ethyl sulfate spectrum. For the other salts, a partial term analysis is given but no polarization data are available. The vibrations of NO_3^- and $SO_4^=$ ions were fairly prominent in the spectrum. Severin also identified some of the crystal field levels of the ground state. Meehan and Nutting[111] had previously done this for the sulfate octahydrate. Both obtained 0, 19.6, 41.5, and 86.4 cm^{-1}.

(14) Thulium (Tm^{+++}, $Z = 69$, $4f^{12}$, $^3H_6$). The spectrum of this ion has received attention because of its analogy to that of Pr^{+++}. Bethe and Spedding[113] calculated its spectrum by using the analogy to the La II spectrum. Thus all the term values of La II were multiplied by a factor of 3.3. The complete matrices of spin-orbit interaction were solved for several ζ_{4f} values near an estimated value, and 2800 cm^{-1} was found to give best agreement. The calculated spectrum seems to be in fairly good agreement with the observed one.[83,114] Nevertheless, there has been no detailed work on this spectrum, and no assignments of J values have been made on the basis of the crystal field effect or the Zeeman effect.

Gobrecht[115] has reported an emission spectrum for Tm^{+++} dissolved in K_2SO_4 and in CaO. Many bands are reported, and assignments are made, but no detailed work has been done to test them.

(15) Ytterbium (Yb^{+++}, $Z = 70$, $4f^{13}$, $^2F_{7/2}$). The sharp spectrum of Yb^{+++} in solids consists of the $^2F_{7/2} \rightarrow {}^2F_{5/2}$ transition and the accompanying crystal field and lattice vibrational transitions. It is observed in $YbCl_3 \cdot 6H_2O$, in absorption as two sharp lines, 10297.13 and 10282.32 cm^{-1}

[112] H. Severin, *Ann. Physik* [6] **1,** 41 (1947).
[113] H. A. Bethe and F. H. Spedding, *Phys. Rev.* **52,** 454 (1937).
[114] H. Gobrecht, *Ann. Physik* [6] **7,** 88 (1950).
[115] H. Gobrecht, *Ann. Physik* [5] **31,** 600 (1938).

plus a number of fainter ones.[116] The two sharp lines are two of the three crystal field components expected for $J = \frac{5}{2}$. Since they are so close together, the third component could be several hundred cm^{-1} away, and might be difficult to pick out of the many weak components. The latter are presumably caused by lattice vibrations.

(16) Yb^{++} ($4f^{14}$, $^1S_0$). As in the case of Sm^{++} and Eu^{++}, the Yb^{++} ion has broad, moderately strong absorption bands. In this case, there is no possibility of a $4f$-$4f$ transition, but a $4f$-$5d$ transition is perfectly possible.

Butement[101] studied Yb^{++} in $SrCl_2$ and $BaCl_2$ solid solutions and found a weak fluorescence peak at 24,880 cm^{-1} in the former, and a stronger fluorescence peak at 23,200 cm^{-1} in the latter. Two absorption bands occurred in aqueous solution, 28,500 cm^{-1} ($f = 0.0051$) and 41,000 cm^{-1} ($f = 0.0275$). Stronger absorption occurred farther into the ultraviolet. The upper state of the fluorescence is presumably the same as the upper state of the weaker of the two absorption bands, and the 5000 cm^{-1} gap between the absorption and fluorescence peaks is due to band width.

These excited states are probably states of the $4f^{13}\,5d$ configuration.

d. *Interaction of the Ion with the Lattice*

There are several manifestations of the interaction between ion and lattice. The crystal field splitting of atomic levels is the most obvious one. The levels are also displaced from their positions in the free ion. This latter effect is less spectacular but perhaps is capable of being interpreted so as to give useful information. A possible interpretation of level displacements would be the one suggested by Owen for the $3d$ transition series (Section 1).

The level displacements may also be caused by second-order crystal field perturbations.

Since up to the present, the rare earth ions have been studied only in a very limited range of environments, and since the free ion term values are not known, it is not easy to investigate the origin of the term displacements. The data on Pr^{+++} surrounded by F, O, and Cl are shown in Table XVII. The term energies given are weighted averages over the crystal field components. The term energy is, however taken from the lowest crystal field component of the ground state. The average energy of the crystal field components of $^3H_4$ is given, but only in one case is this accurately known. In order to compare the positions of multiplet components in different salts, the position of $^3H_4$ must be subtracted in each case. More accurate comparisons are the differences between terms, all of whose crystal field components have been observed. Several term

[116] G. H. Dieke and H. M. Crosswhite, *J. Opt. Soc. Am.* **46**, 885 (1956).

differences are given in the table. The term differences change from one salt to another by as much as 300–400 cm^{-1} in some cases. This is probably too great to be caused by second order effects of the crystal field, and must be regarded as evidence for variations in the screening effects of the ligands or of some other effect not considered in the theory of the static crystal field.

TABLE XVII. COMPARISON OF ENERGY LEVELS OF Pr^{+++} IN DIFFERENT SALTS
The energy levels given are weighted averages over crystal field components.

	$PrCl_3$[a]	$Pr_2Mg_3(NO_3)_{12}\cdot 24H_2O$[b]	PrF_3[a]
$^3P_2$	22231 (cm^{-1})	22669 (cm^{-1})	22687 (cm^{-1})
$^3P_1$	21077	21448	21514
$^3P_0$	20476	20846	20923
$^1D_2$	16728	16906	16861
$^1G_4$	9784		9992
$^3F_4$	6778[c]		(6583)[e]
$^3H_4$	94[c]	(220)[d]	(250)[e]
$^3P_2$-$^3P_0$	1755	1823	1764
$^3P_0$-$^3H_4$	20382	20626	20673
$^1G_4$-$^3F_4$	3006	—	3409
$^3P_1$-$^1D_2$	4349	4542	4653

[a] Data of E. V. Sayre, K. Sancier, and S. Freed, *J. Chem. Phys.* **23,** 2060, 2066 (1955), except for $^3F_4$ and $^3H_4$ levels.
[b] Data of A. M. and K. H. Hellwege, *Z. Physik* **135,** 92 (1953).
[c] Data of G. H. Dieke and R. Sarup, *J. Chem. Phys.* **29,** 741 (1958).
[d] Estimated from data of ref. *b* and theoretical splitting pattern of B. R. Judd, *Proc. Roy. Soc. (London)* **A232,** 458 (1955).
[e] Estimated by scaling crystal field splitting of $PrCl_3$ (ref. *c*) to fit known PrF_3 levels given in ref. *a*.

A third effect of the lattice is the excitation of vibrations. It has been studied in detail for (Pr and $Nd)_2Zn_3(NO_3)_{12}\cdot 24H_2O$.[79,117] Both lattice vibrations and NO_3^- ion vibrations were found. An especially simple transition to investigate is $^3H_4 \rightarrow {}^3P_0$ in Pr^{+++}, which at low temperatures has only one pure electronic origin. Some 42 lines due to lattice vibrations appeared between 74 and 246 cm^{-1} to higher frequencies of the electronic origin. The normal vibrations of NO_3^- appeared split into 6 lines for each nondegenerate mode and into at least 11 for doubly degenerate modes. Similar effects are found in infrared and Raman spectra, and are caused by coupling of several NO_3^- groups; the number of nonequivalent NO_3^- groups closely coupled together must be six or a multiple of six in this case. The lines are broader than the pure electronic lines; this must be caused by the weaker couplings within the crystal.

[117] K. H. Hellwege, *Z. Physik* **113,** 192 (1939).

It was shown that totally symmetric NO_3^- vibrations appear predominantly with the same polarization as the electronic origin, and that certain components of nontotally symmetric vibrations appeared in the opposite polarization. The latter appear by virtue of vibrational-electronic interaction, which would be expected to provide a transition mechanism analogous to that found in molecular spectra. Vibrations most probably mix *d* orbitals into the *f* orbitals in producing this effect.

The low-frequency lattice vibrations also produce vibrational-electronic mixing since they are seen in both polarizations.

e. Conclusions

It seems probable that within a fairly short time, most of the lower lying energy levels of the rare earth ions will be identified. The basis for this prediction is that excellent experimental work is now being carried out, particularly in Dieke's laboratory. Furthermore, recent theoretical work of Elliott *et al.*[118] has considerably extended the calculations of energy levels. Their methods should make further advances in this direction possible.

In another recent paper, Judd[119] has shown that the crystal field in the double nitrates, and presumably in other salts having symmetrical 9-coordination of the rare earth ion, is well approximated by an icosahedral field. The only terms in an icosahedral potential are V_6^0, V_6^3, and V_6^6. It is found in practice that these are the largest terms. The smaller terms may be explained as distortions of the icosahedron. It seems probable therefore that the icosahedral field will assume the same importance for the spectra of rare earth ions that the cubic field now enjoys in relation to the spectra of the transition metal ions.

With this increase in our understanding of the spectra, it should be possible to use them as sensitive indicators of the symmetry of crystal fields. Experiments of this kind have already been undertaken by Freed and co-workers[120] for the determination of the symmetry of the coordination sphere about rare earth ions in solution.

4. Spectra of Transition Metal Ions

a. Chemical and Physical Properties of the Ions

There are three groups of transition metals beginning with one of the elements Ti, Zr, or Hf. The known oxidation states of these elements and

[118] J. P. Elliott, B. R. Judd, and W. A. Runcimen, *Proc. Roy. Soc. (London)* **A240,** 509 (1957).

[119] B. R. Judd, *Proc. Roy. Soc. (London)* **A241,** 122 (1957).

[120] E. V. Sayre, D. G. Miller, and S. Freed, *J. Chem. Phys.* **26,** 109 (1957).

their crystal radii where known are given in Table XVIII. The ions, either simple or complex, which are commonly observed in water solution are italicized in the table. The ability of an ion to exist in water is a measure of its stability and ease of preparation. The crystal radius, also given in the table, is one of the most important factors determining the coordination number of an ion and its solubility in a foreign crystal lattice.

A number of useful generalizations concerning the chemistry of transition metal compounds may be made from Table XVIII. The occurrence of a large number of stable ions having no *d* electrons indicates that the rare gas configurations confer especial stability, as is found in other areas of the periodic table. The elements of the first transition group are most stable in the +2 and +3 valence states, while in the second and third series they are most stable in the +3, +4, and higher states. Of the ions which form octahedral complexes, those having three and six *d* electrons are particularly stable.

The stable ionization or oxidation states are determined both by the nature of the surroundings of the ion and by the properties of its *d* orbitals. If the environment has low electronegativity, and the metal ion has a stable unfilled *d* orbital, an electron transfer may occur from environment to ion. The 3*d* orbitals appear to be more stable than the 4*d* and 5*d* orbitals, so that such a process is more likely to occur in the first transition group than in the higher ones. This is consistent with the greater stability of the high oxidation states of the latter. Conversely it is easier for the ions of the second and third transition groups to lose electrons than it is for those of the first. Of the negative ions, fluoride ions have the least ability to gain or to lose an electron, so that transition metal fluorides are stable over a very wide range of oxidation levels. Oxides and chlorides are stable over a somewhat narrower range, whereas sulfides and iodides exist within a very narrow range of oxidation levels.

Among the lower oxidation states of each group, the most common coordination number is six, but four and eight are also known. It is more accurate to consider the higher oxidation levels as molecules consisting of the metal atom and its nearest neighbors, rather than as highly charged ions. There is no sharp dividing line, of course, and in any oxidation state, a metal ion and its neighbors always have some of the properties of a molecule, or complex ion. In terms of crystal structure, molecules are formed at and above the +5 oxidation level, and such molecules as $CrO_4^{=}$ and MnO_4^- are recognized as separate entities in crystals.

The transition from ionic to molecular structure may also be recognized in the electronic spectra. The spectra of the +2 and +3 ions are adequately explained by crystal field theory. Only relatively minor modifications are necessary to take account of the molecular orbitals formed

TABLE XVIII.[a] THE KNOWN OXIDATION STATES OF THE TRANSITION METAL IONS, AND THEIR CRYSTAL RADII FOR COORDINATION NUMBER SIX

n Sp.	Trans. group	Oxidation state (2 = M^{++} etc.) 1	2	3	4	5	6	7	8
0	I		*Ca*(1.06)	*Sc*(0.83)	*Ti*(0.64)	*V*(0.4)	*Cr*(0.37)	*Mn*[0.46]	
^1S	II		*Sr*(1.27)	*Y*(1.06)	*Zr*(0.87)	*Nb*[0.70]	*Mo*[0.62]	*Tc*	*Ru*
	III		*Ba*(1.43)	*La*(1.22)	*Hf*(0.86)	*Ta*(0.73)	*W*(0.68)	*Re*	*Os*
1	I			*Ti*(0.69)	*V*(0.61)	*Cr*	Mn		
^2D	II			Zr	Nb	*Mo*		Ru	
	III					*W*	Re		
2	I		Ti(0.80)	V(0.66)	Cr	Mn	Fe		
^3F	II		Zr	Nb	Mo(0.68)		Ru		
	III				W(0.68)	Re			
3	I		*V*(0.71)	*Cr*(0.64)	Mn(0.52)				
^4F	II		Nb?	*Mo*	Tc		Rh		
	III			*W*	Re		Ir		
4	I		*Cr*(D)	*Mn*(0.70,*D*)					
	II				*Ru*(0.65)		Pd?		
^5D	III			*Re*	*Os*		Pt		
5	I		*Mn*(0.90)	*Fe*(0.67)	Co?				
^6S	II		*Tc*	*Ru*(0.69)	Rh				
	III		Re	*Os*	*Ir*(0.66)				
6	I	Mn	*Fe*(0.83)	*Co*(0.65)	Ni				
^5D	II		Ru	*Rh*	Pd				
	III	Re	Os	*Ir*	*Pt*				
7	I		*Co*(0.80)						
	II		Rh						
^4F	III		Ir	Pt					
8	I		*Ni*(0.78)	Cu					
^3F	II	Rh	*Pd*(0.50)	Ag					
	III	Ir	*Pt*	*Au*					
9	I	Ni	*Cu*(D)						
^2D	II		Ag						
	III	Pt	Au						
10	I	*Cu*[0.96]	*Zn*(0.83)	*Ga*(0.62)					
^1S	II	*Ag*(1.13)	*Cd*(1.03)	*In*(0.92)					
	III	*Au*[1.37]	*Hg*(1.12)	*Tl*(1.05)					

[a] Most of the radii given are those of Goldschmidt, tabulated by T. Moeller in the book "Inorganic Chemistry," pp. 140–142. Wiley, New York, 1952. Values in square brackets are values calculated by Pauling, which are also tabulated in Moeller's book. The letter *D* means the ion has a strongly distorted environment, usually tetragonal, and the radius has little meaning. *n* = number of *d* electrons; Sp. = spectroscopic state. The commonest oxidation states are given in italics.

between the central ion and its surroundings. By way of contrast, ions of the six-valent state have spectra similar to molecular spectra in that they show sharp and prominent vibrational structure.

b. Spectra of Individual Ions

The spectra of transition metal ions have been studied extensively in aqueous solution both with and without complex-forming substances. A comparison of the spectra of some crystalline hydrates with their aqueous solutions is shown in Fig. 20.

Evidence that these are crystal field spectra are the facts that the absorption strengths are very low ($f \sim 10^{-6}$), and the positions of the bands are given quite well by crystal field theory for an octahedral field. In Fig. 20 the calculated positions of bands are given by the heavy vertical lines: one band has been chosen to provide a Dq value. The parameter Dq could be found, for example, by choosing the abscissa of the Tanabe and Sugano diagrams Figs. 4 through 10 such that the best fit of the spectrum is obtained.

From the close similarity of the spectra in solution to those of the solid hydrates and the fact that the metal ion is octahedrally coordinated by water molecules in the solid state, one concludes that the aqueous ions are octahedrally coordinated. This coordination octahedron may be replaced in solution by more strongly bound groups such as NH_3 molecules. Orgel has reviewed and interpreted the spectra of $3d$ ions in solution, and has shown how Dq varies throughout a series of different ligating groups.[10] Since these groups, or ones similar to them, are found in the crystalline state it is of interest to examine his results. He found that Dq increases from left to right in the sequence:

(increasing $Dq \rightarrow$) I^-, Br^-, Cl^-, F^-,
H_2O, oxalate, pyridine, NH_3, ethylene diamine, NO_2^-.

A further indication that these spectra are crystal field spectra, and therefore highly localized on the metal ion, is that Dq changes by less than a factor of 1.5 throughout the series. The small difference between the energies of the observed levels in the most extreme cases shows that the nature of the surrounding atoms is not the dominating factor, and that perturbation of the metal ion levels is a valid approach to the interpretation of the spectra.

In the following sections we shall discuss the spectra of ions belonging to each of the d^n configurations in each transition group. The emphasis will be placed upon spectra observed in the solid state, although the spectra of some of the ions have been observed only in solution.

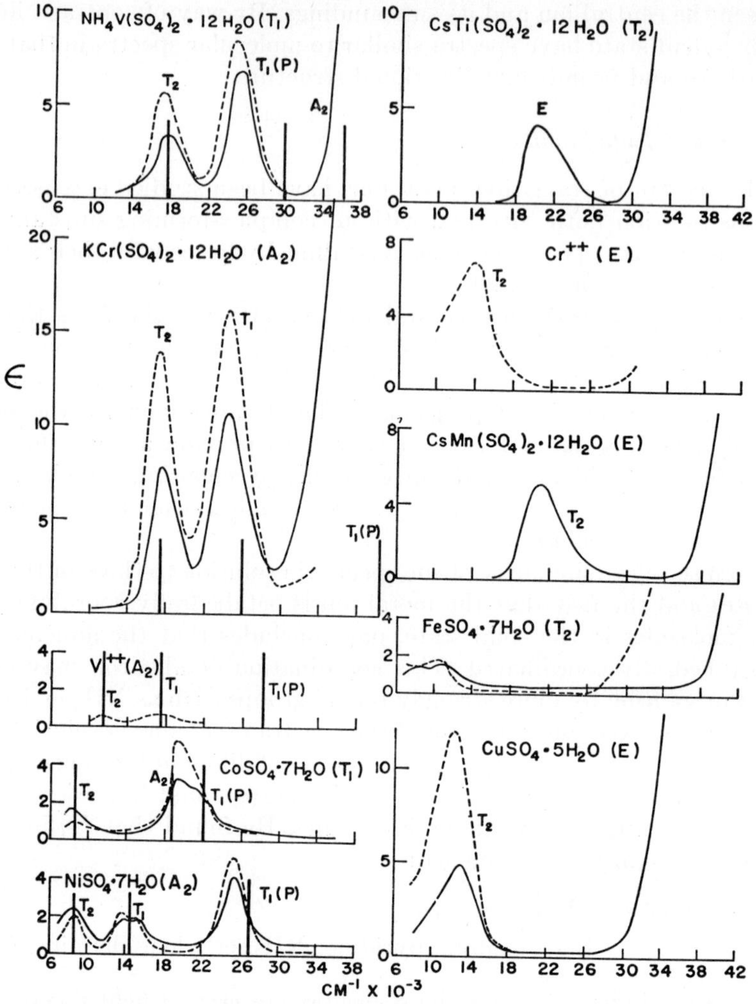

FIG. 20. Absorption spectra of the hydrated ions of the 3*d* transition series. The aqueous solution spectra are shown as dotted lines and the crystalline hydrate spectra as solid lines. The ground state symmetry in the group O_h is given next to the formula for the ion or compound. The excited state symmetry symbols are given next to each absorption band. The theoretical positions of the excited states are shown as solid vertical lines (Holmes and McClure;[49] Hartmann and Schlaefer[121,132]).

3d^1. The $^2T_{2g}$ state of Ti(III) is the ground state in an octahedral field. Hartmann and Schlaefer[121] interpreted the broad absorption band in the visible at 20,300 cm^{-1} as the $^2T_{2g} \rightarrow {}^2E_g$ transition. Spectra of aqueous solutions and of solid $CsTi(SO_4)_3 \cdot 12H_2O$ were observed. Ilse and Hartmann[7] developed the theory of the d^1 ion in an octahedral and in a tetragonal field, and showed that the spectrum of the complex ion $Ti(H_2O)_4Cl_2^+$ could be interpreted on the basis of the tetragonal field theory.

Jorgensen[122] has compared the spectra of several ions having the d^1 configuration, namely, Ti(III), V(IV), Mo(V), W(V), and Nb(IV). Only Ti(III) forms octahedral complexes. The V(IV) ion forms complex ions

TABLE XIX.[a] CHARACTER TABLE FOR D_4 AND C_{4v}

D_4 C_{4v}	E E	$2C_4$ $2C_4$	C_4^2 C_4^2	$2C_2'$ $2\sigma_v$	$2C_2''$ $2\sigma_d$	
A_1	1	1	1	1	1	z (for C_{4v})
A_2	1	1	1	−1	−1	z (for D_4)
B_1	1	−1	1	1	−1	
B_2	1	−1	1	−1	1	
E	2	0	−2	0	0	x, y

[a] *Notes to table:* C_4 = four fold rotation about z-axis; σ_v = 100 and 010 reflection planes; σ_d = 110 and $1\bar{1}0$ reflection planes; C_2' = two fold axes, 100, 010; C_2'' = two fold axes, 110, $1\bar{1}0$.

having a strongly axial tetragonal distortion. The aqueous ions are probably VO^{++} with water molecules attached at octahedral positions. The compound VCl_4, however, appears to have a tetrahedral structure. The band at about 9000 cm^{-1} [123] was identified by Orgel[10] as the $^2E \rightarrow {}^2T_2$ transition. The Dq value is 900 cm^{-1} if this is correct. This value is small for ions having a charge of four, and is therefore characteristic of the Dq values for tetrahedral coordination.

4d^1. The complex ions formed by Mo(V) in solution have not been positively identified, but Jorgensen[122] believes that $MoOCl_5^=$ predominates in about 5*M* HCl solution. Jorgensen interprets the spectra of VO^{++} and of $MoOCl_5^=$ in terms of the diagram of Fig. 21, which shows the effect of a tetragonal distortion superimposed upon a cubic field. Table XIX gives the character table for C_{4v}, the symmetry group of these complex ions. The left of Fig. 21 shows the effect of an axial tetragonal component, appropriate to the d^1 ions, while the right of the diagram

[121] H. Hartmann and H. L. Schlaefer, *Z. Physik Chem. (Leipzig)* **B197,** 116 (1951).
[122] C. K. Jorgensen, *Acta Chem. Scand.* **11,** 73 (1957).
[123] A. G. Whittaker and D. M. Yost, *J. Chem. Phys.* **17,** 188 (1949).

shows the effect of a square tetragonal component. The latter, as we shall see, is appropriate for d^9 systems when the order of the levels is reversed. The diagram of Fig. 2 illustrates qualitatively why the E_g level splits more than the T_{2g} level, and why the doubly degenerate component of T_{2g} increases in energy relative to the nondegenerate component under an axial tetragonal distortion.

The observed transitions in VO^{++} and $MoOCl_5^=$ are thought to be $^2B_2 \rightarrow {}^2E$ and $^2B_2 \rightarrow {}^2B_1$.

3d^9. In the configuration complementary to d^1, the cubic field levels are reversed, and ^2E_g is the lower state. It is particularly susceptible

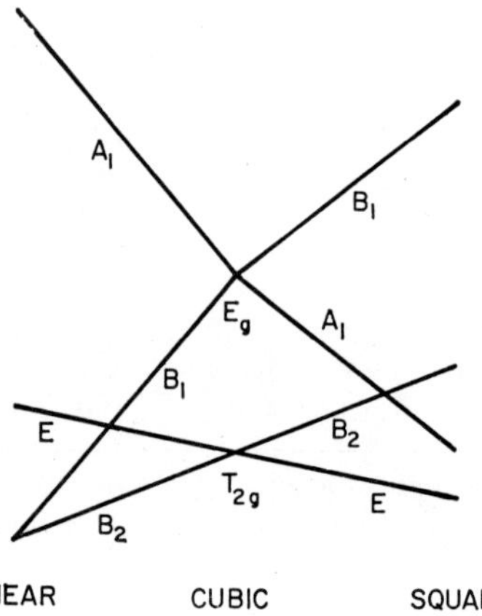

FIG. 21. Energy levels of a single *d*-electron in fields of octahedral and tetragonal symmetry. The right side of the diagram applies to the case in which the polar ligands are farther away (or have a lower charge) than the equatorial ligands of the octahedron. The left side is for the opposite case. When the diagram is inverted, it applies to the d^9 configuration. The levels are given designations appropriate for the symmetry groups C_{4v} and D_4 (Table XIX).

toward splitting by the Jahn-Teller effect, and the tetragonal splitting diagram, Fig. 21 must be applied in order to interpret the spectra. Since the transitions involve a "hole" rather than an electron, the highest level of Fig. 21 is the ground state. Orgel and Dunitz[124] have collected evidence which shows that the copper II ion in crystals is rarely if ever found in a regular octahedral environment, and it is concluded that the Jahn-Teller effect is responsible for this. The type of distortion is that in which the two polar ligands move out from the Cu^{++} leaving the latter "square" coordinated.

[124] L. E. Orgel and J. Dunitz, *Nature* **179,** 462 (1957); M. A. Hepworth, K. H. Jack, and R. S. Nyholm, *ibid.* **179,** 211 (1957).

The effect of a tetragonal distortion of the environment on the spectrum of Cu^{++} has been studied by Belford.[125] He also gives a detailed account of crystal field calculations, including the use of molecular orbitals, which should be quite helpful for anyone who encounters crystal field theory for the first time. He studied a series of copper chelates chosen in a way which provided "crystal fields," or more accurately, ligand fields, of varying symmetry. The results show that the E_g state in a cubic field is split to such an extent by the tetragonal field that its upper component falls among the components of T_{2g}. Thus the ligand field is more nearly square planar than octahedral, corresponding to the right of Fig. 21. Similar results obtained by a different method were reported by Holmes and McClure[49] who examined the spectra of single $CuSO_4{\cdot}5H_2O$ crystals in polarized light.

The variation in frequency and shape of the absorption band of copper II combined in various complexes in solution is described by Jorgensen.[126]

Tetrahedral copper complexes exist. The structure of the nearly tetrahedral $CuCl_4^{=}$ ion was shown by Felsenfeld[127] to be determined by the compromise between the extra crystal field energy gained through a distortion from tetrahedral toward square planar, and the concomitantly increasing repulsion between the Cl^- ions. The E level is the higher one in this case. Since the complex is distorted from T_d into D_{2d} symmetry, the E level must split into two non-degenerate classes, $A_1 + B_1$. The crystal field spectrum must therefore contain two bands, and since Dq should be small, these should lie in the infrared. In the spectrum of solid Cs_2CuCl_4 published by Helmholtz and Kruh,[128] there is a suggestion of a weak band below 10,000 cm^{-1}. The other bands which are seen are much stronger, and are probably charge transfer bands.

$4d^9$, $5d^9$. No crystal field spectra have been reported for Ag^{++} or Au^{++}. AgF_2 and AgO are the only stable simple Ag^{++} salts.

$3d^2$. The theory of the d^2 and d^8 configurations in cubic fields has been worked out in great detail by Liehr and Ballhausen.[41] They included the spin-orbit interaction in these calculations and obtained numerical solutions, for useful ranges of three parameters namely F_2, Dq and λ. It was assumed that $F_2 = 14F_4$. Theoretical work for fields of symmetry lower than cubic has been carried out by Ballhausen and Jorgensen[129] and

[125] L. Belford, M. Calvin, and G. Belford, *J. Chem. Phys.* **26,** 1165 (1957).

[126] C. K. Jorgensen, *Acta Chem. Scand.* **19,** 887 (1956).

[127] G. Felsenfeld, *Proc. Roy. Soc.* (*London*) **A236,** 506 (1956).

[128] L. Helmholz and R. F. Kruh, *J. Am. Chem. Soc.* **74,** 1176 (1952).

[129] C. Ballhausen and C. K. Jorgensen, *Kgl. Danske Videnskab. Selskab Mat.-fys. Medd.* **29,** No. 14 (1955).

Maki,[130] with neglect of spin orbit interaction. The theory for d^2 in a trigonal field was worked out by Pryce and Runciman[130a] in connection with an experimental investigation of V^{+++} in corundum, Al_2O_3. The theoretical work, of course, applies equally well to the d^8 problem, and may be used to interpret the spectra of Ni^{++} compounds, for example.

The only $3d^2$ ion which has received experimental attention is V^{+++}. The ions Ti^{++} and Cr^{++++} are known, however. Another $3d^2$ ion, $MnO_4^{=}$(MnV) will be discussed in Section 5, as it apparently displays only a charge transfer spectrum, rather than a crystal field spectrum.

There are three spin-allowed transitions from the ground state for a d^2 ion. For V^{+++} in corundum, the crystal field is strong, $Dq/C = \frac{3}{4}$; hence almost all the crossings in Fig. 3 lie to the weak-field side, and it is better to describe the states by reference to their strong-field parents. The experimental values for the triplet-triplet transitions[130a,b] and their theoretical values, assuming $Dq = 1860$, $F_4 = 70$, $F_2 = 14F_4$,[41] are given below:

		Experimental	Theoretical
$^3T_1(t^2) \rightarrow$	$^3T_2(et)$	17,400 cm^{-1}	17,400
	$^3T_1(et)$	25,200	25,400
	$^3A_2(e^2)$	34,500	35,800

The assumption $F_2 = 14F_4$ is based upon experience with atomic spectra, so that the three pieces of data are fitted adequately with two parameters.

The same parameters fit the intercombination bands well. The effect of spin-orbit interaction on the energies is almost negligible, although it has been taken into account in the theoretical results quoted below[41] using $\zeta_{3d} = 130$ cm^{-1}. The three observed intercombinations are very

		Experimental	Theoretical
$^3T_1(t^2) \rightarrow$	$^1T_2$, $^1E(t^2)$	8770, 9660 cm^{-1}	9100
	$^1A_1(t^2)$	21,000	20,000
	$^1T_2(et)$		27,600
	$^1T_1(et)$		29,800

sharp lines because there is no change of configuration during the transition. On the other hand, the singlets of *et* are expected to be very broad and hence difficult to observe as transitions from the ground state.

In the above comparison between observed and calculated levels of

[130] G. Maki, *J. Chem. Phys.* **28**, 651 (1958).
[130a] M. H. L. Pryce and W. A. Runciman, *Discussions Faraday Soc.* **26**, 34 (1958).
[130b] W. Low, *Z. physik. Chem.* (*Frankfurt*) **13**, 107 (1957).

V^{+++}, the trigonal field of the corundum crystal has been ignored. Pryce and Runciman have obtained a value of the trigonal field parameter in corundum from the splitting of the $^3T_1(t^2)$ ground state. In a trigonal field, this state splits into $^3A_2 + {}^3E(C_{3v})$, with ^3E about 1200 cm^{-1} above $^3A_2$. The latter splits into a lower S = 0 and an upper S = ±1 state 8 cm^{-1} above, because of second order spin-orbit coupling in the trigonal field. The splitting of 8 cm^{-1} equals $(1.25\lambda)^2/v$, where v is the trigonal field parameter and λ is $\zeta/2$, the spin orbit coupling parameter. The value found for v was +1200 cm^{-1}. There are actually two trigonal field parameters, but v is responsible for all of the first order splitting. The ^3E component of the cubic field ground state has not been observed, and its position is only inferred from the $^3A_2$ splitting observed both in spin resonance experiments and in the optical work of Pryce and Runciman.

(The parameter v may be defined as follows: write the second power trigonal potential as $H(xy + xz + yz)$ and the fourth power part as $I \sum_{\text{cyclic}} xy(-6z^2 + x^2 + y^2)$. Then $v = \frac{6}{7}(H\overline{r^2} - \frac{4}{3}I\overline{r^4})$. The first order splitting of the $^3T_1$ ground state and of the $^3T_2$ state is v, and that of the $^3T_1(et)$ state is $\frac{1}{2}v$.)

It is interesting to compare the value of the trigonal field parameter derived by Sugano and Tanabe[130c] for Cr^{+++} in corundum with the value for V^{+++} in corundum. They found a value of v of about 1050 cm^{-1}. This is the same as Pryce and Runciman's value within experimental error, a very encouraging result, since the trigonal potential is provided by the same crystal in each case. There is a possibility that more refined measurements will reveal differences due to the local distortion of the crystal brought about by the mismatch of the ionic radii between host and guest ions.

The other trigonal field parameter for V^{+++} in corundum may possibly be obtained from the intensity ratios among the absorption bands, as has been shown by Ballhausen.[131] The special circumstance which makes this possible is the following. In the strong field limit, the $^3T_1$ ground state of V^{+++} belongs to the t^2 configuration. There is only one $^3A_2$ state, and it belongs to the e^2 configuration. In the strong field limit therefore, the $^3T_1 \rightarrow {}^3A_2$ transition is a two-electron jump, $t^2 \rightarrow e^2$, and therefore forbidden. There are several factors which cause configuration mixing, thus making the transition allowed but weak. One is the departure from the strong-field limit because of the magnitude of the electron correlation terms. These are found to be too small. The mixing produced by spin orbit interaction is negligible also. The trigonal field is the only effect having the right order of magnitude. Accurate experimental results have

[130c] S. Sugano and Y. Tanabe, *J. Phys. Soc. Japan* **13,** 880 (1958).
[131] C. J. Ballhausen, *Z. physik. Chem. (Frankfurt)* **17,** 246 (1958).

not yet been obtained, but the calculated intensity is approximately correct.

Hartmann and Schlaefer[132] studied the spectrum of V^{+++} in alums in which the ion is surrounded by a nearly regular octahedron of water molecules. Figure 20 shows this spectrum compared to the calculated one. The onset of very strong absorption prevents the highest crystal field band from being observed. The strong absorption undoubtedly corresponds to a charge transfer process involving the surrounding molecules, and Hartmann[133] pointed out that its onset varies in a series of $V(X)_6^{+++}$ complexes with the ionization potential of X.

4d^2. The Nb^{+++} ion appears to have crystal field bands in the visible region.[134,135] Its spectrum has been studied in solution only, and there is even some question as to the authenticity of the oxidation state.

5d^2. Jorgensen[135] reports that W^{++++} in 12M HCl has a band at 19,100 cm^{-1} having an extinction coefficient of 600. This value is high for the crystal field spectrum of an octahedral complex, and low for a charge transfer band, and it remains uninterpreted.

3d^8. The theory of the d^8 configuration in a cubic field is the same as that of d^2 if the sign of Dq is changed. The comprehensive paper of Liehr and Ballhausen[41] applies specifically to this case also. Ranges of parameter values appropriate for Ni^{++} are used in their calculations. The spectrum of the only other $3d^8$ ion, Cu^{+++}, is not known.

The spectrum of $Ni(H_2O)_6^{++}$ in $NiSO_4 \cdot 7H_2O$ crystals is shown in Fig. 26. Charge transfer bands do not interfere with the observation of crystal field bands in this case. The energies of the three transitions, ${}^3A_2 \rightarrow {}^3T_2$, ${}^3T_1(F)$, ${}^3T_1(P)$ are given quite accurately by crystal field theory with $Dq = 860$ cm^{-1} and with B the same as in the free ion.

A more complete test of the theory is provided by the data of Low[136] for Ni^{++} dissolved in single crystals of MgO. The positions of almost all of the excited singlets as well as the triplets were reported. The observed bands and their assignments are:

${}^3A_2 \rightarrow {}^3T_2(F)$	8600 cm^{-1}
${}^3T_1(F)$, ${}^1E(D)$	13,700 − 14,700
${}^1T_2(D)$	21,550
${}^3T_1(P)$	24,500
${}^1A_1(G)$	25,950
${}^1T_1(G)$	28,300
${}^1T_2$, ${}^1E(G)$	34,500

132 H. Hartmann and L. Schlaefer, *Z. Naturforsch.* **6a,** 754, 760 (1951).
133 H. Hartmann, *Z. Naturforsch.* **6a,** 781 (1951).
134 D. Cozzi and S. Vivarelli, *Z. anorg. Chem.* **279,** 165 (1955).
135 C. K. Jorgensen, Report to the 10th Solvay Council, Brussels, May 1956.
136 W. Low, *Phys. Rev.* **109,** 247 (1958).

The positions of all these bands are given to within about 500 cm^{-1} by the crystal field splitting of the ^3F, ^3P, ^1D and ^1G states if the separation of the last three from ^3F are reduced by 2500 cm^{-1} from their values in the free Ni^{++} ion. This reduction amounts to a 10 to 15% decrease in F_2 and F_4. A value of $Dq = 860$ cm^{-1} gives the best fit of the levels.

Some questions about the details of the above assignments are raised by Liehr and Ballhausen.[41] With the inclusion of spin-orbit coupling, the interaction between $^1E(D)$ and the E component of $^3T_1(F)$ can be treated. Since these lie very close together, they mix appreciably. The positions of these and the three other components of the $^3T_1$ multiplet were calculated and were found to lie within a range of 3000 cm^{-1}. The double peak at 13,700 − 14,700 cm^{-1} is due to the multiplet splitting of $^3T_1$ and the $^1E(D)$ state is a weak and as yet unobserved band, according to these calculations.

Details of the spectra such as those produced by spin-orbit splitting appear to be resolvable at temperatures well below the Debye temperature in some crystals. Vibrational bands complicate these low temperature spectra, however. Pappalardo[136a] has observed a series of narrow (4 cm^{-1}) bands near the origin of the $^3A_2 \rightarrow {}^3T_1(P)$ transition in $NiSiF_6 \cdot 6H_2O$ at 4°K. The spacings of these bands are in the neighborhood of 40 cm^{-1}, and must correspond to lattice modes of the crystal. The writer has observed similar fine structure near the origins of the transitions to the states $^1T_2(D)$ and $^3T_1(F)$.[137] These bands probably arise through vibrational-electronic coupling with lattice modes of a symmetry such as to remove the selection rule against the transition. Their low intensity is a consequence of their weak coupling to the $Ni(H_2O)_6^{++}$ complex. Much stronger transitions are produced by vibrations within the complex.

Lattice fine structure has been observed in the spectra of V^{+++},[130a] Cr^{+++},[137a] Mn^{+++}, Co^{+++} and Ni^{+++}[137] in corundum solution (see Fig. 22). In the cases of V^{+++} and Cr^{+++} the intervals are about 200 cm^{-1}, and therefore are close to the infrared frequency of 194 cm^{-1} assigned by Krishnan[138] to the E_u class of D_{3d}, the symmetry group of the corundum lattice. For an impurity site, this vibration would belong to E of C_{3v}. If this vibration is correctly assigned to a doubly degenerate class, its appearance in the visible absorption spectra of transition metal ions in corundum is certain evidence for an upper state in which the axis of threefold symmetry has been destroyed. A vibrational series of an E-mode cannot appear in the spectrum as a progression of single quanta

[136a] R. Pappalardo, *Nuovo cimento* **6,** 392 (1957).

[137] D. S. McClure, unpublished.

[137a] B. N. Grechushnikov and P. P. Feofilov, *J. Exptl. Theoret. Phys.* (*USSR*) **29,** 384 (1955); transl. in *Soviet Phys. JETP* **2,** 330 (1956).

[138] R. S. Krishnan, *Proc. Indian Acad. Sci.* **26A,** 450 (1947).

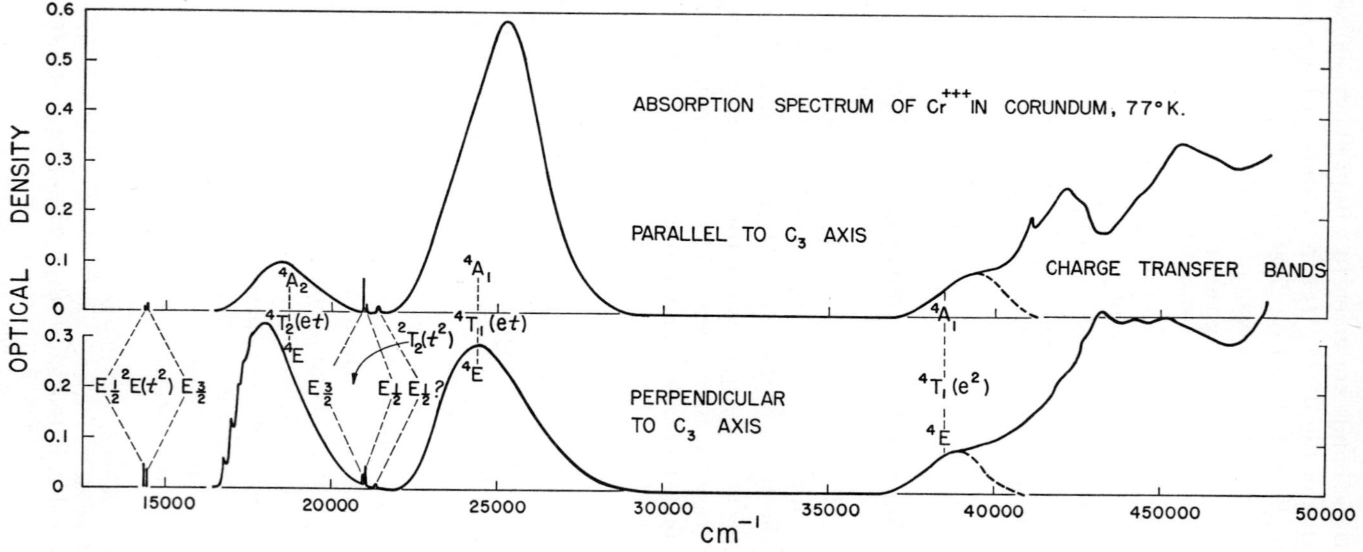

FIG. 22. Polarized absorption spectrum of Cr^{+++} in corundum (Al_2O_3) at 77°K. Both the cubic field and trigonal field designations of the upper levels are given. The ground state for all transitions is $^4A_2(t^3)$. The configuration designation is given incorrectly for each excited state; for the correct description add one t electron to each configuration shown in the figure. To convert optical density, ρ, to molar extinction coefficient, ϵ (where $\epsilon = \rho/cd$, c = concentration of Cr^{+++} in moles/liter, d = thickness of sample in centimeters) multiply ρ by 150. The "charge transfer bands" are so-called because the crystal field states are all assigned. They are very weak for charge transfer bands and may be some of the forbidden transitions discussed in Section 2.

if the upper state has the same symmetry as the ground state. The loss of symmetry may be ascribed to the Jahn-Teller effect. Judging from the unsymmetrical band shapes of the vibrational series, the Jahn-Teller splitting parameter is small compared to the vibrational quantum. Perturbation formulas developed for small Jahn-Teller effects therefore appear to be applicable.[(36,37a)]

The occurrence of an emission spectrum of Ni^{++} in MgO shows that the lattice-electron coupling must be quite small. Kroger, Vink and Van den Boomgaard[139] have observed an emission band having its origin at 21,270 cm^{-1} at low concentrations and low temperatures. The transition thus appears to be ${}^1T_1(D) \rightarrow {}^3A_2(F)$. The rate constant for the exponential decay is 8×10^4 sec^{-1}, a reasonable value for a spin forbidden transition. It is remarkable that radiationless processes do not compete successfully here, since as a rule, only the lower excited electronic states of a molecule or complex ion have any probability of emitting.

The optical spectrum of a single crystal of NiO has been studied by Newman and Chrenko.[139a] The crystal field spectrum is very similar to that of NiO in MgO(Dq = 910, all "zero crystal field levels" a few hundred cm^{-1} lower than in MgO). An absorption band at 0.24 ev, which is not expected to appear in the spectrum of isolated Ni^{++} ions, is thought to be related to the antiferromagnetic state of the crystal. Pratt and Coelho[139b] have also examined the spectra of some antiferromagnetic oxides. There is a possibility that spectral differences above and below the Néel temperature will lead to new knowledge about antiferromagnetic coupling.

Several other references to the spectra of Ni^{++} in octahedral fields are: $NiCl_2{\cdot}6H_2O$;[140] NiO;[140a,b] $NiSO_4{\cdot}7H_2O$ and $NiSO_4{\cdot}6H_2O$.[141]

Ni^{++} dissolved in ZnO should substitute at lattice points, and thus be in tetrahedral sites. The absorption spectrum of this solid solution may be explained by crystal field theory.[140a] The appropriate energy level diagram is the one for d^2, Fig. 4. The value found for Dq is 465 cm^{-1}. The low value seems to be characteristic of tetrahedral coordination, as first pointed out by Orgel and as discussed in Sections 1c and 2c.

The theory of two d-electrons in a tetragonal field has been given in detail by Ballhausen and Jorgensen.[129] It is of particular interest in

[139] F. A. Kröger, H. J. Vink, and J. van den Boomgaard, *Physica* **18,** 77 (1952).
[139a] R. Newman and R. Chrenko, *Bull. Am. Phys. Soc.* [2] **4,** 53 (1959).
[139b] G. W. Pratt, Jr., and R. Coelho, *Bull. Am. Phys. Soc.* [2] **4,** 53 (1959).
[140] J. Gielessen, *Ann. Physik* [5] **22,** 537 (1935).
[140a] D. S. McClure, *J. Phys. Chem. Solids* **3,** 311 (1957).
[140b] W. P. Doyle and G. A. Lonergan, *Discussions Faraday Soc.* **26,** 27 (1958).
[141] H. Hartmann and H. Muller, *Discussions Faraday Soc.* **26,** 49 (1958).

connection with the square planar complexes occurring among the ions of the d^8 configuration. These complexes are diamagnetic, and therefore are in singlet ground states. One way of accounting for their existence is that the lowest ^1E state of an octahedral complex becomes split by a tetragonal field more than the $^3A_2$ ground state, and in a sufficiently strong field, ^1E may become the lowest state.

The spectrum reported for $Ni(CN)_4^=$ (square) apparently is a charge transfer spectrum. Yamada[142] has reported the spectra of single crystals of several square planar complexes in compounds such as $BaNi(CN)_4 \cdot 4H_2O$, $K_2Pd(CN)_4 \cdot 3H_2O$ and K_2PdCl_4. The latter compound ($4d^8$) shows a weak band at 16,700 cm^{-1} which is several times more strongly absorbing when the electric vector is parallel to the $PdCl_4^=$ plane than when it is perpendicular. This could be a crystal field transition appearing through vibrational electronic coupling involving a degenerate vibration of the complex. The other spectra of these compounds are charge transfer bands.

3d³. The d^3 configuration in an octahedral field is especially stable chemically because it corresponds to a half filled t_2 shell. Oxidation produces a state with less stabilization from the crystal field and reduction, a state destabilized by electron pairing energy. The ground state is an orbital singlet in octahedral fields. The commonest ion having the d^3 configuration Cr^{+++}, has received much study. Figure 20 shows the spectrum of chrome alum. The highest frequency peak of this spectrum has been observed but is not shown in the figure.

The value of Dq determined for the alum (using Fig. 5) from the $^4A_2 - {}^4T_2$ separation is 1750 cm^{-1}, and the position of the $^4A_2 \rightarrow {}^4T_1(F)$ transition is calculated to be 26,000 cm^{-1}, in reasonable agreement with the observed peak at 24,700 cm^{-1}. The $^4A_2 \rightarrow {}^4T_1(P)$ transition is predicted to lie at 39,500 cm^{-1}, and was reported by Tsuchida[143] to lie at 38,000 cm^{-1}.

The sharp absorption and emission bands of ruby, Al_2O_3:Cr, have attracted the attention of many investigators.[130c,144–146] The absorption spectrum of this solid solution is shown in Fig. 22. Sharp absorption bands are seen near 14,400 and 21,000 cm^{-1}. The 14,400 band is also observable in emission. These bands are due to transitions from the $^4A_{2g}$ ground state to excited doublet states, while the strong broad bands are transitions to quartets. Their positions are given quite well by the

[142] S. Yamada, *J. Am. Chem. Soc.* **73,** 1182 (1951).
[143] R. Tsuchida, *J. Chem. Soc. Japan* **13,** 388, 426, 471 (1938).
[144] O. Deutschbein, *Ann. Physik* [5] **14,** 712, 729 (1932); *ibid.* [5] **20,** 828 (1934).
[145] S. Sugano and I. Tsujikawa, *J. Phys. Soc. Japan* **13,** 899 (1958).
[146] G. H. Dieke, unpublished work.

theory for the cubic field. Recent work of Sugano, Tanabe, and Tsujikawa[130c,145] has resulted in a clarification of the details of the entire spectrum. Their principal results are as follows:

The effect of the trigonal field of the corundum crystal (approximately C_{3v}) at the Cr^{+++} site is to split the upper quartet states in the following way, where we use the strong field designation of the states:

$$\begin{array}{lll} & {}^4E & K/2 \\ {}^4T_2(t^2e) & & \\ & {}^4A_1 & -K \\ & {}^4E & K/2 \\ {}^4T_1(t^2e) & & \\ & {}^4A_2 & -K \\ & {}^4E & K \\ {}^4T_1(te^2) & & \\ & {}^4A_2 & -2K \end{array}$$

where the splitting parameter K equals $-v/3$, the parameter used in describing the V^{+++} spectrum in the last section. The intensity of absorption of these bands polarized parallel and perpendicular to the C_3 axis was calculated assuming that the non-centrosymmetric part of the crystal field induces the transitions. The polarization ratios so calculated gave results which were in the right order, but predicted more extreme polarization effects than those observed (see Fig. 22). The results were considered good enough to permit the assignment of the components of the bands to E and A representations of C_{3v}. Then from the observed splitting of about 500 cm^{-1}, the parameter K was determined to be -350 cm^{-1} (or $v = +1050$ cm^{-1}).

The next step was to calculate the splittings of the doublet states ${}^2E(t^3)$, ${}^2T_1(t^3)$, and ${}^2T_2(t^3)$ in terms of the trigonal field parameter K and the spin-orbit interaction parameter ζ. The first two of these states are possible assignments for the 14,400 cm^{-1} lines, while the third is undoubtedly the assignment for the 21,000 cm^{-1} lines. Since the doublet states observed all belong to the t^3 configuration, a half filled shell, there are no first order splittings. Configuration interaction was ignored in the calculations. The state ${}^2E$ splits only by coupling to the ${}^2T_2$ state through the combined effect of the trigonal field and the spin-orbit interaction. Using the value of the energy difference between ${}^2E$ and ${}^2T_2$, 6500 cm^{-1}, and the trigonal field parameter K, a value of ζ of 140 cm^{-1} was found from the 30 cm^{-1} splitting of the ${}^2E$ state. With the parameters K and ζ, the splitting pattern of the ${}^2T_2$ state was calculated. Three Kramers doublets should appear, but only two have been identified. There is a line at 21357 cm^{-1} on the short wave side of the prominent doublet 20993, 21068 cm^{-1} which could be the missing component, but then the splitting pattern calculated would be wrong. Unfortunately this line is too broad to permit observation of its Zeeman effect in ordinary magnetic fields. The omission of configuration interaction from the splitting calculations may be a serious one.

The assignments of the lines were made more certain by the calculation of their intensity ratios. The lines borrow their intensity from the bands through spin-orbit interaction. Using the observed absorption strength of the bands parallel and perpendicular to C_3, the calculated absorption anisotropy of the lines was found to agree with the observed anisotropy if the previous assignments based on the splitting calculation are used. The anisotropy of the $^2T_2$ state is shown in Fig. 22. The assignments of the lines was again verified by their Zeeman effects.

TABLE XX. THE SPECTRUM OF Cr^{+++} IN SEVERAL CRYSTALS

Crystal	$^4T_2(t^2e)F$	$^4T_1(t^2e)F$	$^4T_1(te^2)P$
Cr_2O_3	16,800	23,000	29,500
10.9% Cr_2O_3-Al_2O_3	18,300	24,500	>32,000 (edge of band)
8.4% Cr_2O_3-Al_2O_3	18,350	25,000	>36,000 (edge of band)
~0% Cr_2O_3-Al_2O_3 (ruby)	18,150	25,730	39,100
MgO	16,200	22,700	29,700
$KCr(SO_4)_2 \cdot 12H_2O$ (Chrome alum)	17,500	24,700	38,000

The Cr^{+++} ion is one of the few trivalent ions in which all of the strong crystal field bands may be observed. This is due to the particular stability of the d^3 configuration and the resulting absence of low energy charge transfer processes. The three quartet—quartet transitions have been observed in a number of solid compounds and in solid solutions. There is some disagreement over the interpretation of these spectra at present because the $^4T_1(^4P)$ state does not always appear where crystal field theory says it should.

The problem is illustrated by Table XX in which the three bands are recorded for Cr^{+++} in several environments. The $^4T_2$ and $^4T_1(F)$ bands change rather little as the environment of the Cr^{+++} ion is changed. Orgel[147] has explained these changes in terms of the compression of the ion by the crystal. He points out, for example, that when Cr^{+++} is substituted for Al^{+++} in Al_2O_3, it must enter a site which is 0.08 A too small for it, and Dq must become larger as a consequence. This explains the

[147] L. E. Orgel, *Nature* **179,** 1348 (1957).

shift of the bands in ruby to higher frequencies from their positions in Cr_2O_3. Presumably, in MgO, where Cr^{+++} substitutes as a trivalent ion for a much larger divalent ion there is no compression by the crystal, and in agreement with this reasoning the spectrum in MgO appears to be similar to that in Cr_2O_3.[148]

The $^4T_1(P)$ state moves much more than it should with changes in Dq. Its position is also about 9000 cm^{-1} lower in Cr_2O_3 and in MgO than is predicted by crystal field theory. The mixtures of Cr_2O_3 with Al_2O_3 are particularly interesting, but unfortunately the $^4T_1(P)$ peak was not located exactly in the experiments of Ritschl and Müller.[149] In fact these workers did not resolve $^4T_1(F)$ and $^4T_1(P)$ in Cr_2O_3. These peaks have been resolved by the use of a different experimental method.[140a]

If the data in Table XX are correctly interpreted the $^4T_1(P)$ state is much more susceptible to changes in the environment than are the other states. This behavior is not predicted by crystal field theory for a cubic field, and it is doubtful if the trigonal field components can explain such a large effect. The correct explanation of these data may reveal that extension of crystal field theory must be made for the trivalent ions.

There are several other $3d^3$ ions. The spectrum of divalent vanadium ion is known in water solution (Fig. 20) and in some glasses. The spectrum needs fuller investigation but there is reasonably good evidence for a Dq value of 1180 in an octahedral field of water molecules.[49]

In a strong octahedral field the $^4A_2(t_3) \rightarrow {}^2E(t_3)$ separation is nearly independent of field strength. This transition should therefore occur in the tetravalent manganese ion near the value for the free ion. The emission spectrum of Mn(IV) is well known. According to the recent work of Lorenz and Prener[150] the first band of Mn^{+4} in Li_2TiO_3 solid solution (where it is on Ti^{+4} lattice sites and therefore in an octahedral field) is 14800 cm^{-1}. This is only 400 cm^{-1} from the $^4A_2 \rightarrow {}^2E$ band in ruby.

The other crystal field bands of Mn^{++++} have not been reported, and may be obscured by charge transfer bands, since the charge of this ion is rather high.

$4d^3$. Among the higher transition groups there are several ions having d^3 configurations. Cozzi and Vivarelli's data[134] on Nb^{++} in HCl solution show a strong band at 22,700 cm^{-1}, which Jorgensen[135] interprets as the $^4A_2 \rightarrow {}^4T_2(F)$ transition of an octahedral complex.

Solid Mo(III) compounds of known structure exist. Griffiths and Owen[151] reported the spin-resonance of K_3MoCl_6 in solid solution with

[148] W. Low, *Phys. Rev.* **105**, 801 (1957).
[149] R. Ritschl and R. Müller, *Z. Physik* **133**, 237 (1952).
[150] M. R. Lorenz and J. S. Prener, *J. Chem. Phys.* **25**, 1013 (1956).
[151] See refs. in K. D. Bowers and J. Owen, *Rept. Progr. in Phys.* **18**, 304 (1955).

$K_3InCl_6{\cdot}2H_2O$ where Mo^{+++} is probably octahedrally coordinated by Cl^- ions. They found the ground state to be an orbital singlet with $S = \frac{3}{2}$. Hartmann and Schmidt[152] report bands at 14,800 (^2E); 19,200 ($^4T_2$) and 24,200 cm^{-1} ($^4T_1(F)$) for the $MoCl_6^{\equiv}$ complex in water solution. The resulting Dq value is 1920 cm^{-1}, about 15% larger than for the corresponding Cr^{+++} complex, but otherwise the spectra are remarkably similar.

5d^3. Tungsten in the third transition group forms solid colored W^{+++} compounds[153] whose absorption spectra are unknown. The spectrum of $W_2Cl_9^{-3}$ in water solution is known.[135,153] The bands are weak enough to be crystal field bands but they have not been fully interpreted.

Rhenium IV as $ReCl_6^{=}$ and $ReBr_6^{=}$ have been studied in solution by Jorgensen.[154] The first weak band found in the visible corresponds to the $^4A_2 \rightarrow {}^2E$ transition found in Cr^{+++}, Mn^{++++} and Mo^{+++}, and appears at almost the same energy. The strong bands whose energy depends upon Dq were not located definitely and probably are obscured by the charge transfer absorption. Since Dq is larger and B is smaller in the third transition group, the value of Dq/B is quite large, and this complex must be close to the strong field limit near the right of the Tanabe and Sugano diagram, Fig. 5.

Os V is not known. Ir(VI) is present in a gaseous fluoride IrF_6. Its gold-yellow color is probably due to charge transfer bands rather than to crystal field bands. Its ground state is still likely to be the $^4A_2$ of d^3, however. Its infrared spectrum has been studied[155] but there are no reports on the electronic spectrum.

3d^7. The commonest representative of the d^7 configuration is Co^{++} whose spectrum in $Co(H_2O)_6^{++}$ is shown in Fig. 20. Ballhausen and Jorgensen[156] have applied crystal field theory to both octahedrally and tetrahedrally coordinated Co^{++}. The Dq value for the octahedrally coordinated hydrate is such that the $^4T_1(P)$ and $^4A_2(F)$ states are nearly coincident (see Fig. 9). Fine structure appearing in this composite peak at low temperatures has been attributed to splitting the degeneracy of the $^4T_1(P)$ state as well as to the presence of the $^4A_2(F)$ state.[10,49]

The optical and paramagnetic resonance spectra of Co^{++} in MgO have been studied and interpreted in great detail by Low.[157] All of the quartet states and four of the doublets given in Table III were observed. The

[152] H. Hartmann and M. Schmidt, *Z. physik. Chem.* (*Frankfurt*) [N.S.] **11,** 234 (1957).
[153] R. A. Laudise and R. C. Young, *J. Am. Chem. Soc.* **77,** 5288 (1955).
[154] C. K. Jorgensen, *Acta Chem. Scand.* **9,** 710 (1955).
[155] M. C. Mattraw, N. J. Hawkins, D. R. Carpenter, and W. W. Sabol, *J. Chem. Phys.* **23,** 985 (1955).
[156] C. J. Ballhausen and C. K. Jorgensen, *Acta Chem. Scand.* **9,** 397 (1955).
[157] W. Low, *Phys. Rev.* **109,** 256 (1958).

agreement of the positions of the excited states with the predictions of crystal field theory was found to be good. So far there is no definite evidence from the temperature dependence of the optical spectrum for the low-lying multiplet components of Co^{++}. In recent detailed work by Pappalardo[158] on $CoBr_2 \cdot 6H_2O$ and $CoCl_2 \cdot 6H_2O$ some of the bands show very strong temperature dependence in the region between 20 and 77°K, suggesting the existence of a component some 100 cm^{-1} above the ground state.

The splitting diagram for Co^{++} in a tetrahedral field, is Fig. 5, the one used for d^3 in octahedral fields. The spectrum of Co^{++} dissolved in ZnO may be understood in terms of this diagram.[140a] The cobalt undoubtedly substitutes at Zn^{++} lattice sites where it is surrounded by oxygen tetrahedra. Since a center of symmetry is absent, the extinction coefficient is greater than for octahedral coordination by a factor of roughly 100. The great intensity difference indicates appreciable participation of the p-orbitals in the transition and may cast some doubt on the appropriateness of crystal field theory. Nevertheless the two peaks found in ZnO-CoO(1%) may be interpreted as the ${}^4A_2 \rightarrow {}^4T_1(F)$ and ${}^4A_2 \rightarrow {}^4T_1(P)$ transitions. They require $Dq = 370$ cm^{-1} and an F-P separation of 12,000 cm^{-1}. Since the F-P interval is 14,500 cm^{-1} in the free ion this value is entirely reasonable. The ${}^4A_2 \rightarrow {}^4T_2(F)$ transition has not been located but should lie at 3700 cm^{-1}.

Another example of the spectrum of tetrahedral cobalt is that of Cs_3CoCl_5[159,160] in which the tetrahedral ions $CoCl_4^{=}$ have been identified. The strong ${}^4A_2 \rightarrow {}^4T_1(P)$ visible peak occurs at slightly greater wavelengths than in ZnO-CoO, and the infra-red peak, ${}^4A_2 \rightarrow {}^4T_1(F)$ occurs at 6300 cm^{-1}. Thus Dq is 350 cm^{-1}, slightly less than in ZnO. Many sharp bands are found to shorter wavelengths of the ${}^4A_2 \rightarrow {}^4T_1(P)$ peak. The two most prominent of these may be interpreted as ${}^4A_2 \rightarrow {}^2A_1({}^2G)$ (18,940 cm^{-1}) and ${}^4A_2 \rightarrow {}^2T_2$ (mainly ${}^2G$) (21,980 cm^{-1}). Crystal field theory appears to be as accurate for tetrahedral complexes as for octahedral ones in these two examples. The Dq values are very low, in fact less than the value predicted by an electrostatic point charge model from the relation Dq (tetrahedral) $= -\frac{4}{9}Dq$ (octahedral), which may be derived from Eq. (1.4).

Low has shown that crystal field theory accounts reasonably well for the spectrum of Co^{++} dissolved in CaF_2.[161] The Co^{++} substitutes for Ca^{++}, and is therefore surrounded by eight F^- ions arranged at the corners of a

[158] R. Pappalardo, unpublished.
[159] S. Yamada and R. Tsuchida, *Bull. Chem. Soc.* (*Japan*) **27,** 436 (1954).
[160] L. Katzin, *J. Am. Chem. Soc.* **76,** 3089 (1954).
[161] R. Stahl-Brada and W. Low, *Phys. Rev.* **113,** 775 (1959).

cube. The crystal field therefore has the same sign as the tetrahedral field so that Fig. 5 applies, although it also has a center of symmetry. The *p-d* mixing is not the important factor in this case that it could be in a tetrahedral field.

The spectrum of Co^{++} in CaF_2 is analogous to that of Cr^{+++} in an octahedral field. The Dq value is remarkably low however, only 340 cm^{-1}; so that the components of the ^4F state all lie in the infra-red. The low Dq value may be caused by the large difference between the crystal radii of Ca^{++} (1.06 A) and Co^{++} (0.80 A).

The d^7 ions of the second and third transition groups are unstable and difficult to prepare. It is probably for this reason that their spectra have not been reported.

3d⁴. The d^4 configuration is another chemically unstable one. Trivalent d^4 ions may lose one charge and attain the d^5 configuration which is especially stable in weak crystal fields. Or they may acquire an extra positive charge to reach the d^3 configuration which is especially stable in strong crystal fields.

Mn^{+++} and Cr^{++} are the only two d^4 ions in the first transition group. The spectra of the octahedrally coordinated hydrated ions are shown in Fig. 20. They consist of a single broad peak corresponding to the transition between components of the ^5D state; $^5E \rightarrow {}^5T_2$. Since the E state is lowest, these ions are analogous to the Cu^{++} ion, and a large Jahn-Teller distortion results. The somewhat abnormally large Dq values (Table VII) are explained in this way. It also appears that most known Mn^{+++} and Cr^{++} compounds have structures which may be described as distortions of more common structures.[124] There is very little information on the spectra of these two ions.

4d⁴. Ru^{+4} and Rh^{+5} in the second transition group form ions having the d^4 configuration. They probably owe their stability to the formation of non-octahedral complexes. The spectra appear to be charge transfer spectra.[135]

5d⁴. A part of the spectrum of Re(III) in solution is known, but it has not been interpreted.[135]

3d⁶. In a weak octahedral crystal field the ground state of the d^6 configuration arises from the ^5D state, as it does for d^4. Since Dq is opposite in sign from d^4 the $^5T_2$ state should be lowest. The Fe^{++} ion in most of its compounds and a few compounds of Co^{+++} have this ground state. In all others the lowest $^1A_1$ state (there are five) becomes the ground state. This must happen, according to Fig. 8, for $Dq/B \gtrsim 2$.

When $^5T_2$ is the ground state the spectrum is mainly due to the $^5T_2 \rightarrow {}^5E$ transition. The Jahn-Teller effect therefore occurs principally in the upper state, and since the Franck-Condon principle requires the absorption transition to occur in the spatial configuration of the lower state the spectrum might be expected to show some evidence for the excitation of unsymmetrical vibrations. The theoretical problem of the dynamical Jahn-Teller effect, of which this is one example, has been solved by Pryce *et al.*[36] This work is too recent to have been compared to observed spectra in detail. The spectrum of $Fe(H_2O)_6^{++}$ in crystals (Fig. 20) is an extraordinarily broad diffuse band showing two maxima. The breadth shows that many vibrations are excited. Two maxima are predicted to appear under certain conditions as a consequence of the dynamical Jahn-Teller effect.[36]

Intercombination bands have been observed in the $Fe(H_2O)_6^{++}$ spectrum.[42a] Both the spectrum, and the absence of paramagnetism of $Fe(CN)_6^{\equiv}$ indicate a ${}^1A_1$ groundstate.

The diamagnetic Co^{+++} compounds have received a great deal of attention from chemists, and there is a voluminous literature on the solution spectra of its various complexes.[162,163] The polarized spectra of several monoclinic crystals containing cobalt complex ions have been obtained recently by Yamada *et al.*[164] and interpreted in some detail by Ballhausen and Moffitt.[165] These authors studied crystalline trans-Co (ethylene diamine)$(Cl)_2^+$ in which Cl^- ions are at opposite corners of the octahedron about the Co^{+++} ion and the four $—NH_2$ groups of the two ethylene diamines occupy the four equatorial positions. The complex thus has a strong tetragonal distortion. The two lowest excited singlet states of an octahedral complex arise from transfer of an electron from a t_2 orbital into an e orbital, hence they are ${}^1T_1$ and ${}^1T_2$, as shown in Fig. 8. These states are each split by the tetragonal field into two.

The magnitude of this splitting has not been definitely determined. The spectrum observed at room temperature consists of an unpolarized band at 6200 A and a band at 4300 A polarized predominantly perpendicular to the Cl-Co-Cl direction. Ballhausen and Moffitt interpreted the two bands as the components of the ${}^1T_1$ state split apart by the tetragonal field, the lower state being E_g (of D_{4h}) and the upper A_{2g}. The polarizations are explained if it is assumed that the transitions appear only through vibrational-electronic interaction. The E_g state appears in the spectrum through the intercession of e_u (parallel to C_4) and a_{2u} or

162 See papers by M. Linhard and collaborators, e.g., *Z. anorg. u. allgem. Chem.* **271,** 101 (1952).

163 J. C. Bailar, "The Chemistry of the Coordination Compounds." Reinhold, New York, 1956.

164 S. Yamada, A. Nakahara, Y. Shimura, and R. Tsuchida, *Bull. Chem. Soc.* (*Japan*) **28,** 222 (1955); also extensive references to previous work.

165 C. J. Ballhausen and W. Moffitt, *J. Inorg. & Nuclear Chem.* **3,** 178 (1956).

b_{2u} (perpendicular to C_4) vibrations. The A_{2g} state appears perpendicular to C_4 with the help of e_u modes, but as there are no stretching or bending modes of the type a_{1u}, the parallel transition cannot occur. There is an alternative interpretation, however. The tetragonal splitting may be practically unobservable; and the observed states could be the $^1T_1$ and $^1T_2$ states. The polarizations would be the observed ones if the symmetry were D_4 instead of D_{4h}. The two possibilities can be distinguished by observing the temperature dependence of the transition strength to see if vibrations actually participate (Section 1g).

4d⁶, 5d⁶. Many complex ions in the higher transition groups have the d^6 configuration and most of them have the $^1A_1$ ground state. Three such examples are Rh(III) in the 4*d* series and Ir(III) and Pt(IV) complexes in the 5*d* series. These ions have been studied in solution by Jorgensen.[166] Rh(III) and Ir(III) are analogous to Co(III). The *Dq* value is greater for Rh(III) than for Co(III) by a factor of about 1.54, and for Ir(III) by about 1.90, for the same ligand. The singlet-triplet mixing is increased in the heavier elements, and intercombinations are more probable. Jorgensen has found several spin-forbidden bands.

Jorgensen[166] has reported the solution spectra of some Ru^{++} complexes. This ion is analogous to Fe^{++}, but it is not certain that the spectra may be interpreted analogously, and its ground state is not known.

The tetravalent ion Pt(IV) was found to have low-lying charge transfer bands, as might have been expected because of its high charge.[166] Some of the relatively weak crystal field bands are observed in the spectra of $PtCl_6^=$ and $PtBr_6^=$ ions before the onset of the charge-transfer bands.

3d⁵. The half-filled *d* shell is especially stable when the crystal field is weak, because both the *e* and *t* orbitals may then be half-filled, and the pairing energy is zero. In strong octahedral fields the *e* orbitals become less available and the chemical stability of the ion toward reduction to the d^6 ion is decreased. In weak fields the ground state is derived from the ^6S state of the free ion. Since there may be only one sextet state in d^5, all excited states are either quartets or doublets. In strong fields the ground state is derived from the t^5 configuration, and is therefore $^2T_2$, as shown in Fig. 7.

In a half-filled shell, the one-electron operators in the Hamiltonian have no first-order effect on the states. Thus in weak field theory in which the free ion levels are the basis functions, the crystal field is a second-order perturbation. Therefore all the lines in Fig. 7 meet the ordinate with zero slope. The energy diagram is the same for tetrahedral, octahedral, and cubal fields.

[166] C. K. Jorgensen, *Acta Chem. Scand.* **10,** 518 (1956).

The best known example of the d^5 configuration is Mn^{++}. The spectrum of MnF_2 shown in Fig. 23 is typical of the crystal field spectra of Mn^{++} compounds.[167] The Mn^{++} ion is surrounded by a fairly regular octahedron of F^- ions in this crystal. Certain bands in the spectrum become quite sharp at low temperatures as shown in the lower part of Fig. 23. The theoretical splitting pattern for $Dq = 740$ cm^{-1} is shown for comparison with the spectrum. The agreement is remarkably good when the $^6S - {}^4G$ separation is treated as an additional parameter, as suggested by the molecular orbital theory, Section 2a. All of the observed excited states are quartets.

The sharpness of the spectra was explained by Orgel[12] who pointed out that the band widths in the Mn^{++} spectrum are proportional to $d(h\nu)/d(Dq)$ where $h\nu$ is the transition frequency. The change of $d(h\nu)/d(Dq)$ is a measure of the change in the relative occupation of the e and t orbitals resulting from the transition. In fact, $d(h\nu)/d(Dq) = 4N_t - 6N_e$, where N_e and N_t are the numbers of electrons in the e and t orbitals. Since the e orbitals are σ antibonding, and the t orbitals practically nonbonding, a change in electron distribution between them results in a change in the equilibrium configuration of the local environment. As a result of the action of the Frank-Condon principle such a change gives rise to broadening of the spectrum.

The breadths of the bands in the spectrum of MnF_2, Fig. 23, are quite well explained by this theory. The slopes of the Tanabe-Sugano diagram, Fig. 7 are obviously greatest for the broadest bands of Fig. 23. The very sharp band at 25,250 cm^{-1} corresponds to the horizontal line on Fig. 7, for which $d(h\nu)/d(Dq) = 0$.

The spectrum of Mn^{++} is especially important because of the wide use of manganese phosphors, many of which depend upon the presence of the divalent form.[168] The emitting level of Mn^{++} is the $^4T_{1g}(^4G)$ state, the first excited state shown in Fig. 7, for the region of Dq/B near unity. The variation in color from one phosphor to another depends upon the value of Dq and the $^6S - {}^4G$ separation appropriate to the environment of the Mn^{++} ion. The $^6S - {}^4G$ separation may be interpreted, as Owen[18] has done, in terms of the extent to which the d orbitals of Mn^{++} form molecular orbitals with their surroundings (see Section 2a). In $Mn(H_2O)_6^{++}$, the reduction of this interval from the free-ion value is 1900 cm^{-1}, or seven per cent. The splitting of the ^4G level may amount to as much as 7000 cm^{-1}, so that the chief color changes are due to changes in Dq.

Further modifications of the color of the emitted light are produced by the band width in absorption and emission. Thus there is a total of

[167] J. W. Stout, *J. Chem. Phys.*, in press.

[168] F. A. Kröger, "Some Aspects of the Luminescence of Solids." Elsevier, New York, 1948.

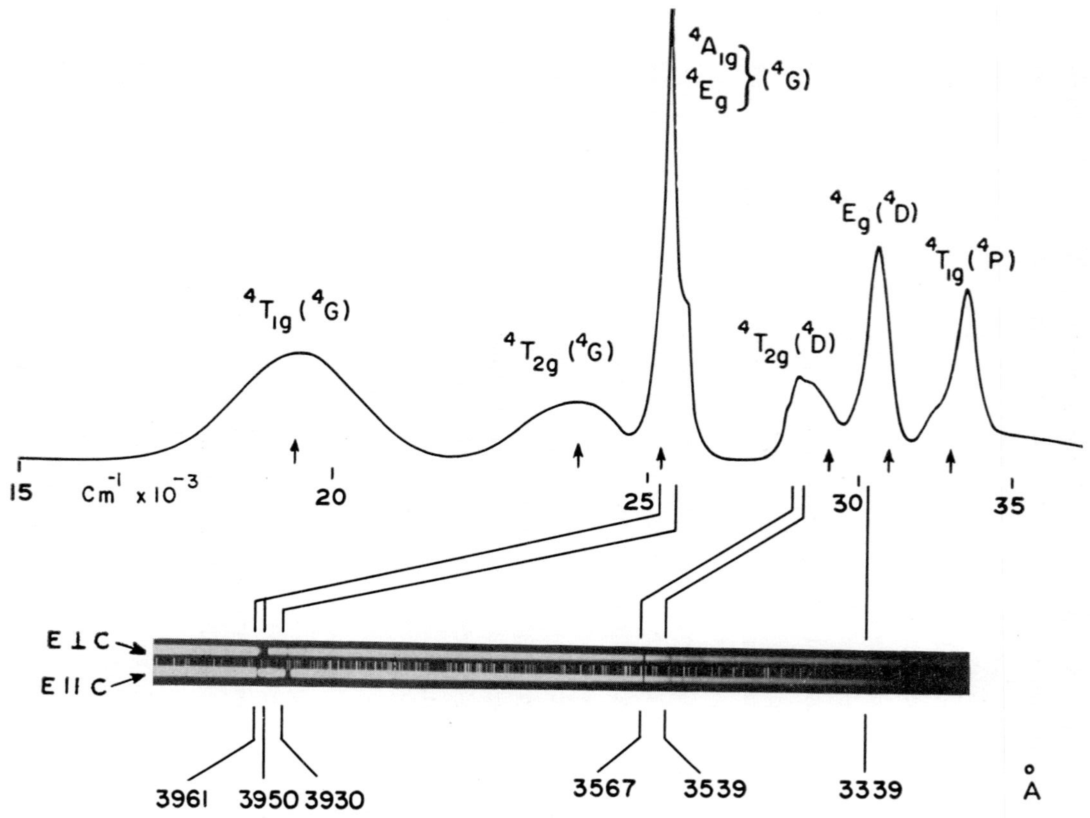

FIG. 23. The absorption spectrum of MnF_2. The upper spectrum is a low-resolution record made at room temperature. The lower spectrum was photographed at 20°K using a single crystal and polarized light. The calculated positions of the peaks are shown by the short arrows. The parameters used in the calculation are: $Dq = 740$ cm^{-1}; and the $^6S - {}^4G$ separation was reduced by 1500 cm^{-1} from the value for the free ion (J. W. Stout and J. Bigeleisen, unpublished). *Note added in proof:* Professor Stout plans to publish this figure (*J. Chem. Phys.*, in press). Some of the wavelengths given here differ slightly from his final values.

at least three parameters which determine the wavelength of the emission maximum.

The values of N_t and N_e may be obtained from the eigenvectors of the secular equation for the $^4T_{1g}$ states, shown in Table V. In the $^6A_1(^6S)$ ground state, $N_e = 2$ and $N_t = 3$. The distribution of electrons among the e and t orbitals in the lowest $^4T_{1g}$ state is the same as this in zero crystal field, and changes to $N_e = 1$; $N_t = 4$ in a strong crystal field. Most of the change occurs between $Dq = 0$ and $Dq = 500$ cm^{-1}. These results mean that in the usual range of Dq values, one less e orbital, or one less σ antibonding orbital is occupied in the excited state. The surroundings of the Mn^{++} ion should therefore move inward slightly in the equilibrium position of the excited state. However, since the e shell is occupied by only one electron, the equilibrium configuration must be unsymmetrical. As pointed out by Orgel,[169] the Mn^{++} ion in the first $^4T_{1g}$ state must resemble Cr^{++} in its ground state where a large Jahn-Teller effect results in an appreciable distortion of the octahedral surroundings. The band width in absorption and emission therefore cannot arise from a totally symmetric configuration coordinate, as has been assumed in the past.[170,171]

A systematic study of the emission of Mn^{++} ion substituted at lattice sites of various perovskite crystals was carried out by Klasens *et al.*[172] They discovered that there is a roughly linear relationship between the lattice constant of the crystal into which the Mn^{++} is introduced and the wavelength of maximum emission. When the space within the crystal available to the ion is decreased, the value of Dq increases, and the $^4T_{1g}$ state moves to lower energy, as Fig. 7 shows it must. The authors did not study the absorption spectra, however, and therefore it is impossible to disentangle the three factors determining the emission wavelength. However, since the band width of the emitted light does not change greatly from one solid solution containing Mn^{++} to another, the experiments must illustrate principally how Dq changes. When interpreted in this way, the data show that Dq increases 66 cm^{-1} for a 0.1 A decrease in the lattice constant when Dq is near 700 cm^{-1}.

The emission lifetimes of Mn^{++} phosphors have not been fully explained. Since the emitting transition is quartet to sextet, some spin-orbit interaction is necessary in order to break down the multiplicity selection rule. From Table VIII the ζ value is seen to be about 335 cm^{-1}, and therefore appreciable sextet-quartet mixing could occur. There must be further

[169] L. E. Orgel, *J. Chem. Phys.* **23,** 1958 (1955).
[170] C. C. Klick and J. H. Schulman, *J. Opt. Soc. Am.* **42,** 910 (1952).
[171] F. E. Williams, *Brit. J. Appl. Phys.* Suppl. 4 (Luminescence), 597 (1954).
[172] H. A. Klasens, P. Zalm, and F. O. Huysman, *Philips Research Repts.* **8,** 441 (1953).

destruction of selection rules by vibrations or by the environmental symmetry, since all the states are g states. There is some possibility, however, that at least in the longer-lived emitters (10^{-1} sec), the transitions are magnetic dipole rather than electric dipole, so that $g \rightarrow g$ transitions would be permitted.

Gielessen[140] has reported the low-temperature spectra of many Mn^{++} compounds. These spectra seem never to have been interpreted in detail since the work was done before the full development of crystal field theory. Pappalardo[173] has recently studied the low-temperature spectra of four hydrated manganous salts in the form of single crystals, and has made a partial interpretation of some of the features observed.

The isoelectronic ion Fe^{+++} has a crystal field spectrum similar to that of Mn^{++}, but charge transfer bands permit the observation of only the lowest three transitions. The Dq value of 1400 cm^{-1} is extraordinarily small for a trivalent ion. The transitions in Fe^{+++} compounds are difficult to observe because, in addition to being weak, they are excessively broad except for the ${}^6A_1({}^6S) \rightarrow {}^4A_1, {}^4E({}^4G)$ doublet, seen at 24,500 cm^{-1} in $Fe(H_2O)_6^{+++}$.[42a,49]

4d^5, *5d*^{5}. The hexachloro complexes of the ions Ru^{+++} and Ir^{++++} have been studied by Jorgensen[174] who reports only charge transfer bands. Evidently the crystal field is very strong, and optical excitation probably results in the addition of an electron to the metal ion to form the chemically stable d^6 shell. The spin resonance of these substances has been observed[151] and it shows that the maximum electron pairing has occurred (right side of Fig. 7) to give the ${}^2T_2$ ground state.

5. Charge Transfer Spectra

The term charge transfer spectrum could be applied with some justification to many observed spectra. There is a large literature on charge transfer spectra of organic complexes, some of which has been reviewed by Orgel.[175] The photochemists have long recognized that charge transfer processes in solution lead to photo-oxidation or reduction. Orgel's review and one by Rabinowitch[57] cover this field. Exciton spectra in ionic crystals are also thought of as due to charge transfer processes. This section will be limited chiefly to a discussion of charge transfer spectra in transition metal compounds. It will be shown that the experimental data justify such a classification for many of the spectral bands observed in these substances. Either the initial or the final state responsible for this type of charge transfer spectrum is localized mainly on the atoms

[173] R. Pappalardo, *Phil. Mag.* [8] **2,** 1397 (1957).
[174] C. K. Jorgensen, *Acta Chem. Scand.* **10,** 578 (1956).
[175] L. E. Orgel, *Quart. Revs. (London)* **8,** 422 (1954).

surrounding a central metal atom. The electronic transition process is thus one step less localized than a crystal field transition.

It is useful first to review the systems in which charge transfer spectra are observed. In one large class we may place molecules or molecular ions containing a transition metal surrounded by electronegative atoms. This class may be further subdivided according to the symmetry of the molecule. Two important subclasses are the octahedral MX_6^{n-} ions and the tetrahedral MX_4^{n-} ions. These molecular ions preserve their identity in solution, and have consequently been studied widely by chemists. Most of the information on their spectra comes from data taken on solutions. The octahedral ions are formed with singly charged anions or uncharged ligating groups having at least one nonbonding pair of electrons. Examples are halides, NH_3, H_2O, NO_2^-, CN^-. The tetrahedral ions form with the higher valence states of the metal and doubly negative anions such as $O^=$.

One of the most important physical principles determining the molecular structure appears to be the balance between the attraction of the central ion for the ligands, and the repulsion of the ligands for each other. Thus the divalent ions as a rule form octahedral complexes only with uncharged ligands, or with ligands such as CN^- in which the charge is spread over two or more atoms. Even the latter are very unstable in water solution. Trivalent ions of the first and higher transition series form octahedral CN^- complexes, but octahedral halide complexes only become relatively stable (in water) in the second and third transition groups. The divalent ions in some cases form tetrahedral halide and oxide complexes. There are no octahedral oxide complexes, even though oxygen is commonly found with this coordination type in solids. The reason is simply that other cations in addition to the central ion are needed to stabilize such a structure.

Statements as to the existence or nonexistence of a complex ion are often based upon whether or not the ion may exist in water solution. Such a statement is actually a comparison of the stability of the complex ion with the corresponding hydrated ion or its hydrolysis product.

Charge transfer spectra are observed in the solid state whether or not molecular ions are present. The band width, which is not a factor in the spectra of the molecular ions, may be an important feature of the spectra of the solids. Crystals containing well-defined molecular ions must have very small band widths, although no definitive experimental work on this point has appeared. When the molecular ions are poorly defined or nonexistent, the widths of the individual charge transfer bands may increase. In some cases this appears to lead to loss of all the details of the spectrum.

Data on the charge transfer spectra of solids are thus often difficult to interpret. They are also more difficult to obtain than are data for solutions. For these reasons, there is much less to say about the spectra of solids than about spectra of solutions. The solution data thereby become all the more important, since it is only through them that it is possible to develop a semiempirical understanding of charge transfer spectra at present.

a. Octahedral Coordination

The spectra of octahedral complex ions of transition metals may be understood from Fig. 16. The crystal field bands are the transitions $2t_{2g} \rightarrow 2e_g$, and the allowed charge transfer bands are the transitions $1t_{2u} \rightarrow 2t_{2g}$, $2e_g$. The various complex ions differ in the number of electrons in the $2t_{2g}$ orbital in the ground state, and in the relative energies of the ligand and central ion orbitals.

There is a great deal of information on Co III complex ions (Section 2c). In the ground state the $2t_{2g}$ orbital is filled. The two spin-permitted transitions to the configuration $(2t_{2g})^5(2e_g)^1$ give rise to two weak "crystal field" bands at low energy, corresponding to $Dq/B \approx 2$ and the charge transfer transition $t_{2u} \rightarrow 2e_g$ gives rise to intense higher energy bands. The $\pi t_{1g} \rightarrow de_g$ transition is forbidden, but should be stronger than the crystal field transitions.

The spectra of the ions $Co(NH_3)_6^{+++}$, $[Co(NH_3)_5Cl]^{++}$, and

$$[Co(NH_3)_4Cl_2]^+ \text{ (both } cis \text{ and } trans\text{)}$$

have been interpreted recently in these terms.[176] Not all of the charge transfer bands of $Co(NH_3)_6^{+++}$ can be observed because of their high energy. When a Cl^- ion is substituted for an NH_3 molecule on the octahedron about Co^{++}, a pair of strong bands appear in the near ultraviolet. They are interpreted as charge transfer bands which appear because it is easier to remove an electron from Cl^- than from NH_3. This interpretation is supported by the change in energy of the transitions with changing halogen. It is shown in Fig. 24 that substitution of Br^- or I^- for Cl^- reduces the transition energy by up to a volt. The striking difference between the behavior of the crystal field bands and the charge transfer bands toward substitution is also shown in Fig. 24. The crystal field bands are hardly affected by the substitution. This behavior is good evidence for the correctness of the interpretation of both types of spectra. In the detailed interpretation of the spectra, one must take account of

[176] K. Nakamoto, J. Fujita, M. Kobayashi, and R. Tsuchida, *J. Chem. Phys.* **27**, 439 (1957).

the splitting produced by the unsymmetrical array of ligands in the halide-substituted complexes.

In the second and third transition groups, Rh^{+++} and Ir^{+++} are isoelectronic with Co^{+++}. Their crystal field bands occur in the near ultraviolet rather than in the visible because Dq is 54 and 90% greater for these ions respectively than for Co^{+++} (Section 4). The charge transfer bands are also higher in energy because the $2e_g$ orbital is higher. The first charge transfer band of $[Rh(NH_3)_5I]^{++}$ occurs at 36,000 cm^{-1} (Jorgensen[177]) while that of the corresponding cobalt ion[176] is at 25,000 cm^{-1}.

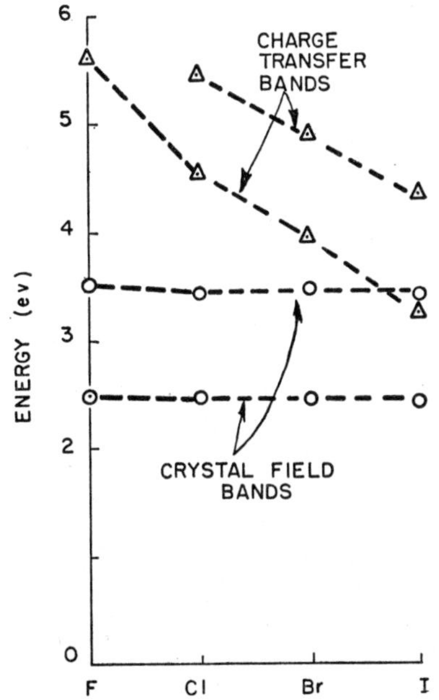

FIG. 24. Change in the energy of the excited states of $Co(NH_3)_5X^{++}$ as X changes from F to Cl, etc.

The charge transfer transitions of the d^6 ions occur at high energy because the t_{2g} orbitals are filled. From another point of view, these ions have high chemical stability toward oxidation or reduction, and therefore the charge transfer energies are high. The d^5 ions, however, have lower energy charge transfer bands because a t_{2g} orbital is available as the final state of the transition.

Jorgensen[178] has studied the charge transfer spectra of the octahedral

[177] C. K. Jorgensen, *Acta Chem. Scand.* **10,** 500 (1956).
[178] C. K. Jorgensen, *Acta Chem. Scand.* **10,** 518 (1956).

complex d^5 ions Fe III, Ru III, Rh IV, Ir IV. A perfect comparison of these with d^6 ions cannot be made because they have not all been studied with the same ligands. An idea as to the order of magnitude of the effect of having a hole in the t_2 shell may be obtained from the fact that the first charge transfer band of the d^5 ion $IrBr_6^{=}$ occurs at 14,000 cm^{-1}, while that of the d^6 ion $IrBr_6^{\equiv}$ occurs above 30,000 cm^{-1}. If the orbitals did not change their energies when the electron is added to $IrBr_6^{=}$, the difference between the charge transfer bands would be the average energy of the crystal field bands of $IrBr_6^{\equiv}$, or approximately 24,000 cm^{-1}. This appears to be in the correct range.

Another such example is afforded by the comparison of $Fe(CN)_6^{\equiv}$, a d^5 ion, with $Fe(CN)_6^{\equiv}$, a d^6 ion. Both have the strong crystal field ground states, so that the diagram of Fig. 16 applies to them. The first charge transfer bands occur at 24,000 and 38,000 cm^{-1}, respectively, an increase of 14,000 cm^{-1}.[179] This increase is probably smaller than the $t_{2g} - e_g$ separation in $Fe(CN)_6^{\equiv}$, but there are no reliable data on this point. The published data on the other $X(CN)_6^{\equiv}$, $X(CN)_6^{\equiv}$ pairs are not reliable enough for further comparisons of this type among the cyanides. All of the d^6 ions have their first charge transfer bands at quite high energies, however. Table XXI illustrates this.[179] The table also

TABLE XXI. FIRST CHARGE TRANSFER BANDS OF d^6 CYANIDES

$3d^6$	$Fe(CN)_6^{\equiv}$	38,000 cm^{-1}	$Co(CN)_6^{\equiv}$	45,000	50,000 cm^{-1}
$4d^6$	Ru	40,000	Rh	40,000	
$5d^6$	Os	42,000	Ir	45,000	

indicates an increase in the energy of charge transfer as the principle quantum number of the d-shell increases.

The data on some high-energy spectra of transition metal halide crystals have been reviewed by Pick in the Landolt-Börnstein Tabellen (p. 882 in ref. 179). The spectra have not been interpreted in detail, but they may probably be identified as charge transfer spectra. The change in energy with change in halogen atom is about the same as in Fig. 24. As the atomic number of the metal atom is increased from Mn to Co to Ni, the spectral energy decreases. The same trend is also apparent in the charge transfer spectra of a series of isoelectronic tetrahedral ions, as shown in Fig. 26 as will be discussed presently.

Parkinson and Williams[180] calculated the energy of the charge transfer process in MnF_2 by assuming a classical model, and used the result to interpret the two absorption bands found by them at 7.66 and 6.0 ev. A

[179] Landolt-Börnstein Tabellen, [6] Vol. 1, Part 3, Molekeln II, p. 236 (1951).
[180] W. W. Parkinson, Jr., and F. E. Williams, *J. Chem. Phys.* **18,** 534 (1950).

classical model of the crystal implies that the ions are localized at the lattice points, and have no extension in space. A charge transfer process therefore implies doing work against the Madelung potential and the electron affinity of the halogen, and getting back only the second ionization potential of the metal. The value of the transition energy calculated in this way for MnF_2 was 18 ev. Since polarization corrections could reduce this very little, Parkinson and Williams concluded that the absorption processes at 7.6 and 6.0 ev are not charge transfers but possibly $3d \rightarrow 4s$ or $4p$ transitions within the Mn^{++} ion.

This interpretation conflicts with our previous one based on empirical evidence. The apparently analogous transitions in Co and Ni halides are as low as 4 ev. Their energies vary strongly with the halide, so that even if an internal $d \rightarrow s$ or p transition in the metal ion is involved, the electrons of the halide ions must also participate strongly.

A correct interpretation may contain aspects of the foregoing two. However, a calculation of the transition energy of a charge transfer process assuming that an electron is transferred from one ion to another against the full Madelung potential, certainly cannot be correct. One must take account of the nonlocalization, overlap, and mixing of the electronic orbitals taking part in the transition. These factors all tend to reduce the transition energy.

b. Tetrahedral Coordination

We have already mentioned the series of isoelectronic tetrahedral ions MnO_4^-, $CrO_4^=$, $VO_4^{\equiv}$, $TiO_4^{\equiv}$ in which the nuclear charge decreases from first to last, and there are no occupied d orbitals in the ground state (Section 2c). The most complete experimental work on these ions has been done by Teltow.[59] Most of his work was done with the ion dissolved in single crystals of an isoelectronic substance. The diluents were, for example, $KClO_4$ for $KMnO_4$, K_2SO_4 for K_2CrO_4, and $Na_3PO_4 \cdot 12H_2O$ for $Na_3VO_4 \cdot 12H_2O$. The diluents contain only ions having completed shells, and their absorption does not begin until the vacuum ultraviolet region. The ions Teltow studied have dimensions very similar to those of the corresponding diluents.

The absorption spectra of MnO_4^- and $CrO_4^=$ are characteristic molecular spectra. There are several pure electronic transitions upon which are superimposed progressions of the totally symmetric M-O stretching vibration of approximately 800 cm^{-1}. Two strong electronic transitions in the visible and near ultraviolet and several weak ones have been found. The vibration frequencies, band widths, and electronic origins are remarkably independent of the substance used for the dilution. On the other hand, valuable information about the electronic spectra was obtained by changing the symmetry of the crystal in which the ion was dissolved. The perturbations from the surroundings result in splitting the degenerate electronic states in a way characteristic of their symmetry properties.

The interpretation of the spectra is due to Wolfsberg and Helmholz.[55] These authors carried out the molecular orbital treatment outlined in Section 2c in order to find the approximate excitation energies of the various levels. Then they used the observed crystal perturbations to verify the assignments of the electronic origins. A further check on the assignments was provided by the comparison between calculated and observed intensities.

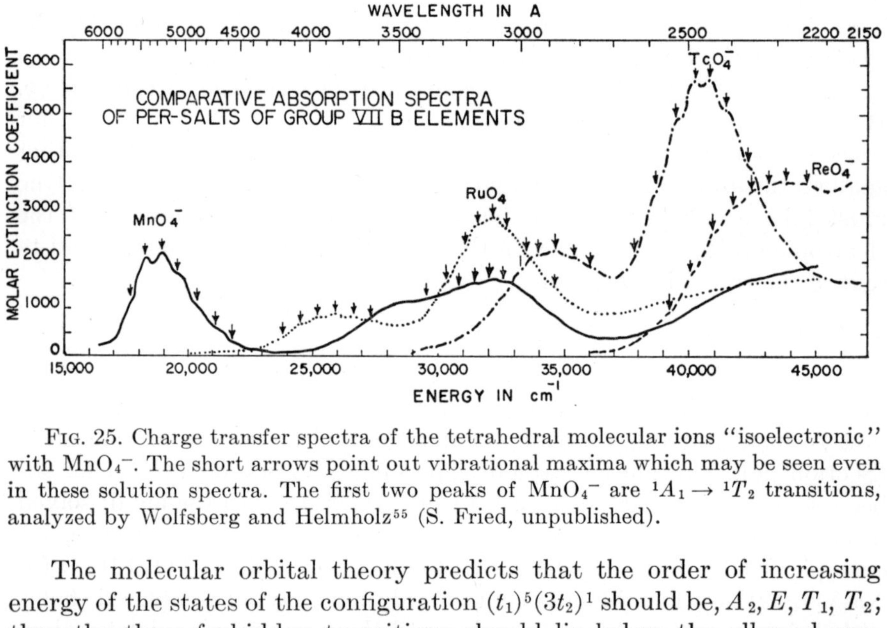

FIG. 25. Charge transfer spectra of the tetrahedral molecular ions "isoelectronic" with MnO_4^-. The short arrows point out vibrational maxima which may be seen even in these solution spectra. The first two peaks of MnO_4^- are ${}^1A_1 \rightarrow {}^1T_2$ transitions, analyzed by Wolfsberg and Helmholz[55] (S. Fried, unpublished).

The molecular orbital theory predicts that the order of increasing energy of the states of the configuration $(t_1)^5(3t_2)^1$ should be, A_2, E, T_1, T_2; thus the three forbidden transitions should lie below the allowed one. Many weak bands are found in the spectrum of MnO_4^- between 13,000 and 18,000 cm^{-1}, while the strong absorption, shown in Fig. 25 begins at about 18,000 cm^{-1}. When MnO_4^- is dissolved in $KClO_4$, the ion is at a site of symmetry C_s, and all degeneracy must be removed. The components of the T_2 state move apart about 30 cm^{-1}, and the corresponding transitions from the ground state are polarized in accordance with the site-group selection rules. The allowed ${}^1A_1 \rightarrow {}^1T_2$ transition should split in C_s into ${}^1A' \rightarrow {}^1A'$ (two such transitions polarized in the a-c plane) and ${}^1A' \rightarrow {}^1A''$ (polarized along the b axis), as is actually found. The configuration assignments may be verified only by observation of the forbidden transitions. There is quite definite evidence for two states, possibly E and T_1 lying below the T_2 state. No evidence for the presence of a nondegenerate state was found. The configuration change for these transitions cannot, therefore, be determined.

The forbidden transitions apparently appear mainly because of vibrational-electronic interaction (Section 1g). Their electronic origins are missing in almost all cases, showing that the direct environmental perturbation of the intensities is very small.

Some of the excited states of these ions are degenerate, and must be subject to the dynamical Jahn-Teller effect.[36] The spectra have not been interpreted in enough detail as yet to show what the size of the effect is. Furthermore there should also be an excitation exchange degeneracy in the pure crystals which would lead to band splitting in case there are several molecules per unit cell. This has not been identified either.

The energy of excitation of the first allowed transition increases by 8000 cm^{-1} in going from MnO_4^- to $CrO_4^=$ and apparently is greater still in $VO_4^{\equiv}$. Teltow found one broad peak at 37,000 cm^{-1} in $Na(VO_4, PO_4) \cdot 12H_2O$ which could reasonably be ascribed to this transition. The absence of sharp vibrational structure shows that $VO_4^{\equiv}$ either undergoes a large distortion in its first excited state (via the Jahn-Teller effect), or is much less isolated from the lattice than are MnO_4^- and $CrO_4^=$ ions. In either case the observed breadth of the spectrum suggests that $VO_4^{\equiv}$ is a much less tightly bound molecular unit than MnO_4^- and $CrO_4^=$, as we concluded in Section 2e.

There has been no experimental work on the second and third transition groups comparable to Teltow's detailed work in the first. The effect of increasing principal quantum number on the spectra of the isoelectronic ions MnO_4^-, TcO_4^-, ReO_4^- is shown in Fig. 25. The spectrum of the isoelectronic RuO_4 molecule is also shown in this figure. It will be recalled that the lower oxidation states of the $4d$ and $5d$ groups are not as stable toward oxidation as those of the corresponding $3d$ ions. The increase in energy of the charge transfer levels with increasing atomic number, shown in Fig. 25, parallels this chemical effect, and in fact is very similar to it if the previously outlined interpretation of charge transfer spectra is correct. Either to reduce the ion chemically or to excite the charge transfer level, requires transfer of an electron from the surroundings to an orbital largely localized on the metal ion, and this process apparently requires more energy as the principal quantum number increases. Jorgensen has related this fact to the increasing number of nodes in the d function.[181]

[181] C. K. Jorgensen, "Energy Levels of Complexes and Gaseous Ions." Jul. Gjellerups Forlag, Copenhagen, 1957.

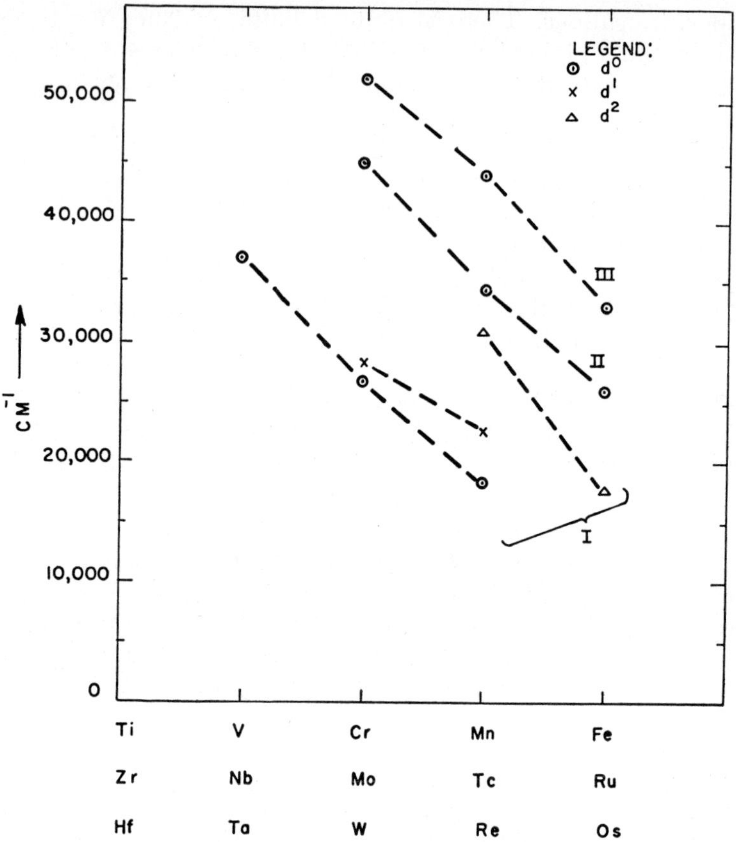

FIG. 26. Positions of the peak corresponding to the first allowed ($^1A_1 \rightarrow {}^1T_2$) transition in XO_4^{n-} ions. Data for ions of the three transition periods (I, II, III) are given. Note that the transition energy decreases with atomic number in the same period, but increases with the number of the period, and with the number of added "*d* electrons."

Considerable additional data on the spectra of tetrahedral molecular oxy-ions have been reported by Symons and co-workers.[182,183] They prepared the little-known ions $CrO_4^{\equiv}$, $MnO_4^{=}$, $MnO_4^{\equiv}$, and $FeO_4^{=}$, and recorded their optical spectra in water solution and their spin resonance spectra in various solid solutions. These ions have one or two *d* electrons in their ground states. Their lowest transition was interpreted to be a charge transfer in the reverse direction from the ones we have considered

[182] A. Carrington, D. S. Schonland, and M. C. R. Symons, *J. Chem. Soc.* p. 659 (1957).
[183] A. Carrington, D. J. E. Ingram, D. S. Schonland, and M. C. R. Symons, *J. Chem. Soc.* p. 4710 (1956).

previously, namely from the $3t_2$ orbital which is mainly $3d$ to the $2a_1$ orbital predominantly on oxygen. The first allowed transition is raised in energy relative to an ion having an empty $3t_2$ shell.

The energy of the first allowed transition for the tetrahedral ions is plotted against atomic number in Fig. 26. The three effects which we have mentioned, namely those of nuclear charge, principal quantum number, and number of d electrons present, are illustrated in this figure.

McGlynn and Kasha[184] have reviewed the spectral data on the oxyanions, including the types of ions we have been considering as well as those containing only nonmetal atoms. Many of these spectra were interpreted as "$n \rightarrow \pi$" transitions, where "n" means a nonbonding oxygen orbital and "π" means a π orbital of the entire complex. The experimental data which support this assignment are the moderately low intensity ($f = 0.01 - 0.1$) and the polarization. These transitions are partly of the charge transfer type and partly of the type localized on the ligands.

c. *Conclusions*

The occurrence of charge transfer spectra in a wide variety of cases is well substantiated. In no case, however, has sufficient work been carried out, either experimental or theoretical, to establish the validity of the one-configuration theory. It appears that careful study of the tetrahedral ions, such as MnO_4^-, $MnO_4^=$, and $CrO_4^=$, offers the best opportunity to identify energy states, and to observe the effects of configuration interaction.

6. Spectra of the Post-Transition Metal Ions

The ions of some of the metals beyond the ends of the transition groups have low-lying excited energy levels (below 10 ev). The ground states of these ions and the ground and low excited configurations are given in Table XXII.

The energy levels corresponding to these configurations are known for most of the ions in the gaseous state.[65,65a] The spectra of the ions in many solid solutions and in the pure salts are also known. The correspondence between the spectra of the free ions and those in the solid state is sufficiently close to show that the atomic levels are of major importance in the solid. The absence of low-energy absorption spectra in the analogous compounds of the pretransition metal ions is further confirmation of this idea, since these ions have no bound excited levels below 10 ev.

In spite of the agreement in order of magnitude between levels of the

[184] S. P. McGlynn and M. Kasha, *J. Chem. Phys.* **24**, 481 (1956).

free ions, and those of the bound ions, the energy levels of the latter must be highly perturbed by the anions of the crystal lattice. Such perturbations must be considerably greater than those acting on the *d* shells of the transition metal ions. The *d* shells are not directly involved in bonding, whereas the outer *s* and *p* orbitals of the post-transition metal ions are the actual bonding orbitals. Even if one assumes that the

TABLE XXII

	Ground state	Excited state
Cu^+, Ag^+, Au^+ Zn^{++}, Cd^{++}, Hg^{++}	$^1S_0$, $(n-1)d^{10}$	$(n-1)d^9ns$, $(n-1)d^9np$
Zn^+, Cd^+, Hg^+	$^2S_{\frac{1}{2}}(n-1)d^{10}ns$	$(n-1)d^{10}np$
Zn^0, Cd^0, Hg^0 Ga^+, In^+, Tl^+ Ge^{++}, Sn^{++}, Pb^{++} As^{+++}, Sb^{+++}, Bi^{+++}	$^1S_0(n-1)d^{10}ns^2$	$(n-1)d^{10}nsnp$

crystal is "completely ionic," the anions must polarize the outer orbitals of the cations strongly.

The spectra which apparently correspond to $s^2 \rightarrow sp$ transitions have been studied extensively in solids and will be discussed first.

a. The Transitions $s^2 \rightarrow sp$

The ions of the sixth period having the s^2 ground configuration have received more study than the fourth or fifth period ions. The fourth period ions are somewhat unstable chemically, and have consequently received the least attention.

There has been considerable theoretical work in recent years on the KCl:Tl system following the original work of Seitz.[185] It has been reviewed by Klick and Schulman in Vol. 5 of this series,[186] and will not be discussed in detail here.

In the remainder of this section we shall be concerned with the experimental evidence bearing upon the interpretation of the spectra.

(*1*) *Hg*. The un-ionized mercury atom is isoelectronic with Tl^+, etc., and its transitions are well known in the vapor phase.

Prener *et al.*[187] found that mercury could be atomically dispersed by absorption in a zeolite, and that the resulting preparation had an absorp-

[185] F. Seitz, *J. Chem. Phys.* **6**, 150 (1938).
[186] C. C. Klick and J. H. Schulman, *Solid State Phys.* **5**, 97 (1957).
[187] J. S. Prener, R. E. Hanson, F. E. Williams, *J. Chem. Phys.* **21**, 759 (1953).

tion band in the neighborhood of the lower resonance transition of the free atom, 2537 A. A broad luminescence band was also observed.

(*2*) *Tl*$^+$. Figure 27 shows how the energy levels of the Tl^+ ion depend upon the alkali halide in which it is dissolved. Most of this information is taken from the early work done under Hilsch.[188] It also shows the levels

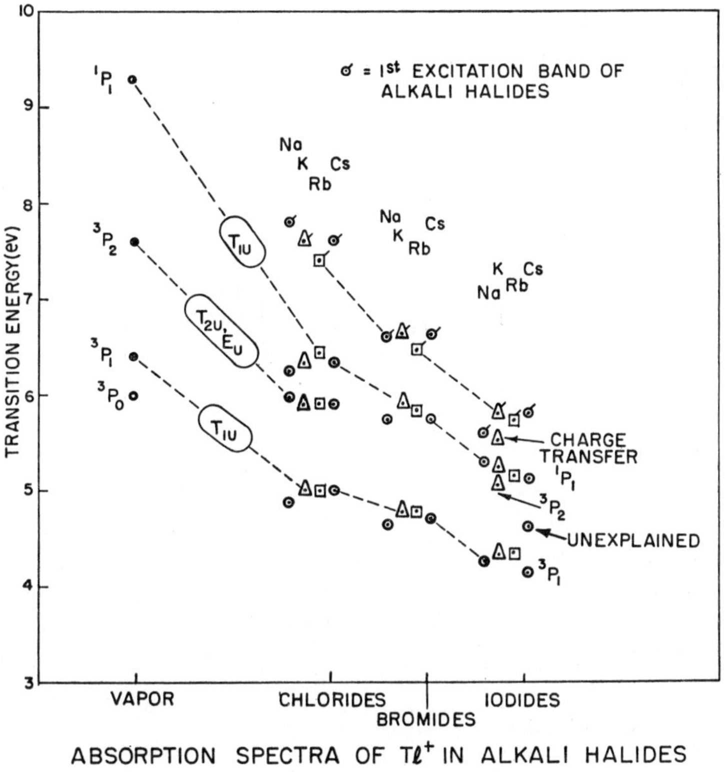

FIG. 27. Absorption peaks of Tl^+ ion in the alkali halides. The energy states of the free Tl^+ are shown at left, and the first exciton band of the alkali halides is also shown for reference.

of the free ion to which the absorption peaks have been attributed, and the positions of the first exciton levels of the alkali halides. The absorption coefficients to the three thallium peaks are in qualitative agreement with their attribution to the corresponding levels of the free ion. In the

[188] R. Hilsch, *Z. Physik* **44,** (1927); M. Forro, *ibid.* **58,** 613 (1929); E. Koch, *ibid.* **57,** 638 (1929); **59,** 378 (1930); W. Von Meyeren, *ibid.* **61,** 321 (1930); W. Bunger, *ibid.* **66,** 311 (1930); W. Bunger and W. Flechsig, *ibid.* **67,** 42 (1931); **69,** 637 (1932); R. Hilsch, *Physik. Z.* **38,** 1031 (1937).

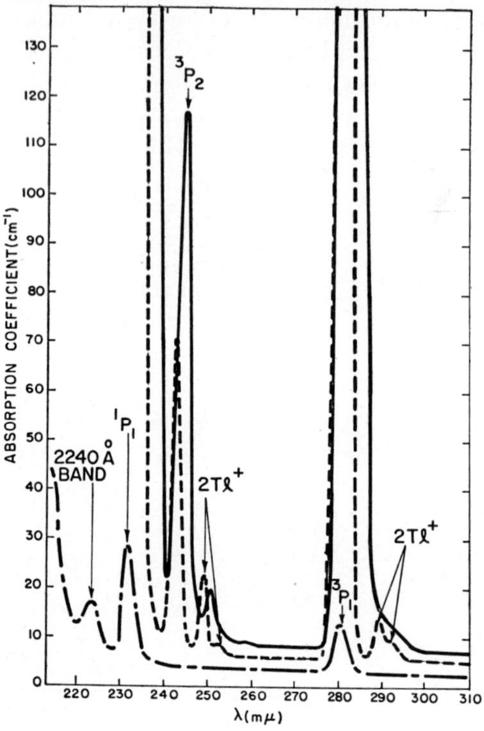

FIG. 28. Absorption spectrum of Tl^+ in KI, after Yuster and Delbecq.[189]

a. – – – – (lowest curve) Tl^+ concentration 1.2×10^{-4} mole per cent; 4°K. Only the strongest bands, namely $^3P_1$, $^1P_1$ and the 2240 A band are visible. b. ——— Tl^+ concentration 3.7×10^{-2} mole per cent; 77°K. The $^3P_2$ band appears strongly, and evidences of bands due to pairs of neighboring Tl^+ ions are present. c. — — — Same concentration as *b*, but 4°K. The $^3P_2$ band is considerably weaker than at 77°K and four bands due to Tl^+ pairs are resolved.

notation appropriate to a cubic field these levels are

$$^3P_0 \equiv A_{1u},\ ^3P_1 \equiv T_{1u};\ ^3P_2 \equiv T_{2u},\ E_u;\ ^1P_1 \equiv T_{1u}.$$

The spin-orbit product is used because of the large spin-orbit interaction.

Figure 28 shows the absorption spectrum of KI:Tl, from the work of Yuster and Delbecq.[189] The peaks in the spectrum are identified by the atomic level to which they are supposed to correspond.

Transitions to the $^3P_1$ and $^1P_1$ states are fully allowed. The ratio of transition probabilities is about 1:5 because the spin-orbit mixing is not complete. The same ratio is observed for KCl:Tl as for KI:Tl. Knox

[189] P. Yuster and C. Delbecq, *J. Chem. Phys.* **21**, 892 (1953).

and Dexter,[190] and later Williams *et al.*[191] have attempted to calculate this intensity ratio, using atomic wave functions without taking explicit account of the anions. The latter authors obtained good agreement with experiment by using the singlet-triplet separation found in the crystal, and the spin-orbit coupling parameters of the free ion. Thus the effect of the crystal was taken into account empirically in an indirect way. These results offer some evidence that the atomic part of the wave functions is the more important part. Williams *et al.* also calculated the transition probability ratio for In^+ in KCl which has a spectrum similar to that of Tl^+ in KCl. The very much weaker transition to the state attributed to $^3P_1$ of In^+ is in agreement with the lower spin-orbit coupling constant in In^+ compared to Tl^+. The calculated transition probability ratio (1:25) agrees with the one observed.

The transitions to $^3P_2(E_u$ and $T_{2u})$ and $^3P_0(A_{1u})$ are forbidden by electronic selection rules for the cubic group (Table I). Vibrations having the symmetries E_g and T_{2g} are the permissible vibrations of an octahedron which may cause transitions to the T_{2u} state. Transitions to the E_u state are caused by the T_{2g} type vibration, whereas transitions to the A_{1u} state are not induced by any vibration of an octahedral array of atoms. Rotational oscillations of the entire octahedron in the crystal are of the correct symmetry to induce the $A_{1g} \rightarrow A_{1u}$ transition, but in order to be effective the p orbitals would have to overlap appreciably with atoms beyond the nearest neighbor shell. The fact that $^3P_0$ has not been observed in absorption is therefore understandable in terms of the Seitz model.

One component of $^3P_2$ has been observed in several mixed crystal systems. Yuster and Delbecq[189] showed that the intensity of absorption to the weak 2440 A Tl^+ band in KI (Fig. 28) is strongly dependent on temperature. Since it is between the bands previously identified as $^1P_1$ and $^3P_1$ and has the temperature dependence expected for a vibrationally induced transition, it is undoubtedly one of the states associated with $^3P_2$. This band had been identified previously by its intensity and position alone.[185]

From the fact that the intensity of the $^3P_2$ band increases from $f = 0.0042$ to $f = 0.011$ when the temperature rises from 4°K to 77°K,[190] one can infer that vibrational quanta on the order of 10–50 cm^{-1} are involved in the transition. Useful information as to the vibrational frequency spectrum in the vicinity of the Tl^+ ion might be obtained by a more detailed study of the temperature dependence of this band.

The two states of the $^3P_2$ band, T_{2u} and E_u, are accidentally degenerate according to crystal field theory, since the orbital part of the wave func-

[190] R. S. Knox and D. L. Dexter, *Phys. Rev.* **104,** 1245 (1956).
[191] F. E. Williams, B. Segall, and P. D. Johnson, *Phys. Rev.* **108,** 46 (1957).

tion contains harmonics of the first order, and the potential contains harmonics of the fourth or higher orders. Crystal field theory could not apply very well to this case however and some splitting would be expected. For example, the *p*-functions must be admixed with *f*- and higher harmonics through the polarizing effect of the anions. Such a splitting has not been observed, and it is not known whether the observed band contains both components, or if one is hidden by the nearby $^1P_1$ band. If this problem could be resolved experimentally, it would provide important information about the interaction between the Tl^+ and the surrounding anions.

One further interesting detail brought out by Yuster and Delbecq's work is the presence of three partially resolved peaks in the $^1P_1$ transition and the absence of any such resolution in the $^3P_1$ transition. The three peaks could conceivably be a result of the dynamical Jahn-Teller effect.[36] A similar case is observed in the Fe^{++} spectrum, discussed in Section 4.

Klick and Compton[192a] have recently found that if a single crystal of KCl:Tl at 4°K is illuminated with polarized light in any region of the main absorption band ($^1S_0 \rightarrow {}^3P_1$) the emitted light is also polarized. The maximum polarization of the emission is observed when the electric vector of the exciting light is oriented along a cube axis: the emitted light is also preferentially polarized along the same cube axis. The polarization is not observable at temperatures as high as 77°K.

Since $^3P_1$ is the emitting state, Klick and Compton suggest that the polarization is brought about by the removal of the three-fold degeneracy through the Jahn-Teller effect. The observed distortion direction is in disagreement, however, with the theoretical predictions of Opik and Pryce.[35] There is also the problem of understanding how the polarization can be preserved between the act of absorption and the act of emission, since a large amount of vibrational energy must be lost in the interim.

Recent work on the absorption spectrum[192b] has revealed an unusual temperature dependence of the two main bands of Tl^+ in KCl. Between 77°K and 295°K they become about 40% broader on the low energy side, as well as becoming assymmetric. Further recent information on these transitions is provided by the work of Eppler and Drickamer[192c] on the pressure shift of the absorption bands. Pressure causes a red-shift in most cases. A sudden shift occurs at the phase transition between the f.c.c. and b.c.c. structures in the neighborhood of 20,000 atm for the potassium halides. This is a blue shift for KCl:Tl and KBr:Tl and a red shift for KI:Tl.

[192a] C. C. Klick and W. D. Compton, *J. Phys. Chem. Solids* **7**, 170 (1958).
[192b] D. A. Patterson, *Phys. Rev.* **112**, 296 (1958).
[192c] R. A. Eppler and H. G. Drickamer, *J. Phys. Chem. Solids* **6**, 180 (1958).

Although statements have been made that these new experimental facts are not at variance with the Williams-Seitz model for Tl^+ in the alkali halides, this center appears to be more complicated than envisioned by the model. The possibility that it is unsymmetrical even in the ground state might be explored. Such a possibility is suggested by the structure of TlF, which is a tetragonally distorted NaCl-type structure.

From the evidence presented earlier for the assignments of the absorption bands of Tl^+ (and In^+) in the alkali halides, one must conclude that the assignments are essentially correct. One may now ask: to what extent do the anions participate in the atomic transitions, and where are the bands corresponding to charge transfer transitions?

Yuster and Delbecq observed a band in KI:Tl at 2240 A, between the $^1P_1$ peak and the first exciton band of KI. This, they suggest is a band similar to the α and β bands of the alkali halides; specifically it is thought to be due to absorption in the anions surrounding the Tl^+, whereas the α and β bands arise from absorption in the anions surrounding an anion vacancy or an F center respectively.

The fact that the 2240 A band appears only 0.22 ev above the $^1P_1$ band means that the excited states of both transitions may be quite similar. The 2330 A band could depend more on the Tl^+ wave functions, and the 2240 A band depend more on the anion wave functions. The anion states and the Tl^+ states should be considered as two types of electron configuration which must become mixed to produce the final states. A reasonable starting point for a quantitative calculation would seem to be Overhauser's treatment of excitons in the alkali halides[193] coupled with the Williams-Seitz model for the Tl^+ ion.[185,186]

The trends shown in Fig. 27 permit one to make some qualitative statements about the effect of the anions on the spectra. The two principal observations are that the Tl^+ transitions occur at lower energy, and the over-all multiplet width $= (^1P_1 - {}^3P_1)$ is reduced as the ion is taken from vacuum into crystals containing progressively heavier halide ions. At the same time, as shown in Fig. 27, the lowest exciton level of the alkali halides moves to lower energies, and in the iodides it is within 0.7 ev of the $^1P_1$ band of Tl^+. The perturbation of the Tl^+ levels by the exciton levels therefore increases markedly from the chlorides to the iodides. The reduction in multiplet width $(^1P_1 - {}^3P_1)$ could be ascribed to a reduction in the effective spin-orbit interaction and electrostatic interaction parameters due to the spreading of the atomic orbitals of Tl^+ over the surrounding anions. Such an explanation is qualitatively similar to the one applied in Section 2 to the transition metal ions.

(3) Pb^{++} and Bi^{+++}. The spectrum of Pb^{++} is known in some of the

[193] A. W. Overhauser, *Phys. Rev.* **101,** 1702 (1956).

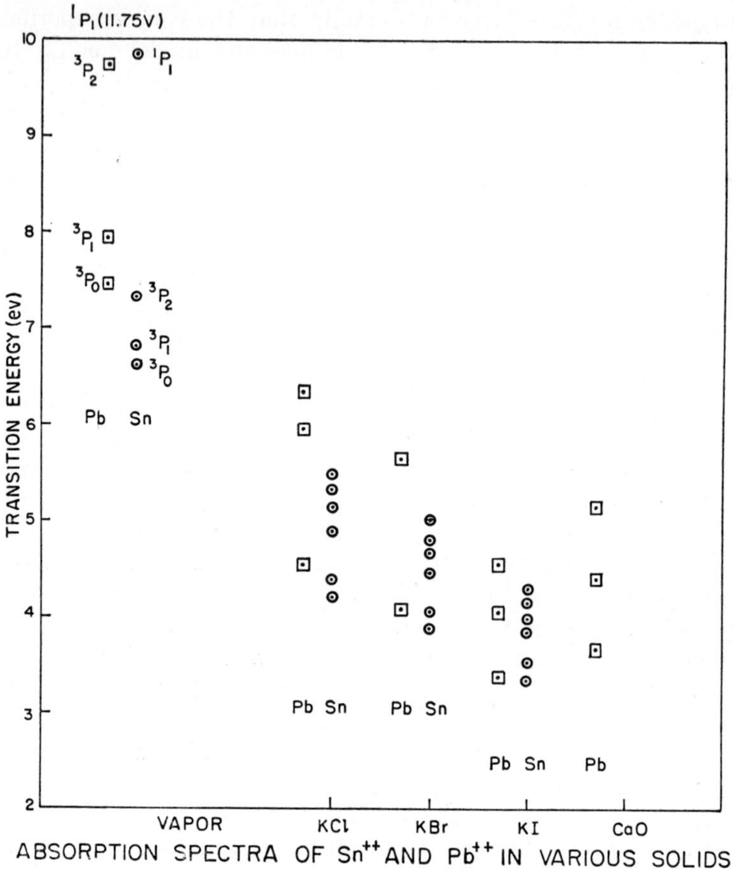

FIG. 29. Absorption peaks of Pb^{++} and Sn^{++} in various solids. The levels of the free ions are shown to the left. The four upper peaks for Sn^{++} are attributed to $^1P_1$ and the lower two to $^3P_1$. The highest peak for Pb^{++} is attributed to $^1P_1$, and the lowest to $^3P_1$.

alkali halides[189] and alkaline-earth oxides and sulfides.[194] It has not been studied in as much detail as Tl^+; hence the assignments of the observed bands are not as certain. The energy levels as observed in several solids are shown in Fig. 29. The spectra in the alkali halides are remarkably similar to those of Tl^+. They occur at almost the same energies, and consist of two strong bands having an intensity ratio of about 3:1, the higher energy band being the stronger one. These may be the $^1P_1$ and $^3P_1$

[194] J. Ewles, *Proc. Roy. Soc.* (*London*) **A167,** 34 (1938); J. Ewles and C. Curry, *Proc. Phys. Soc.* (*London*) **A63,** 708 (1950); J. Ewles and N. Lee, *J. Electrochem. Soc.* **100,** 392, 399, 402 (1953).

bands. In a few cases a weak band appears as a shoulder on the low-energy side of the $^1P_1$ band, and could be identified as the $^3P_2$ band. Forro[195] showed that this band increases in strength at high temperatures in the KCl:Pb system, and that the $^3P_1$ band is practically independent of temperature. This behavior is in accordance with the assignments. The separation between the $^1P_1$ and $^3P_1$ bands is about 30% larger than in Tl^+, as would be expected from the larger spin-orbit coupling constants of Pb^{++}. Although the states of the free Pb^{++} ion lie 1–3 volts higher than the corresponding states of Tl^+, the absorption of the Pb^{++} occurs at about the same frequency as that of Tl^+ in the alkali halides. The decrease in energy of absorption in going from chlorides to iodides is also greater for Pb^{++} than for Tl^+.

Fredericks and Scott[196] have recently shown that Pb^{++} in KCl migrates at high temperatures in an electric field as if it were a negative ion. They conclude that it must exist in the crystal as a complex ion such as $PbCl_6^{\equiv}$. This is not unexpected since with its charge of +2 the Pb^{++} ion may pull the halide ions out of their equilibrium positions against the attractive forces of the nonovalent alkali metal ions. Opposing the displacement of the halide ions is the effect of the ionic radius of the Pb^{++} ion (1.32 A). When replacing an alkali ion smaller than itself, the Pb^{++} may not be able to pull the halide ions out of their equilibrium positions to the same extent as when it replaces a larger alkali ion.

Another factor that affects the spectrum of the Pb^{++} ion in the alkali halides is the association of the ion with vacancies. Burstein and his associates[197] showed that the spectrum of NaCl:Pb is very sensitive to its thermal history, and could show that bands appearing at the sides of the main absorption at 2730 A were due to Pb-vacancy pairs or to Pb-Pb pairs. They also found that KCl:Pb is *not* sensitive to its thermal history. Since Pb^{++} and K^+ have nearly the same crystal radii (1.32 A), and Na^+ has a smaller one (0.98 A), they suggest that the aggregation of vacancies and Pb^{++} ions is favored by a state of strain in the lattice. In this work and in another investigation of the NaCl:Pb system,[198] the excitation and emission spectra were studied in detail and were found to be more sensitive to the presence of defects than are the usual absorption spectra taken at room temperature.

All of the evidence favors the view that the states of the Pb^{++} ion in alkali halides are more intimately mixed with the states of the lattice

[195] M. Forro, *Z. Physik* **56**, 543 (1929).
[196] W. J. Fredericks and A. B. Scott, *J. Chem. Phys.* **28**, 249 (1958).
[197] E. Burstein, J. J. Oberly, B. W. Henvis, and J. W. Davisson, *Phys. Rev.* **81**, 459 (1951).
[198] J. H. Schulman, R. J. Ginther, and C. C. Klick, *J. Opt. Soc. Am.* **40**, 854 (1950).

than for the case of Tl^+. There should therefore be a number of "trapped exciton"-like states similar to the 2240 A band of KI:Tl^+; one could also consider them as molecular charge transfer states similar to those considered in Sections 2 and 5. These states have not been reported.

The spectroscopic study of Pb^{++} in oxides and sulfides is far from being complete. The excitation spectra given by Ewles[194] show where the ultraviolet absorption bands are, but are ambiguous because of insufficient information on the preparation of the samples. The points on the diagram of Fig. 29 are therefore somewhat doubtful. Some of them may correspond to the optical absorption by pairs of Pb^{++} ions in the lattice. The bands for the oxide seem to fall between those of the bromide and iodide.

Both Pb^{++} and Bi^{+++} in the alkaline earth oxides and sulfides show vibrational structure in absorption and emission. These spectra have been studied by Ewles and co-workers,[194] and by Runciman.[199] The latter has given an analysis of the data which appears more tenable than that of Ewles *et al.* In emission there is found for SrO:Bi a series of vibrational bands given by

$$\nu = 25717 - 493v_1 - 195v_2 \qquad v_1 = 0, 1, 2, 3 \ldots \quad v_2 = 0, 1$$

whereas in absorption the same system shows bands given by

$$\nu = 27230 + 430v_1 + 215v_2 \qquad v_1 = 0, 1, 2, 3 \ldots \quad v_2 = 0, 1.$$

In the emission spectrum, ground-state frequencies appear, while in absorption those of the excited state appear. The frequencies seem reasonable for the vibrations of an octahedral BiO_6 molecule in which the oxygen atoms are vibrating. The appearance of two frequencies is interesting, because there is only one totally symmetric mode of an octahedron; thus one of the frequencies may be nontotally symmetric, unless it involves the next neighbor shell around the Bi^{+++} ion. If the transition is the fully allowed $A_{1g}(^1S_0) \rightarrow T_{1u}(^3P_1)$, the only reason for the appearance of a nontotally symmetric mode in the spectrum would be the occurrence of the Jahn-Teller effect, which is predicted to occur in the upper state.

The fact that the absorption and emission spectra do not have the same origin casts doubt on the interpretation of the spectrum as a $^1S_0 \rightarrow {}^3P_1$ transition. The apparent origins differ by 1513 cm^{-1}. Either they belong to different electronic states or some weak bands have not been found. Contrary to opinions expressed elsewhere,[199] the origins of

[199] W. A. Runciman, *Proc. Phys. Soc.* **A68**, 647 (1955).

absorption and emission must coincide if the transitions are the same. Without further study to settle the question of the electronic assignments, little can be done with the published data on the vibrational intervals. However, the mere appearance of discrete vibrations shows that the equilibrium displacement upon excitation is much less in these systems than, for example, in KCl:Tl.

(4) In^+, Sn^{++}. A broad survey of the absorption and emission spectra of Sn^{++} in alkali halides has been made by Huniger and Rudolph.[200] The main features of the spectrum are again two peaks in the ultraviolet having an intensity ratio of about 10:1, with the higher energy peak the stronger. This intensity ratio is higher than for In^+ (25:1)[191] as would be expected from the increase of the spin-orbit interaction parameters. The transitions are strong, with $f \approx 1$. These features persist throughout all of the halides.

A striking pecularity of the Sn^{++} spectra is the "fine" structure: the $^1P_1$ peak consists of four well-resolved subpeaks and the $^3P_1$ peak of two. The positions of these peaks are shown in Fig. 29. The spectrum of Sn^{++} in KI is practically identical to that in CaI_2, so it is not likely that the structure is caused by the presence of a defect near the Sn^{++} ion. The subpeaks are separated by about 1500 cm^{-1}, which is definitely too high for vibrational structure. Furthermore the intervals are not strongly dependent upon either the halide ion or the alkali metal ion. As in the case of Tl^+, a possible explanation is the occurrence of the Jahn-Teller effect in the upper state.

b. *The Transitions $d^{10} \rightarrow d^9s$*

The low-energy transitions in Cu^+ and Ag^+ occur in the vapor at the frequencies shown in Table XXIII.[65a]

TABLE XXIII

	d^9s				d^9p	
	$^3D_3$	$^3D_2$	$^3D_1$	$^1D_2$	^3P, ^3F, ^3D	^1F, ^1P, ^1D
Cu^+	21,925 cm^{-1}	22,844	23,995	26,261	66,415–71,490	71,916–73,592
Ag^+	39,164	40,741	43,739	46,045	80,172–86,883	89,130–90,883

All of the lower transitions are forbidden either for reasons of symmetry (even → even) or multiplicity (singlet → triplet).

When these ions are present in the alkali halides they give rise to absorption bands between 2000 and 3500 A. The absorption coefficients

[200] M. Huniger and J. Rudolph, *Z. Physik* **117**, 81 (1940).

are smaller by one or two orders of magnitude than those of Pb^{++} and Tl^{+} in alkali halides. If these bands are to be assigned to atomic-like transitions, their low intensity is in agreement with the forbidden nature of all the lowest transitions of the free atoms.

Most of the experimental work dates back to 1927–1931.[201] It would be desirable to have spectra extending over a greater wavelength range and to have more low-temperature studies. However, some interpretation of the spectra is possible on the basis of existing data.

MacMahon and Forro[201] showed that the strength of the transitions in Cu^{+} and Ag^{+} in NaCl and KCl are temperature dependent. Forro showed also that they obey Beer's law, and that we are therefore dealing with absorption in single ions. The temperature dependence suggests a vibrational perturbation of a forbidden transition, and therefore indicates that the $d^{10} \rightarrow d^{9}s$ transitions are involved. Most of the spin-forbidden $d^{10} \rightarrow d^{9}p$ transitions are likely to be strong enough because of the spin-orbit coupling that vibrations would not cause appreciable intensification.

There do not seem to be any correlated studies of the excitation and emission spectra of Cu^{+} or Ag^{+} in alkali halides. These would be desirable to help answer the question of whether or not the lowest excited state is the one seen in absorption near 2500 A. However, note two recent references.[202,203]

The excitation spectra of copper-containing alumino-silicate phosphors, and the corresponding emission spectra were published by Claffy and Schulman.[204] The excitation spectrum is very similar to the absorption spectrum of Cu^{+} in NaCl even though in the alumino-silicate lattice, the Cu^{+} ion must be surrounded by oxygen atoms.

The peak of the emission spectrum occurs at 4500 A, and is displaced so far from the absorption spectrum that it is not clear that the same transition is involved in both spectra. Thus it is possible that some lower energy states of the Cu^{+} ion are present, but are not observed in the excitation spectrum.

If the spectra actually arise from the $d^{10} \rightarrow d^{9}s$ transitions, one problem is to explain why they shift toward higher energies upon entering the solid. The prominent absorption peaks of Cu^{+} and Ag^{+} in NaCl lie at 39,200 cm^{-1} and 47,700 cm^{-1} respectively. In each case the transition requires about two volts more energy in the solid than in the vapor. This

[201] A. Smakula, *Z. Physik* **45,** 1 (1927); A. M. MacMahon, *ibid.* **52,** 336 (1928); M. Forro, *ibid.* **56,** 235 (1929); **58,** 613 (1929).
[202] E. Boesman and W. Dekeyser, *Physica* **24,** 52 (1958).
[203] M. L. Kats, *Optika y Spekt.* **3,** 602 (1957).
[204] E. W. Claffy and J. H. Schulman, *J. Electrochem. Soc.* **98,** 409 (1951).

fact also favors the hypothesis as to the atomic origin of the transitions: an atom in a $(n-1)d^{10}$ configuration should fit more easily into a crystal lattice than one in the more extended configuration $(n-1)d^9ns$, so that its ground state is stabilized more than the excited state.

c. The Transitions $d^{10}s \rightarrow d^{10}p$

No examples of the transitions $d^{10}s \rightarrow d^{10}p$ have come to the writer's attention. The ions having the ground configuration $d^{10}s$ form dimers such as Hg_2^{++} when combined with halides and in water solution. Chemical reactions which would be expected to lead to complex Hg^+ ions lead instead to Hg^{++} and Hg.

d. Spectra of Ion Pairs

Yuster and Delbecq[189] have observed absorption transitions due to pairs of Tl^+ ions in KI:Tl. These absorption bands vary in strength with the square of the Tl^+ concentration and are first observable at about 0.01 mole per cent. Four such bands are pointed out in Fig. 28; two apparently related to $^1P_1$ and two related to $^3P_1$.

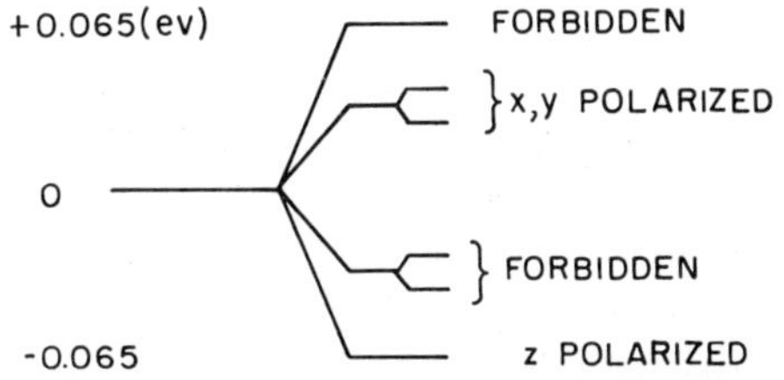

FIG. 30. Hypothetical energy level splitting diagram for two Tl^+ ions at neighboring cation sites in KI lattice. The z direction is long the line joining the atoms.

In some of the earlier work[188] it was shown that phosphorescence only occurs when the concentration of Tl^+ in the alkali halide is large enough for the formation of pairs in adjacent cation positions. The interpretation of the phosphorescence is discussed by Seitz.[185]

The absorption spectrum of pairs has not been fully interpreted, and in fact further experimental work will probably be needed before this can be done. If we use the simple model of the dipole-dipole excitation exchange which has been applied with success to molecular crystals[45a] a theoretical splitting pattern may be derived for a pair of Tl^+ ions. Such a pattern is shown for the $^1P_1$ state of Tl^+ in Fig. 30. It is assumed that the two Tl^+ ions are in undisplaced nearest neighbor K^+ positions, 4.98 A apart, and that the transition moment per Tl^+ ion as calculated from Yuster and Delbecq's f number of 0.80 is 0.75 A electrons. It is also assumed that the electronic properties of the Tl^+ ions are not affected by their proximity to each other. The lowest transition is allowed, according to the theory, and is polarized along the $Tl^+ - Tl^+$ direction. The 2500 A band could be identified with this transition. However, its separation of 0.432 ev from the parent ${}^1P_1$ state is almost seven times larger than the calculated value, 0.065 ev. Such a discrepancy is large enough to indicate that the simple model used to calculate the effect is incorrect. Some of the more complex possibilities which must therefore be considered are the displacement of the Tl^+ ions toward each other, and the direct involvement of the halogen ions in the transition. In order to learn something about these effects, the experiments of Yuster and Delbecq should be carried out with other alkali halides.

New absorption bands appearing as the concentration is raised have also been observed in the system KCl:Cu.[201] The lowest concentrations at which these are observed are on the order of 0.1 mole per cent. It has not been shown that their intensity depends on the square of the concentration, and no precise spectroscopic studies have been made. Evidence for absorption bands due to pairs of Pb^{++} ions has been found[197] but no accurate spectral work has been reported.

e. Conclusions

The lowest energy spectra which occur both in the transition and post-transition ions in solids arise from electron jumps occurring principally within the metal ion. The analogy between the two groups has been illustrated for the low oxidation levels of the metal where *s* or *d* electrons are available to make a transition. An analogy may also exist for the higher oxidation levels in which no electron shells of the metal ion are available to make low-energy transitions. In the case of the transition metal ions the high oxidation levels gave rise to what are termed "charge transfer spectra." These ions exist in the form of molecules rather than as discrete ions. The same is true in the higher oxidation levels of the post transition metal ions. Molecules such as $SbO_4^{\equiv}$, $SnCl_6^{=}$ are commonly occurring chemical species. The spectra of such molecular ions have not been studied in detail, but they may be expected to resemble the charge transfer spectra of the transition metal ions. Orbital diagrams such as those of Figs. 16 and 17 could be drawn for these cases. One major difference between the charge transfer processes in the two cases is that the *s* orbital of the metal ion will be the acceptor, rather than the *d* orbital as in the transition metal ions. This will result in a simpler spectrum.

Throughout this article we have concentrated on the problem of finding a zeroth-order approximation in terms of which the spectra may be understood qualitatively, and which may serve as a starting point for calculation. This was a straightforward problem for molecular crystals, and for the *f*- and *d*-electron states of rare earth and transition metal ions. It has been shown that the methods chosen for these systems allow refined calculations to be made which explain many details of the spectra. This point has not been reached as yet for the states we have identified as charge transfer states, nor for the $s \rightarrow p$ or $d \rightarrow s$ transitions of post-transition metal ions. This is the area in which new knowledge of the spectra of localized ions or molecular ions will be of the most help in understanding the exciton spectra of pure solids. The approach to these problems outlined here should lead to satisfactory solutions.

ACKNOWLEDGMENTS

I wish to thank Drs. Tanabe and Sugano for the use of the figures and secular equations from their 1954 paper. I am also indebted to Drs. S. Freed, J. W. Stout, S. Fried, P. Yuster and C. Delbecq for permission to use figures from their papers.

Author Index

The numbers in parentheses are footnote numbers and are inserted to enable the reader to locate a cross reference when the author's name does not appear at the point of reference in the text.

C

I

J

Y

Z

Subject Index

H

I

J

K

L